Praise for *Stochastic Local Search: Foundations and Applications*

'Hoos and Stützle, two major players in the field, provide us with an excellent overview of stochastic local search. If you are looking for a book that covers all the major meta-heuristics, gives you insight into their working and guides you in their application to a wide set of combinatorial optimization problems, this is the book.'
Marco Dorigo, Université Libre de Bruxelles

'*Stochastic Local Search: Foundations and Applications* provides an original and synthetic presentation of a large class of algorithms more commonly known as meta-heuristics. Over the last 20 years, these methods have become extremely popular, often representing the only practical approach for tackling so many of the hard combinatorial problems that are encountered in real-life applications. Hoos and Stützle's treatment of the topic is comprehensive and covers a variety of techniques, including simulated annealing, tabu search, genetic algorithms and ant colony optimization, but a main feature of the book is its proposal of a most welcome unifying framework for describing and analyzing the various methods.'
Michel Gendreau, Université de Montréal

'Local search algorithms are often the most practical approach to solving constraint satisfaction and optimization problems that admit no fast deterministic solution. This book is full of information and insights that would be invaluable for both researchers and practitioners.'
Henry Kautz, University of Washington

'This extensive book provides an authoritative and detailed exposition for novices and experts alike who need to tackle difficult decision or combinatorial optimization problems. The chapters span fundamental theoretical questions such as, "When and why do heuristics work well?" but also more applied aspects involving, for instance, the comparison of very different algorithms. The authors are university faculty members and leading players in their research fields; our communities will enjoy in particular their book's valuable teaching material and a "complete" bibliography of the state of the art for the field.'
Olivier Martin, Université Paris-Sud, Orsay

'The authors provide a lucid and comprehensive introduction to the large body of work on stochastic local search methods for solving combinatorial problems. The text also covers a series of carefully executed empirical studies that provide significant further insights into the performance of such methods and show the value of an empirical scientific methodology in the study of algorithms. An excellent overview of the wide range of applications of stochastic local search methods is included.'
Bart Selman, Cornell University

'Stochastic local search is a powerful search technique for solving a wide range of combinatorial problems. If you only want to read one book on this important topic, you should read Hoos and Stützle's. It is a comprehensive and informative survey of the field that will equip you with the tools and understanding to use stochastic local search to solve the problems you come across.'

Toby Walsh, Cork Constraint Computation Centre
University College Cork

'This book provides remarkable coverage and synthesis of the recent explosion of work on randomized local search algorithms. It will serve as a good textbook for classes on heuristic search and metaheuristics as well as a central reference for researchers. The book provides a unification of a broad spectrum of methods that enables concise, highly readable descriptions of theoretical and experimental results.'

David L. Woodruff, University of California, Davis

STOCHASTIC LOCAL SEARCH
FOUNDATIONS AND APPLICATIONS

STOCHASTIC LOCAL SEARCH
FOUNDATIONS AND APPLICATIONS

Holger H. Hoos
Department of Computer Science
University of British Columbia
Canada

Thomas Stützle
Department of Computer Science
Darmstadt University of Technology
Germany

AMSTERDAM • BOSTON • HEIDELBERG • LONDON
NEW YORK • OXFORD • PARIS • SAN DIEGO
SAN FRANCISCO • SINGAPORE • SYDNEY • TOKYO

ELSEVIER   MORGAN KAUFMANN PUBLISHERS IS AN IMPRINT OF ELSEVIER

MORGAN KAUFMANN PUBLISHERS

Senior Editor	Denise Penrose
Publishing Services Manager	Simon E. M. Crump
Editorial Coordinator	Emilia Thiuri
Editorial Assistant	Valerie Witte
Cover Design	Gary Ragaglia and Holger H. Hoos
Cover Image	"Antelope Canyon" © Thomas Morse/Chuck Place Photo
Text Design	Rebecca Evans and Associates
Composition	Kolam USA
Technical Illustration	Dartmouth Publishing, Inc.
Copyeditor	Lori Newhouse
Proofreader	Calum Ross
Indexer	Robert Swanson
Interior printer	The Maple-Vail Book Manufacturing Group
Cover printer	Phoenix Color

Morgan Kaufmann Publishers is an imprint of Elsevier.
500 Sansome Street, Suite 400, San Francisco, CA 94111

This book is printed on acid-free paper.

Library of Congress Cataloging-in-Publication Data
Application submitted

ISBN: 1-55860-872-9

For information on all Morgan Kaufmann publications, visit our Web site at *www.mkp.com*.

Printed and bound by CPI Group (UK) Ltd, Croydon, CR0 4YY
Transferred to Digital Print 2011

[Science] ... is not a steady march from ignorance to knowledge.
It's more like mountaineering expedition.
On the way up an unscaled peak,
climbers will gain some altitude on one route,
then find it's a dead end.
They'll spot a better one, backtrack a little and move on.
The fact that they sometimes have to take
a step backward for every two steps forward
doesn't mean they are wasting their time.
It means that inching up an uncharted mountain is tough work.
When you step back, though, and take a look at the overall picture
— a long view from the upper slopes of the mountain —
it turns out in hindsight that the path was clear.

(Michael D. Lemonick, Science Writer)

THIS BOOK IS DEDICATED TO OUR PARENTS

DIESES BUCH IST UNSEREN ELTERN GEWIDMET

Eva-Marie Hoos & Hans-Helmut Hoos

Berta Stützle & Günther Stützle

About the Authors

Holger H. Hoos is an Assistant Professor at the Computer Science Department of the University of British Columbia (Canada). His Ph.D. thesis on stochastic local search algorithms for computationally hard problems in artificial intelligence, completed in 1998 at Darmstadt University of Technology (Germany), received the 'Best Dissertation Award 1999' of the German Informatics Society. He has been working on the design and empirical analysis of stochastic local search algorithms since 1994, and his research in this area has been published in book chapters, journal articles and at major conferences in AI and OR.

Holger's research interests are currently focused on topics in artificial intelligence, bioinformatics, empirical algorithmics and computer music. At the University of British Columbia, he is a founding member of the Bioinformatics, Empirical & Theoretical Algorithmics Laboratory (BETA-Lab), a member of the Laboratory for Computational Intelligence (LCI), and a faculty associate of the Peter Wall Institute for Advanced Studies.

Thomas Stützle is an Assistant Professor at the Computer Science Department of Darmstadt University of Technology (Germany). He received an M.Sc. degree in Industrial Engineering and Management Science at the University of Karlsruhe and a Ph.D. from the Computer Science Department of Darmstadt University of Technology. He was a postgraduate fellow at the Department of Statistics and Operations Research, Universidad Complutense de Madrid and a Marie Curie Fellow at IRIDIA, Université Libre de Bruxelles. Thomas has been involved in several EU funded projects on the study of stochastic local search techniques and his research is published in various journals, book chapters and conferences in OR and AI. His current research focuses on the further development of SLS methods, search space analysis, the automatisation of the design and the tuning of SLS algorithms, and new hybridisation schemes for the effective solution of hard combinatorial problems.

Contents

PROLOGUE

Imagine you visit a friend in the beautiful city of Augsburg in southern Germany. It is summer and you set yourselves the challenge to visit all 127 'Biergärten' (beer gardens) on a single day. (If you don't like beer, or if you don't have friends in Augsburg, consider visiting all coffee shops in Vancouver, Canada.) Can this be done? If so, which route should you take? Clearly, your chances of reaching your goal may depend on finding a short round trip that takes you to all 127 places.

As you arrive at Biergarten No. 42, your friend gives you the following puzzle, offering to pay for all your drinks if you can solve it before the night is over: 'Last week my friends Anne, Carl, Eva, Gustaf and I went out for dinner every night, Monday to Friday. I missed the meal on Friday because I was visiting my sister and her family. But otherwise, every one of us had selected a restaurant for a particular night and served as a host for that dinner. Overall, the following restaurants were selected: a French bistro, a sushi bar, a pizzeria, a Greek restaurant, and the Brauhaus. Eva took us out on Wednesday. The Friday dinner was at the Brauhaus. Carl, who doesn't eat sushi, was the first host. Gustaf had selected the bistro for the night before one of the friends took everyone to the pizzeria. Tell me, who selected which restaurant for which night?'

There are various approaches for solving these problems. Given the huge number of possible round trips through the Biergärten, or assignments of weekdays, hosts, and restaurants, systematic enumeration (i.e., trying out all possibilities) is probably not a realistic option. Some people would take a more sophisticated approach and eliminate certain assignments or partial tours through careful reasoning, while systematically searching over the remaining alternatives.

1

But most of us would probably take a rather different approach in practice: starting with a rough and somewhat arbitrary first guess, small changes are repeatedly performed on a given tour or assignment, with the goal of improving its quality or of getting closer to a feasible solution. This latter type of approach is known as *stochastic local search (SLS)* and plays a very important role in solving combinatorial problems like the ones illustrated above. (It may be noted that the logical puzzle and the shortest round trip problem can be seen as instances of the Propositional Satisfiability and Travelling Salesman Problems, which will be more formally introduced in Chapter 1 and used throughout this book.)

Why Stochastic Local Search?

There are many reasons for studying stochastic local search (SLS) methods. As illustrated above, SLS is closely related to a very natural approach in human problem solving. Many SLS methods are surprisingly simple, and the respective algorithms are rather easy to understand, communicate and implement. Yet, these algorithms can often solve computationally hard problems very effectively and robustly. SLS methods are also typically quite general and flexible. The same SLS methods have been found to work well for a broad range of different combinatorial problems, and existing algorithms can often be modified quite naturally and easily to solve variants of a given problem. This makes SLS methods particularly attractive for solving real-world problems, which are often not completely or correctly specified at the beginning of a project and may consequently undergo numerous revisions or modifications before all relevant aspects of the given application situation are captured.

Another reason for the popularity of SLS lies in the fact that this computational approach to problem solving facilitates an explorative approach to algorithm design. Furthermore, as we will discuss in more detail in Chapter 2, many prominent and successful SLS methods are inspired by natural phenomena, which gives them an additional intellectual appeal.

For these (and many other) reasons, SLS methods are among the most prominent and successful techniques for solving computationally hard problems in many areas of computer science (specifically artificial intelligence) and operations research; they are also widely used for solving combinatorial problems in other disciplines, including engineering, physics, management science and bioinformatics.

The academic interest in SLS methods can be traced back to the beginnings of modern computing. In operations research, local search algorithms were developed and described in the 1950s, and in artificial intelligence, SLS methods have been studied since the early days of the field, in the 1960s. To date, the study of SLS algorithms falls into the intersection of algorithmics, statistics, artificial intelligence, operations research, and numerous application areas. At the same time, SLS methods play a prominent role in these fields and are rapidly becoming part of the respective mainstream academic curricula.

About this Book

'Stochastic Local Search: Foundations and Applications' was primarily written for researchers, students and practitioners with an interest in efficient heuristic methods for solving hard combinatorial problems. In particular, it is geared towards academic and industry researchers in computer science, artificial intelligence, operations research, and engineering, as an introduction to and overview of the field or as a reference text; towards graduate students in computer science, operations research, mathematics, or engineering, as well as towards senior undergraduate students with some background in computer science and mathematics, as primary or supplementary text for a course, or for self-study; and towards practitioners, who need to solve combinatorial problems for practical applications, as a reference text or as an introduction to and overview of the field.

The main goal of this book is to provide its readers with important components of a scientific approach to the design and application of SLS methods, and to present them with a broad, yet detailed view on the general concepts and specific instances of SLS methods, including aspects of their development, analysis, and application. More specifically, we aim to give our readers access to detailed knowledge on the most prominent and successful SLS techniques; to facilitate an understanding of the relationships, the characteristic similarities and differences between existing methods; to introduce and discuss basic and advanced aspects of the empirical analysis of SLS algorithms; and to give hands-on knowledge on the application of some of the most widely used SLS methods to a variety of combinatorial problems.

Stochastic search algorithms are being studied by a large number of researchers from different communities, many of which have quite different views on the topic or specific aspects of it. While striving for a balanced and objective presentation, this book provides a view on stochastic local search that is based on our background and experience. This is reflected, for instance, in the specific choice of our formal definition of stochastic local search (Chapter 1), in the GLSM model for hybrid SLS methods (Chapter 3), the extensive and in-depth coverage of empirical analysis and search space structure (Chapters 4 and 5), as well as in the selection of algorithms and problems we cover in varying degree of detail (particularly in Chapters 9 and 10). There are rational reasons for most

– if not all – of these choices; nevertheless, in many cases, equally defensible alternative decisions could have been made.

Clearly, some topics would benefit from broader and deeper coverage. However, even relatively large book projects are subject to certain resource limitations in both time and space, and it is our hope that our choices of the material and its presentation will make this book useful for the previously stated purposes.

Structure and Supplementary Materials

The main body of this book consists of two parts. **Part 1**, which comprises Chapters 1 to 5, covers the foundations of the study of stochastic local search algorithms, including:

- fundamental concepts, definitions, and terminology (Chapter 1),

- an introduction to a broad range of important SLS methods and their most relevant variants (Chapter 2),

- a conceptual and formal model that facilitates the development and understanding of hybrid SLS methods (Chapter 3),

- a methodical approach for the empirical analysis of SLS methods and other randomised algorithms (Chapter 4), and

- features and properties of the spaces searched by SLS algorithms and their impact on SLS behaviour (Chapter 5).

The material from the first two chapters provides the basis for all other aspects of SLS algorithms covered in this book; Chapters 1 and 2 should therefore be read before any other chapters and in their natural sequence. Chapters 3, 4, and 5 are quite independent from each other and expand the foundations of SLS in different directions. Chapter 3 complements Chapter 2; since it discusses some of the more complex SLS methods in a different light, it can be very useful for reviewing and deepening the understanding of these practically very relevant methods. The scope of Chapter 4 extends substantially beyond the empirical analysis of SLS algorithms; although most of the material covered in the subsequent chapters does not directly depend on the concepts and methods from Chapter 4, we strongly believe that anyone involved in the design and application of SLS algorithms should be familiar at least with the basic issues and approaches discussed there. Chapter 5 in some sense covers the most advanced material presented in this book; it should be useful to readers interested in a deeper knowledge of the factors and reasons underlying SLS behaviour and performance, but reading it is not a prerequisite to understanding any of the material covered in the other chapters.

Part 2 comprises Chapters 6 to 10, which present, in varying degree of scope and detail, SLS algorithms for a number of well-known and widely studied combinatorial problems. Except for Chapter 7, which should be read after Chapter 6 since it builds on much of the material covered there, all chapters of this second part are basically independent of each other and can be studied in any combination and order. Chapters 6 to 8 provide a reasonable coverage of the most prominent and successful SLS methods for the respective problems and discuss the respective algorithms in a relatively detailed way. Chapters 9 and 10 are of a more introductory nature; their focus lies on a small number of SLS algorithms for the respective combinatorial problems that have been selected primarily based on their performance and general interest. In particular, the five main sections of Chapter 10 are independent of each other and can be studied in any combination and order.

'In Depth' Sections.   Additional, clearly marked 'In Depth' sections are included in various chapters. These provide additional material that expands or complements the main body of the respective chapter, often at a more technical or detailed level. These sections are generally not required for understanding the main text, but in many cases they should be helpful for obtaining a deeper understanding of important concepts and issues.

Further Readings.   Towards the end of each chapter, a 'Further Readings and Related Work' section provides additional references and pointers to literature on related topics. In the case of subjects for which there is a large body of literature, these represent only a small selection of references deemed especially relevant and/or accessible by the authors. These references should provide good starting points for the reader interested in a broader and deeper knowledge of the respective topic.

Chapter Summaries.   Each chapter closes with a summary section that briefly reviews the most relevant concepts and ideas covered in the respective chapter. The purpose of this summary is to provide the reader with a high-level overview of the material presented in the chapter, and to point out connections (and differences) between the respective concepts and approaches. Together with the chapter introductions and exercises, these summaries facilitate rapid reviewing of previously studied or known material.

Exercises. Each chapter is accompanied by a collection of exercises, classified according to their degree of difficulty as 'easy', 'medium' and 'hard'. This classification is only approximate and does not necessarily reflect the anticipated amount of time needed for producing a solution; although an exercise marked as 'easy' may be relatively straightforward to solve, it may still require a substantial amount of time until the details of the solution are worked out and written down. The exercises cover the material presented in the respective chapter and are intended to facilitate a deeper understanding of the subject matter. They include theoretical questions as well as hands-on implementation and experimentation exercises.

References and Bibliography. References to the technical and research literature are provided throughout the book, particularly in the previously mentioned 'Further Readings and Related Work' sections. These give rise to an extensive bibliography that covers much of the most relevant literature on SLS algorithms and related topics, with a particular emphasis on recent publications.

Glossary and Index. The glossary contains brief explanations of important technical terms useful throughout the book. In conjunction with the extensive and thoroughly compiled index, the glossary particularly facilitates using this book as a reference book or for self-study.

Webpage and Supplementary Materials. Supplementary materials are provided from the book webpage at `www.sls-book.net`. These include slide sets that may be useful in the context of courses that use the book as a primary or supplementary text (see also Section 'Suggested Uses' below), as well as reference implementations of some of the SLS algorithms discussed in this book (needed for some of the hands-on exercises and useful for further practical experience) and some educational tools, for example, for the empirical analysis of SLS behaviour.

Suggested Uses

This book was designed for various types of uses. As a whole, it is intended to be used as a reference book for researchers and practitioners or as the primary text for a specialised graduate or upper-level undergraduate course on stochastic search algorithms; furthermore, parts of it can be used as primary reading or supplementary material for modules of more general courses in artificial intelligence, algorithms, operations research, combinatorial problem solving, empirical methods in computer science, etc. The following specific suggestions reflect our own experience, including the use of parts of this book by students, researchers, and course instructors at the University of British Columbia (Vancouver, Canada) and Darmstadt University of Technology (Darmstadt, Germany).

General introduction to SLS methods, particularly for self-study. Chapters 1 and 2; Sections 3.1 to 3.3 and 3.6; Sections 4.1 to 4.3 and 4.6; Section 5.8; any one or two sections from Chapter 10. For more advanced self-study, the remaining materials can be added as desired; particularly the remaining sections of Chapters 4 as well as Chapters 6 and 8 are highly recommended.

Graduate Course on SLS methods/stochastic search. Chapters 1 and 2; Sections 3.1 to 3.3 and 3.6; Chapter 4; Sections 5.1 to 5.3 and 5.8; Chapters 6 and 7 without the sections on CSP and MAX-CSP; Chapter 8; and any two sections from Chapter 10. Depending on the precise format, focus and level of the course, this selection may be expanded in various ways, for example, by additionally covering Section 9.1 and any one other section from Chapter 9. For a general course on stochastic search methods, an additional module on randomised systematic search algorithms should be included (a sample set of slides for such a module is available from `www.sls-book.net`).

SLS Module(s) in a general AI course. Parts of Chapters 1 and 2; Sections 3.1 to 3.3 and 3.6; Sections 4.1 to 4.3 and 4.6; parts of Chapter 6; and possibly parts of Chapters 8, 9, or 10. The selections from Chapters 1, 2, 6 and 8 to 10 will naturally be based on the prerequisite knowledge of the students as well as the format, level and other modules of the course. A minimal subset for a module of about two lectures in an undergraduate course would mainly take parts of Chapters 1 and 2 and illustrate the working principles of SLS methods using example applications described in Part 2.

SLS Module(s) in a general algorithms course. Parts of Chapters 1 and 2; Sections 3.1 to 3.3 and 3.6; Sections 4.1 to 4.3 and 4.6; Sections 5.1 to 5.3 and 5.8; parts of Chapters 6 and 8; and possibly one or more sections from Chapter 10. The precise balance between these components will naturally depend on the exact nature of the course, particularly on its focus on theoretical or practical aspects of problem solving. In the context of strongly practically oriented algorithms courses, the in-depth sections in Chapters 4, 6 and 8 may be of particular interest.

SLS Module(s) in a discrete optimisation course. Parts of Chapters 1 and 2; Sections 3.1 to 3.3 and 3.6; Chapter 4; parts of Chapter 8 and 9; and any one or two sections from Chapter 10. Additional material, particularly from Chapters 6 and 7, can be used to further expand and complement this selection.

Parts of this book can also be used as primary or supplementary material for specialised graduate courses on SAT, CSP, TSP, scheduling and empirical methods in computing.

The Making of SLS:FA

The process of creating this book is in many ways related to the subject material discussed therein. Not unlike the fundamental approach of local search, it involved navigating a huge space of possibilities in an iterative manner. This process was initiated in 1998, when both, H. H. and T. S. were finishing their Ph.D. theses at the Computer Science Department of Darmstadt University of Technology, and the idea of combining materials from both theses into a comprehensive book on Stochastic Local Search first arose. Five years and about 650 pages later, we reached the end of this search trajectory. The result of a myriad of construction, perturbation and evaluation steps is this book. Interestingly and perhaps not too surprisingly, both, the writing process and its end result turned out to be very different from what we had originally imagined.

Although it would be hard to precisely define the objective(s) being optimised through the writing process, it took us through many situations that closely resemble those of a stochastic local search algorithm trying to solve a challenging instance of a hard combinatorial problem. There were phases of rapid progress and stagnation; we encountered (and overcame) numerous local minima; and along the way, we had to make many decisions based on very limited local information, various forms of heuristic guidance, and some degree of experience.

Random, or at least completely unforeseen and unpredictable, factors played a large role in this local search process. Rather trivial sources of randomness, such as hardware and software glitches, were complemented by more fundamental stochastic influences, such as the random thoughts and ideas that on warm summer nights seem to preferably lurk around the Biergärten, always looking for a receptive mind, or the random person sticking their head into the office door, causing the more organised ideas to fly apart in a hurry. Without these random influences, and the circumstances conducive to them, this book could not have been created in its present form.

At the same time, this book has been shaped by many other factors and influences. These include the places and circumstances under which part of the work was done. (Some of the more interesting places where parts of the book have been written include a log cabin on Sechelt Inlet, the beautiful and tranquil Nitobe Garden, a grassy spot near the top of Whyte Islet in Howe Sound, and the wild and remote inlets of the Pacific Northwest, onboard the Nautilus Explorer.) More importantly, they include a huge and diverse amount of interaction with friends and family, mentors, colleagues, students and our publishers, who provided crucial guidance, diversification, evaluation and general support. Finally, especially during the final phase of the process, our work on this book was largely driven by Hofstadter's Law: 'It always takes longer than you expect, even when you take into account Hofstadter's Law.' [Hofstadter, 1979], the significance and effects of which can hardly be overestimated.

As a consequence, it would be foolish to believe that our stochastic local search process has led us into a global optimum. However, we feel that, largely thanks to the previously mentioned factors and influences, in the process of creating this book we managed to avoid and escape from many low-quality local optima, and achieved an end result that we hope will be useful to those who study it. In this context, we are deeply grateful towards those who contributed directly and indirectly to this work, and who provided us with guidance and support in our local — and global — search.

High-level guidance is of central importance in any effective search process; in our case, there are several people who played a key role in shaping our approach to scientific research and who provided crucial support during various stages of our academic careers. First and foremost, we thank Wolfgang Bibel, our former advisor and 'Doktorvater', for providing a highly supportive and stimulating academic environment in which we could freely pursue our research interests, and whose encouragement and substantial support was highly significant in getting this project underway. Furthermore, H. H. gratefully acknowledges the ongoing and invaluable support from his academic mentors and colleagues, Alan Mackworth and Anne Condon, who also played an important role during the early stages of writing this book. T. S. would especially like to thank Marco Dorigo for the pleasure of joint research and for his support in many senses.

On the other side, we have received more specific guidance on the contents of this book from a number of colleagues, students and fellow SLS researchers. Their detailed comments led to improvements in various parts of this book and helped to significantly reduce the number of errors. (Obviously, the responsibility for those errors that we managed to hide well enough to escape their vigilance rests solely with us.) In this context, we especially thank (in alphabetical order) Markus Aderhold, Christian Blum, Marco Chiarandini, Anne Condon, Irina Dumitrescu, Frank Hutter, David Johnson, Olivier Martin, Luis Paquete, Marco Pranzo, Tommaso Schiavinotto, Kevin Smyth, Dan Tulpan and Maxwell Young. We also acknowledge helpful comments by Craig Boutilier, Rina Dechter, Jin-Kao Hao, Keld Helsgaun, Kalev Kask, Henry Kautz, Janek Klawe, Lucas Lessing, Elena Marchiori, David Poole, Rubén Ruiz García, Alena Shmygelska and Dave Tompkins. Special thanks go to David Woodruff, Toby Walsh, Celso Ribeiro and Peter Merz, whose detailed comments provided valuable guidance in improving the presentation of our work.

In addition, we gratefully acknowledge the interesting and stimulating discussions on the topics of this book that we shared with many of our co-authors, colleagues, students and fellow researchers at TUD and UBC, as well as at conferences, workshops, tutorials and seminars. It is their encouragement, enthusiasm and continuing interest that provided much of the background and motivation for this work.

The staff at Morgan Kaufmann, Elsevier, Kolam and Dartmouth Publishing have been instrumental in the realisation of this book in many ways; we deeply appreciate their expertise and friendly support throughout the various stages of this project. We are particularly grateful to Denise Penrose, Senior Editor at Morgan Kaufmann, whose enthusiasm for this project and patience in dealing with the adverse effects of Hofstadter's Law (as well as with her authors' more peculiar wishes and ideas) played a key role in creating this book. Simon Crump, Publishing Services Manager at Elsevier, and Jamey Stegmaier, Project Manager at Kolam USA, have been similarly instrumental during the production stages, and we gratefully acknowledge their help and support. We also thank Jessica Meehan and her team at Dartmouth Publishing, who produced many of the figures, as well as Lori Newhouse and Calum Ross for copyediting and proofreading the book, and Robert Swanson for creating the index. Many thanks also to Emilia Thiuri and Valerie Witte, for their help during the draft stages, throughout the reviewing process and during production, and to Brian Grimm, marketing manager at Morgan Kaufmann, for substantially increasing the visibility of our work. H. H. also wishes to thank Valerie McRae for her help with proofreading the manuscript in various draft stages, and for much appreciated moral and administrative support.

Finally, we thank our families who provided the stable and stimulating environment that formed the starting point of our personal and intellectual development, and who shape and accompany the trajectories of our lives in a unique and special way. H. H. expresses his deepest gratitude to Sonia and Jehannine for being his partners in adventure, joy, and sorrow, and his parents, siblings and extended family for their affection and diversifying influence. T. S. especially thanks his wife Maria José for sharing her life with him, Alexander for all his curiosity and love, and his parents for their continuous care and support.

This book has been shaped by many factors and influences, but first and foremost it is the product of our joint research interests and activities, which co-evolved over the past seven years into an immensely fruitful and satisfying collaboration and, more importantly, into a close friendship.

part I

Foundations

The machine does not isolate us
from the great problems of life
but plunges us more deeply into them.

—Antoine de Saint-Exupéry, Pilot & Writer

1 Introduction

This introductory chapter provides the background and motivation for studying stochastic local search algorithms for combinatorial problems. We start with an introduction to combinatorial problems and present SAT, the satisfiability problem in propositional logic, as well as TSP, the travelling salesman problem, as the central problems used for illustrative purposes throughout the first part of this book. This is followed by a short introduction to computational complexity. Next, we discuss and compare various fundamental search paradigms, including the concepts of systematic and local search, after which we formally define and discuss the notion of stochastic local search, one of the practically most important and successful approaches for solving hard combinatorial problems.

1.1 Combinatorial Problems

Combinatorial problems arise in many areas of computer science and other disciplines in which computational methods are applied, such as artificial intelligence, operations research, bioinformatics and electronic commerce. Prominent examples are tasks such as finding shortest or cheapest round trips in graphs, finding models of propositional formulae or determining the 3D-structure of proteins. Other well-known combinatorial problems are encountered in planning, scheduling, time-tabling, resource allocation, code design, hardware design and genome sequencing. These problems typically involve finding groupings, orderings or assignments of a discrete, finite set of objects that satisfy certain conditions or constraints. Combinations of these *solution components* form the potential solutions of a combinatorial problem. A scheduling problem, for instance, can be

13

seen as an assignment problem in which the solution components are the events to be scheduled, and the values assigned to events correspond to the time at which these occur. This way, typically a huge number of *candidate solutions* can be obtained; for most combinatorial optimisation problems, the space of potential solutions for a given problem instance is at least exponential in the size of that instance.

Problems and Solutions

At this point, it is useful to clarify the distinction between *problems* and *problem instances*. In this book, by 'problem', we mean abstract problems (sometimes also called *problem classes*), such as 'for any given set of points in the Euclidian plane, find the shortest round trip connecting these points'. In this example, an *instance* of the problem would be to find the shortest round trip for a specific set of points in the plane. The *solution* of such a problem instance would be a specific shortest round trip connecting the given set of points. The solution of the abstract problem, however, is an algorithm that, given a problem instance, determines a solution for that instance. Generally, problems can be defined as sets of problem instances, where each instance is a pair of input data and solution data. This is an elegant mathematical formalisation; however, in this book we will define problems using a slightly less formal, but more intuitive (yet precise), representation.

For instances of combinatorial problems, we draw an important distinction between candidate solutions and solutions. *Candidate solutions* are potential solutions that may possibly be encountered during an attempt to solve the given problem instance; but unlike solutions, they may not satisfy all the conditions from the problem definition. For our shortest round trip example, typically any valid round trip connecting the given set of points, regardless of length, would be a candidate solution, while only those candidate round trips with minimal length would qualify as solutions. It should be noted that while the definition of any combinatorial problem states clearly what is considered a solution for an instance of this problem, the notion of candidate solution is not always uniquely determined by the problem definition, but can already reflect a particular approach for solving the problem. As an example, consider the variant of the shortest round trip problem in which we are only interested in trips that visit each given point exactly once. In this case, candidate solutions could be either arbitrary round trips which do not necessarily respect this additional condition, or the notion of candidate solution could be restricted to round trips that visit no point more than once.

Decision Problems

Many combinatorial problems can be naturally characterised as *decision problems*: for these, the solutions of a given instance are specified by a set of logical conditions. As an example of a combinatorial decision problem, consider the *Graph Colouring Problem*: given a graph G and a number of colours, find an assignment of colours to the vertices of G such that two vertices that are connected by an edge are never assigned the same colour. Other prominent combinatorial decision problems include finding satisfying truth assignments for a given propositional formula (the *Propositional Satisfiability Problem, SAT*, which we revisit in more detail in Section 1.2) or scheduling a series of events such that a given set of precedence constraints is satisfied. For any decision problem, we distinguish two variants:

> the *search variant*, where, given a problem instance, the objective is to find a solution (or to determine that no solution exists);

> the *decision variant*, in which for a given problem instance, one wants to answer the question whether or not a solution exists.

These variants are closely related because algorithms solving the search variant can always be used to solve the decision variant. Interestingly, for many combinatorial decision problems, the converse also holds: algorithms for the decision variant of a problem can be used for finding actual solutions.

Optimisation Problems

Many practically relevant combinatorial problems are optimisation problems rather than decision problems. *Optimisation problems* can be seen as generalisations of decision problems, where the solutions are additionally evaluated by an *objective function* and the goal is to find solutions with optimal objective function values. The objective function is often defined on candidate solutions as well as on solutions; the objective function value of a given candidate solution (or solution) is also called its *solution quality*. For the Graph Colouring Problem mentioned previously, a natural optimisation variant exists, where a variable number of colours is used and the goal is, given a graph, to find a colouring of its vertices, using only a minimal (rather than a fixed) number of colours.

Any combinatorial optimisation problem can be stated as a *minimisation problem* or as a *maximisation problem*, depending on whether the given objective function is to be minimised or maximised. Often, one of the two formulations is more natural, but algorithmically, minimisation and maximisation problems are treated equivalently. In this book, for uniformity and formal convenience, we generally formulate optimisation problems as minimisation problems. For each combinatorial optimisation problem, we distinguish two variants:

> the *search variant*: given a problem instance, find a solution with minimal (or maximal, respectively) objective function value;

> the *evaluation variant*: given a problem instance, find the optimal objective function value (i.e., the solution quality of an optimal solution).

Clearly, the search variant is the more general of these, since with the knowledge of an optimal solution, the evaluation variant can be solved trivially. Additionally, for each optimisation problem, we can define:

> *associated decision problems*: given a problem instance and a fixed solution quality bound b, find a solution with an objective function value smaller than or equal to b (for minimisation problems; greater than or equal to b for maximisation problems) or determine that no such solution exists.

Many combinatorial optimisation problems are defined based on an objective function as well as on logical conditions. In this case, candidate solutions satisfying the logical conditions are called *feasible* or *valid*, and among those, *optimal solutions* can be distinguished based on their objective function value. While the use of logical conditions in addition to an objective function often leads to more natural formulations of a combinatorial optimisation problem, it should be noted that the logical conditions can always be integrated into the objective function in such a way that the feasible candidate solutions correspond to the solutions of an associated decision problem (i.e., to candidate solutions with bounded solution quality).

As we will see throughout this book, many algorithms for decision problems can be extended to related optimisation problems in a rather natural way. However, such simple extensions of algorithms that work well on certain decision problems are not always effective for finding optimal or near-optimal solutions of the corresponding optimisation problems, and consequently, different algorithmic methods need to be considered for this task.

1.2   **Two Prototypical Combinatorial Problems**

In the following, we introduce two well-known combinatorial problems which will be used throughout the first part of this book for illustrating algorithmic techniques and approaches. These are the Propositional Satisfiability Problem (SAT), a prominent combinatorial decision problem which plays a central role in several areas of computer science, and the Travelling Salesman Problem (TSP), one of the most extensively studied combinatorial optimisation problems. Besides their prominence and well established role in algorithm development, both problems have the advantage of being conceptually simple, which facilitates the development, analysis and presentation of algorithms and algorithmic ideas. Both will be discussed in more detail in Part 2 of this book (see Chapters 6 and 8).

The Propositional Satisfiability Problem (SAT)

Roughly speaking, the Propositional Satisfiability Problem is, given a formula in propositional logic, to decide whether there is an assignment of truth values to the propositional variables appearing in this formula under which the formula evaluates to 'true'. In the following, we present a formal definition of SAT. While the details of this definition may not be crucial for comprehending the restricted forms of the problem used in the remainder of this book, they are important for a deeper understanding of the nature and properties of the general SAT problem.

Propositional logic is based on a formal language over an alphabet comprising propositional variables, truth values and logical operators. Using logical operators, propositional variables and truth values are combined into propositional formulae which represent propositional statements. Formally, the syntax of propositional logic can be defined in the following way:

DEFINITION 1.1 **Syntax of Propositional Logic**

$S := V \cup C \cup O \cup \{(,)\}$ *is the* alphabet of propositional logic, *with* $V := \{x_i \mid i \in \mathbb{N}\}$ *denoting the countable infinite set of* propositional variables, $C := \{\top, \bot\}$ *the set of* truth values *(or propositional constants)* true *and* false, *and* $O := \{\neg, \wedge, \vee\}$ *the set of* propositional operators negation *('not')*, conjunction *('and') and* disjunction *('or')*.

The set of propositional formulae *is characterised by the following inductive definition:*

- *the truth values* $\top$ *and* $\bot$ *are propositional formulae;*

- *each propositional variable* $x_i \in V$ *is a propositional formula;*

- *if* F *is a propositional formula, then* $\neg F$ *is also a propositional formula;*

- *if F_1 and F_2 are propositional formulae, then $(F_1 \wedge F_2)$ and $(F_1 \vee F_2)$ are also propositional formulae.*

Only strings obtained by a finite number of applications of these rules are propositional formulae.

Remark: Often, additional binary operators, such as '$\leftarrow$' (implication) and '$\leftrightarrow$' (equivalence), are used in propositional formulae. These can be defined based on the operators from Definition 1.1; hence, including them into our propositional language does not increase its expressiveness.

Assignments are mappings from propositional variables to truth values. Using the standard interpretations of the logical operators on truth values, assignments can be used to evaluate propositional formulae. Hence, the semantics of propositional logic can be defined as follows:

DEFINITION 1.2 **Semantics of Propositional Logic**

The variable set $Var(F)$ of formula F is defined as the set of all variables appearing in F.

A variable assignment of formula F is a mapping $a : Var(F) \mapsto \{\top, \bot\}$ of the variable set of F to the truth values. The set of all possible variable assignments of F is denoted by $Assign(F)$.

The value $Val(F, a)$ of formula F under assignment a is defined inductively based on the syntactic structure of F:

- $Val(\top, a) := \top$

- $Val(\bot, a) := \bot$

- $Val(x_i, a) := a(x_i)$

- $Val(\neg F_1, a) := \neg Val(F_1, a)$

- $Val(F_1 \wedge F_2, a) := Val(F_1, a) \wedge Val(F_2, a)$

- $Val(F_1 \vee F_2, a) := Val(F_1, a) \vee Val(F_2, a)$

The truth values '$\top$' and '$\bot$' represent logical truth and falsehood, respectively; the operators '$\neg$' (negation), '$\wedge$' (conjunction) and '$\vee$' (disjunction) are defined by the following truth tables:

$\neg$	
$\top$	$\bot$
$\bot$	$\top$

$\wedge$	$\top$	$\bot$
$\top$	$\top$	$\bot$
$\bot$	$\bot$	$\bot$

$\vee$	$\top$	$\bot$
$\top$	$\top$	$\top$
$\bot$	$\top$	$\bot$

Remark: There are many different notations for the truth values '$\top$' and '$\bot$', including '0' and '1', '-1' and '$+1$', 'T' and 'F', as well as 'TRUE' and 'FALSE'. Likewise, the propositional operators '$\neg$', '$\wedge$' and '$\vee$' are often denoted '$-$', '$*$' and '$+$', or 'NOT', 'AND' and 'OR'.

Because the variable set of a propositional formula is always finite, the complete set of assignments for a given formula is also finite. More precisely, for a formula containing n variables there are exactly 2^n different variable assignments.

Considering the values of a formula under all possible assignments, the fundamental notion of satisfiability can be defined in the following way:

DEFINITION 1.3 **Satisfiability**

A variable assignment a is a model *of formula F if, and only if, $Val(F, a) = \top$; in this case we say that a satisfies F.*

A formula F is called satisfiable *if, and only if, there exists at least one model of F.*

Based on the notion of satisfiability, we can now formally define the SAT problem.

DEFINITION 1.4 **The Propositional Satisfiability Problem**

Given a propositional formula F, the Propositional Satisfiability Problem (SAT) *is to decide whether or not F is satisfiable.*

Obviously, SAT can be seen as a combinatorial decision problem, where variable assignments represent candidate solutions and models represent solutions. As for any combinatorial decision problem, we can distinguish a decision variant and a search variant: in the former, only a yes/no decision regarding the satisfiability of the given formula is required; in the latter, also called the *model-finding variant*, in case the given formula is satisfiable, a model has to be found.

Often, logical problems like SAT are studied for syntactically restricted classes of formulae. Imposing syntactic restrictions usually facilitates theoretical studies and can also be very useful for simplifying the design and analysis of algorithms. *Normal forms* are types of syntactically restricted formulae such that for an arbitrary formula F there is always at least one semantically equivalent formula F' in normal form. Thus, each normal form induces a subclass of propositional formulae which is as expressively powerful as full propositional logic. The two most commonly used normal forms, CNF and DNF, are introduced in the following definition.

DEFINITION 1.5 **Normal Forms**

A literal *is a propositional variable (called a* positive literal*) or its negation (called a* negative literal*). Formulae of the syntactic form* $c_1 \wedge c_2 \wedge \ldots \wedge c_m$ *are called* conjunctions, *while formulae of the form* $d_1 \vee d_2 \vee \ldots \vee d_m$ *are called* disjunctions.

A propositional formula F *is in* conjunctive normal form (CNF), *if, and only if, it is a conjunction over disjunctions of literals. In this context, the disjunctions are called* clauses. *A CNF formula* F *is in* k-CNF, *if, and only if, all clauses of* F *contain exactly* k *literals.*

A propositional formula F *is in* disjunctive normal form (DNF), *if, and only if, it is a disjunction over conjunctions of literals. In this case, the conjunctions are called* clauses. *A DNF formula* F *is in* k-DNF, *if, and only if, all clauses of* F *contain exactly* k *literals.*

EXAMPLE 1.1 **A Simple SAT Instance**

Let us consider the following propositional formula in CNF:

$$
\begin{aligned}
F := &(\neg x_1 \vee x_2) \\
\wedge\, &(\neg x_2 \vee x_1) \\
\wedge\, &(\neg x_1 \vee \neg x_2 \vee \neg x_3) \\
\wedge\, &(x_1 \vee x_2) \\
\wedge\, &(\neg x_4 \vee x_3) \\
\wedge\, &(\neg x_5 \vee x_3)
\end{aligned}
$$

For this formula, we obtain the variable set $Var(F) = \{x_1, x_2, x_3, x_4, x_5\}$; consequently, there are $2^5 = 32$ different variable assignments. Exactly one of these, $x_1 = x_2 = \top, x_3 = x_4 = x_5 = \bot$, is a model, rendering F satisfiable.

The Travelling Salesman Problem (TSP)

The motivation behind the Travelling Salesman Problem (also known as Travelling Salesperson Problem) is the problem faced by a salesperson who needs to visit a number of customers located in different cities and tries to find the shortest round trip accomplishing this task. In a more general and abstract formulation, the TSP is, given a directed, edge-weighted graph, to find a shortest cyclic path that visits every node in this graph exactly once. In order to define this problem formally, we first introduce the notion of a Hamiltonian cycle:

DEFINITION 1.6 **Path, Hamiltonian Cycle**

Let $G := (V, E, w)$ be an edge-weighted, directed graph where $V := \{v_1, v_2, \ldots, v_n\}$ is the set of $n = \#V$ vertices, $E \subseteq V \times V$ the set of (directed) edges, and $w : E \mapsto \mathbb{R}^+$ a function assigning each edge $e \in E$ a weight $w(e)$.

A path in G is a list $(u_1, u_2, \ldots, u_k)$ of vertices $u_i \in V$ $(i = 1, \ldots, k)$, such that any pair (u_i, u_{i+1}), $i = 1, \ldots, k-1$, is an edge in G. A cyclic path in G is a path for which the first and the last vertex coincide, i.e., $u_1 = u_k$ in the above notation.

A Hamiltonian cycle in G is a cyclic path p in G that visits every vertex of G (except for its starting point) exactly once, i.e., $p = (u_1, u_2, \ldots, u_n, u_1)$ is a Hamiltonian cycle in G if, and only if, $n = \#V$, and $\{u_1, u_2, \ldots, u_n\} = V$.

The weight of a path p can be calculated by adding up the weights of the edges in p:

DEFINITION 1.7 **Path Weight**

For a given edge-weighted, directed graph and a path $p := (u_1, \ldots, u_k)$ in G, the path weight *$w(p)$ is defined as $w(p) := \sum_{i=1}^{k-1} w((u_i, u_{i+1}))$.*

Now, the TSP can be formally defined in the following way:

DEFINITION 1.8 **The Travelling Salesman Problem**

Given an edge-weighted, directed graph G, the Travelling Salesman Problem (TSP) *is to find a Hamiltonian cycle with minimal path weight in G.*

Often, the TSP is defined in such a way that the underlying graphs are always complete graphs, that is, any pair of vertices is connected by an edge, because for any TSP instance with an underlying graph G that is not complete, one can always construct a complete graph G' such that the TSP for G' has exactly the same solutions as the one for G. (This is done by choosing the edge weights for edges missing in G high enough that these edges can never occur in an optimal solution.) In the remainder of this book we will always assume that TSP instances are specified as complete graphs. Under this assumption, the Hamiltonian cycles in a given graph correspond exactly to the cyclic permutations of the underlying vertex set.

Interesting subclasses of the TSP arise when the edge weighting function w has specific properties. The following definition covers some commonly used cases:

DEFINITION 1.9 **Asymmetric, Symmetric and Euclidean TSP Instances**

A TSP instance is called symmetric *if, and only if, the weight function w of the underlying graph is symmetric, that is, if for all $v, v' \in V$, $w((v, v')) = w((v', v))$; if w is not symmetric, the instance is called* asymmetric. *The Travelling Salesman Problem for asymmetric instances is also called the* Asymmetric TSP (ATSP).

A symmetric TSP instance satisfies the triangle inequality if, and only if, $w((u_1, u_3)) \leq w((u_1, u_2)) + w((u_2, u_3))$ for any triples of different vertices u_1, u_2 and u_3. A TSP instance is metric *if, and only if, the vertices in the given graph correspond to points in a metric space such that the edge weight between any two vertices corresponds to the metric distance between the respective points. A TSP instance is called* Euclidean *if, and only if, the vertices correspond to the points in a Euclidean space and if the weight function w is a Euclidean distance metric. Finally, TSP instances for which the vertices are points on a sphere and the weight function w represents geographical (great circle) distance are called* geographic.

EXAMPLE 1.2 **A Sample (Geographic) TSP Instance**

Figure 1.1 shows a geographic TSP instance with 16 vertices. The vertices of the underlying graph correspond to 16 locations Ulysses is reported to have visited on his odyssey, and the edge weights represent the geographic distances between these locations. The figure also shows the optimal solution, that is, the shortest round trip (length 6 859 km). This tour is calculated based on direct air distances, and can only be travelled by a 'modern Ulysses' using an aircraft. This TSP instance has been first described by Grötschel and Padberg [1993]; it can be found as `ulysses16.tsp` in the TSPLIB Benchmark Library [Reinelt, 2003].

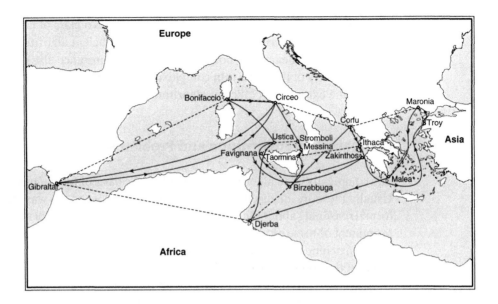

Figure **1.1** A graphic representation of the geographic TSP instance 'ulysses16' and its optimal solution (dashed line); the solid line and arrows indicate the sequence in which Homer's Ulysses supposedly visited the 16 locations. See Example 1.2 for details.

1.3 **Computational Complexity**

A natural way for solving most combinatorial decision and optimisation problems is, given a problem instance, to search for solutions in the space of its candidate solutions. For that reason, these problems are sometimes also characterised as *search problems*. However, for a given instance of a combinatorial problem, the set of candidate solutions is very large, typically at least exponential in the size of that instance. For example, given a SAT instance with 100 variables, typically all 2^{100} different truth assignments are considered candidate solutions. This raises the following question: 'Is it possible to search such vast spaces efficiently?' More precisely, we are interested in the time required for solving an instance of a combinatorial problem as a function of the size of this instance.

Questions like this lie at the core of computational complexity theory, a well-established field of computer science with considerable impact on other areas. In the context of this book, complexity theory plays a role, because the primary field of application of stochastic local search algorithms is a class of computationally very hard combinatorial problems, for which no efficient algorithms are known (where efficient means polynomial run-time w.r.t. instance size). Moreover, to date a majority of the experts in complexity theory believe that for fundamental reasons the existence of efficient algorithms for these problems is impossible.

Complexity of Algorithms and Problems

The complexity of an algorithm is defined on the basis of formal machine models. Usually, these are idealised, yet universal models, designed in a way that facilitates formal reasoning about their behaviour. One of the first, and still maybe the most prominent of these model is the Turing machine. For Turing machines and other formal machine or programming models, computational complexity is defined in terms of the space and time requirements of computations.

Complexity theory usually deals with problem classes (generally countable sets of problem instances) instead of single instances. For a given algorithm or machine model, the complexity of a computation is characterised by the functional dependency between the size of an instance and the time and space required to solve this instance. Here, instance size is defined as the length of a reasonably concise description; hence, for a SAT instance, its size corresponds to the length of the propositional formula (written in linear form), while the size of a TSP instance is typically proportional to the size of the underlying graph.

For reasons of analytical tractability, many problems are formulated as decision problems, and time and space complexity are analysed in terms of the worst-case asymptotic behaviour. Given a suitable definition of the computational complexity of an algorithm for a specific problem, the complexity of the problem itself can be defined as the complexity of the best algorithm for this problem. Because generally time complexity is the more restrictive factor, problems are often categorised into complexity classes with respect to their asymptotic worst-case time complexity.

$\mathcal{NP}$-hard and $\mathcal{NP}$-complete Problems

Two particularly interesting complexity classes are $\mathcal{P}$, the class of problems that can be solved by a *deterministic* machine in polynomial time, and $\mathcal{NP}$, the class of problems that can be solved by a *nondeterministic* machine in polynomial time. (Note that nondeterministic machines are *not* equivalent to machines that make random choices; they are hypothetical machines which can be thought of as having the ability to make correct guesses for certain decisions.) Of course, every problem in $\mathcal{P}$ is also contained in $\mathcal{NP}$, because deterministic calculations can be emulated on a nondeterministic machine. However, the question whether also $\mathcal{NP} \subseteq \mathcal{P}$, and consequently $\mathcal{P} = \mathcal{NP}$, is one of the most prominent open problems in computer science. Since many extremely application-relevant problems are in $\mathcal{NP}$, but possibly not in $\mathcal{P}$ (i.e., no polynomial-time deterministic algorithm is known), this so-called $\mathcal{P}$ *vs* $\mathcal{NP}$ *Problem* is not only of theoretical interest. For these computationally hard problems, the best algorithms known so far have exponential time complexity. Therefore, for growing problem size, the problem instances become quickly intractable, and even tremendous advances in hardware design have little effect on the size of the problem instances solvable with state-of-the-art technology in reasonable time.

Many of these hard problems from $\mathcal{NP}$ are closely related and can be translated into each other in polynomial deterministic time (these translations are also called *polynomial reductions*). A problem that is at least as hard as any other problem in $\mathcal{NP}$ (in the sense that each problem in $\mathcal{NP}$ can be polynomially reduced to it) is called $\mathcal{NP}$-*hard*. Thus, $\mathcal{NP}$-hard problems in some sense can be regarded as at least as hard as every problem in $\mathcal{NP}$. But they do not necessarily have to belong to the class $\mathcal{NP}$ themselves, as their complexity may actually be higher. $\mathcal{NP}$-hard problems that are contained in $\mathcal{NP}$ are called $\mathcal{NP}$-*complete*; in a certain sense, these problems are the hardest problems in $\mathcal{NP}$.

The SAT problem, introduced in Section 1.2, is the prototypical $\mathcal{NP}$-complete problem. Historically, it was the first problem for which $\mathcal{NP}$-completeness was established [Cook, 1971]. $\mathcal{NP}$-completeness of SAT can directly be proven by encoding the calculations of a Turing machine M for an $\mathcal{NP}$ problem into a propositional formula whose models correspond to the accepting computations of M. Furthermore, it is quite easy to show that SAT remains $\mathcal{NP}$-complete when restricted to CNF or even 3-CNF formulae (see, e.g., Garey and Johnson [1979]). On the other hand, SAT is decidable in linear time for DNF, for 2-CNF [Cook, 1971] and for Horn formulae [Dowling and Gallier, 1984].

Our second example problem, the TSP, is known to be $\mathcal{NP}$-hard [Garey and Johnson, 1979]. The same holds for many special cases, such as Euclidean TSPs and even TSPs in which all edge weights are either one or two. In all of these cases, the associated decision problem for optimal solution quality is $\mathcal{NP}$-complete. However, there exist a number of polynomially solvable special cases of the TSP, such as fractal TSP instances that are generated by so-called *Lindenmayer Systems* [Moscato and Norman, 1998] or specially structured Euclidean instances where, for example, all vertices lie on a circle; for an extensive overview of polynomially solvable special cases of the TSP we refer to Burkard et al. [1998b] and Gilmore et al. [1985].

Besides SAT and TSP, many other well-known combinatorial problems are $\mathcal{NP}$-hard or $\mathcal{NP}$-complete, including the Graph Colouring Problem, the Knapsack Problem, as well as many scheduling and timetabling problems, to name just a few [Garey and Johnson, 1979]. It should be noted that for $\mathcal{NP}$-complete combinatorial decision problems, the search and decision variants are equally hard in the sense that if one could be solved deterministically in polynomial time, the same would apply to the other. This is the case because any algorithm for the search variant also solves the decision variant; and furthermore, given a decision algorithm and a specific problem instance, a solution (if existent) can be constructed by iteratively fixing solution components and deciding solubility of the resulting, modified instance (this approach requires only a polynomial number of calls to the decision algorithm). In the same sense, for $\mathcal{NP}$-hard optimisation problems, the search and evaluation variants are equally hard. Furthermore, if either of these variants could be solved efficiently (i.e., in polynomial time on a deterministic machine), all decision variants could be solved efficiently as well; and if all decision variants could be solved efficiently, the same would hold for the search and evaluation variant.

One fundamental result of complexity theory states that it suffices to find a polynomial time deterministic algorithm for one single $\mathcal{NP}$-complete problem to prove that $\mathcal{NP} = \mathcal{P}$. This is a consequence of the fact that all $\mathcal{NP}$-complete problems can be encoded into each other in polynomial time. Today, most experts believe that $\mathcal{P} \neq \mathcal{NP}$; however, so far all efforts of finding a proof for this inequality have been unsuccessful, and there has been some speculation that today's mathematical methods might be too weak to solve this fundamental problem.

Not All Combinatorial Problems are Hard

Although many combinatorial problems are $\mathcal{NP}$-hard, it should be noted that not every computational task that can be formulated as a combinatorial problem is inherently difficult. A well-known example for a problem that, at the first glance, might seem to require searching an exponentially large space of candidate solutions is the *Shortest Path Problem*: given an edge-weighted graph G (where all edge weights are positive) and two vertices u, v in G, find the shortest route from u to v, that is, the path with minimal total edge weight. Fortunately, this shortest path problem can be solved efficiently; in particular, a simple recursive scheme for calculating all pairwise distances between u and any other vertex in the given graph, known as Dijkstra's algorithm [Dijkstra, 1959], can find shortest paths in quadratic time w.r.t. the number of vertices in the given graph. In general, there are many other combinatorial problems that can be solved by polynomial-time algorithms. In many cases, these efficient algorithms are based on a general method called *dynamic programming* (cf. [Bertsekas, 1995]).

Practically Solving Hard Combinatorial Problems

Nevertheless, many practically relevant combinatorial problems, such as scheduling and planning problems, are $\mathcal{NP}$-hard and therefore generally not efficiently solvable to date (and possibly, if $\mathcal{NP} \neq \mathcal{P}$, not efficiently solvable at all). However, being $\mathcal{NP}$-complete or $\mathcal{NP}$-hard does not mean that it is impossible for a problem to be solved efficiently. Practically, there are at least three ways of dealing with these problems:

- find an application relevant subclass of the problem that can be solved efficiently;
- use efficient approximation algorithms;
- use stochastic approaches.

Regarding the first strategy, we have to keep in mind that $\mathcal{NP}$-hardness is a property of an entire problem class Π, whereas in practice, often only instances from a certain subclass $\Pi' \subseteq \Pi$ occur. In general, Π' need not be $\mathcal{NP}$-hard, that is, while for Π an efficient algorithm might not exist, it may still be possible to find an efficient algorithm for the subclass Π'; as an example consider the SAT problem for 2-CNF formulae, which is polynomially solvable.

Furthermore, $\mathcal{NP}$-hardness results characterise the worst-case complexity of a problem, and typical problem instances may be much easier to solve. Formally, this can be captured by the notion of *average case complexity*; although average case complexity results are typically significantly harder to prove and are hence much rarer than worst-case results, empirical studies suggest that for many $\mathcal{NP}$-hard problems, typical or average case instances can be solved reasonably efficiently. The same applies to the time complexity of concrete algorithms for combinatorial problems; a well-known example is the Simplex Algorithm for linear optimisation, which has worst-case exponential time complexity [Klee and Minty, 1972], but has been empirically shown to achieve polynomial run-times (w.r.t. problem size) in the average case.

In the case of an $\mathcal{NP}$-hard optimisation problem that cannot be narrowed down to an efficiently solvable subclass, another option is to accept suboptimal solutions. Formally, the degree of suboptimality of a solution quality q is typically expressed in the form of the *approximation ratio*, defined as q/q^* for a minimisation problem, and q^*/q for a maximisation problem, where q^* is the optimal solution quality for the given problem instance.

For a given optimisation problem we can then consider *associated approximation problems*, in which the objective is to find solutions with an approximation ratio bounded from above by a given constant $r > 1$. Often, as r is increased, the computational complexity of these approximation problems decreases to the point where they become practically solvable. In some cases, allowing a relatively small margin from the optimal solution quality renders the problem deterministically solvable in polynomial time. In other cases, the approximation problem remains $\mathcal{NP}$-hard, while for practically occurring problem instances, suboptimal solutions of acceptable quality can be found in reasonable time.

For example, it is well known that the general TSP for instances with arbitrary edge weights is not efficiently approximable to any constant factor, that is, there is no deterministic algorithm that is guaranteed to find solutions of quality within a constant factor of the optimum for a given problem instance in polynomial time [Sahni and Gonzalez, 1976]. Yet, for instances satisfying the triangle inequality, Christofides' polynomial-time construction algorithm guarantees an approximation ratio of at most 1.5 [Christofides, 1976]. Furthermore, in the case of Euclidean TSP instances, a *polynomial time approximation scheme* exists, that is, there are algorithms that find solutions for arbitrary approximation ratios larger than one in polynomial time w.r.t. instance size [Arora, 1998].

Sometimes, however, even reasonably efficient approximation methods cannot be devised, or the problem is a decision problem, to which the notion of approximation cannot be applied at all. In these cases, one further option is to focus on probabilistic rather than deterministic algorithms. At first glance, this idea seems to be appealing. After all, according to the definition of the complexity class $\mathcal{NP}$, at least $\mathcal{NP}$-complete problems can be efficiently solved by (hypothetical) nondeterministic machines. But this, of course, is of little practical use, since it is unlikely that such idealised machines can be built; and for an actual probabilistic algorithm there is typically merely a small chance that it can solve the given problem in polynomial time. In practice, the success probability of such an algorithm can be arbitrarily small. Nevertheless, in numerous cases, probabilistic algorithms have been found to be considerably more efficient on $\mathcal{NP}$-complete or $\mathcal{NP}$-hard problems than the best deterministic methods available. In other cases, probabilistic methods and deterministic methods complement each other in the sense that for certain types of problem instances one or the other has been found to be superior. SAT and TSP, the two combinatorial problems introduced previously, are amongst the most fundamental and best-known problems in this category.

Finally, it should be noted that even truly exponential scaling of run-time with instance size does not necessarily rule out solving practically relevant problem instances. For theoretical purposes, complexity analysis typically focuses on

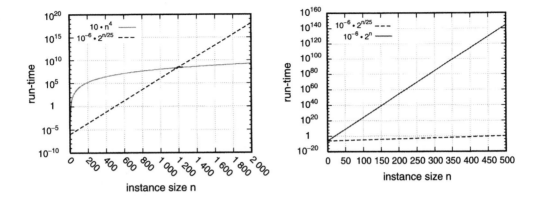

Figure 1.2 *Left:* Polynomial-time algorithms are not always better than exponential-time algorithms; in this example, for problem instances of size smaller than 1 187, the exponential-time algorithm performs better than the polynomial-time algorithm. *Right:* Performance differences between two exponential-time algorithms; in this example, one algorithm can solve instances up to size $n = 500$ in the same time required by the other algorithm for instances of size 20.

asymptotic behaviour, and for exponential scaling, constants (such as the base of the exponential) are mostly not considered. In practice, however, these constants are obviously extremely important, especially, when the size of problem instances that need to be solved has reasonable upper bounds. Consider, for example, an algorithm A with time complexity $10^{-6} \cdot 2^{n/25}$ (where n is the problem size), and another algorithm B with time complexity $10 \cdot n^4$ (see also Figure 1.2). Of course, for big problem instances, here about $n > 1\,187$, A becomes quickly dramatically more costly than B. However, for $n \leq 1\,100$, A is much more efficient than B (for $n = 100$, the performance ratio is larger than 10^8 in favour of the exponential-time algorithm). It is important to keep in mind that exponential complexity should be avoided whenever possible, and does eventually, as instance size grows, render the application of an algorithm infeasible. However, for many problems where exponential time complexity is unavoidable (unless $\mathcal{P} = \mathcal{NP}$), some algorithms, though exponential in time complexity, or even incomplete, can still be dramatically more efficient than others and hence make it feasible to solve the problem for practically interesting instance sizes (see Figure 1.2, right side). This is where heuristic guidance, combined with randomisation and probabilistic decisions (all of which are central issues of this book), can make the difference.

IN DEPTH SOME ADVANCED CONCEPTS IN COMPUTATIONAL COMPLEXITY

Besides the fundamental notions of $\mathcal{NP}$-hardness and $\mathcal{NP}$-completeness discussed above, there are a number of other concepts from complexity theory that are of interest in the context of combinatorial problems. This in-depth section briefly covers some of these advanced concepts which, although not essential for understanding the material presented in the main text, are particularly relevant to the topics discussed later in this book.

The instances of many combinatorial problems contain numbers, such as the edge weights in the TSP. For some of these problems, there are algorithms whose run-time is bounded from above by a polynomial function over $|\pi|$, the size of a given problem instance π, and $\text{Max}(\pi)$, the maximum absolute value of any number occurring in π; such algorithms are called *pseudo-polynomial*. Note that the run-time of a pseudo-polynomial algorithm may be exponential in the size of π, and hence not have polynomial time complexity. (The reason for this is that the maximum absolute value of any number in π can be exponential in $|\pi|$.) $\mathcal{NP}$-complete decisions problems for which a pseudo-polynomial algorithms is known are called *weakly $\mathcal{NP}$-complete* or *$\mathcal{NP}$-complete in the weak sense*, while those that remain $\mathcal{NP}$-complete, even if run-time is measured as a function of $\text{Max}(\pi)$ and $\text{Max}(\pi)$ is bounded from above by a polynomial in $|\pi|$, are called *strongly $\mathcal{NP}$-complete* or *$\mathcal{NP}$-complete in the strong sense*. A prominent example of an $\mathcal{NP}$-complete problem for which a pseudo-polynomial algorithm is known is the Knapsack Problem; examples for strongly $\mathcal{NP}$-complete problems include TSP and the Set Covering Problem (see Chapter 10, Section 10.3) [Garey and Johnson, 1979].

The notion of $\mathcal{NP}$-hardness applies to decision and optimisation problems alike. Slightly more technically as above, $\mathcal{NP}$-hardness can be defined as follows: A problem Π is *$\mathcal{NP}$-hard* if the existence of a polynomial time algorithm for Π implies the existence of a polynomial time algorithm for all $\mathcal{NP}$-complete problems (and hence, for all problems in $\mathcal{NP}$). Conversely, a problem Π is *$\mathcal{NP}$-easy* if the existence of a polynomial-time algorithm for any $\mathcal{NP}$-complete problem implies the existence of a polynomial time algorithm for Π. While strictly speaking, the notion of $\mathcal{NP}$-completeness only applies to decision problems, the relationship between the complexity of many hard optimisation problems and their associated decision problems can be formally captured by the concept of $\mathcal{NP}$-equivalence: an optimisation problem is *$\mathcal{NP}$-equivalent* if, and only if, it is $\mathcal{NP}$-hard and $\mathcal{NP}$-easy. It is well-known and relatively easy to show that the TSP is $\mathcal{NP}$-equivalent, while its associated decision problem is $\mathcal{NP}$-complete [Garey and Johnson, 1979].

As discussed previously, one of the strategies for solving hard combinatorial optimisation problems more efficiently is to settle for suboptimal solution qualities. The following concepts capture the computational complexity of such approximations more precisely.

A *polynomial-time approximation scheme (PTAS)* for a combinatorial optimisation problem is an algorithm that is guaranteed to achieve an approximation ratio of $r = 1+\epsilon$ for any given $\epsilon > 0$ in time bounded from above by a polynomial function of the size of the given problem instance π. A polynomial-time approximation scheme whose run-time is also at most polynomial in $1/\epsilon$ is called a *fully-polynomial-time approximation scheme (FPTAS)*. The class of problems that have polynomial-time approximation schemes and fully-polynomial-time approximation schemes are called $\mathcal{PTAS}$ and $\mathcal{FPTAS}$, respectively. An example of a problem in $\mathcal{PTAS}$ is the Euclidean TSP [Arora, 1998], while the Knapsack Problem is known to be in $\mathcal{FPTAS}$ [Ibarra and Kim, 1975].

The complexity class $\mathcal{APX}$ comprises all optimisation problems for which there exists an algorithm that is guaranteed to find a solution within a constant factor of the optimal solution quality of any given instance. Note that $\mathcal{APX}$ contains $\mathcal{PTAS}$. A problem Π in $\mathcal{APX}$ is *$\mathcal{APX}$-complete* if the existence of a polynomial-time approximation scheme for Π implies the existence of polynomial-time approximation schemes for all problems in $\mathcal{APX}$. Hence, $\mathcal{APX}$-complete problems are the hardest $\mathcal{APX}$ problems, as

$\mathcal{NP}$-complete problems are the hardest problems in $\mathcal{NP}$. Prominent examples for $\mathcal{APX}$-complete problems are metric TSP and MAX-SAT, a well-known optimisation variant of SAT (see also Chapter 95) [Ausiello et al., 1999] .

1.4   Search Paradigms

Basically all computational approaches for solving hard combinatorial problems can be characterised as search algorithms. The fundamental idea behind the search approach is to iteratively generate and evaluate candidate solutions; in the case of combinatorial decision problems, evaluating a candidate solution means to decide whether it is an actual solution, while in the case of an optimisation problem, it typically involves determining the respective value of the objective function. Although for $\mathcal{NP}$-hard combinatorial problems the time complexity of finding solutions can grow exponentially with instance size, evaluating candidate solutions can often be done much more efficiently, that is, in polynomial time. For example, for a given TSP instance, a candidate solution would correspond to a round trip visiting each vertex of the given graph exactly once, and its objective function value can be computed easily by summing up the weights associated with all the edges used for that round trip.

Generally, the evaluation of candidate solutions much depends on the given problem, and is often rather straightforward to implement. The fundamental differences between search algorithms are in the way in which candidate solutions are generated, which can have a very significant impact on the algorithms' theoretical properties and practical performance. In this context, general mechanisms can be defined that are applicable to a broad range of search problems. Consequently, in the remainder of this section, we discuss various search paradigms based on their underlying approaches to generating candidate solutions.

Perturbative *vs* Constructive Search

Candidate solutions for instances of combinatorial problems are composed of *solution components*, such as the assignments of truth values to individual propositional variables (atomic assignments) in the case of SAT. Hence, given candidate solutions can easily be changed into new candidate solutions by modifying one or more of the corresponding solution components. This can be characterised as perturbing a given candidate solution, and hence we classify search algorithms that rely on this mechanism for generating the candidate solutions to be tested as *perturbative search methods*. Applied to SAT, perturbative search would start with one or more complete truth assignments and then at each step generate other truth assignments by changing the truth values of a number of variables in each such assignment.

While for perturbative approaches, the search typically takes place directly in the space of candidate solutions, it can sometimes be useful to also include *partial candidate solutions* in the search space, that is, candidate solutions in which one or more solution components are missing. Examples for such partial candidate solutions are partial truth assignments for a SAT instance which leave the truth values of some propositional variables unspecified, and partial round trips for a TSP instance, which correspond to paths in the corresponding graph that visit a subset of the vertices and can be extended into Hamiltonian cycles by adding additional edges.

The task of generating (complete) candidate solutions by iteratively extending partial candidate solutions can be formulated as a search problem in which typically the goal is to obtain a 'good' candidate solution, where for optimisation problems, the goodness corresponds to the value of the objective function. Algorithms for solving this type of problem are called *constructive search methods* (or *construction heuristics*).

As a simple example, consider the following method for generating solution candidates for a given TSP instance. Start at a randomly chosen vertex in the graph, and then iteratively follow an edge with minimal weight connecting the current vertex to one of the vertices that has not yet been visited. This method generates a path that, by adding the starting vertex as a final element to the corresponding list, can be easily extended into a Hamiltonian cycle in the given graph, that is, a candidate solution for the TSP instance. This simple construction heuristic for the TSP is called the *Nearest Neighbour Heuristic*; on its own, it typically does not generate candidate solutions with close-to-optimal objective function values, but it is commonly and successfully used in combination with perturbative search methods (this will be discussed in more detail in Chapter 8).

Systematic *vs* Local Search

A different, and more common, classification of search approaches is based on the distinction between systematic and local search: *Systematic search algorithms* traverse the search space of a problem instance in a systematic manner which guarantees that eventually either a (optimal) solution is found, or, if no solution exists, this fact is determined with certainty. This typical property of algorithms based on systematic search is called *completeness*. *Local search algorithms*, on the other hand, start at some location of the given search space and subsequently move from the present location to a neighbouring location in the search space, where each location has only a relatively small number of neighbours, and each of the moves is determined by a decision based on local knowledge only. Typically, local search algorithms are *incomplete*, that is, there is no guarantee that an existing solution is eventually found, and the fact that no solution exists can never be determined with certainty. Furthermore, local search methods can visit the same location within the search space more than once. In fact, many local search algorithms are prone to getting stuck in some part of the search space which they cannot escape from without using special mechanisms, such as restarting the search process or performing some type of diversification steps.

As an example for a simple local search method for SAT, consider the following algorithm: given a propositional formula F in CNF over n propositional variables, randomly pick a variable assignment as a starting point. Then, in each step, check whether the current variable assignment satisfies F. If not, randomly select a variable, and change its truth value from $\perp$ to $\top$ or vice versa. Terminate the search when a model is found, or after a specified number of search steps have been performed unsuccessfully. This algorithm is called *Uninformed Random Walk* and will be revisited in Section 1.5.

To obtain a simple systematic search algorithm for SAT, we modify this local search method in the following way. Given an ordering of the n propositional variables, with each variable assignment a we uniquely associate a number k between 0 and $2^n - 1$ such that digit i of the binary representation of k is 1 if, and only if, assignment a assigns $\top$ to propositional variable i. Our systematic search algorithm starts with the variable assignment setting all propositional variables to $\perp$, which corresponds to the number 0. Then, in each step we move to the variable assignment obtained by incrementing the numerical value associated with the current assignment by one. The procedure terminates when the current assignment satisfies F or after $2^n - 1$ of these steps. Obviously, this procedure searches the space of all variable assignments in a systematic way and will either return a model of F or terminate unsuccessfully after $2^n - 1$ steps, in which case we can be certain that F is unsatisfiable.

Local Search = Perturbative Search?

Local search methods are often, but not always based on perturbative search. The Uninformed Random Walk algorithm for SAT introduced previously is a typical example of a perturbative local search algorithm, because in each search step we change the truth value assigned to one variable, which corresponds to a perturbation of a candidate solution. However, local search can also be used for constructive search processes. This is exemplified by the Nearest Neighbour

Heuristic for the TSP introduced earlier in this section, where vertices are iteratively added to a given partial tour based on the weight of the edges leading to vertices adjacent to the last vertex on that tour. Clearly, this process corresponds to a constructive local search on the given graph. Generally, construction heuristics can be interpreted as constructive local search methods, and as we will see in Chapter 2, there are some prominent examples of SLS methods based on constructive local search.

In many cases, constructive local search can be combined with perturbative local search. A typical example is the use of the Nearest Neighbour Heuristic for generating the starting points for a perturbative local search algorithm for the TSP. Another interesting example is Ant Colony Optimisation [Dorigo and Di Caro, 1999], which can be seen as a perturbative search method where in each step one or more constructive local searches are performed. (See also Chapter 2, Section 2.4.)

Interestingly, perturbative search, although naturally associated with local search methods, can also provide the basis for systematic search algorithms. As an example, let us consider the systematic variant of the Uninformed Random Walk algorithm for SAT presented on page 33. The steps of this search algorithm correspond to perturbations of complete variable assignments; consequently, the algorithm can be considered a perturbative systematic search method. As this example shows, perturbative search methods can be complete. It should be noted, however, that we are presently not aware of any perturbative systematic search methods that achieve competitive performance on any hard combinatorial problem.

Constructive Search + Backtracking = Systematic Search

Another interesting relationship can be established between constructive search methods and systematic search algorithms. Let us once more consider our prototypical example for constructive search, the Nearest Neighbour Heuristic for the TSP. If we modify this algorithm such that in each step of the construction process the given partial tour can be extended with arbitrary neigbours of its last vertex, it is clear that the constructive search method thus obtained can in principle find the optimal solution to any given TSP instance. Hence, an algorithm which could systematically enumerate all such constructions would obviously be guaranteed to solve arbitrary TSP instances optimally (given sufficient time), that is, it would be complete.

Such a complete algorithm for the TSP can be obtained easily by combining the Nearest Neighbour Heuristic with *backtracking*. At each choice point of the construction algorithm (including the initial vertex), a list of all alternative choices is kept. Once a complete tour has been generated, the search process 'backtracks' to the most recent choice point at which unexplored alternatives exist, and the constructive search is resumed there using an alternate vertex at this point. This backtracking process first tries alternate choices for recent decisions (which are deep in the corresponding search tree), and once all alternatives are explored for a given choice point, revisits earlier choices. In this latter case, all subsequent choice points are newly generated, that is, in our example, from that point on, we first use the Nearest Neighbour Heuristic to generate another complete tour, and then recursively continue to revise the choices made in this process.

Visiting all solutions by means of a backtrack search method leads to an algorithm with at least exponential time complexity, which becomes rapidly infeasibly even for relatively small problem instances. Fortunately, in many situations it is possible to prune large parts of the corresponding search tree which can be shown to not contain any solutions. For example, in the case of the TSP, the search on a given branch can be terminated if the length of the current partial tour plus a lower bound on the length of the completion of the tour exceeds the shortest tour found in the search so far. This type of algorithm is called *branch & bound* or *A* search* in the operations research and artificial intelligence communities, respectively.

For SAT, one can easily devise a backtrack algorithm that searches a binary search tree in which each node corresponds to assigning a truth value to one variable, which is then fixed for the subtree beneath that node. This tree can be pruned considerably by using *unit propagation*, a technique that propagates the logical consequences of particular atomic variable assignments down the search tree and effectively eliminates subtrees from the search that cannot contain a model of the given formula. Unit propagation is one of the key techniques used in all state-of-the-art systematic search algorithms for SAT.

In general, systematic backtracking is a recursive mechanism which can be used to build a complete search algorithm on top of a constructive search method. This approach can be applied to basically any constructive search algorithm. Moreover, many prominent and successful systematic search algorithms can be decomposed into a constructive search method and some form of backtracking. It should be noted that the construction methods used in this context need not be as 'greedy' as the Nearest Neighbour Heuristic. Furthermore, although many well-known systematic search algorithms are deterministic, it is possible to combine randomised construction heuristics with backtracking in order to obtain stochastic systematic search algorithms (see, e.g., Gomes et al. [1998]).

There is also some flexibility in the backtracking mechanisms, which do not have to revisit choices in the simple recursive manner indicated above; in fact, as long as there is a reasonably compact representation of all unexplored candidate solutions, essentially any strategy that guarantees to eventually evaluate all of these leads to a complete search algorithm. In particular, this allows the order in which decisions are revisited to be randomised or dynamically changed based on search progress — approaches which provide the basis for some of the best known systematic search algorithms for combinatorial problems such as SAT.

Advantages and Disadvantages of Local Search

It might appear that due to their incompleteness, local search algorithms are generally inferior to systematic methods. But as will be shown later, this is not the case. Firstly, many problems are of a constructive nature and their instances are known to be solvable. In this situation, the goal of any search algorithm is to generate a solution rather than just to decide whether one exists. This holds in particular for optimisation problems, such as the Travelling Salesman Problem (TSP), where the actual problem is to find a solution of sufficiently high quality, but also for underconstrained decision problems, which are not uncommon in practice. Obviously, the main advantage of a complete algorithm — its ability to detect that a given problem instance has no solution — is not relevant for finding solutions of solvable instances.

Secondly, in a typical application scenario the time to find a solution is often limited. Examples for such real-time problems can be found in virtually all application domains. Actually one might argue that almost every real-world problem involving interaction with the physical world, including humans, has real-time constraints. Common examples are real-time production scheduling, robot motion planning and decision making, most game playing situations, and speech recognition for natural language interfaces. In these situations, systematic algorithms often have to be aborted after the given time has been exhausted, which, of course, renders them incomplete. This is particularly problematic for certain types of systematic optimisation algorithms that search through spaces of partial solutions without computing complete solutions early in the search (this is the case for many dynamic programming algorithms); if such a systematic algorithm is aborted prematurely, usually no solution candidate is available, while in the same situation local search algorithms typically return the best solution found so far.

Ideally, algorithms for real-time problems should be able to deliver reasonably good solutions at any point during their execution. For optimisation problems this typically means that run-time and solution quality should be positively correlated; for decision problems one could guess a solution when a time-out occurs, where the accuracy of the guess should increase with the run-time of the algorithm. This so-called *any-time property* of algorithms is usually difficult to achieve, but in many situations the local search paradigm is naturally suited for devising any-time algorithms.

Generally, systematic and local search algorithms are somewhat complementary in their applications. An example for this can be found in Kautz and Selman's work on solving SAT-encoded planning problems, where a fast local search algorithm is used for finding solutions whose optimality is proven by means of a systematic search algorithm [Kautz and Selman, 1996]. As we will discuss later in more detail, local search algorithms are often advantageous in certain situations, particularly if reasonably good solutions are required within a short time, if parallel processing is used and if the knowledge about the problem domain is rather limited. In other cases, particularly when provably optimal solutions are required, time constraints are less important and some knowledge about the problem domain can be exploited, systematic search may be the better choice. There is also some evidence that for certain problems, different types of instances are more effectively solved using local or systematic search methods, respectively. Unfortunately, to date the general question of when to prefer local search over systematic methods and vice versa remains mostly unanswered.

1.5 Stochastic Local Search

Many widely known and high-performance local search algorithms make use of randomised choices in generating or selecting candidate solutions for a given combinatorial problem instance. These algorithms are called *stochastic local search (SLS) algorithms*, and they constitute one of the most successful and widely used approaches for solving hard combinatorial problems.

SLS algorithms have been used for many years in the context of combinatorial optimisation problems. Among the most prominent algorithms of this kind we find the Lin-Kernighan Algorithm for the Travelling Salesman Problem [Lin and Kernighan, 1973], as well as general methods such as Evolutionary Algorithms (see, e.g., Bäck [1996]) and Simulated Annealing [Kirkpatrick et al., 1983] (these SLS methods will be presented and discussed in Chapter 2). More recently, it has become evident that stochastic local search algorithms can also be very successfully applied to the solution of $\mathcal{NP}$-complete decision problems such as the Graph Colouring Problem (GCP) [Hertz and de Werra, 1987; Minton et al., 1992] or the Satisfiability Problem in propositional logic (SAT) [Selman et al., 1992; Gu, 1992; Selman et al., 1994].

A General Definition of Stochastic Local Search

As outlined in the previous section, local search algorithms generally work in the following way. For a given instance of a combinatorial problem, the search for solutions takes place in the space of candidate solutions. Note that this search space may include partial candidate solutions, as required in the context of constructive search algorithms. The local search process is started by selecting an initial candidate solution, and then proceeds by iteratively moving from one candidate solution to a neighbouring candidate solution, where the decision on each search step is based on a limited amount of local information only. (See also Figure 1.3.) In stochastic local search algorithms, these decisions as well as the search initialisation can be randomised. Furthermore, the search process may use additional memory, for example, for storing a limited number of recently visited candidate solutions. Formally, a stochastic local search algorithm can be defined in the following way:

DEFINITION 1.10 **Stochastic Local Search Algorithm**

> *Given a (combinatorial) problem* Π, *a stochastic local search algorithm for solving an arbitrary problem instance* π ∈ Π *is defined by the following components:*

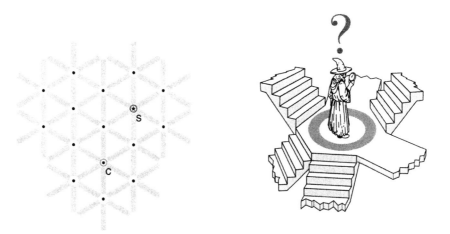

Figure 1.3 Illustration of stochastic local search. *Left:* 'Bird's-eye view' of a search space region; *s* marks a solution and *c* the current search position; neighbouring candidate solutions are connected by lines. *Right:* In each step, the search process moves to a neighbouring search position that is chosen based on local information only; here, the elevation of search positions indicates a heuristic value that is used for selecting the search steps to be performed.

- *the* search space $S(\pi)$ *of instance* π*, which is a finite set of* candidate solutions $s \in S$ *(also called* search positions, locations, configurations *or* states*)*;

- *a* set of (feasible) solutions $S'(\pi) \subseteq S(\pi)$;

- *a* neighbourhood relation *on* $S(\pi)$, $N(\pi) \subseteq S(\pi) \times S(\pi)$;

- *a finite set of* memory states $M(\pi)$*, which, in the case of SLS algorithms that do not use memory, may consist of a single state only;*

- *an* initialisation function $init(\pi) : \emptyset \mapsto \mathcal{D}(S(\pi) \times M(\pi))$*, which specifies a probability distribution over initial search positions and memory states;*

- *a* step function $step(\pi) : S(\pi) \times M(\pi) \mapsto \mathcal{D}(S(\pi) \times M(\pi))$ *mapping each search position and memory state onto a probability distribution over its neighbouring search positions and memory states;*

- *a* termination predicate $terminate(\pi) : S(\pi) \times M(\pi) \mapsto \mathcal{D}(\{\top, \bot\})$ *mapping each search position and memory state to a probability distribution over truth values (* $\top$ *= true,* $\bot$ *= false), which indicates the probability with which the search is to be terminated upon reaching a specific point in the search space and memory state.*

In the above, $\mathcal{D}(S)$ *denotes the set of probability distributions over a given set* S*, where formally, a probability distribution* $D \in \mathcal{D}(S)$ *is a function* $D : S \mapsto \mathbb{R}_0^+$ *that maps elements of* S *to their respective probabilities.*

Remark: In this definition, all components depend on the given problem instance π. Formally, these could be defined as (higher-order) functions mapping the given problem instance onto the corresponding search space, solution set, etc. While this is a straightforward extension of the definition as given above, for increased readability, we specify the components instantiated for a given problem instance; furthermore, we will often omit the formal reference to the problem instance, by writing S instead of $S(\pi)$, etc.

Any neighbourhood relation $N(\pi)$ can be equivalently specified in the form of a function $N : S(\pi) \mapsto 2^{S(\pi)}$ that maps candidate solutions $s \in S$ to the sets of their respective direct neighbours $N(s) := \{s' \in S \mid N(s, s')\} \subseteq S$; the set $N(s)$ is called the *neighbourhood set*, or just the *neighbourhood*, of s.

The combination of search position and memory state forms the state of the SLS algorithm, or *search state*. In the simplest case, search states solely consist of the respective candidate solution, and no additional memory is used;

this is formally captured by $M(\pi) := \{m_0\}$, where m_0 is an arbitrary constant. If additional memory is used, the memory state can consist of multiple independent attributes, that is, $M(\pi) := M_1 \times M_2 \times \ldots \times M_{l(\pi)}$ for some instance dependent constant $l(\pi)$. Although $M(\pi)$ can, in principle, represent a number of memory states that is exponential in the size of the given problem instance, typically it has a compact (i.e., polynomially bounded) representation. The memory state can be used to represent information that the algorithm is using to control the search process, such as the temperature parameter in Simulated Annealing or the tabu status of solution components in Tabu Search (cf. Chapter 2), but also simple book keeping mechanisms such as an iteration counter that can be used, for instance, in the context of a restart mechanism.

As an alternative to the initialisation and step functions, one can also specify *initialisation* and *step procedures* that draw an element from the probability distributions $init(\pi)()$ and $step(\pi)(s, m)$ for a given search position s and memory state m. The same holds for the termination predicate. (The notation $step(\pi)(s, m)$ and $init(\pi)()$ reflects the fact that these components are formally defined as higher order functions. For example, $step$ is instantiated through the first argument π into an instance specific step function that has s and m as its two arguments.) In the remainder of this book, we will use both types of definitions interchangeably, where $init(\pi)$, $step(\pi, s, m)$ and $terminate(\pi, s, m)$, when used in algorithm outlines, represent the procedures realising the probabilistic selection from the corresponding probability distributions. In cases where no additional memory is used, that is, $\#M(\pi) = 1$, we will often write $step(\pi, s)$ and $terminate(\pi, s)$ instead of $step(\pi, s, m)$ and $terminate(\pi, s, m)$.

Based on the components of the definition, the algorithm outlines in Figures 1.4 and 1.5 specify the semantics of stochastic local search algorithms for the search variants of decision and optimisation problems, respectively. The only major difference between the two versions is that for optimisation problems, the best candidate solution found so far, the so-called *incumbent solution*, is being memorised and returned upon termination of the algorithm (if it is a feasible solution); in this context, the objective function f for the given problem is used to determine the quality of candidate solutions. Furthermore, for decision problems, the termination condition is typically satisfied as soon as a solution is found, that is, $s \in S'$. In the case of optimisation problems, however, finding a feasible solution $s \in S'$ is typically not a sufficient termination criterion; in fact, many SLS algorithms for optimisation problems search through spaces containing feasible solutions only, that is, $S' = S$.

It may be noted that any SLS algorithms realises a Markov process; in particular, the behaviour of an SLS algorithm from a given search state (s, m) does not depend on any aspects of the search history that lead to that state, except for the information captured in s and m.

procedure *SLS-Decision*(π)
    input: *problem instance* $\pi \in \Pi$
    output: *solution* $s \in S'(\pi)$ **or** $\emptyset$
    $(s, m) := init(\pi, m);$
    while not *terminate*(π, s, m) **do**
        $(s, m) := step(\pi, s, m);$
    end
    if $s \in S'(\pi)$ **then**
        return s
    else
        return $\emptyset$
    end
end *SLS-Decision*

Figure 1.4 General outline of a stochastic local search algorithm for a decision problem Π.

procedure *SLS-Minimisation*(π')
    input: *problem instance* $\pi' \in \Pi'$
    output: *solution* $s \in S'(\pi')$ **or** $\emptyset$
    $(s, m) := init(\pi', m);$
    $\hat{s} := s;$
    while not *terminate*(π', s, m) **do**
        $(s, m) := step(\pi', s, m);$
        if $f(\pi', s) < f(\pi', \hat{s})$ **then**
            $\hat{s} := s;$
        end
    end
    if $\hat{s} \in S'(\pi')$ **then**
        return $\hat{s}$
    else
        return $\emptyset$
    end
end *SLS-Minimisation*

Figure 1.5 General outline of a stochastic local search algorithm for a minimisation problem Π' with objective function f; $\hat{s}$ is the incumbent solution, that is, the best candidate solution found at any time during the search so far.

EXAMPLE 1.3   **A Simple SLS Algorithm for SAT**

For a given SAT instance, that is, a CNF formula F, we define the search space as *Assign*(F), the set of all possible variable assignments of F. Obviously, the set of solutions is then given by the set of all models (satisfying assignments) of F. A frequently used neighbourhood relation is the so-called *one-flip neighbourhood*, which defines two variable assignments to be direct neighbours if, and only if, they differ in the truth value of exactly one variable, while agreeing on the assignment of the remaining variables. Formally, this can be written in the following way: for all $a, a' \in Assign(F)$, $N(a, a')$ if, and only if, there exists $v' \in Var(F)$, such that $Val(v', a) \neq Val(v', a')$ and for all $v \in Var(F) - \{v\}$, $Val(v, a) = Val(v, a')$. The search mechanism we will specify in the following does not use any memory, and hence we define $M := \{0\}$.

As an initialisation function, let us consider an 'uninformed' random selection realised by a uniform distribution over the entire search space. This initialisation function randomly selects any assignment of F with equal probability. Formally, it can be written as $init()(a', m) := init()(a') := 1/\#S = 1/2^n$, where $a' \in S$ is an arbitrary variable assignment of F and n is the number of variables appearing in F. (Note that formally, $init() = init(F)()$ is a probability distribution and $init()(a')$ denotes the probability of a' under the distribution $init()$. According to our earlier convention, we omit the problem instance from the notation of $init$ and $step$ when it is clear from the context.) Analogously, we can define a step function that maps any variable assignment a to the uniform distribution over all its neighbouring assignments. Formally, if $N(a) := \{a' \in S \mid N(a, a')\}$ is the set of all assignments neighbouring to a, the step function can be defined as $step(a, m)(a', m) := step(a)(a') := 1/\#N(a) = 1/n$.

This SLS algorithm is called *uninformed random walk*; as one might imagine, it is quite ineffective, since it does not provide any mechanism for steering the search towards solutions of the given problem instance.

Neighbourhoods and Neighbourhood Graphs

Generally, the choice of an appropriate neighbourhood relation is crucial for the performance of an SLS algorithm and often, this choice needs to be made in a problem specific way. Nevertheless, there are standard types of neighbourhood relations which form the basis for many successful applications of stochastic local search. One of the most widely used types of neighbourhood relations is the so-called *k-exchange neighbourhoods*, in which two candidate solutions are neighbours if, and only if, they differ in at most k solution components.

The neighbourhood used in the simple SAT algorithm from Example 1.3 (as well as in most state-of-the-art SLS algorithms for SAT) is a 1-exchange neighbourhood. For the TSP, one could define a k-exchange neighbourhood such that

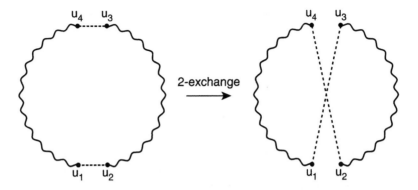

Figure 1.6 Schematic view of a single SLS step based on the standard 2-exchange neighbourhood relation for the TSP.

from a given candidate round trip, all its direct neighbours can be reached by changing the positions of at most k vertices in the corresponding permutation. However, this neighbourhood relation was found to be inferior to a different type of k-exchange neighbourhood, where the edges of the given graphs are viewed as the solution components, and two candidate round trips are k-exchange neighbours if, and only if, one can be obtained from the other by removing at most k edges and rewiring the resulting partial tours [Reinelt, 1994]. Figure 1.6 illustrates two tours that are neighbours under this latter 2-exchange neighbourhood.

Any neighbourhood relation N induces a (directed) graph on the underlying search space S; in this *neighbourhood graph* G_N, two vertices s, s' are connected by an edge (s, s') if, and only if, $(s, s') \in N$. Formally, for a given problem instance π, the neighbourhood graph induced by search space $S(\pi)$ and neighbourhood relation $N(\pi)$ is defined as $G_N(\pi) := (S(\pi), N(\pi))$. Many important properties of the neighbourhood relation are reflected directly in the neighbourhood graph. For instance, most standard neighbourhood relations (such as the k-exchange neighbourhoods introduced above) are symmetric, that is, $\forall s, s' \in S : (N(s, s') \Leftrightarrow N(s', s))$; this means, that the neighbourhood graph is symmetric in its edges and essentially corresponds to an undirected graph. This situation is relevant in practice, because it is a necessary precondition for an SLS algorithm's ability to directly reverse search steps. The degree of each vertex in the neighbourhood graphs corresponds to the size of its neighbourhood. In many cases, in particular for k-exchange neighbourhoods, all vertices of G_N have the same degree, (i.e., the underlying neighbourhood graph is regular). Another important property of the neighbourhood graph is its diameter, $diam(G_N)$, which gives a worst-case lower bound on the number of search steps required for reaching (optimal) solutions from arbitrary points in the search space. Neighbourhood graphs and their properties are further discussed in Chapter 5.

Search Strategies, Steps and Trajectories

Typically, the first three components of our definition of an SLS algorithm, the search space, solution set and neighbourhood relation, depend very much on the problem being solved. Together, these components provide the basis for solving a given problem using stochastic local search. But based on a given definition of a search space, solution set and neighbourhood relation, a wide range of *search strategies*, specified by the definition of initialisation and step functions, can be devised. To some extent, such search strategies can be independent from the underlying search space, solution set, and neighbourhood, and consequently can be studied and presented separately from these. In this context, the following concepts are often useful:

DEFINITION 1.11   **Search Steps and Search Trajectories**

Let Π be a (combinatorial) problem, and let $\pi \in \Pi$ be an arbitrary instance of Π. Given an SLS algorithm A for Π according to Definition 1.10, a search step *(also called* move *) is a pair $(s, s') \in S \times S$ of neighbouring search positions such that the probability for A moving from s to s' is greater than 0, that is, $N(s, s')$ and $step(s)(s') > 0$.*

A search trajectory *is a finite sequence $(s_0, s_1, \ldots, s_k)$ of search positions s_i $(i = 0, \ldots, k)$ such that for all $i \in \{1, \ldots, k\}$, (s_{i-1}, s_i) is a search step and the probability of initialising the search at s_0 is greater than zero, that is, $init()(s_0, m) > 0$ for some $m \in M$.*

For the simple SLS algorithm for SAT introduced in Example 1.3, each search step is an arbitrary pair of neighbouring variable assignments, and a search trajectory is a sequence of variable assignments in which each pair of successive elements is neighbouring; obviously such a trajectory corresponds to a sequence of search steps. In general, any search trajectory corresponds to a walk in the neighbourhood graph.

Uninformed SLS: Random Picking and Random Walk

The two (arguably) simplest SLS strategies are Uninformed Random Picking and Uninformed Random Walk. Both do not use memory and are based on an initialisation function that returns the uniform distribution over the entire search space. SLS algorithms based on this initialisation function randomly select any element of the search space S with equal probability as a starting point for the search process.

For *Uninformed Random Picking*, a complete neighbourhood relation is used (i.e., $N = S \times S$) and the step function maps each point in S to a uniform distribution over all its neighbours, that is, every point in S. Effectively, this strategy randomly samples the search space, drawing a new candidate solution in every step.

Uninformed Random Walk uses the same initialisation function, but for a given, arbitrary neighbourhood relation $N \subseteq S \times S$, its step function returns the uniform distribution over the set of neighbours of the given candidate solution, which implements a uniform, random selection from that neighbourhood in each step. Obviously, for the complete neighbourhood relation, this coincides with Uninformed Random Picking; for more restricted neighbourhoods it leads to a strategy that slightly more resembles the intuitive notion of local search.

As one might imagine, both of these uninformed SLS strategies are quite ineffective, since they do not provide any mechanism for steering the search towards solutions. Nevertheless, as we will see later, in combination with more directed search strategies, both Uninformed Random Picking and variants of Uninformed Random Walk play a role in the context of preventing or overcoming premature stagnation in complex and much more effective SLS algorithms.

Evaluation Functions

To improve on the simple uninformed SLS strategies discussed above, a mechanism is needed to guide the search towards solutions. For a given instance π of a decision problem, this can be achieved using an *evaluation function* $g(\pi)(s)$: $S(\pi) \mapsto \mathbb{R}$ that maps each search position onto a real number in such a way that the global optima of π correspond to the solutions of π. In the following, we will use the notation $g(\pi, s)$ instead of $g(\pi)(s)$ in the context of algorithm outlines, and we will often write $g(s)$ when the problem instance π is clear from the context; analogously, we use $f(\pi, s)$ and $f(s)$ instead of $f(\pi)(s)$ to denote objective function values.

The evaluation function is used for assessing or ranking candidate solutions in the neighbourhood of the current search position. The efficacy of the guidance thus provided depends on properties of the evaluation function and its integration into the search mechanism being used. Typically, the evaluation function is problem specific, and its choice is to some degree dependent on the search space, solution set and neighbourhood underlying the SLS approach under consideration.

In the case of SLS algorithms for combinatorial optimisation problems, the objective function characterising the problem is often used as an evaluation function, such that the values of the evaluation function correspond directly to the quantity to be optimised. However, sometimes different evaluation functions can provide more effective guidance towards high-quality or optimal solutions. For example, in the case of unweighted MAX-SAT, an optimisation variant of SAT in which the objective is to maximise the number of satisfied clauses, local search algorithms with better theoretical approximation guarantees can be obtained when using a specific evaluation function different from the number of clauses satisfied by a given assignment [Khanna et al., 1994] (see also Chapter 7).

For combinatorial decision problems, sometimes evaluation functions are naturally suggested by the objective functions of optimisation variants, but often there is more than one obvious choice of an evaluation function. In the case of SLS algorithms for SAT, the following evaluation function g is often used: given a formula F in CNF and an arbitrary variable assignment a of F, $g(F, a)$ is defined as the number of clauses of F that are unsatisfied under a. Obviously, the models of F correspond to the global minima of g and are characterised by $g(F, a) = 0$. It may be noted that this evaluation function corresponds to the objective function of the previously mentioned unweighted MAX-SAT problem.

> *Remark:* In the literature, often no distinction is made between an objective function and an evaluation function. To minimise potential confusion between the definition of the problem to be solved (which, in case of an optimisation problem, includes an objective function) and the definition of an SLS algorithm for solving this problem (which might make use of an evaluation function different from the problem's objective function), we systematically distinguish between the two concepts in this book.

Generally, through the use of an evaluation function whose global optima correspond to the (optimal) solutions, decision problems and optimisation problems can be treated analogously. However, for a decision problem, the result of the SLS algorithm is generally useless, unless it is a global optimum of the evaluation function and hence corresponds to a solution. For optimisation problems, suboptimal solutions (usually local minima) can be very useful — in which case the respective evaluation function should guide the algorithm to high-quality solutions as effectively as possible (which might complicate or conflict with providing effective guidance towards optimal solutions).

> *Remark:* In the literature, the evaluation function is often treated as an integral part of the definition of an SLS algorithm. Although it is technically possible to define SLS algorithms using the concept of an evaluation function instead of that of a step function, the resulting definitions would capture the concept of stochastic local search less naturally and would lead to unnecessarily complex or imprecise representations of certain SLS algorithms.

These difficulties specifically arise for SLS algorithms that use multiple or dynamically changing evaluation functions (such techniques are prominent and successful in various domains). Using our definition, in many cases the concept of an evaluation function still provides a useful and convenient means for structuring the definition of step functions.

Iterative Improvement

One of the most basic SLS algorithms using an evaluation function is *Iterative Improvement*. Given a search space S, solution set S', neighbourhood relation N and evaluation function g, Iterative Improvement starts from a randomly selected point in the search space, and then tries to improve the current candidate solution w.r.t. g. The initialisation function is typically the same as in Uninformed Random Picking, that is, for arbitrary $s \in S$, $init(s) := 1/\#S$. Furthermore, if for a given candidate solution s, $I(s)$ is the set of all neighbouring candidate solutions $s' \in N(s)$ for which $g(s') < g(s)$, then the step function can be formally defined as:

$$step(s)(s') := \begin{cases} 1/\#I(s) & \text{if } s' \in I(s) \\ 0 & \text{otherwise} \end{cases}$$

This SLS strategy is also known as *iterative descent* or *hill-climbing*, where the latter name is motivated by the application of Iterative Improvement to maximisation problems. Note that in the case where none of the neighbours of a candidate solution s realises an improvement w.r.t. the evaluation function, $step(s)$ is not a probability distribution. Hence, when using this step function, the search process is terminated as soon as this case is encountered — an obviously unsatisfying mechanism which we will revisit shortly.

EXAMPLE 1.4 **Iterative Improvement for SAT**

Using the same definition for the search space, solution set, neighbourhood relation, and set of memory states as in Example 1.3 (page 41*f.*), we consider the evaluation function g which maps each variable assignment a to the number of clauses of the given formula F that are unsatisfied under a. Iterative Improvement then starts the search at a randomly selected variable assignment (like Uninformed Random Walk, see Example 1.3), and in each step, it randomly selects one of the assignments that leave fewer clauses unsatisfied than the current candidate solution. Since according to the definition of the neighbourhood relation, each search step corresponds to flipping the truth value associated with one of the variables appearing in F, Iterative Improvement can be seen as always performing variable flips that increase the overall number of satisfied clauses.

To efficiently implement iterative improvement algorithms, evaluation function values are typically maintained using *incremental updates* (also called *delta evaluations*) after each search step. This is done by calculating the effects of the differences between the current candidate solution s and a neighbouring candidate solution s' on the evaluation function value. Since in many cases, the evaluation function value of a candidate solution consists of independent contributions of its individual solution components (or of small subsets of solution components), this can often be achieved by solely considering the contributions of those solution components that are not common to s and s'. For example, in the case of the TSP, where the solution components correspond to the edges of the given graph, when using the standard 2-exchange neighbourhood, neighbouring round trips p and p' differ in two edges. Given $w(p)$, the weight of p, the weight $w(p')$ can be incrementally computed by subtracting the weight of the edges contained in p but not in p' and adding the weight of the edges contained in p' but not in p. Note how in this example, using incremental updating, the computation of $w(p')$ requires at most four arithmetic operations, regardless of the number n of vertices in the given graph, compared to n arithmetic operations if $w(p')$ is computed from scratch.

Local Minima

In our definition of Iterative Improvement, the step function is not well-defined for candidate solutions that do not have any improving neighbours. A candidate solution with this property corresponds to a *local minimum* of the evaluation function g. Formally, this is captured in the following definition:

DEFINITION I.12   **Local Minimum, Strict Local Minimum**

Given a search space S, a solution set $S' \subseteq S$, a neighbourhood relation $N \subseteq S \times S$ and an evaluation function $g : S \mapsto \mathbb{R}$, a local minimum is a candidate solution $s \in S$ such that for all $s' \in N(s)$, $g(s) \leq g(s')$. We call a local minimum s a strict local minimum if for all $s' \in N(s)$, $g(s) < g(s')$. (Local maxima and strict local maxima can be defined analogously.)

Note that under this definition, global minima of the evaluation function are also considered to be local minima. Intuitively, local minima, and even more so, strict local minima, are positions in the search space from which no single search step can achieve an improvement w.r.t. the evaluation function. In cases where an SLS algorithm guided by an evaluation function encounters a local minimum that does not correspond to a solution, this algorithm can 'get stuck'. This happens, for example, when an Iterative Improvement algorithm is defined in such a way that it terminates (or just stays at the same candidate solution) when a local optimum is encountered.

There are no general (non-trivial) theoretical bounds on the solution quality of local optima for arbitrary combinatorial optimisation problems. While such bounds have been proven for specific problems (e.g., the Euclidean TSP [Chandra et al., 1999]), general guarantees can only be given for complete neighbourhood relations, in which case any local minimum is also a global minimum. Yet, the size of such complete neighbourhoods is typically exponential w.r.t. instance size, and therefore they cannot be searched reasonably efficiently in practice. However, typical instances of combinatorial optimisation problems can be empirically shown to have high-quality local optima which often can be found reasonably efficiently by high-performance SLS algorithms.

Computational Complexity of Local Search

While empirically, local minima of basically any instance of a combinatorial optimisation problem can be found reasonably fast, theoretically, in most cases the number of steps needed by an iterative improvement algorithm to find a local optimum cannot be bounded by a polynomial. However, any local search algorithm should at the very least be able to execute individual local search steps efficiently. This idea gives rise to the complexity class $\mathcal{PLS}$ [Johnson et al., 1988]. Intuitively, $\mathcal{PLS}$ is the class of problems for which a local search algorithm exists in which initial positions and search steps as well as the evaluation function values of search positions can always be computed in polynomial time (w.r.t. instance size) on a deterministic machine. This means that local optimality can be verified efficiently or, in case a candidate solution is not locally optimal, a neighbouring solution of better quality can be generated in polynomial time. Note that this theoretical concept does not include any statement on the number of local search steps required for reaching a local optimum.

Analogously to the notion of $\mathcal{NP}$-completeness, the class of $\mathcal{PLS}$-complete problems is defined in such a way that it captures the hardest problems in $\mathcal{PLS}$. If for any of these problems local optima can be found in polynomial time, the same would hold for all problems in $\mathcal{PLS}$. It is conjectured that the class of polynomial local search problems is a strict subset of $\mathcal{PLS}$ and hence, in the worst case superpolynomial run-time may be required by any algorithm to find local minima of a $\mathcal{PLS}$-complete problem. The first well-known combinatorial optimisation problem that was shown to be $\mathcal{PLS}$-complete is the partitioning of weighted graphs under the Kernighan-Lin neighbourhood [Kernighan and Lin, 1970]. The TSP under the neighbourhood induced by a variant of the Lin-Kernighan Algorithm [Lin and Kernighan, 1973], one of the most efficient local search algorithm for the TSP, has also been shown to be $\mathcal{PLS}$-complete [Papadimitriou, 1992]. Furthermore, $\mathcal{PLS}$-completeness has been shown for Iterative Improvement algorithms for the TSP that are based on the standard k-exchange neighbourhood with sufficiently large $k > 3$ [Krentel, 1989], while the question of $\mathcal{PLS}$-completeness when using 2- or 3-exchange neighbourhoods remains open [Johnson and McGeoch, 1997; Yannakakis, 1997].

Escape Strategies

In many cases, local minima are quite common (this will be further discussed in Chapter 5), and for optimisation problems, the corresponding candidate solutions are typically not of sufficiently high quality. Consequently, techniques for avoiding or escaping from local minima are of central importance in SLS algorithm design, and a large number of such mechanisms have been proposed and evaluated in the literature. Many of these are discussed in detail or mentioned in passing in the following chapters, and specifically the next chapter introduces some of the most prominent and successful approaches for avoiding search stagnation due to local minima. Therefore, we restrict the present discussion to two very simple methods.

One straightforward way of modifying Iterative Improvement such that local minima are dealt with more reasonably is to simply reinitialise the search process whenever a local minimum is encountered. While this simple *restart strategy* can work reasonably well when the number of local minima is rather small or restarting the algorithm is not very costly (in terms of overhead cost for initialising data structures, etc.), in many cases this technique is rather ineffective. Alternatively, one can relax the improvement criterion and, when a local minimum is encountered, perform a randomly chosen *non-improving step*. This can be realised as a uniform random selection among all neighbours of the current search position (which corresponds to an Uninformed Random Walk step), or it can be done by randomly selecting one of the neighbours that result in the lowest increase in evaluation function value (this corresponds to a 'mildest ascent step' and is closely related to a variant of Iterative Improvement that will be discussed in more detail in Chapter 2).

For neither of those latter two mechanisms is there any guarantee that the search algorithm effectively escapes from arbitrary local minima, because the nature of a local minimum can be such that after any such 'escape step', the only improving step available leads directly back into the same local minimum.

Furthermore, in the case of non-strict local minima, minimally worsening steps
will lead to walks in so-called *plateaus* — regions of neighbouring candidate
solutions with identical evaluation function values. Such plateaus can be very
extensive (cf. Chapter 5), and it can be difficult to decide whether the search process is trapped in a plateau region that does not allow any further improvement
without an effective escape mechanism.

Intensification *vs* Diversification

As we will show in more detail in later chapters, the strong randomisation of
local search algorithms, that is, the utilisation of stochastic choice as an integral
part of the search process, can lead to significant increases in performance and
robustness. However, with this potential comes the need to balance randomised
and goal-directed components of the search strategy, a trade-off which is often
characterised as 'diversification *vs* intensification'. *Intensification* refers to search
strategies that aim to greedily improve solution quality or the chances of finding
a solution in the near future by exploiting, for instance, the guidance given by the
evaluation function. *Diversification* strategies try to prevent search stagnation
by making sure that the search process achieves a reasonable coverage when
exploring the search space and does not get trapped in relatively confined regions
that do not contain (sufficiently high-quality) solutions. In this sense, Iterative
Improvement is an intensification strategy, while Uninformed Random Walk is a
diversification strategy, and as we will see in the next chapter, both strategies can
be combined into an SLS approach called *Randomised Iterative Improvement*,
which typically shows improved performance over both pure search methods.

A large variety of techniques for combining and balancing intensification
and diversification strategies has been proposed, and to some extent these will
be presented and discussed in the remainder of this book. While the resulting
SLS algorithms often perform very well in practice, typically their behaviour
is not well understood. The successful application of these algorithms is often
based on intuition and experience rather than on theoretically or empirically
derived principles and insights, particularly when it comes to the trade-off between diversification and intensification. While in this context, problem specific
knowledge is often (if not typically) crucial for achieving peak performance and
robustness, a solid understanding of the various types of SLS methods, combined
with detailed knowledge of their properties and characteristics is of at least equal
importance.

The latter is especially relevant in cases where one of the reasons for applying
SLS algorithms is a lack of sufficient specific knowledge about the problem to
be solved; in this situation, where specialised algorithms are typically not available, SLS algorithms are attractive because they often allow solving the problem

reasonably efficiently using fairly generic and easily implementable algorithms. More importantly, for many hard combinatorial problems, such generic SLS methods can also be quite naturally extended with or adapted based on problem-specific knowledge as it becomes available. The specialised SLS algorithms thus obtained are often amongst the best-known techniques for solving these problems, and specifically for large instances of optimisation problems or under tight constraints on time and other computational resources, in many cases they represent the only known methods for finding (high-quality) solutions in practice.

In the following chapters, we will introduce and discuss a broad range of SLS algorithms, covering many state-of-the-art generic SLS methods. Our discussion will focus on underlying general properties and design principles, such as the combination of search strategies and methods for balancing intensification and diversification aspects of search. Later, we will show in detail how these general methods are applied and adapted to specific combinatorial problems, yielding high-performance or state-of-the-art algorithms for solving these problems.

IN DEPTH RANDOMNESS AND PROBABILISTIC COMPUTATION

Implementations of randomised algorithms almost always realise all random choices and decisions by means of a *pseudo-random number generator (PRNG)* [Knuth, 1997]. PRNGs are provided by basically all modern programming environments; they are based on deterministically generated cycles of integers, from which floating point numbers can be obtained by appropriate scaling. PRNGs should satisfy the following conditions:

- The generated sequence of numbers should be serially uncorrelated, that is, n-tuples from the sequence should be independent of one another.

- The generator should have a long period; while ideally it should not cycle at all, in practice, the repetition should occur only after a very large set of numbers has been generated.

- The sequence should be uniform and unbiased, that is, equal fractions of generated numbers should fall into equal intervals.

- The algorithm underlying the PRNG and its implementation should be as efficient as possible.

To date, most PRNGs are based on three types of methods: linear congruential generators (LCGs), lagged Fibonacci generators (LFGs) and the Mersenne Twister (MT) [Matsumoto and Nishimura, 1998]. Especially the Mersenne Twister presents an interesting alternative to standard PRNGs, which are typically based on linear congruential generators. The MT 19 937 version of the Mersenne Twister has a period of $2^{19\,937} - 1$ and has been shown to generate sequences of excellent quality. It is also very efficient (according to empirical measurements it is up to four times faster than the standard rand() function in C/C++); furthermore, implementations in many programming languages are freely and publicaly available [Matsumoto, 2003].

Basically all PRNGs produce uniformly distributed numbers. In most SLS algorithms, random decisions involve uniform or biased choices from a finite set; these can be easily implemented using a uniform random number generator. Sometimes, however, it is desirable to sample from a different type of distribution, such as a normal or exponential distribution. This can be achieved by appropriately chosen transformations of the output of a uniform random number generator, such as the well-known Box-Muller transformation, which generates a pair of normally distributed random values from a pair of uniformly distributed values [Box and Muller, 1958].

As an alternative to using pseudo-random numbers generated by PRNGs, true random numbers can be used; these can be obtained in various ways, all of which are based on sampling and processing a source of entropy, such as the decay of radioactive material, atmospheric noise from a radio, etc. (Of course, whether these physical phenomena are truly random is ultimately unclear; however, for all practical purposes, they appear to be random according to the commonly used criteria.) Random numbers from such sources are freely and publicly available from several websites (cf. Haahr [2003]; Walker [2003]); furthermore, there are various commercial hardware devices that can be plugged into a standard PC (empirical studies suggest that not all of these consistently produce high-quality random number sequences). Compared to efficient PRNGs, these true random number generators are very slow, which can be a serious drawback in the context of heavily randomised algorithms, such as many SLS algorithms or Monte Carlo simulations, which may require millions of random numbers per second.

It is well known that for implementing certain types of probabilistic algorithms, especially Monte-Carlo simulations of physical systems, it is of crucial importance to use a high-quality (pseudo-)random number generator. In the case of SLS algorithms, the issue is less clear. There is no empirical evidence that true random number sequences offer any advantages over the sequences obtained from state-of-the-art pseudo-random number generators, and given the general availability and efficiency of the latter, there is no reason to use lower quality PRNGs or true random number sources.

It may be noted that by implementing a probabilistic algorithm using a PRNG, what is implemented is effectively a derandomised, entirely deterministic version of the algorithm. As previously mentioned, with current hardware this has the advantage of higher efficiency; it also greatly facilitates debugging, since any run of the algorithm can be perfectly reproduced by initialising the PRNG with the same seed. This raises the question whether this derandomisation can result in a loss of computational power or efficiency.

From a theoretical point of view, SLS algorithms, like all algorithms that use randomised decisions, are based on a probabilistic model of computation, such as the probabilistic Turing machine [de Leeuw et al., 1955; Gill, 1977; Rabin, 1976], which can be seen as a variant of a conventional deterministic Turing machine that has access to an arbitrary number of random bits, each of which is zero or one with probability $1/2$, that can be used at arbitrary points in the computation.

Note that there is an important difference between nondeterministic and probabilistic machine models. Nondeterministic models — which are used, for example, in the definition of the complexity class $\mathcal{NP}$ — intuitively can be seen as having the ability to make nondeterministic guesses such that computation time is minimised. (Alternatively, nondeterministic machines can be viewed as pursuing all possible paths of computation simultaneously, such that only the shortest of these potential computations determines the run-time.) Probabilistic machine models, on the other hand, can be seen as making actual randomised choices; consequently, each possible path of computation corresponds to a set of such mutually independent random choices and has a probability associated with it. This gives rise to a probability distribution over computation paths and consequently, over run-times.

Probabilistic models of computation and randomised algorithms are of substantial interest in complexity theory. For decision problems, one can distinguish between three types of randomised algorithms: depending on the probability of giving incorrect 'yes' and 'no' answers, respectively, there are algorithms with zero, one-sided and two-sided error. SLS algorithms, as considered in this book, are generally of the first type, which is also known as the class of Las Vegas algorithms. (These are formally defined and further discussed in Chapter 4.) The class of problems that can be solved by probabilistic algorithms with zero error probability in expected run-time that is polynomial in the size of the given input is known as $\mathcal{ZPP}$. Another prominent probabilistic complexity class is $\mathcal{BPP}$, the class of problems that can be solved in polynomial time (w.r.t. the size of the given input) in the worst case by a probabilistic algorithm with two-sided error probability bounded from above by $1/2 - \epsilon$ for some $\epsilon > 0$ [Papadimitriou, 1994].

While it is known that $\mathcal{ZPP} \subseteq \mathcal{BPP}$, many questions regarding the relationships between these probabilistic complexity classes and other complexity classes, in particular $\mathcal{P}$ and $\mathcal{NP}$, remain open. It is rather easy to see that $\mathcal{P} \subseteq \mathcal{ZPP}$ (and hence also $\mathcal{P} \subseteq \mathcal{BPP}$); furthermore, it is known that $\mathcal{ZPP} \subseteq \mathcal{NP}$. Assuming $\mathcal{P} \neq \mathcal{NP}$, it is not known whether $\mathcal{ZPP}$ is a proper superset of $\mathcal{P}$ or a proper subset of $\mathcal{NP}$. Furthermore, the relationship between $\mathcal{BPP}$ and $\mathcal{NP}$ is unknown. Interestingly, it is strongly believed that $\mathcal{P} = \mathcal{BPP}$ (see, for example, Kabanets [2002]), and hence also $\mathcal{P} = \mathcal{ZPP}$, which suggests that from a theoretical point of view, the use of true randomisation may not substantially improve our ability of solving hard combinatorial problems. The fact that empirically, the derandomisation of probabilistic algorithms by use of high-quality PRNGs typically does not appear to result in performance losses is consistent with this belief and, in fact, is commonly seen as additional support for it. There is little doubt, however, that the typical properties of true and pseudo-random number sequences stated above are crucial for the excellent performance of many SLS methods and other probabilistic algorithms.

1.6 Further Readings and Related Work

Due to the introductory nature of this chapter, there is a huge body of literature related to the concepts presented here. Introductions to combinatorial problems and search methods can be found in many modern or classic textbooks on combinatorial optimisation, operations research or artificial intelligence (such as Aarts and Lenstra [1997], Lawler [1976], Nemhauser and Wolsey [1988], Papadimitriou and Steiglitz [1982], Poole et al. [1998], Reeves [1993b], Rayward-Smith et al. [1996], Russel and Norvig [2003], etc.); details on heuristic search can also be found in Pearl [1984]. For a slightly different definition of combinatorial optimisation problems we refer to the classical text by Papadimitriou and Steiglitz [1982]. A detailed discussion of complexity theory, $\mathcal{NP}$-completeness and

$\mathcal{NP}$-hard problems can be found in Garey and Johnson [1979], Papadimitriou [1994] or Reischuk [1990].

For a general reference to recent research on the Propositional Satisfiability Problem we refer the interested reader to the book edited by van Maaren, Gent and Walsh [2000], and to the overview article by Gu et al. [1997]. For details on and a large number of variants of the TSP we refer to the now classical book edited by Lawler et al. [1985] or the monograph by Reinelt [1994]. For a detailed account of the state-of-the-art in TSP solving with SLS algorithms up to 1997, the book chapter by Johnson and McGeoch [1997] is the best reference; results of more recent variants are collected in a book chapter by Johnson and McGeoch [2002] and on the web pages for the 8th DIMACS Challenge on the TSP [Johnson et al., 2003a]. Regarding stochastic local search methods for SAT, early work includes the studies by Selman et al. [1992] and Gu [1992], while some of the better performing algorithms have been presented by McAllester et al. [1997]. For an overview and comparison of the best-performing SLS algorithms for SAT up to the year 2000 we refer to Hoos and Stützle [2000a]. Further details on state-of-the-art algorithms for SAT and TSP as well as further references for these problems can be found in Chapters 6 and 8 of this book.

1.7 **Summary**

This chapter started with a brief introduction to *combinatorial problems* and distinguished between two main types of problems, *decision* and *optimisation problems*. We introduced the *Propositional Satisfiability Problem (SAT)* and the *Travelling Salesman Problem (TSP)* as two prototypical combinatorial problems. Both problems are conceptually simple and easy to state, which facilitates the design and analysis of algorithms. At the same time, they are computationally hard and appear at the core of many real-world applications; hence, these problems pose a constant challenge for the development of new algorithmic techniques for solving hard combinatorial problems. Many combinatorial problems, including SAT and TSP, are $\mathcal{NP}$-*hard*; consequently, there is little hope for finding algorithms with better than exponential worst-case behaviour. However, this does not imply that all instances of these problems are intrinsically hard. Interesting or application-relevant subclasses of hard combinatorial problems can be efficiently solvable. For many optimisation problems, there are efficient *approximation algorithms* that can find good solutions reasonably efficiently. Additionally, *stochastic algorithms* can help in solving combinatorial problems more robustly and efficiently in practice.

Next, we discussed various *search paradigms* and highlighted their relations and properties. We distinguished *perturbative local search methods*, which operate on fully instantiated candidate solutions, from *constructive search algorithms*, which iteratively extend partial candidate solutions. Combinations of constructive search algorithms with *backtracking* lead to *complete, systematic search methods* that are traditionally known as *tree search* or *refinement search techniques*. *Local search algorithms*, which move between candidate solutions based on local information only, have the advantage of being easily applicable to a broad range of combinatorial problems, for many of which they have been shown to be the most effective solution methods. Furthermore, they are typically rather easy to implement and often have attractive any-time properties. But these advantages come at a price. Local search algorithms are typically *incomplete* and, particularly in the case of *stochastic local search methods*, they are generally difficult to analyse – an issue that will be addressed in more detail in Chapters 4 and 5.

Finally, we gave a general definition of *stochastic local search algorithms* that covers both, perturbative as well as constructive methods within a unified framework. Based on this definition we introduced and discussed a number of simple SLS strategies such as *Iterative Improvement*, which forms the basis of many of the more complex SLS methods presented in the next chapter.

Exercises

1.1 [**Easy**] Consider the following *Graph Colouring Problem*: Given a graph $G := (V, E)$ with vertex set V and edge relation E, assign a minimal number of colours $c_1, c_2, \ldots, c_k$ to the vertices such that two vertices that are connected by an edge in E are never assigned the same colour.

Show how this problem fits the (informal) definition of a combinatorial problem from Section 1.1, and state the different decision and optimisation variants.

1.2 [**Medium**] Recall the puzzle from the prologue:

Last week my friends Anne, Carl, Eva, Gustaf and I went out for dinner every night, Monday to Friday. I missed the meal on Friday because I was visiting my sister and her family. But otherwise, every one of us had selected a restaurant for a particular night and served as a host for that dinner. Overall, the following restaurants were selected: a French bistro, a sushi bar, a pizzeria, a Greek restaurant, and the Brauhaus. Eva took us out on Wednesday. The Friday dinner was at the Brauhaus. Carl, who doesn't eat sushi, was the first host. Gustaf had selected the bistro for the night before one of the friends took everyone to the pizzeria. Tell me, who selected which restaurant for which night?

Formalise this puzzle as a SAT instance.

1.3   [**Hard**]   Consider the problem of finding a Hamiltonian cycle in a given (undirected) graph (cf. Definition 1.6, page 21). Show how the known result that the Hamiltonian Cycle Problem is $\mathcal{NP}$-hard implies the $\mathcal{NP}$-hardness of the TSP for graphs in which all edge weights are equal to one or two.

(Hint: You need to show that any polynomial-time deterministic TSP algorithm could be used for solving (suitably encoded) instances of the Hamiltonian Cycle Problem.)

1.4   [**Hard**]   Consider the following argument. For the Euclidean TSP, given an arbitrary approximation ratio $r > 1$, there exists a deterministic algorithm that achieves that ratio in polynomial run-time w.r.t. the number of vertices, n. Hence, the associated decision problems for arbitrary solution quality bounds can be solved by a deterministic algorithm with run-time polynomial in n, which implies that the search variant of the Euclidean TSP is also efficiently solvable. (Note that this conclusion is in direct contradiction with the known result that the Euclidean TSP is $\mathcal{NP}$-hard.) Why is this argument flawed?

(Hint: Think carefully about the nature of the solution quality bounds.)

1.5   [**Easy**]   Given an arbitrary TSP instance G, does the Nearest Neighbour Heuristic (see Section 1.4, page 31*ff.*) always return the same solution, that is, does G have a uniquely defined nearest neighbour tour? (Justify your answer.)

1.6   [**Easy**]   Consider the following recursive algorithm for SAT:

```
procedure DP-SAT(F,A)
    input: propositional formula F, partial truth assignment A
    output: true or false

    if A satisfies F then
        return true
    end
    if ∃ unassigned variable in A then
        randomly select variable x that is unassigned in A;
        A' := A extended by x := ⊤;
        A" := A extended by x := ⊥;
    end
    if DP-SAT (F,A') = true or DP-SAT (F,A") = true then
        return true
```

```
           else
                 return false
           end
     end DP-SAT
```

Which search paradigm does this algorithm implement and which of the properties discussed in Section 1.4 does it possess?

1.7   [*Medium*]   Design a simple complete stochastic local search algorithm for SAT and show how it fits Definition 1.10 (page 38*f.*). Show that your algorithm is complete and discuss the practical importance of this completeness result.

1.8   [*Medium*]   Consider the following, alternative definition of a stochastic local search algorithm.

Given a (combinatorial) problem Π, a stochastic local search algorithm for solving an arbitrary problem instance $\pi \in \Pi$ is defined by the following components:

- *a (directed) search graph* $G(\pi) := (V, E)$, *where the elements of V are the candidate solutions of π and the arcs in E connect any candidate solution to those candidate solutions that can be reached in one search step;*

- *an evaluation function* f_π, *which assigns a numerical value* $f_\pi(s)$ *to each candidate solution s and whose global maxima correspond to the (optimal) solutions of π;*

- *an initialisation procedure* $init(\pi)$, *which determines a candidate solution at which the search process is started;*

- *an iteration procedure* $iter(\pi)$, *which for any candidate solution s selects a candidate solutions* s' *such that* $(s, s') \in E$;

- *a termination function* $terminate(\pi)$, *which for a given candidate solution determines whether the search is to be terminated (this function can make use of a random number generator and a limited amount of memory on earlier events in the search process).*

Is this definition equivalent to Definition 1.10 (page 38*f.*), that is, does it cover the same class of algorithms? Discuss the differences between the definitions and try to decide which one is better.

1.9   [*Medium*]   Consider the decision variant of the Graph Colouring Problem as described in Exercise 1.1. Design an iterative improvement algorithm for this problem.

1.10 **[Easy]**   Consider the following *Conflict-Directed Random Walk* algorithm for SAT:

procedure *CDRW-SAT* (*F*)
    input: *CNF formula F*
    output: *model of F* **or** 'no solution found'
    a := randomly chosen assignment of the variables in formula *F*;
    while not (*a* is a model of *F*) **do**
        c := randomly chosen clause in *F* that is unsatisfied under *a*;
        v := randomly chosen variable from *c*;
        a := *a* with *v* flipped;
    end
    return 'no solution found';
end *CDRW-SAT*

Let N be the neighbourhood relation under which assignments a and a' are direct neighbours if, and only if, a' can be reached from a in a single search step according to this algorithm. Is N symmetric? (Justify your answer.)

2 SLS METHODS

Stochastic Local Search (SLS) is a widely used approach to solving hard combinatorial optimisation problems. Underlying most, if not all, specific SLS algorithms are general SLS methods that can be applied to many different problems. In this chapter we present some of the most prominent SLS methods and illustrate their application to hard combinatorial problems, using SAT and TSP as example domains.

The techniques covered here range from simple iterative improvement algorithms to complex SLS methods, such as Ant Colony Optimisation and Evolutionary Algorithms. For each of these SLS methods, we motivate and describe the basic technique and discuss important variants. Furthermore, we identify and discuss important characteristics and features of the individual methods and highlight relationships between them.

2.1 Iterative Improvement (Revisited)

In Chapter 1, Section 1.5, we introduced Iterative Improvement as one of the simplest, yet reasonably effective SLS methods. We have pointed out that one of the main limitations of Iterative Improvement is the fact that it can, and often does, get stuck in local minima of the underlying evaluation function. Here, we discuss how using larger neighbourhoods can help to alleviate this problem without rendering the exploration of local neighbourhoods prohibitively expensive.

Large Neighbourhoods

As pointed out before, the performance of any stochastic local search algorithm depends significantly on the underlying neighbourhood relation and, in particular, on the size of the neighbourhood. Consider the standard k-exchange neighbourhoods introduced in Chapter 1, Section 1.5. It is easy to see that for growing k, the size of the neighbourhood (i.e., the number of direct neighbours for each given candidate solution), also increases. More precisely, for a k-exchange neighbourhood, the size of the neighbourhood is in $O(n^k)$, that is, the neighbourhood size increases exponentially with k.

Generally, larger neighbourhoods contain more and potentially better candidate solutions, and hence they typically offer better chances for finding locally improving search steps. They also lead to neighbourhood graphs with smaller diameters, which means that an SLS trajectory can potentially more easily explore different regions of the underlying search space. In a sense, the ideal case would be a neighbourhood relation for which any locally optimal candidate solution is guaranteed to be globally optimal. Neighbourhoods which satisfy this property are called *exact*; unfortunately, in most cases exact neighbourhoods are exponentially large with respect to the size of the given problem instance, and searching an improving neighbouring candidate solution may take exponential time in the worst case. (Efficiently searchable exact neighbourhoods exist in a few cases; for example, the Simplex Algorithm in linear programming is an iterative improvement algorithm that uses a polynomially searchable, exact neighbourhood, and is hence guaranteed to find a globally optimal solution.)

This situation illustrates a general tradeoff: using larger neighbourhoods might increase the chance of finding (high-quality) solutions of a given problem in fewer local search steps when using SLS algorithms in general and Iterative Improvement in particular; but at the same time, the time complexity for determining improving search steps is much higher in larger neighbourhoods. Typically, the time complexity of an individual local search step needs to be polynomial w.r.t. the size of the given problem instance. However, depending on problem size, even quadratic or cubic time per search step might already be prohibitively high if the instance is very large.

Neighbourhood Pruning

Given the tradeoff between the benefits of using large neighbourhoods and the associated time complexity of performing search steps, one attractive idea for improving the performance of Iterative Improvement and other SLS algorithms is to use large neighbourhoods but to reduce their size by never examining neighbours that are unlikely to (or that provably cannot) yield any improvements in evaluation function value. While in many cases, the use of large neighbourhoods is only practically feasible in combination with such pruning, the same pruning techniques can be applied to relatively small neighbourhoods, where they can lead to substantial improvements in SLS performance.

For the TSP, one such pruning technique that has been shown to be useful in practice is the use of *candidate lists*, which for each vertex in the given graph contain a limited number of their closest direct neighbours, ordered according to increasing edge weight. The search steps performed by an SLS algorithm are then limited to consider only edges connecting a vertex i to one of the vertices in i's candidate list. The use of such candidate lists is based on the intuition that high-quality solutions will be likely to include short edges between neighbouring vertices (cf. Figure 1.1, page 23). In the case of the TSP, pruning techniques have shown significant impact on local search performance not only for large neighbourhoods, but also for rather small neighbourhoods, such as the standard 2-exchange neighbourhood.

Other neighbourhood pruning techniques identify neighbours that provably cannot lead to improvements in the evaluation function based on insights into the properties of a given problem. An example for such a pruning technique is described by Nowicki and Smutnicki [1996a] in their tabu search approach to the Job Shop Problem, which will be described in Chapter 9.

Best Improvement *vs* First Improvement

Another method for speeding up the local search is to select the next search step more efficiently. In the context of iterative improvement algorithms, the search step selection mechanism that implements the step function from Definition 1.10 (page 38*f.*) is also called *pivoting rule* [Yannakakis, 1990]; the most widely used pivoting rules are the so-called best improvement and first improvement strategies described in the following.

Iterative Best Improvement is based on the idea of randomly selecting in each search step one of the neighbouring candidate solutions that achieve a maximal improvement in the evaluation function. Formally, the corresponding step function can be defined as follows: given a search position s, let $g^* := \min\{g(s') \mid s' \in N(s)\}$ be the best evaluation function value in the neighbourhood of s. Then $I^*(s) := \{s' \in N(s) \mid g(s') = g^*\}$ is the set of maximally improving neighbours of s, and we define $step(s)(s') := 1/\#I^*(s)$ if $s' \in I^*(s)$, 0 otherwise. Best Improvement is also called *greedy hill-climbing* or *discrete gradient descent*. Note that Best Improvement requires a complete evaluation of all neighbours in each search step.

The *First Improvement* neighbour selection strategy tries to avoid the time complexity of evaluating all neighbours by performing the first improving step encountered during the inspection of the neighbourhood. Formally, Iterative First Improvement is best defined by means of a step procedure rather than a step function. At each search position s, the First Improvement step procedure evaluates the neighbouring candidate solutions $s' \in N(s)$ in a particular fixed order, and the first s' for which $g(s') < g(s)$, that is, the first improving neighbour encountered, is selected. Obviously, the order in which the neighbours are evaluated can have a significant influence on the efficiency of this strategy. Instead of using a fixed ordering for evaluating the neighbours of a given search position, random orderings can also be used. For fixed evaluation orderings, repeated runs of Iterative First Improvement starting from the same initial solution will end in the same local optimum, while by using random orderings, many different local optima can be reached. In this sense, random-order First Improvement inherently leads to a certain diversification of the search process. The following example illustrates the variability of the candidate solutions reached by random-order First Improvement.

EXAMPLE 2.1 **Random-Order First Improvement for the TSP**

In this example, we empirically study a random-order first improvement algorithm for the TSP that is based on the 2-exchange neighbourhood. This algorithm always starts from the same initial tour, which visits the vertices of the given graph in their canonical order (i.e., in the order $v_1, v_2, \ldots, v_n, v_1$). Furthermore, when initialising the search, a random permutation of the integers from 1 to n is generated, which determines the order in which the neighbhourhood is scanned in each search step. (This permutation remains unchanged throughout the search process.) As usual for simple iterative improvement methods, the search is terminated when a local minimum of the given evaluation function (here: weight of the candidate tour) is encountered.

This algorithm was run 1 000 times on pcb3038, a TSP instance with 3 038 vertices available from the TSPLIB benchmark library. For each of these runs, the length of the final, locally optimal tour (i.e., the weight of the corresponding path in the graph) was recorded. Figure 2.1 shows the cumulative distribution of the percentage deviations of these solution quality values from the known optimal solution. (The cumulative distribution function specifies for each relative solution quality value q on the x-axis the relative frequency with which a solution quality smaller or equal to q is obtained.) Clearly, there is a large degree of variation in the qualities of the 1 000 tours produced by our random-order iterative first improvement algorithm. The average tour length is 8.6% above the known optimum, while the 0.05-

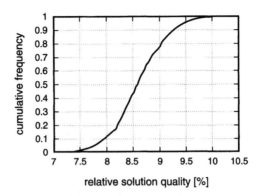

Figure **2.1** Cumulative distribution of the solution quality returned by a random-order first improvement 2-exchange algorithm for the TSP on TSPLIB instance pcb3038, based on 1 000 runs of the algorithm.

and 0.95-quantiles of this solution quality distribution can be determined as 7.75 % and 9.45 % above the optimum.

Based on the shape of this empirical distribution, it can be conjectured that the solution quality data follow a normal distribution. This hypothesis can be tested using the Shapiro-Wilk test [Shapiro and Wilk, 1965], a statistical goodness-of-fit test that specifically checks whether given sample data are normally distributed. In this example, the test accepts the hypothesis that the solution quality data follow a normal distribution with mean 8.6 and standard deviation 0.51 at a p-value of 0.2836. (We refer to Chapter 4 for more details on statistical tests). Normally distributed solution qualities occur rather frequently in the context of SLS algorithms for hard combinatorial problems, such as the TSP.

As for large neighbourhoods, in the context of pivoting rules there is a trade-off between the number of search steps required for finding a local optimum and the computation time for each search step. Search steps in first improvement algorithms can often be computed more efficiently than in best improvement algorithms, since in the former case, typically only a small part of the local neighbourhood is evaluated, especially as long as there are multiple improving search steps from the current candidate solution. However, the improvement obtained by each step of First Improvement is typically smaller than for Best Improvement and therefore, more search steps have to be performed in order to reach a local optimum. Additionally, Best Improvement benefits more than First Improvement from the use of caching and updating mechanisms for evaluating neighbours efficiently.

Remark: Besides First Improvement and Best Improvement, iterative improvement algorithms can use a variety of other pivoting rules. One example is *Random Improvement*, which randomly selects a candidate solution from the set $I(s) := \{s' \in N(s) \mid g(s') < g(s)\}$; this selection strategy can be implemented as First Improvement where a new random evaluation ordering is used in each search step. Another example is the *least improvement* rule, which selects an element from $I(s)$ that minimally improves the current candidate solution.

Variable Neighbourhood Descent

Another way to benefit from the advantages of large neighbourhoods without incurring a high time complexity of the search steps is based on the idea of using standard, small neighbourhoods until a local optimum is encountered, at which point the search process switches to a different (typically larger) neighbourhood, which might allow further search progress. This approach is based on the fact that the notion of a local optimum is defined relative to a neighbourhood relation, such that if a candidate solution s is locally optimal w.r.t. a neighbourhood relation N_1 it need not be a local optimum for a different neighbourhood relation N_2. The general idea of changing the neighbourhood during the search has been systematised by the *Variable Neighbourhood Search (VNS)* framework [Mladenović and Hansen, 1997; Hansen and Mladenović, 1999].

VNS comprises a number of algorithmic approaches including *Variable Neighborhood Descent (VND)*, an iterative improvement algorithm that realises the general idea behind VNS in a very straightforward way. In VND, k neighbourhood relations $N_1, N_2, \ldots, N_k$ are used, which are typically ordered according to increasing size. The algorithm starts with neighbourhood N_1 and performs iterative improvement steps until a local optimum is reached. Whenever no further improving step is found for a neighbourhood N_i and $i + 1 \leq k$, VND continues the search in neighbourhood N_{i+1}; if an improvement is obtained in N_i, the search process switches back to N_1, from where the search is continued as previously described. An algorithm outline for VND is shown in Figure 2.2. In general, there are variants of this basic VND method that switch between neighbourhoods in different ways. It has been shown that Variable Neighbourhood Descent can considerably improve the performance of iterative improvement algorithms both w.r.t. to the solution quality of the local optima reached, as well as w.r.t. the time required for finding (high-quality) solutions compared to using standard Iterative Improvement in large neighbourhoods [Hansen and Mladenović, 1999].

It may be noted that apart from VND, there are several other variants of the general idea underlying Variable Neighbourhood Search. Some of these — in

procedure $VND(\pi', N_1, N_2, \ldots, N_k)$

    input: *problem instance $\pi' \in \Pi'$, neighbourhood relations $N_1, N_2, \ldots, N_k$*

    output: *solution $\hat{s} \in S'(\pi')$* **or** $\emptyset$

    $s := init(\pi');$

    $\hat{s} := s;$

    $i := 1;$

    repeat

        find best candidate solution s' in neighbourhood $N_i(s);$

        if $g(s') < g(s)$ **then**

            $s := s';$

            if $f(s) < f(\hat{s})$ **then**

                $\hat{s} := s;$

            end

            $i := 1;$

        else

            $i = i + 1;$

        end

    until $i > k$

    if $\hat{s} \in S'$ **then**

        return $\hat{s}$

    else

        return $\emptyset$

    end

end *VND*

Figure 2.2 Algorithm outline for Variable Neighbourhood Descent for optimisation problems; note that the evaluation function g is used for checking whether the search has reached a local minimum, while the objective function of the given problem instance, f, is used for detecting improvements in the incumbent candidate solution. For further details, see text.

particular, Basic VNS and Skewed VNS [Hansen and Mladenović, 2002] — are conceptually closely related to Iterated Local Search, a hybrid SLS method that will be discussed later in this chapter (cf. Section 2.3).

Variable Depth Search

A different approach to selecting search steps from large neighbourhoods reasonably efficiently is to compose more complex steps from a number of steps in small, simple neighbourhoods. This idea is the basis of *Variable Depth Search*

(VDS), an SLS method introduced first by Kernighan and Lin for the Graph Partitioning Problem [1970] and the TSP [1973]. Generally, VDS can be seen as an iterative improvement method in which the local search steps are variable length sequences of simpler search steps in a small neighbourhood. Constraints on the feasible sequences of simple steps help to keep the time complexity of selecting complex steps reasonably low. (For an algorithm outline of VDS, see Figure 2.3.)

As an example for a VDS algorithm, consider the *Lin-Kernighan (LK) Algorithm for the TSP*. The LK algorithm performs iterative improvement using complex search steps each of which corresponds to a sequence of 2-exchange steps. The mechanism underlying the construction of a complex step can be

procedure *VDS*(π')

    input: *problem instance* $\pi' \in \Pi'$

    output: *solution* $\hat{s} \in S'(\pi')$ **or** $\emptyset$

    $s := init(\pi')$;

    $\hat{s} := s$;

    while not *terminate* (π', s) **do**

        $t := s$;

        $\hat{t} := t$;

        repeat

            $t := selectBestFeasibleNeighbour(\pi', t)$;

            if $f(t) < f(\hat{t})$ **then**

                $\hat{t} := t$;

            end

        until *terminateConstruction* $(\pi', t, \hat{t})$;

        $s := \hat{t}$;

        if $f(s) < f(\hat{s})$ **then**

            $\hat{s} := s$;

        end

    end

    if $\hat{s} \in S'$ **then**

        return $\hat{s}$

    else

        return $\emptyset$

    end

end *VDS*

Figure 2.3 Algorithm outline for Variable Depth Search for optimisation problems; for details, see text.

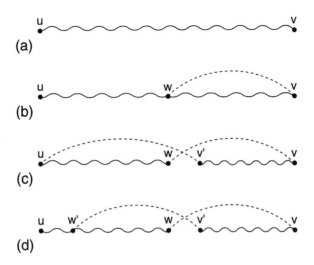

Figure 2.4 Schematic view of a Lin–Kernighan exchange step: (a) shows a Hamiltonian path, (b) a possible δ-path, (c) the next Hamiltonian path (which is closed by introducing the left dashed edge) and (d) indicates a next possible δ-path.

understood best by considering a sequence of Hamiltonian paths, that is, paths that contain each vertex in the given graph G exactly once. Figure 2.4a shows an example in which a Hamiltonian path between nodes u and v is obtained from a valid round trip by removing the edge (u, v). Let us fix one of the endpoints in this path, say u; the other endpoint is kept variable. We can now introduce a cycle into this Hamiltonian path by adding an edge (v, w) (see Figure 2.4b). The resulting subgraph can also be viewed as a spanning tree of G with one additional edge; it is called a *δ-path*. The cycle in this δ-path can be broken by removing a uniquely defined edge (w, v') incident to w, such that the result is a new Hamiltonian path that can be extended to a Hamiltonian cycle (and hence a candidate solution for the TSP) by adding an edge between v' and the fixed endpoint u (this is the dashed edge (v', u) in Figure 2.4c). Alternatively, a different edge can be added, leading to a new δ-path as indicated in Figure 2.4d.

Based on this fundamental mechanism, the LK algorithm computes complex search steps as follows: Starting with the current candidate solution (a Hamiltonian cycle) s, a δ-path p of minimal path weight is determined by replacing one edge as described above. If the Hamiltonian cycle t obtained from p by adding a (uniquely defined) edge has weight smaller than s, then t (and its weight) is memorised. The same operation is now performed with p as a starting point, and iterated until no δ-path can be obtained with weight smaller than that of the best Hamiltonian cycle found so far. Finally, the minimal weight Hamiltonian cycle s' found in this iterative process provides the end point of a complex search step.

Note that this can be interpreted as a sequence of 1-exchange steps that alternate between δ-paths and Hamiltonian cycles.

In order to limit the time complexity of constructing complex search steps, VDS algorithms use two types of restrictions, *cost restrictions* and *tabu restrictions*, on the selection of the constituting simple search steps. In the case of the LK algorithm, any edge that has been added cannot be removed and any edge that has been removed cannot be introduced any longer. This tabu restriction has the effect that a candidate sequence for a complex step is never longer than n, the number of vertices in the given graph. The original LK algorithm also uses a number of additional mechanisms, including a limited form of backtracking, for controlling the generation of complex search steps; as a consequence, the final tour returned by the algorithm is guaranteed to be optimal w.r.t. the standard 3-exchange neighbourhood. Along with other details of the LK algorithm, these mechanisms are described in Chapter 8, Section 8.2.

VDS algorithms have been used with considerable success for solving a number of problems other than the TSP, including the Graph Partitioning Problem [Kernighan and Lin, 1970], the Unconstrained Binary Quadratic Programming Problem [Merz and Freisleben, 2002] and the Generalised Assignment Problem [Yagiura et al., 1999].

Dynasearch

Like VDS, *Dynasearch* is an iterative improvement algorithm that tries to build a complex search step based on a combination of simple search steps [Potts and van de Velde, 1995; Congram et al., 2002; Congram, 2000]. However, differently from VDS, Dynasearch requires that the individual search steps that compose a complex step are mutually independent. Here, independence means that the individual search steps do not interfere with each other with respect to their effect on the evaluation function value and the feasibility of candidate solutions. In particular, any dynasearch step from a feasible candidate solution s is guaranteed to result in another feasible candidate solution, and the overall improvement in evaluation function value achieved by the dynasearch step can be obtained by summing up the effects of applying each individual search step to s.

As an example for this independence condition, consider a TSP instance and a specific Hamiltonian cycle $t = (u_1, \ldots, u_n, u_{n+1})$, where $u_{n+1} = u_1$. A 2-exchange step involves removing two edges (u_i, u_{i+1}) and (u_j, u_{j+1}) from t (without loss of generality, we assume that $1 \leq i$ and $i + 1 < j \leq n$). Two 2-exchange steps that remove edges (u_i, u_{i+1}), (u_j, u_{j+1}) and (u_k, u_{k+1}), (u_l, u_{l+1}), respectively, are independent if, and only if, either $j < k$ or $l < i$. An example of a pair of independent 2-exchange steps is given in Figure 2.5. Any set of independent steps can be executed in parallel, leading to an overall

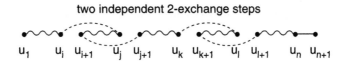

Hamiltonian cycle ($u_1, ..., u_n, u_{n+1}$)

two independent 2-exchange steps

Figure **2.5** Example of a pair of independent 2-exchange steps that can potentially form a dynasearch step.

improvement equal to the sum of the improvements achieved by the simple component steps and to a feasible candidate solution, here: another Hamiltonian cycle in the given graph.

The neighbourhood explored by Dynasearch consists of the set of all possible complex search steps; it can be shown that in general this neighourhood can be of exponential size w.r.t. to the size of the underlying simple neighbourhoods. However, through the use of a dynamic programming algorithm it is possible to find the best possible complex search step in polynomial time. (Roughly speaking, the key principle of Dynamic Programming is to iteratively solve a sequence of increasingly large subproblems that leads to a solution of the given problem, exploiting independence assumptions such as the one described previously [Bertsekas, 1995].) Only in the worst case, one complex dynasearch step consists of a single simple step. Although Dynasearch is a very recent local search technique, it has already shown very promising performance on several combinatorial optimisation problems, such as the Single Machine Total Weighted Tardiness Problem [Congram et al., 2002] (we discuss a dynasearch algorithm for this well-known $\mathcal{NP}$-hard scheduling problem in Chapter 9, Section 9.2), the TSP and the Linear Ordering Problem [Congram, 2000].

2.2 'Simple' SLS Methods

In the previous section, we introduced several ways of extending simple exchange neighbourhoods that can significantly enhance the performance of Iterative Improvement and prevent this algorithm from getting stuck in very low-quality local optima. Another way of addressing the same problem is to modify the step function, such that for a fixed and fairly simple neighbourhood, the search process can perform worsening steps which help it to escape from local optima. As mentioned in Chapter 1, Section 1.5, the simplest technique for achieving this is to use randomised variants of Iterative Improvement or a restart strategy that re-initialises the search process whenever it gets stuck in a local optimum. In this section, we will discuss a number of different methods that achieve the same effect often in a more efficient and robust way. These are simple in the sense that they essentially perform only one type of search step, while later in this chapter, we will discuss hybrid SLS algorithms, which combine various different types of search steps, as well as population-based SLS methods.

Randomised Iterative Improvement

One of the simplest ways of extending iterative improvement algorithms such that worsening steps can be performed is to sometimes select a neighbour at random, rather than an improving neighbour, within the individual search steps. Such uninformed random walk steps may be performed with a fixed frequency, such that the alternation between improvement steps and random walk steps follows a deterministic pattern. Yet, depending on the improvement strategy used, this may easily lead to a situation in which the effect of the random walk steps are immediately undone in subsequent improvement steps, leading to cycling behaviour and preventing an escape from given local optima. Therefore, it is preferable to probabilistically determine in each search step whether to apply an improvement step or a random walk step. Typically, this is done by introducing a parameter $wp \in [0,1]$, called *walk probability* or *noise parameter*, that corresponds to the probability of performing a random walk step instead of an improvement step.

The resulting algorithm is called *Randomised Iterative Improvement (RII)*. Like Iterative Improvement, it typically uses a random initialisation of the search, as described in Chapter 1, Section 1.5. Its step function can be written as $step_{RII}(s)(s') := wp \cdot step_{URW}(s)(s') + (1 - wp) \cdot step_{II}(s)(s')$, where $step_{URW}(s)(s')$ is the step function for uninformed random walk and $step_{II}(s)(s')$ is a variant of the step function for the Iterative Improvement Algorithm (see Section 1.5) that differs only in that a minimally worsening neighbour is selected if the set $I(s)$ of strictly improving neighbours is empty. As shown in Figure 2.6, the RII step function is typically implemented as a two level choice, where first a probabilistic decision is made on which of the two types of search steps is to be applied, and then the corresponding search step is performed. Obviously, there is no need to terminate this SLS algorithm as soon as a local optimum is encountered. Instead, the termination predicate can be realised in various ways. One possibility is to stop the search after a limit on the CPU time or the number of search steps has been reached; alternatively, the search may be terminated when a given number of search steps has been performed without achieving any improvement.

```
procedure step-RII(π, s, wp)
    input: problem instance π, candidate solution s, walk probability wp
    output: candidate solution s'

    u := random([0, 1]);
    if (u ≤ wp) then
        s' := step_URW(π, s);
    else
        s' := step_II(π, s);
    end
    return s'
end step-RII
```

Figure 2.6 Standard implementation of the step function for Randomised Iterative Improvement; *random*([0, 1]) returns a random number between zero and one using a uniform probability distribution.

A beneficial consequence of using a probabilistic decision on the type of local search performed in each step is the fact that arbitrarily long sequences of random walk steps (or improvement steps, respectively) can occur, where the probability of performing r consecutive random walk steps is wp^r. Hence, there is always a chance to escape even from a local optimum that has a large 'basin of attraction' in the sense that many worsening steps may be required to ensure that subsequent improvement steps have a chance of leading into different local optima. In fact, for RII it can be proven that, when the search process is run long enough, eventually a (optimal) solution to any given problem instance is found with arbitrarily high probability. (More details on this proof can be found in the in-depth section on page 155*ff.*)

EXAMPLE 2.2 **Randomised Iterative Improvement for SAT**

RII can be very easily applied to SAT by combining the uninformed random walk algorithm presented in Example 1.3 (page 41*f.*) and an iterative improvement algorithm like that of Example 1.4 (page 47*f.*), using the same search space, solution set, neighbourhood relation and initialisation function as defined there. The only difference is that here, we will apply a best improvement local search algorithm instead of the simple descent method from Example 1.4: In each step, the best improvement algorithm flips a variable that leads to a maximal increase in the evaluation function. Note that such a best improvement algorithm need not terminate at a local optimum, because in this situation the maximally improving variable flip is a perfectly valid worsening step (more precisely: a least worsening step). The step function for RII is composed of the two step functions for this greedy improvement algorithm and for uninformed random walk as described previously: With probability wp, a random neighbouring solution is returned, otherwise with probability $1 - wp$, a best improvement step is applied. We call the resulting algorithm GUWSAT.

Interestingly, a slight variation of the GUWSAT algorithm for SAT from Example 2.2, called GSAT with Random Walk (GWSAT), has been proven rather successful (see also Chapter 6, page 269*f.*). The only difference between GUWSAT and GWSAT is in the random walk step. Instead of uninformed random walk steps, GWSAT uses 'informed' random walk steps by restricting the random neighbour selection to variables occurring in currently unsatisfied clauses; among these variables, one is chosen according to a uniform distribution. When GWSAT was first proposed, it was among the best performing SLS algorithms for SAT. Yet, apart from this success, Randomised Iterative Improvement is rather rarely applied. This might be partly due to the fact that it is such a simple extension of Iterative Improvement, and more complex SLS algorithms often achieve better performance. Nevertheless, RII certainly deserves attention as a simple and generic extension of Iterative Improvement that can be generalised easily to more complex SLS methods.

Probabilistic Iterative Improvement

An interesting alternative to the mechanism for allowing worsening search steps underlying Randomised Iterative Improvement is based on the idea that the probability of accepting a worsening step should depend on the respective deterioration in evaluation function value such that the worse a step is, the less likely it would be performed. This idea leads to a family of SLS algorithms called *Probabilistic Iterative Improvement (PII)*, which is closely related to Simulated Annealing, a widely used SLS method we discuss directly after PII. In each search step, PII selects a neighbour of the current candidate solution according to a given function $p(g, s)$, which determines a probability distribution over neighbouring candidate solutions of s based on their respective evaluation function values. Formally, the corresponding step function can be written as $step(s)(s') := p(g, s)$.

Obviously, the choice of the function $p(g, s)$ is of crucial importance to the behaviour and performance of PII. Note that both Iterative Improvement, as defined in Chapter 1, Section 1.5, and Randomised Iterative Improvement can be seen as special cases of PII that are obtained for particular choices of $p(g, s)$. Generally, PII algorithms for which $p(g, s)$ assigns positive probability to all neighbours of s have properties similar to RII, in that arbitrarily long sequences of worsening moves can be performed and (optimal) solutions can be found with arbitrarily high probability as run-time approaches infinity.

EXAMPLE 2.3   **PII / Constant Temperature Annealing for the TSP**

The following, simple application of PII to the TSP illustrates the underlying approach and will also serve as a convenient basis for introducing the more general SLS method of Simulated Annealing. Given a TSP instance represented by a complete, edge-weighted graph G, we use the set of all vertex permutations as search space, S, and the same set as our set of feasible candidate solutions, S'. (This simply means that we consider each Hamiltonian cycle in G as a valid solution.) As the neighbourhood relation, N, we use a reflexive variant of the 2-exchange neighbourhood defined in Chapter 1, Section 1.5, which for each candidate solution s contains s itself as well as all Hamiltonian cycles that can be obtained by replacing two edges in s.

The search process uses a simple randomised initialisation function that picks a Hamiltonian cycle uniformly at random from S. The step function is implemented as a two-stage process, in which first a neighbour $s' \in N(s)$ is selected uniformly at random, which is then accepted according to the following probability function:

$$
p_{accept}(T, s, s') := \begin{cases} 1 & \text{if } f(s') \leq f(s) \\ \exp\left(\frac{f(s) - f(s')}{T}\right) & \text{otherwise} \end{cases} \tag{2.1}
$$

This acceptance criterion is known as the *Metropolis condition*. The parameter T, which is also called *temperature*, determines how likely it is to perform worsening search steps: at low temperature values, the probability of accepting a worsening search step is low, while at high temperature values, the algorithm accepts even drastically worsening steps with a relatively high probability. As for RII, various termination predicates can be used for determining when to end the search process.

This algorithm corresponds to a simulated annealing algorithm in which the temperature is being kept constant at T. In fact, there exists some evidence suggesting that compared to more general simulated annealing approaches, this algorithm performs quite well, but in general, the determination of a good value for T may be difficult [Fielding, 2000].

Simulated Annealing

Considering the example PII algorithm for the TSP, in which a temperature parameter T controls the probability of accepting worsening search steps, one rather obvious generalisation is to allow T to vary over the course of the search process. Conceptually, this leads to a family of SLS algorithms known as *Simulated Annealing (SA)*, which was proposed independently by Kirkpatrick, Gelatt and Vecchi [1983], and Cerný [1985]. SA was originally motivated by the annealing of solids, a physical process in which a solid is melted and then cooled down slowly in order to obtain perfect crystal structures, which can be modelled as a state of minimum energy (also called ground state). To avoid defects (i.e., irregularities) in the crystal, which correspond to meta-stable states in the model, the cooling needs to be done very slowly.

The idea underlying SA is to solve combinatorial optimisation problems by a process analogous to the physical annealing process. In this analogy, the candidate solutions of the given problem instance correspond to the states of the physical system, the evaluation function models the thermodynamic energy of the solid, and the globally optimal solutions correspond to the ground states of the physical system.

Like PII, Simulated Annealing typically starts from a random initial solution. It then performs the same general type of PII steps as defined in Example 2.3, where in each step first a neighbour s' of s is randomly chosen (*proposal mechanism*), and then an acceptance criterion parameterised by the temperature parameter T is used to decide whether the search accepts s' or whether it stays at s (see Figure 2.7). One standard choice for this acceptance criterion is a probabilistic choice according to the Metropolis condition (see Equation 2.1, page 75), which was also used in an early article on the simulation of the physical annealing process [Metropolis et al., 1953], where the parameter T corresponded to the actual

> **procedure** *step-SA*(π, s, T)
>
>   **input:** *problem instance π, candidate solution s, temperature T*
>   **output:** *candidate solution s''*
>
> $s' := proposal(\pi, s)$;
> $s'' := accept(\pi, s, s', T)$;
>   **return** s''
>
> **end** *step-SA*

Figure 2.7 Standard step function for Simulated Annealing; *proposal* randomly selects a neighbour of s, *accept* chooses probabilistically between s and s', dependent on temperature T.

physical temperature. Throughout the search process, the temperature is adjusted according to a given *annealing schedule* (often also called *cooling schedule*).

Formally, an annealing schedule is a function that for each run-time t (typically measured in terms of the number of search steps since initialisation) determines a temperature value $T(t)$. Annealing schedules are commonly specified by an initial temperature T_0, a temperature update scheme, a number of search steps to be performed at each temperature and a termination condition.

In many cases, the initial temperature T_0 is determined based on properties of the given problem instances such as the estimated cost difference between neighbouring candidate solutions [Johnson et al., 1989; van Laarhoven and Arts, 1987]. Simple geometric cooling schedules in which temperature is updated as $T := \alpha \cdot T$ have been shown to be quite efficient in many cases [Kirkpatrick et al., 1983; Johnson et al., 1989]. The number of steps performed at each temperature setting is often chosen as a multiple of the neighbourhood size.

Simulated Annealing can use a variety of termination predicates; a specific termination condition often used for SA is based on the *acceptance ratio*, that is, the ratio of proposed steps to accepted steps. In this case, the search process is terminated when the acceptance ratio falls below a certain threshold or when no improving candidate solution has been found for a given number of search steps.

EXAMPLE 2.4 **Simulated Annealing for the TSP**

The PII algorithm for the TSP specified in Example 2.3 (page 75) can be easily extended into a Simulated Annealing algorithm (see also Johnson and McGeoch [1997]). The search space, solution set and neighbourhood relation are defined as in Example 2.3. We also use the same initialisation and step functions, where *propose*(π, s) randomly selects a neighbour of s, and *accept*(π, s, s', T) probabilistically accepts s' depending on T, using the Metropolis condition. The temperature T is initialised such that only 3% of the proposed steps are not accepted, and updated according to a geometric cooling schedule with $\alpha = 0.95$; for each temperature value, $n \cdot (n-1)$ search steps are performed, where n is the size (i.e., number of vertices) of the given problem instance. The search is terminated when for five consecutive temperature values no improvement of the evaluation function has been obtained, and the acceptance rate of new solutions has fallen below 2%.

Compared to standard iterative improvement algorithms, including 3-opt local search (an iterative improvement method based on the 3-exchange neighbourhood on edges) and the Lin-Kernighan Algorithm, the SA algorithm presented in Example 2.4 performs rather poorly. By using additional techniques, including neighbourhood pruning (cf. Section 2.1), greedy initialisation, low temperature starts and look-up tables for the acceptance probabilities, significantly improved results, which are competitive with those obtained by the Lin-Kernighan Algorithm, can be obtained. Greedy initialisation methods, such as starting with a nearest neighbour tour, help SA to find high-quality candidate solutions more rapidly. To avoid that the beneficial effect of a good initial candidate solution is destroyed by accepting too many worsening moves, the initial temperature is set to a low value. The use of look-up tables deserves particular attention. Obviously, calculating the exponential function in Equation 2.1 (page 75) is computationally expensive compared to the evaluation of one neighbouring solution obtained by one 2-exchange step. By using a precomputed table of values of the function $\exp(\Delta/T)$ for a range of argument values Δ/T and by looking up the accetance probabilities $\exp((f(s) - f(s'))/T)$ from that table, a very significant speedup (in our example about 40%) can be achieved [Johnson and McGeoch, 1997].

A feature of Simulated Annealing that is often noted as particularly appealing is the fact that under certain conditions the convergence of the algorithm, in the sense that any arbitrarily long trajectory is guaranteed to end in an optimal solution, can be proven [Geman and Geman, 1984; Hajek, 1988; Lundy and Mees, 1986; Romeo and Sangiovanni-Vincentelli, 1991]. However, the practical usefulness of these results is very limited, since they require an extremely slow cooling that is typically not feasible in practice.

Tabu Search

A fundamentally different approach for escaping from local minima is to use aspects of the search history rather than random or probabilistic techniques for accepting worsening search steps. *Tabu Search (TS)* is a general SLS method that systematically utilises memory for guiding the search process [Glover, 1986; 1989; 1990; Hansen and Jaumard, 1990]. The simplest and most widely applied version of TS, which is also called *Simple Tabu Search*, consists of an iterative improvement algorithm enhanced with a form of short-term memory that enables it to escape from local optima.

Tabu Search typically uses a best improvement strategy to select the best neighbour of the current candidate solution in each search step, which in a local optimum can lead to a worsening or plateau step (*plateau steps* are local search steps which do not lead to a change of the evaluation function value). To prevent the local search to immediately return to a previously visited candidate solution and to avoid cycling, TS forbids steps to recently visited search positions. This can be implemented by explicitly memorising previously visited candidate solutions and ruling out any step that would lead back to those. More commonly,

procedure *step-TS*(π, *s*, *tt*)
    input: *problem instance* π, *candidate solution s, tabu tenure tt*
    output: *candidate solution s'*

    $N' := admissibleNeighbours(\pi, s, tt)$;
    $s' := selectBest(N')$;
    return *s'*
end *step-TS*

Figure 2.8 Standard step function for Tabu Search; *admissibleNeighbours*(π, *s*, *tt*) returns the set of admissible neighbours of *s* given the tabu tenure *tt*, *selectBest*(N') randomly chooses an element of N' with maximal evaluation function value.

reversing recent search steps is prevented by forbidding the re-introduction of solution components (such as edges in case of the TSP) which have just been removed from the current candidate solution. A parameter *tt*, called *tabu tenure*, determines the duration (in search steps) for which these restrictions apply. Forbidding possible moves using a tabu mechanism has the same effect as dynamically restricting the neighbourhood $N(s)$ of the current candidate solution *s* to a subset $N' \subset N(s)$ of *admissible neighbours*. Thus, Tabu Search can also be viewed as a dynamic neighbourhood search technique [Hertz et al., 1997].

This tabu mechanism can also forbid search steps leading to attractive, unvisited candidate solutions. Therefore, many tabu search algorithms make use of a so-called *aspiration criterion*, which specifies conditions under which the tabu status of candidate solutions or solution components is overridden. One of the most commonly used aspiration criteria overrides the tabu status of steps that lead to an improvement in the incumbent candidate solution.

Figure 2.8 shows the step function that forms the core of Tabu Search. It uses a function *admissibleNeighbours* to determine the neighbours of the current candidate solution that are not tabu or are tabu but satisfy the aspiration criterion. In a second stage, a maximally improving step is randomly selected from this set of admissible neighbours.

EXAMPLE 2.5 **Tabu Search for SAT**

Using the same definition for the search space, solution set and neighbourhood relation as in Example 1.3 (page 41*f.*), and the same evaluation function as in Example 1.4 (page 47*f.*), Tabu Search can be applied to SAT in a straightforward way. The search starts with a randomly chosen variable assignment. Each search step corresponds to a single variable flip that is selected according to the associated change in the number of unsatisfied clauses and its tabu status. More precisely, in each search step, all variables are considered admissible that either have not been flipped during the least tt steps, or that, when flipped, lead to a lower number of unsatisfied clauses than the best assignment found so far (this latter condition defines the aspiration criterion). From the set of admissible variables, a variable that, when flipped, yields a maximal decrease (or, equivalently, a minimal increase) in the number of unsatisfied clauses is selected uniformly at random. The algorithm terminates unsuccessfully if after a specified number of flips no model of the given formula has been found.

This algorithm is known as *GSAT/Tabu*; it has been shown empirically to achieve very good performance on a broad range of SAT problems (see also Chapter 6). When implementing GSAT/Tabu, it is crucial to keep the time complexity of the individual search steps minimal, which can be achieved by using special data structures and a dynamic caching and incremental updating technique for the evaluation function (this will be discussed in more detail in Chapter 6, in the in-depth section on page 271*ff.*; Chapter 6 also provides a detailed overview of state-of-the-art SLS algorithms for SAT). It is also very important to determine the tabu status of the propositional variables efficiently. This is done by storing with each variable x the search step number it_x when it was flipped last and comparing the difference between the current iteration number it and it_x to the tabu tenure parameter, tt: variable x is tabu if, and only if, $it - it_x$ is smaller than tt.

In general, the performance of Tabu Search crucially depends on the setting of the tabu tenure parameter, tt. If tt is chosen too small, search stagnation may occur; if it is too large, the search path is too restricted and high-quality solutions may be missed. A good parameter setting for tt can typically only be found empirically and often requires considerable fine-tuning. Therefore, several approaches to make the particular settings of tt more robust or to adjust tt dynamically during the run of the algorithm have been introduced.

Robust Tabu Search [Taillard, 1991] achieves an increased robustness of performance w.r.t. the tabu tenure by repeatedly choosing tt randomly from an interval $[tt_{min}, tt_{max}]$. Additionally, Robust Tabu Search forces specific local search moves if these have not been applied for a large number of iterations. For example, in the case of SAT, this corresponds to forcing a specific variable to be flipped if it has not been flipped in the last $k \cdot n$ search steps, where $k > 1$ is a parameter and n is the number of variables in a given formula. (Note that it does not make sense to set k to a value smaller or equal to one in this case.) A variant of Robust Tabu Search is currently amongst the best known algorithms for MAX-SAT, the optimisation variant of SAT (see also Chapter 7).

Reactive Tabu Search [Battiti and Tecchiolli, 1994], uses the search history to adjust the tabu tenure tt dynamically during the search. In particular, if candidate solutions are repeatedly encountered, this is interpreted as evidence that search stagnation has occurred, and the tabu tenure is increased. If, on the contrary, no repetitions are found during a sufficiently long period of time, the tabu tenure is gradually decreased. Additionally, an escape mechanism based on a series of random changes is used to prevent the search process from getting trapped in a specific region of the search space. In Section 10.2 (page 482*ff.*) we will present in detail a reactive tabu search algorithm for the Quadratic Assignment Problem.

Generally, the efficiency of Tabu Search can be further increased by using techniques exploiting a form of *intermediate-term* or *long-term memory* to achieve additional intensification or diversification of the search process. Intensification strategies correspond to efforts of revisiting promising regions of the search space, for example, by recovering *elite candidate solutions*, that is, candidate solutions that are amongst the best that have been found in the search process so far. When recovering an elite candidate solution, all tabu restrictions associated with it can be cleared, in which case the search may follow a different search path. (For an example, we refer to the tabu search algorithm presented in Section 9.3, page 446*ff.*) Another possibility is to freeze certain solution components and to keep them fixed during the search. In the TSP case, this amounts to forcing certain edges to be kept in the candidate solutions seen over a number of iterations.

Diversification can be achieved by generating new combinations of solution components, which can help to explore regions of the search space that have not been visited yet. One way of achieving this is by introducing a few rarely used solution components into the candidate solutions. An example for such a mechanism is the forced execution of search steps, as in Robust Tabu Search. Another possibility is to bias the local search by adding a component to the evaluation function contribution of specific search steps based on the frequency with which these were applied. For a detailed discussion of diversification and intensification techniques that exploit intermediate and long-term memory we refer to Glover and Laguna [1997].

Overall, tabu search algorithms have been successfully applied to a wide range of combinatorial problems, and for many problems they are among best known algorithms w.r.t. the tradeoff between solution quality and computation time [Battiti and Protasi, 2001; Galinier and Hao, 1997; Nowicki and Smutnicki, 1996b; Vaessens et al., 1996]. We will discuss several tabu search algorithms in the second part of this book. Crucial for these successful applications of Tabu Search is often a carefully chosen neighbourhood relation, as well as the use of efficient caching and incremental updating schemes for the evaluation of candidate solutions.

Dynamic Local Search

So far, the various techniques for escaping from local optima discussed in this chapter were all based on allowing worsening steps during the search process. A different approach for preventing iterative improvement methods from getting stuck in local optima is to modify the evaluation function whenever a local optimum is encountered in such a way that further improvement steps become possible. This can be achieved by associating *penalty weights* with individual solution components, which determine the impact of these components on the evaluation function value. Whenever the iterative improvement process gets trapped in a local optimum, the penalties of some solution components are increased. This leads to a degradation in the current candidate solution's evaluation function value until it is higher than the evaluation function values of some of its neighbours (which are not affected in the same way by the penalty modifications), at which point improving moves become available. This general approach provides the basis for a number of SLS algorithms which we collectively refer to as *dynamic local search (DLS)* methods.

Figure 2.9 shows an algorithm outline of DLS. As motivated above, the underlying idea is to find local optima of a dynamically changing evaluation function g' using a simple local search algorithm *localSearch*, which typically performs iterative improvement until a local minimum in g' is found. The modified evaluation function g' is obtained by adding penalties *penalty(i)* to solution components used in a candidate solution s to the original evaluation function value $g(\pi', s)$:

$$g'(\pi', s) := g(\pi, s) + \sum_{i \in SC(\pi', s)} penalty(i),$$

where $SC(\pi', s)$ is the set of solution components of π' used in a candidate solution s.

The penalties *penalty(i)* are initially set to zero and subsequently updated after each subsidiary local search. Typically, *updatePenalties* increases the penalties of some or all the solution components used by the locally optimal candidate solution s' obtained from *localSearch*(π, g', s). Particular DLS algorithms differ in how this update is performed. One main difference is whether the penalty modifications are done in an additive way or in a multiplicative way. In both cases, the penalty modification is typically parameterised by some constant λ, which also takes into account the range of evaluation function values for the particular instance being solved. Additionally, some DLS techniques occasionally decrease the penalties of solution components not used in s' [Schuurmans and Southey, 2000; Schuurmans et al., 2001].

procedure *DLS* (π')

 input: *problem instance* $\pi' \in \Pi'$

 output: *solution* $\hat{s} \in S(\pi')$ **or** $\emptyset$

 $s := init(\pi')$;

 $s := localSearch(\pi', s)$;

 $\hat{s} := s$;

 while not *terminate* (π', s) **do**

  $g' := g + \sum_{i \in SC(\pi', s)} penalty(i)$;

  $s' := localSearch(\pi, g', s)$;

  if ($f(s') < f(\hat{s})$;

   $\hat{s} := s'$;

  end

  $updatePenalties(\pi, s')$;

 end

 if $\hat{s} \in S'$ **then**

  return $\hat{s}$

 else

  return $\emptyset$

 end

end *DLS*

Figure **2.9** Algorithm outline of Dynamic Local Search for optimisation problems; *penalty*(i) is the penalty associated with solution component i, $SC(\pi', s)$ is the set of solution components used in candidate solution s, *localSearch*(π', g', s) is a subsidiary local search procedure using evaluation function g', and *updatePenalties* is a procedure for updating the solution component penalties. (Further details are given in the text.)

Penalising all solution components of a locally optimal candidate solution can cause difficulties if certain solution components that are required for any optimal solution are also present in many other local optima. In this case, it can be useful to only increase the penalties of solution components that are least likely to occur in globally optimal solutions. One specific mechanism that implements this idea uses the solution quality contribution of a solution component i in candidate solution s', $f_i(\pi, s')$, to estimate the utility of increasing *penalty*(i):

$$util(s', i) := \frac{f_i(\pi, s')}{1 + penalty(i)} \tag{2.2}$$

Using this estimate of utility, *updatePenalties* then only increases the penalties of solution components with maximal utility values. Note that dividing the solution quality distribution by $1 + penalty(i)$ avoids overly frequent penalisation of specific solution components by reducing their utility. (This mechanism is used in a particular DLS algorithm called Guided Local Search [Voudouris and Tsang, 1995].)

It is worth noting that in many cases, the solution quality contribution of a solution component does not depend on the current candidate solution. In the case of the TSP, for example, the solution components are typically the edges of the given graph, and their solution quality contributions are given by their respective weights. There are cases, however, where the solution quality contributions of individual solution components are dependent on the current candidate solution s', or, more precisely, on all solution components of s'. This is the case, for example, for the Quadratic Assignment Problem (see Section 10.2, page 477*ff.*), where DLS algorithms typically use approximations of the actual solution cost contribution [Voudouris and Tsang, 1995; Mills et al., 2003].

EXAMPLE 2.6 **Dynamic Local Search for the TSP**

This example follows the first application of DLS to the TSP, as presented by Voudouris and Tsang [1995; 1999], and describes a particular DLS algorithm called *Guided Local Search (GLS)*. Given a TSP instance in the form of an edge-weighted graph G, the same search space, solution set and 2-exchange neighbourhood is used as in Example 2.1 (page 64*f.*). The solution components are the edges of G, and the cost contribution of each edge e is given by its weight, $w(e)$. The subsidiary local search procedure *localSearch* performs first improvement steps in the underlying 2-exchange neighbourhood and can be enhanced by using standard speed-up techniques, which are described in detail in Chapter 8, Section 8.2.

In GLS, the procedure *updatePenalties*(π, s) increments the penalties of all edges of maximal utility contained in candidate solution s by a factor λ, which is chosen in dependence of the average length of good tours; in particular a setting of

$$\lambda := 0.3 \cdot \frac{f(s_{2\text{-}opt})}{n}$$

where $f(s_{2\text{-}opt})$ is the objective function value of a 2-optimal tour, and n is the number of vertices in G, has been shown to yield very good results on a set of standard TSP benchmark instances [Voudouris and Tsang, 1999].

The fundamental idea underlying DLS of adaptively modifying the evaluation function during a local search process has been used as the basis for a number of SLS algorithms for various combinatorial problems. Among the earliest DLS algorithms is the Breakout Method, in which penalties are added to solution components of locally optimal solutions [Morris, 1993]. GENET [Davenport et al., 1994], an algorithm that adaptively modifies the weight of constraints to be satisfied, has directly inspired Guided Local Search, one of the most widely applied DLS methods [Voudouris and Tsang, 1995; 2002]. Closely related SLS algorithms, which can be seen as instances of the general DLS method presented here, have been developed for constraint satisfaction and SAT, where penalties are typically associated with the clauses of a given CNF formula [Selman and Kautz, 1993; Cha and Iwama, 1996; Frank, 1997]; this particular approach is also known as *clause weighting*. Some of the best-performing SLS algorithms for SAT and MAX-SAT (an optimisation variant of SAT) are based on clause weighting schemes inspired by Lagrangean relaxation techniques [Hutter et al., 2002; Schuurmans et al., 2001; Wu and Wah, 2000].

2.3 Hybrid SLS Methods

As we have seen earlier in this chapter, the behaviour and performance of 'simple' SLS techniques can often be improved significantly by combining them with other SLS strategies. We have already presented some very simple examples of such hybrid SLS methods. Randomised Iterative Improvement, for example, can be seen as a hybrid SLS algorithm, obtained by probabilistically combining standard Iterative Improvement and Uninformed Random Walk (cf. Section 2.2, page 72*ff.*). Similarly, many SLS implementations make use of a random restart mechanism that terminates and restarts the search process from a randomly chosen initial position based on standard termination conditions; this can be seen as a hybrid combination of the underlying SLS algorithm and Uninformed Random Picking. In this section, we present a number of well-known and very successful SLS methods that can be seen as hybrid combinations of simpler SLS techniques.

Iterated Local Search

In the previous sections, we have discussed various mechanisms for preventing iterative improvement techniques from getting stuck in local optima of the evaluation function. Arguably one of the simplest and most intuitive ideas for addressing this fundamental issue is to use two types of SLS steps: one for reaching local optima as efficiently as possible, and the other for effectively escaping from local optima. This is the key idea underlying *Iterated Local Search (ILS)* [Lourenço et al., 2002], a SLS method that essentially uses these two types of search steps alternately to perform a walk in the space of local optima w.r.t. the given evaluation function.

```
procedure ILS(π')
    input: problem instance π' ∈ Π'
    output: solution ŝ ∈ S(π') or ∅

    s := init(π');
    s := localSearch(π', s);
    ŝ := s;
    while not terminate(π', s) do
        s' := perturb(π', s');
        s'' := localSearch(π', s');
        if (f(s'') < f(ŝ)) then
            ŝ := s'';
        end
        s := accept(π', s, s'');
    end
    if ŝ ∈ S' then
        return ŝ
    else
        return ∅
    end
end ILS
```

Figure 2.10 Algorithm outline of Iterated Local Search (ILS) for optimisation problems. (For details, see text.)

Figure 2.10 shows an algorithm outline for ILS. As usual, the search process can be initialised in various ways, for example, by starting from a randomly selected element of the search space. From the initial candidate solution, a locally optimal solution is obtained by applying a subsidiary local search procedure *localSearch*. Then, each iteration of the algorithm consists of three major stages: first, a perturbation is applied to the current candidate solution s; this yields a modified candidate solution s' from which in the next stage a subsidiary local search is performed until a local optimum s'' is obtained. In the last stage, an acceptance criterion *accept* is used to decide from which of the two local optima, s or s'', the search process is continued. Both functions, *perturb* and *accept*, can use aspects of the search history, for example, when the same local optima are repeatedly encountered, stronger perturbation steps may be applied. As in the case of most other SLS algorithms, a variety of termination predicates *terminate* can be used for deciding when the search process ends.

The three procedures *localSearch*, *perturb* and *accept* form the core of any ILS algorithm. The specific choice of these procedures has a crucial impact on the performance of the resulting algorithm. As we will discuss in the following, these components need to complement each other for achieving a good tradeoff between intensification and diversification of the search process, which is critical for obtaining good performance when solving hard combinatorial problems.

It is rather obvious that the subsidiary local search procedure, *localSearch*, has a considerable influence on the performance of any ILS algorithm. In general, more effective local search methods lead to better performing ILS algorithms. For example, when applying ILS to the Travelling Salesman Problem, using 3-opt local search (i.e., an iterative improvement algorithm based on the 3-exchange neighbourhood relation) typically leads to better performance than using 2-opt local search, while even better results than with 3-opt local search are obtained when using the Lin-Kernighan Algorithm as a subsidiary local search procedure. While often, iterative improvement methods are used for the subsidiary local search within ILS, it is perfectly possible to use more sophisticated SLS algorithms, such as SA, TS or DLS, instead.

The role of *perturb* is to modify the current candidate solution in a way that will not be immediately undone by the subsequent local search phase. This helps the search process to effectively escape from local optima, and the subsequent local search phase has a chance to discover different local optima. In the simplest case, a random walk step in a larger neighbourhood than the one used by *localSearch* may be sufficient for achieving this goal. There are also ILS algorithms that use perturbations consisting of a number of simple steps (e.g., sequences of random walk steps in a 1-exchange neighbourhood).

Typically, the strength of the perturbation has a strong influence on the length of the subsequent local search phase; weak perturbations usually lead to shorter local search phases than strong perturbations, because the local search procedure requires fewer steps to reach a local optimum. If the perturbation is too weak, however, the local search will often fall back into the local optimum just visited, which leads to search stagnation. At the same time, if the perturbation is too strong, its effect can be similar to a random restart of the search process, which usually results in a low probability of finding better solutions in the subsequent local search phase. To address these issues, both the strength and the nature of the perturbation steps may be changed adaptively during the search. Furthermore, there are rather complex perturbation techniques, such as the one used in Lourenço [1995], which is based on finding optimal solutions for parts of the given problem instance.

The acceptance criterion, *accept*, also has a strong influence on the behaviour and performance of ILS. A strong intensification of the search is obtained if the better of the two solutions s and s'' is always accepted. ILS algorithms using this acceptance criterion effectively perform iterative improvement in the space of local optima reached by the subsidiary local search procecudure. Conversely, if the new local optimum, s'', is always accepted regardless of its solution quality, the behaviour of the resulting ILS algorithm corresponds to a random walk in the space of the local optima of the given evaluation function. Between these extremes, many intermediate choices exist; for example, the Metropolis acceptance criterion known from Simulated Annealing has been used in an early class of ILS algorithms called *Large Step Markov Chains* [Martin et al., 1991]. While all these acceptance criteria are Markovian, that is, they only depend on s and s'', it has been shown that acceptance criteria that take into account aspects of the search history, such as the number of search steps since the last improvement of the incumbent candidate solution, often help to enhance ILS performance [Stützle, 1998c].

EXAMPLE 2.7 **Iterated Local Search for the TSP**

In this example we describe the *Iterated Lin-Kernighan (ILK) Algorithm*, an ILS algorithm that is currently amongst the best performing incomplete algorithms for the Travelling Salesman Problem. ILK is based on the same search space and solution set as used in Example 2.3 (page 75). The subsidiary local search procedure *localSearch* is the Lin-Kernighan variable depth search algorithm (LK) described in Section 2.1 (page 68*ff.*).

Like almost all ILS algorithms for the Travelling Salesman Problem, ILK uses a particular 4-exchange step, called a *double-bridge move*, as a perturbation step. This double-bridge move is illustrated in Figure 2.11; it has the desirable property that it cannot be directly reversed by a sequence of 2-exchange moves as performed by the LK algorithm. Furthermore, it was found in empirical studies that this perturbation is effective independently of problem size. Finally, an acceptance criterion is used that always returns the better of the two candidate solutions s and s''. An efficient implementation of

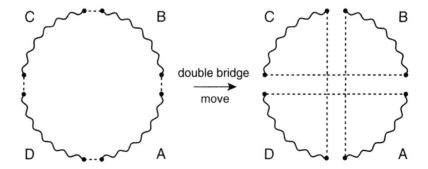

Figure 2.11 Schematic representation of the double-bridge move used in ILK. The four dashed edges on the left are removed and the remaining paths A, B, C, D are reconnected as shown on the right side.

this structurally rather simple algorithm has been shown to achieve excellent performance [Johnson and McGeoch, 1997]. (Details on this and other ILS algorithms for the TSP are presented in Section 8.3, page 384*ff*.)

Generally, ILS can be seen as a straight-forward, yet powerful technique for extending 'simple' SLS algorithms such as Iterative Improvement. The conceptual simplicity of the underlying idea led to frequent re-discoveries and many variants, most of which are known under various names, such as *Large Step Markov Chains* [Martin et al., 1991], *Chained Local Search* [Martin and Otto, 1996], as well as, when applied to particular algorithms, to specific techniques, such as Iterated Lin-Kernighan algorithms [Johnson and McGeoch, 1997]. Despite the fact that the underlying ideas are quite different, there is also a close conceptual relationship between ILS and certain variants of Variable Neighbourhood Search (VNS), such as Basic VNS and Skewed VNS [Hansen and Mladenović, 2002].

ILS algorithms are also attractive because they are typically easy to implement: in many cases, existing SLS implementations can be extended into ILS algorithms by adding just a few lines of code. At the same time, ILS algorithms are currently among the best-performing incomplete search methods for many combinatorial problems, the most prominent application being the Travelling Salesman Problem [Johnson and McGeoch, 1997; Martin and Otto, 1996]. For an overview of various issues arising in the design and implementation of ILS algorithms we refer to Lourenço et al. [2002].

Greedy Randomised Adaptive Search Procedures

A standard approach for quickly finding high-quality solutions for a given combinatorial optimisation problem is to apply a greedy construction search method (see also Chapter 1, Section 1.4) that, starting from an empty candidate solution, at each construction step adds the solution component ranked best according to a heuristic selection function, and to subsequently use a perturbative local search algorithm to improve the candidate solution thus obtained. In practice, this type of hybrid search method often yields much better solution quality than simple SLS methods initialised at candidate solutions obtained by Uninformed Random Picking (see Chapter 1, Section 1.5). Additionally, when starting from a greedily constructed candidate solution, the subsequent perturbative local search process typically takes much fewer improvement steps to reach a local optimum. By iterating this process of greedy construction and perturbative local search, even higher-quality solutions can be obtained.

Unfortunately, greedy construction search methods can typically only generate a very limited number of different candidate solutions. *Greedy Randomised*

```
procedure GRASP(π')
    input: problem instance π' ∈ Π'
    output: solution ŝ ∈ S(π') or ∅

    s := ∅;
    ŝ := s;
    f(ŝ) := ∞;
    while not terminate(π', s) do
        s := construct(π');
        s' := localSearch(π', s);
        if (f(s') < f(ŝ)) then
            ŝ := s';
        end
    end
    if ŝ ∈ S' then
        return ŝ
    else
        return ∅
    end
end GRASP
```

Figure **2.12** Algorithm outline of GRASP for optimisation problems. (For details, see text.)

Adaptive Search Procedures (GRASP) [Feo and Resende, 1989; 1995] try to avoid this disadvantage by randomising the construction method, such that it can generate a large number of different good starting points for a perturbative local search method.

Figure 2.12 shows an algorithm outline for GRASP. In each iteration of the algorithm, first a candidate solution s is generated using a randomised constructive search procedure, *construct*. Then, a local search procedure, *localSearch*, is applied to s, yielding an improved (typically, locally optimal) candidate solution s'. This two-phase process is iterated until a termination condition is satisfied.

In contrast to standard greedy constructive search methods, the constructive search algorithm used in GRASP does not necessarily add a solution component with maximal heuristic value in each construction step, but rather selects randomly from a set of highly ranked solution components. This is done by defining in each construction step a *restricted candidate list (RCL)* and then selecting one of the solution components in the RCL randomly according to a uniform distribution. In GRASP, there are two different mechanisms for defining the RCL: by cardinality restriction or by value restriction. In the case of a cardinality restriction, only the k best-ranked solution components are included in the RCL.

Value restriction allows the number of elements in the RCL to vary. More specifically, let $g(l)$ be the greedy heuristic value of a solution component l, L be the set of feasible solution components, and let $g_{min} := \min\{g(l) \mid l \in L\}$ and $g_{max} := \max\{g(l) \mid l \in L\}$ be the best and worst heuristic values among the feasible components, respectively. Then a component l is inserted into the RCL if, and only if, $g(l) \leq g_{min} + \alpha(g_{max} - g_{min})$. Clearly, the smaller k or α, the greedier is the selection of the next solution component.

The constructive search process performed within GRASP is 'adaptive' in the sense that the heuristic value for each solution component typically depends on the components that are already present in the current partial candidate solution. This takes more computation time than using static heuristic values that do not change during the construction process, but this overhead is typically amortised by the higher quality solutions obtained when using the 'adaptive' search method.

Note that it is entirely feasible to perform GRASP without a perturbative local search phase; the respective restricted variants of GRASP are also known as *semi-greedy* heuristics [Hart and Shogan, 1987]. But in general, the candidate solutions obtained from the randomised constructive search process are not guaranteed to be locally optimal with respect to some simple neighbourhood; hence, even the additional use of a simple iterative improvement algorithm typically yields higher quality solutions with rather small computational overhead. Indeed, for a large number of combinatorial problems, empirical results indicate that the additional local search phase improves the performance of the algorithm considerably.

EXAMPLE 2.8 **GRASP for SAT**

GRASP can be applied to SAT in a rather straightforward way [Resende and Feo, 1996]. The constructive search procedure starts from an empty variable assignment and adds an atomic assignment (i.e., an assignment of a truth value to an individual propositional variable of the given CNF formula) in each construction step. The heuristic function used for guiding this construction process is defined by the number of clauses that become satisfied as a consequence of adding a particular atomic assignment to the current partial assignment.

Let $h(i, v)$ be the number of (previously unsatisfied) clauses that become satisfied as a consequence of the atomic assignment $x_i := v$, where $v \in \{\top, \bot\}$. In each construction step, an RCL is built by cardinality restriction; this RCL contains the k variable assignments with the largest heuristic value $h(i, v)$.

In the simplest case, the current partial assignment is extended by an atomic variable assignment that is selected from the RCL uniformly at random. In Resende and Feo [1996], a slightly more complex assignment strategy is followed. If an unsatisfied clause c exists, in which only one variable is unassigned under the current partial assignment, this variable is assigned the value that renders c satisfied. (This mimics unit propagation, a well-known simplification strategy for SAT that is widely used in complete SAT algorithms.) Only if no such clause exists, a random element of the RCL is selected instead.

After a complete assignment has been generated, the respective candidate solution is improved using a best improvement variant of the iterative improvement algorithm for SAT from Example 1.4 (page 47*f.*). The search process is terminated when a solution has been found or after a given number of iterations has been exceeded.

This GRASP algorithm together with other variants of *construct* was implemented and tested on a large number of satisfiable SAT instances from the DIMACS benchmark suite [Resende and Feo, 1996]. While the results were reasonably good at the time the algorithm was first presented, it is now outperformed by more recent SLS algorithms for SAT (see Chapter 6).

GRASP has been applied to a large number of combinatorial problems, including MAX-SAT, Quadratic Assignment and various scheduling problems; we refer to Festa and Resende [2001] for an overview of GRASP applications. There are also a number of recent improvements and extensions of the basic GRASP algorithm; some of these include reactive GRASP variants in which, for example, the parameter α used in value-restricted RCLs is dynamically adapted [Prais and Ribeiro, 2000], and combinations with tabu search or path relinking algorithms [Laguna and Martí, 1999; Lourenço and Serra, 2002]. For a detailed introduction to GRASP and a discussion of various extensions of the basic GRASP algorithm as presented here, we refer to Resende and Ribeiro [2002].

Adaptive Iterated Construction Search

Considering algorithms based on repeated constructive search processes, such as GRASP, the idea of exploiting experience gained from past iterations for guiding further solution constructions is appealing. One way of implementing this idea is to use weights associated with the possible decisions that are made during the construction process. These weights are adapted over multiple iterations of the search process to reflect the experience from previous iterations. This leads to a family of SLS algorithms we call *Adaptive Iterated Construction Search (AICS)*.

An algorithm outline of AICS is shown in Figure 2.13. At the beginning of the search process, all weights are initialised to some small value τ_0. Each

procedure *AICS*(π')
    input: *problem instance* $\pi' \in \Pi'$
    output: *solution* $\hat{s} \in S(\pi')$ *or* $\emptyset$

    $s := \emptyset$;
    $\hat{s} := s$;
    $f(\hat{s}) := \infty$;
    $w := initWeights(\pi')$;
    while not *terminate*(π', s) **do**
        $s := construct(\pi', w, h)$;
        $s' := localSearch(\pi', s)$;
        if $f(s') < f(\hat{s})$ **then**
            $\hat{s} = s'$;
        end
        $w := adaptWeights(\pi', s', w)$;
    end
    if $\hat{s} \in S'$ **then**
        return $\hat{s}$
    else
        return $\emptyset$
    end
end *AICS*

Figure **2.13** Algorithm outline of Adaptive Iterated Construction Search for optimisation problems. (For details, see text.)

iteration of AICS consists of three phases. First, a constructive search process is used to generate a candidate solution s. Next, an additional perturbative local search phase is performed on s, yielding a locally optimal solution s'. Finally, the weights are adapted based on the solution components used in s' and the solution quality of s'. As usual, various termination conditions can be used to determine when the search process is ended.

The constructive search process uses the weights as well as a heuristic function h on the solution components to probabilistically select components for extending the current partial candidate solution. Generally, h can be chosen to be a standard heuristic function, as used for greedy methods or in the context of tree search algorithms; alternatively, h can be based on lower bounds on the solution quality of s, such as the bounds used in branch & bound algorithms. For AICS, it can be advantageous to implement the solution component selection in such a way that at all points of the construction process, with a small probability, any component solution can be added to the current partial candidate solution, irrespective of its weight and heuristic value.

As in GRASP, the perturbative local search phase typically improves the quality of the candidate solution generated by the construction process, leading to an overall increase in performance. In the simplest case, iterative improvement algorithms can be used in this context; however, it is perfectly possible and potentially beneficial to use more powerful SLS methods that can escape from local optima of the evaluation function. Typically, there is a tradeoff between the computation time used by the local search phase *vs* the construction phase, which can only be optimised empirically and depends on the given problem domain.

The adjustment of the weights, as implemented in the procedure *adaptWeights*, is typically done by increasing the weights that correspond to the solution components contained in s'. In this context, it is also possible to use aspects of the search history; for example, by using the incumbent candidate solution as the basis for the weight update, the sampling performed by the construction and perturbative search phases can be focused more directly on promising regions of the search space.

EXAMPLE 2.9 **A Simple AICS Algorithm for the TSP**

The AICS algorithm presented in this example is a simplified version of Ant System for the TSP by Dorigo, Maniezzo and Colorni [1991; 1996], enhanced by an additional perturbative search phase, which in practice improves the performance of the original algorithm. (Ant System is a particular instance of Ant Colony Optimisation, an SLS method discussed in the following section.) It uses the same search space and solution set as used in Example 2.3 (page 75).

Weights $\tau_{ij} \in \mathbb{R}_0^+$ are associated with each edge (i, j) of the given graph G, and heuristic values $\eta_{ij} := 1/w((i, j))$ are used, where $w((i, j))$ is the weight of edge (i, j). At the beginning of the search process, all edge weights are initialised to a small value, τ_0. The function *construct* iteratively constructs vertex permutations (corresponding to Hamiltonian cycles in G). The construction process starts with a randomly chosen vertex and then extends the partial permutation ϕ by probabilistically selecting a vertex not contained in ϕ according to the following distribution:

$$p_{ij} := \frac{[\tau_{ij}]^\alpha \cdot [\eta_{ij}]^\beta}{\sum_{l \in N'(i)} [\tau_{il}]^\alpha \cdot [\eta_{il}]^\beta} \qquad \text{if } j \in N'(i) \qquad (2.3)$$

and 0 otherwise, where $N'(i)$ is the feasible neighbourhood of vertex i, that is, the set of all neighbours of i that are not contained in the current partial permutation ϕ, and α and β are parameters that control the relative impact of the weights *vs* the heuristic values.

Upon the completion of each construction process, an iterative improvement search using the 2-exchange neighbourhood is performed until a vertex permutation corresponding to a Hamiltonian cycle with minimal path weight is reached.

The adaption of the weights τ_{ij} is done by first decreasing all τ_{ij} by a constant factor and then increasing the weights of the edges used in s' proportionally to the path weight $f(s')$ of the Hamiltonian cycle represented by s', that is, for all edges (i, j), the following update is performed:

$$\tau_{ij} := (1 - \rho) \cdot \tau_{ij} + \Delta(i, j, s') \tag{2.4}$$

where $0 < \rho \leq 1$ is a parameter of the algorithm, and $\Delta(i, j, s')$ is defined as $1/f(s')$ if edge (i, j) is contained in the cycle represented by s' and as zero otherwise.

The decay mechanism controlled by the parameter ρ helps to avoid unlimited increased of the weights τ_{ij} and lets the algorithm 'forget' the past experience reflected in the weights. The specific definition of $\Delta(i, j, s')$ reflects the idea that edges contained in good candidate solutions should be used with higher probability in subsequent constructions. The search process is terminated after a fixed number of iterations.

Different from most of the other SLS methods presented in this chapter, AICS has not (yet) been widely used as a general SLS technique. It is very useful, however, as a general framework that helps to understand a number of recent variants of constructive search algorithms. In particular, various incomplete tree search algorithms can be seen as instances of AICS, including the stochastic tree search algorithm by Bresina [1996], the Squeeky-Wheel Optimisation algorithm by Joslin and Clements [1999], and the Adaptive Probing algorithm by Ruml [2001]. Furthermore, AICS can be viewed as a special case of Ant Colony Optimisation, a prominent SLS method based on an adaptive iterated construction process involving populations of candidate solutions.

2.4 **Population-Based SLS Methods**

All SLS methods we have discussed so far manipulate only one single candidate solution of the given problem instance in each search step. A straightforward extension is to consider algorithms where several individual candidate solutions are simultaneously maintained; this idea leads to the population-based SLS methods discussed in this section. Although in principle, one could consider population-based search methods in which the population size may vary throughout the search process, the population-based SLS methods considered here typically use constant size populations.

Note that population-based SLS algorithms fit into the formal definition of an SLS algorithm (Definition 1.10, page 38*f.*) by considering search positions that are sets of individual candidate solutions. Though interesting in some ways, this view is somewhat unintuitive, and for most practical purposes, it is preferable to think of the search process as operating on sets of candidate solutions for the given problem instance. For example, a population-based SLS algorithm for SAT intuitively operates on a set of variable assignments. In the following, unless explicitly stated otherwise, we will use the term 'candidate solution' in this intuitive sense, rather than to refer to entire populations.

The use of populations offers several conceptual advantages in the context of SLS methods. For instance, a population of candidate solutions provides a straightforward means for achieving search diversification and hence for increasing the exploration capabilities of the search process. Furthermore, it facilitates the use of search mechanisms that are based on the combination of promising features from a number of individual candidate solutions.

Ant Colony Optimisation

Ant Colony Optimisation (ACO) is a population-based SLS method inspired by aspects of the pheromone-based trail-following behaviour of real ants; it was first introduced by Dorigo, Maniezzo and Colorni [1991] as a metaphor for solving hard combinatorial problems, such as the TSP. ACO can be seen as a population-based extension of AICS, based on a population of agents (*ants*) that indirectly communicate via distributed, dynamically changing information, the so-called *(artificial) pheromone trails*. These pheromone trails reflect the collective search experience and are exploited by the ants in their attempts to solve a given problem instance. The pheromone trail levels used in ACO correspond exactly to the weights in AICS. Here, we use the term 'pheromone trail level' instead of 'weight' to be consistent with the literature on Ant Colony Optimisation.

An algorithm outline of ACO for optimisation problems is shown in Figure 2.14. Conceptually, the algorithm is usually thought of as being executed by k ants, each of which creates and manipulates one candidate solution. The search process is started by initialising the pheromone trail levels; typically, this is done by setting all pheromone trail levels to the same value, τ_0. In each iteration of ACO, first a population sp of k candidate solutions is generated by a constructive search procedure *construct*. As in AICS, in this construction process each ant starts with an empty candidate solution and iteratively extends the current partial candidate solution with solution components that are selected probabilistically according to the pheromone trail levels and a heuristic function, h.

procedure *ACO*(π')
    input: *problem instance* $\pi' \in \Pi'$
    output: *solution* $\hat{s} \in S'(\pi')$ **or** $\emptyset$

    $sp := \{\emptyset\}$;
    $\hat{s} := \emptyset$;
    $f(\hat{s}) := \infty$;
    $\tau := initTrails(\pi')$;
    while not *terminate*(π', sp) **do**
        $sp := construct(\pi', \tau, h)$;
        $sp' := localSearch(\pi', sp)$;
        if $f(best(\pi', sp')) < f(\hat{s})$ **then**
            $\hat{s} = best(\pi', sp')$;
        end
        $\tau := updateTrails(\pi', sp', \tau)$;
    end
    if $\hat{s} \in S'$ **then**
        return $\hat{s}$
    else
        return $\emptyset$
    end
end *ACO*

Figure **2.14** Algorithm outline of Ant Colony Optimisation for optimisation problems; *best*(π', sp') denotes the individual from population *sp* with the best objective function value. The use of the procedure *localSearch*(π', sp') is optional. (For details, see text.)

Next, a perturbative local search procedure *localSearch* may be applied to each candidate solution in *sp*; typically, an iterative improvement method is used in this context, resulting in a population *sp'* of locally optimal candidate solutions. If the best of the candidate solutions in *sp'*, *best*(π', sp'), improves on the overall best solution obtained so far, this candidate solution becomes the new incumbent candidate solution. As in GRASP and AICS, this perturbative local search phase is optional, but typically leads to significantly improved performance of the algorithm.

Finally, the pheromone trail levels are updated based on the candidate solutions in *sp'* and their respective solution qualities. The precise pheromone update mechanism differs between various ACO algorithms. A typical mechanism first uniformly decreases all pheromone trail levels by a constant factor (intuitively, this corresponds to the physical process of pheromone evaporation), after which a subset of the pheromone trail levels is increased; this subset and the amount of the increase is determined from the quality of the candidate solutions in sp' and $\hat{s}$, and from the solution components contained in these. As usual, a number of different termination predicates can be used to determine when to end the search process. As an alternative to standard termination criteria based on CPU time or the number of iterations, these can include conditions on the make-up of the current population, sp', such as the variation in solution quality across the elements of sp' or their average distance from each other.

EXAMPLE 2.10 A Simple ACO Algorithm for the TSP

In this example, we present a variant of Ant System for the TSP, a simple ACO algorithm which played an important role as the first application of the ant colony metaphor to solving combinatorial optimisation problems [Dorigo et al., 1991; Dorigo, 1992; Dorigo et al., 1996].

This algorithm can be seen as a slight extension of the AICS algorithm from Example 2.9 (page 94*f*.). The initialisation of the pheromone trail levels is performed exactly like the weight initialisation in the AICS example. The functions *construct* and *localSearch* are straightforward extensions of the ones from Example 2.9 that perform the respective construction and perturbative local search processes for each individual candidate solution independently.

The pheromone trail update procedure, *updateTrails*, is also quite similar to the *adaptWeights* procedure from the AICS example; in fact, it is based on the same update as specified in Equation 2.4 (page 95), but instead of $\Delta(i, j, s')$, now a value $\Delta(i, j, sp')$ is used, which is based on contributions from all candidate solutions in the current population sp' according to the following definition:

$$\Delta(i, j, sp') := \sum_{s' \in sp'} \Delta(i, j, s'), \tag{2.5}$$

where $\Delta(i, j, s')$ is defined as $1/f(s')$ if edge (i, j) is contained in the Hamiltonian cycle represented by candidate solution s' and as zero otherwise. According to this definition, the pheromone trail levels associated with edges which belong to the highest-quality candidate solutions (i.e., low-weight Hamiltonian cycles) and which have been used by the most ants are increased the most. This reflects the idea that heuristically, these edges are most likely to be contained in even better (and potentially optimal) candidate solutions and should therefore be selected with higher probability during future construction phases. The search process is terminated after a fixed number of iterations.

Note how, in terms of the biological metaphor, the phases of this algorithm can be interpreted loosely as the actions of ants that walk the edges of the given graph to construct tours (using memory to ensure that only Hamiltonian cycles are generated as candidate solution) and deposit pheromones to reinforce the edges of their tours.

The algorithm from Example 2.10 differs from the original Ant System (AS) only in that AS did not include a perturbative local search phase. For many (static) combinatorial problems and a variety of ACO algorithms, it has been shown, however, that the use of a perturbative local search phase leads to significant performance improvements [Dorigo and Gambardella, 1997; Maniezzo et al., 1994; Stützle and Hoos, 1996; 1997].

ACO, as introduced here, is typically applied to static problems, that is, to problems whose instances (i) are completely specified before the search process is started and (ii) do not change while solving the problem. (All combinatorial problems covered in this book are static in this sense.) In this case, the construction of candidate solutions, the perturbative local search phase, and the pheromone updates are typically performed in a parallel and fully synchronised manner by all ants. There are, however, different approaches, such as *Ant Colony System*, an ACO method in which ants modify the pheromone trails during the construction phase [Dorigo and Gambardella, 1997]. When applying ACO to dynamic optimisation problems, that is, optimisation problems where parts of the problem instances (such as the objective function) change over time, the distinction between synchronous and asynchronous, decentralised phases of the algorithm becomes very important. This is reflected in the *ACO metaheuristic* [Dorigo et al., 1999; Dorigo and Di Caro, 1999; Dorigo and Stützle, 2004], which provides a general framework for ACO applications to both, static and dynamic combinatorial problems.

ACO algorithms have been applied to a wide range of combinatorial problems. The first ACO algorithm, Ant System, was applied to the TSP and several other combinatorial problems. It has been shown to be capable of solving some non-trivial instances of these problems, but its performance falls substantially short of that of state-of-the-art algorithms. Nevertheless, Ant System can be seen as a proof-of-concept that the ideas underlying ACO can be used to solve combinatorial optimisation problems. Following Ant System, many other Ant Colony Optimisation algorithms have been developed, including Ant Colony System [Dorigo and Gambardella, 1997], $\mathcal{MAX}$–$\mathcal{MIN}$ Ant System [Stützle and Hoos, 1997; Stützle and Hoos, 2000] and the ANTS Algorithm [Maniezzo, 1999]. These algorithms differ in important aspects of the search control and introduced advanced features, such as the use of look-ahead or pheromone trail level updates during the construction phase or diversification mechanisms, such as bounds on the range of possible pheromone trail levels. Some of the most prominent ACO applications are to dynamic optimisation problems, such as routing in telecommunications networks, in which traffic patterns are subject to significant changes over time [Di Caro and Dorigo, 1998]. We refer to the book by

Dorigo and Stützle [2004] for a detailed account of the ACO metaheuristic, different ACO algorithms, theoretical results and ACO applications.

Evolutionary Algorithms

With Ant Colony Optimisation, we saw an example of a population-based SLS method in which the only interaction between the individual elements of the population of candidate solutions is of a rather indirect nature, through the modification of a common memory (namely, the pheromone trails). Perhaps the most prominent example for a type of population-based SLS algorithms based on a much more direct interaction within a population of candidate solutions is the class of *Evolutionary Algorithms (EAs)*.

In a broad sense, Evolutionary Algorithms are a large and diverse class of algorithms inspired by models of the natural evolution of biological species [Bäck, 1996; Mitchell, 1996]. They transfer the principle of evolution through mutation, recombination and selection of the fittest, which leads to the development of species that are better adapted for survival in a given environment, to solving computationally hard problems. Evolutionary algorithms are generally iterative, population-based approaches: Starting with a set of candidate solutions (the initial population), they repeatedly apply a series of three genetic operators, *selection*, *mutation* and *recombination*. Using these operators, in each iteration of an evolutionary algorithm, the current population is (completely or partially) replaced by a new set of candidate solutions; in analogy with the biological inspiration, the populations encountered in the individual iterations of the algorithm are often called *generations*.

The *selection* operator implements a (generally probabilistic) choice of individual candidate solutions either for the next generation or for the subsequent application of the mutation and recombination operators; it typically has the property that fitter individuals have a higher probability of being selected. *Mutation* is based on a unary operation on individuals that introduces small, often random modifications. *Recombination* is based on an operation that generates one or more new individuals (the *offspring*) by combining information from two or more individuals (the *parents*). The most commonly used type of recombination mechanism is called *crossover*; it is originally inspired by a fundamental mechanism in biological evolution of the same name, and essentially assembles pieces from a linear representation of the parents into a new individual. One major challenge in designing evolutionary algorithms is the design of recombination operators that combine parents in such a way that the resulting offspring is likely to inherit desirable properties from their parents, while improving on their parents' solution quality.

Note how Evolutionary Algorithms fit into our general definition of SLS algorithms, when the notion of a candidate solution as used in an SLS algorithm is applied to populations of candidate solutions of the given problem instance, as used in an EA. The concepts of search space, solution set and neighbourhood, as well as the generic functions *init*, *step* and *terminate*, can be easily applied to this population-based concept of a candidate solution. Nevertheless, to keep this description conceptually simple, in this section we continue to present evolutionary algorithms in the traditional way, where the notion of candidate solution refers to an individual of the population comprising the search state.

Intuitively, by using a population of candidate solutions instead of a single candidate solution, a higher search diversification can be achieved, particularly if the initial population is randomly selected. The primary goal of Evolutionary Algorithms for combinatorial problems is to evolve the population such that good coverage of promising regions of the search space is achieved, resulting in high-quality solutions of a given optimisation problem instance. However, pure evolutionary algorithms often seem to lack the capability of sufficient search intensification, that is, the ability to reach high-quality candidate solutions efficiently when a good starting position is given, for example, as the result of recombination or mutation. Hence, in many cases, the performance of evolutionary algorithms for combinatorial problems can be significantly improved by adding a local search phase after applying mutation and recombination [Brady, 1985; Suh and Gucht, 1987; Mühlenbein et al., 1988; Ulder et al., 1991; Merz and Freisleben, 1997; 2000b] or by incorporating a local search process into the recombination operator [Nagata and Kobayashi, 1997]. The class of Evolutionary Algorithms thus obtained is usually called *Memetic Algorithms (MAs)* [Moscato, 1989; Moscato and Norman, 1992; Moscato, 1999; Merz, 2000] or *Genetic Local Search* [Ulder et al., 1991; Kolen and Pesch, 1994; Merz and Freisleben, 1997].

In Figure 2.15 we show the outline of a generic memetic algorithm. At the beginning of the search process, an initial population is generated using function *init*. In the simplest (and rather common) case, this is done by randomly and independently picking a number of elements of the underlying search space; however, it is equally possible to use, for example, a randomised construction search method instead of random picking. In each iteration of the algorithm, recombination, mutation, perturbative local search and selection are applied to obtain the next generation of candidate solutions. As usual, a number of termination criteria can be used for determining when to end the search process.

The recombination function, $recomb(\pi', sp)$, typically generates a number of offspring solutions by repeatedly selecting a set of parents and applying a recombination operator to obtain one or more offspring from these. As mentioned

```
procedure MA(π′)
    input: problem instance π′ ∈ Π′
    output: solution ŝ ∈ S′(π′) or ∅

    sp := init(π′);
    sp := localSearch₁(π′, sp);
    ŝ := best(π′, sp);
    while not terminate(π′, sp) do
        sp′ := recomb(π′, sp);
        sp′ := localSearch₂(π′, sp′);
        sp″ := mutate(π, sp ∪ sp′);
        sp″ := localSearch₃(π′, sp″);
        if f(best(π′, sp′ ∪ sp″)) < f(ŝ) then
            ŝ = best(π′, sp′ ∪ sp″);
        end
        sp := select(π′, sp, sp′, sp″);
    end
    if ŝ ∈ S′ then
        return ŝ
    else
        return ∅
    end
end MA
```

Figure 2.15 Algorithm outline of a memetic algorithm for optimisation problems; $best(\pi', sp)$ denotes the individual from a population sp with the best objective function value. (For details, see text.)

before, this operation is generally based on a linear representation of the candidate solutions, and pieces together the offspring from fragments of the parents; this type of mechanism creates offspring that inherit certain subsets of solution components from their parents. One of the most commonly used recombination mechanisms is the *one-point binary crossover operator*, which works as follows. Given two parent candidate solutions represented by strings $x_1 x_2 \ldots x_n$ and $y_1 y_2 \ldots y_n$, first, a cut point i is randomly chosen according to a uniform distribution over the index set $\{2, \ldots, n\}$. Two offspring candidate solutions are then defined as $x_1 x_2 \ldots x_{i-1} y_i y_{i+1} \ldots y_n$ and $y_1 y_2 \ldots y_{i-1} x_i x_{i+1} \ldots x_n$ (see also Figure 2.16).

One challenge when designing recombination mechanisms stems from the fact that often, simple crossover operators do not produce valid solution candidates. Consider, for example, a formulation of the TSP in which the solution candidates are represented by permutations of the vertex set, written as vectors

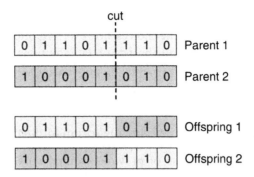

Figure 2.16 Schematic representation of the one-point binary crossover operator.

$(u_1, u_2, \ldots, u_n)$. Using a simple one-point binary crossover operation as the basis for recombination obviously leads to vectors that do not correspond to Hamiltonian cycles of the given graph. In cases like this, either a repair mechanism has to be applied to transform the results of a standard crossover into a valid candidate solution, or special crossover operators have to be used, which are guaranteed to produce valid candidate solutions only. In Chapter 8 we give two examples of high-performance memetic algorithms for the TSP that illustrate both possibilities. An overview of different specialised crossover operators for the TSP can be found in Merz and Freisleben [2001].

The role of function $mutate(\pi, sp \cup sp')$ is to introduce relatively small perturbations in the individuals in $sp \cup sp'$. Typically, these perturbations are of stochastic nature, and they are performed independently for each individual in $sp \cup sp'$, where the amount of perturbation applied is controlled by a parameter called the *mutation rate*. It should be noted that mutation need not be applied to all individuals of $sp \cup sp'$; instead, a subsidiary selection function can be used to determine which candidate solutions are to be mutated. (Until rather recently, the role of mutation compared to recombination for the performance of one of the most prominent types of evolutionary algorithms, *Genetic Algorithms*, has been widely underestimated [Bäck, 1996].)

As in ACO and AICS, perturbative local search is often useful and necessary for obtaining high-quality candidate solutions. It typically consists of selecting some or all individuals in the current population, and then applying an iterative improvement procedure to each element of this set independently.

Finally, the selection function used for determining the individuals that form the next generation sp of candidate solutions typically considers elements of the original population, as well as the newly obtained candidate solutions, and selects from these based on their respective evaluation function values (which, in this context, are usually referred to as *fitness values*). Generally, the selection is done in such a way that candidate solutions with better evaluation function values have a higher chance of 'surviving' the selection process. Many selection schemes involve probabilistic choices; however, it is often beneficial to use elitist strategies, which ensure that the best candidate solutions are always selected. Generally, the goal of selection is to obtain a population with good evaluation function values, but at the same time, to ensure a certain diversity of the population.

EXAMPLE 2.11 **A Memetic Algorithm for SAT**

As in the case of previous examples of SLS algorithms for SAT, given a propositional CNF formula F with n variables, we define the search space as the set of all variable assignments of F, the solution set as the set of all models of F, and a basic neighbourhood relation under which two variable assignments are neighbours if, and only if, they differ exactly in the truth value assigned to one variable (1-flip neighbourhood). As an evaluation function, we use the number of clauses in F unsatisfied under a given assignment.

Note that the variable assignments for a formula with n variables can be easily represented as binary strings of length n by using an arbitrary ordering of the variables and representing the truth values $\top$ and $\bot$ by 1 and 0, respectively. We keep the population size fixed at k assignments.

To obtain an initial population, we use k (independent) iterations of Uninformed Random Picking from the search space, resulting in an initial population of k randomly selected variable assignments. The recombination procedure performs $n/2$ one-point binary crossovers (as defined above) on pairs of randomly selected assignments from sp, resulting in a set sp' of n offspring assignments.

The function $mutate(F, sp \cup sp')$ simply flips μ randomly chosen bits of each assignment in $sp \cup sp'$, where $\mu \in \{1, \ldots, n\}$ is a parameter of the algorithm; this corresponds to performing μ steps of Uninformed Random Walk independently for all $s \in sp \cup sp'$ (see also Section 1.5). For perturbative local search, we use the same iterative best improvement algorithm as in Example 1.4 (page 47f.), which is run until a locally minimal assignment is obtained. The function $localSearch_3(F, sp'')$ returns the set of assignments obtained by applying this procedure to each element in s''; the same function is used for $localSearch_1$ and $localSearch_2$.

Finally, $select(F, sp, sp', sp'')$ applies a simple elitist selection scheme, in which the k best assignments in $sp \cup sp' \cup sp''$ are selected to form the next generation (using random tie-breaking, if necessary). Note that this selection scheme ensures that the best assignment found so far is always included in the new population. The search process is terminated when a model of F is found or a fixed number of iterations has been performed without finding a model.

So far, we are not aware of any Memetic Algorithm or Evolutionary Algorithm for SAT that achieves a performance comparable to state-of-the-art SAT algorithms. However, even when just following the general approach illustrated in this example, there are many alternate choices for the recombination, mutation, perturbative local search and selection procedures, few of which appear to have been implemented and studied so far.

The most prominent type of Evolutionary Algorithms for combinatorial problem solving has been the class of *Genetic Algorithms (GAs)* [Holland, 1975; Goldberg, 1989]. In early GA applications, individual candidate solutions were typically represented as bit strings of fixed length. Using this approach, interesting theoretical properties of certain Genetic Algorithms can be proven, such as the well-known Schema Theorem [Holland, 1975]. Yet, this type of representation has been shown to be disadvantageous in practice for solving certain types of combinatorial problems [Michalewicz, 1994]; in particular, this is the case for permutation problems such as the TSP, which are represented more naturally using different encodings.

Besides Genetic Algorithms, there are two other major approaches based on the same metaphor of Evolutionary Computation: *Evolution Strategies* [Rechenberg, 1973; Schwefel, 1981] and *Evolutionary Programming* [Fogel et al., 1966]. All three approaches have been developed independently and, although all of them originated in the 1960s and 1970s, only in the beginning of the 1990s did researchers become fully aware of the common underlying principles [Bäck, 1996]. These three types of Evolutionary Algorithms tend to be primarily applied to different types of problems: While Genetic Algorithms are typically used for solving discrete combinatorial problems, Evolution Strategies and Evolutionary Programming were originally developed for solving (continuous) numerical optimisation problems. For a detailed discussion of the similarities and differences between these different types of Evolutionary Algorithms and their applications, we refer to Bäck [1996].

2.5   Further Readings and Related Work

There exists a huge amount of literature on the various SLS methods discussed in this chapter. Since it would be impossible to give a reasonably complete list of references, we refer the interested reader to some of the most relevant and accessible literature, and point out books as well as conference and workshop proceedings that will provide additional material and further references.

There are relatively few books that provide a general introduction to and overview of different SLS techniques. One of these is the book on 'modern heuristics' by Michalewicz and Fogel [2000], which is rather focused on Evolutionary Algorithms but also discusses other SLS methods; another one is the book by Sait and Youssef [1999], which includes the discussion of two lesser-known SLS techniques: Simulated Evolution and Stochastic Evolution. For a tutorial-like introduction to some of the SLS techniques covered in this chapter, such as SA, TS or GAs, we refer to the book edited by Reeves [1993b]. More advanced material is provided in the book on local search edited by Arts and Lenstra [1997], which contains expert introductions to individual SLS techniques as well as overviews on the state-of-the-art of applying SLS methods to various combinatorial problems. The Handbook of Metaheuristics [Glover and Kochenberger, 2002] includes reviews of different SLS methods and additional related topics by leading experts on the respective subjects.

There is a large number of books dedicated to individual SLS techniques. This is particularly true for Evolutionary Algorithms, one of the oldest and most developed SLS methods. Currently, the classics in this field are certainly the early books describing these techniques [Holland, 1975; Goldberg, 1989; Schwefel, 1981; Fogel et al., 1966]; the book by Mitchell [1996] offers a good introduction to Genetic Algorithms. Similarly, there exist a number of books dedicated to Simulated Annealing, including Aarts and Korst [1989] or van Laarhoven and Aarts [1987]. For an overview of the literature on SA as of 1988 we refer to Collins et al. [1988]. A tutorial-style overview of SA is given in Dowsland [1993], and a summary of theoretical results and statistical annealing schedules can be found in Aarts et al. [1997]. For a general overview of Tabu Search and detailed discussions of its features, we refer to the book by Glover and Laguna [1997]. This book also covers in detail various more advanced strategies, such as Strategic Oscillation and Path Relinking, as well as some lesser-known tabu search methods. Ant Colony Optimisation is covered in detail in the book by Dorigo and Stützle [2004].

For virtually all of the SLS methods covered in this chapter, large numbers of research articles have been published in a broad range of journals and conference proceedings. Research on some of the most prominent SLS methods is presented at dedicated conferences or workshop series. Again, Evolutionary Algorithms are particularly well represented, with conference series like GECCO (Genetic and Evolutionary Computation Conference), CEC (Congress on Evolutionary Computation) or PPSN (Parallel Problem Solving from Nature) as well as some smaller conferences and workshops dedicated to specific subjects and issues in the general context of Evolutionary Algorithms. Similarly, The ANTS series of workshops (From Ant Colonies to Artificial Ants: A Series of International Workshops on Ant Algorithms) provides a specialised forum for research on Ant Colony Optimisation algorithms and closely related topics. Many of the most recent developments and results in these areas can be found in the respective proceedings.

The Metaheuristics International Conference (MIC) series, initiated in 1995, has a broader scope, including many of the SLS methods described in Sections 2.2 and 2.3 of this chapter. The corresponding post-conference collections of articles [Osman and Kelly, 1996; Voß et al., 1999; Hansen and Ribeiro, 2001; Resende and de Sousa, 2003] are a good reference for recent developments in this general area. An extensive, commented bibliography on various SLS methods can be found in Osman and Laporte [1996].

In the operations research community, papers on SLS algorithms now appear frequently in journals such as the INFORMS Journal on Computing, Operations Research, European Journal of Operational Research and Computers & Operations Research. There even exists one journal, the Journal of Heuristics that is dedicated to research related to SLS methods.

Since the early 1990s, SLS algorithms have also been very prominent in the artificial intelligence community, particularly in the context of applications to SAT, constraint satisfaction, planning and scheduling. The proceedings of major AI conferences, such as IJCAI (International Joint Conference on Artificial Intelligence), AAAI (AAAI National Conference on Artificial Intelligence), ECAI (European Conference on Artificial Intelligence), as well as the proceedings of the CP (Principles and Practice of Constraint Programming) conferences and leading journals in AI, including Artificial Intelligence and the Journal on AI Research (JAIR), contain a large number of articles on SLS algorithms and their application to AI problems (we will provide many of these references in Part II of this book).

There are a number of SLS methods that we did not present in this chapter, some of which are closely related to the approaches we discussed. First, let us mention that there exists a number of other algorithms that make use of large neighbourhoods. Prominent examples are Ejection Chains [Glover and Laguna, 1997] and Cyclic Exchange Neighbourhoods [Thompson and Orlin, 1989; Thompson and Psaraftis, 1993]. For an overview of SLS methods based on very large scale neighbourhoods we refer to Ahuja et al. [2002]. Other SLS methods include: Threshold Accepting, a variant of Simulated Annealing that uses a deterministic acceptance criterion [Dueck and Scheuer, 1990]; Extremal Optimisation, which in each step tries to eliminate 'defects' of the current candidate solution and accepts every new candidate solution independent of its solution quality [Boettcher and Percus, 2000; Boettcher, 2000]; and Variable Neighbourhood Search (VNS), which is based on the fundamental idea of changing the neighbourhood relation during the search process. The VNS framework comprises Variable Neighbourhood Descent (see Section 2.1) as well as various methods that can be seen as special cases of Iterated Local Search (see Section 2.3); other VNS algorithms, however, such as Variable Neighbourhood Decomposition

Search (VNDS) [Hansen and Mladenović, 2001b], differ significantly from the SLS methods discussed in this chapter.

Ant Colony Optimisation is only the most successful example of a class of algorithms that are often referred to as swarm intelligence methods [Bonabeau et al., 1999; Kennedy et al., 2001]. A technique that is inspired by Evolutionary Algorithms is Estimation of Distribution Algorithms (EDA) [Baluja and Caruana, 1995; Larrañaga and Lozano, 2001; Mühlenbein and Paaß, 1996]; these algorithms build and iteratively update a probabilistic model of good candidate solutions that is used to generate populations of candidate solutions. Another population-based SLS method is Scatter Search [Glover, 1977; Glover et al., 2002; Laguna and Martí, 2003], which is similar to Memetic Algorithms, but typically uses a more general notion of recombination and differs in some other details of how the population is handled.

Several of these and other SLS methods are described in the Handbook of Metaheuristics [Glover and Kochenberger, 2002] and in the book New Ideas in Optimisation [Corne et al., 1999].

2.6   **Summary**

At the beginning of this chapter we discussed important details and refinements of *Iterative Improvement*, one of the most fundamental SLS methods. *Large neighbourhoods* can be used to improve the performance of iterative improvement algorithms, but they are typically very costly to search; in this situation, as well as in general, *neighbourhood pruning techniques* and *pivoting rules,* such as *first-improvement neighbour selection,* can help to increase the efficiency of the search process. More advanced SLS methods, such as *Variable Neighbourhood Descent (VND)*, *Variable Depth Search (VDS)* and *Dynasearch* use dynamically changing or complex neighbourhoods to achieve improved performance over simple iterative improvement algorithms. Although these strategies yield significantly better performance for a variety of combinatorial problems, they are also typically more difficult to implement than simple iterative improvement algorithms and often require advanced data structures to realise their full benefit.

Generally, the main problem with simple iterative improvement algorithms is the fact that they get easily stuck in local optima of the underlying evaluation function. By using large or complex neighbourhoods, some poor-quality local optima can be eliminated; but at the same time, these extended neighbourhoods are typically more costly or more difficult to search. Therefore, in this chapter we introduced and discussed various other approaches for dealing with the problem of local optima as encountered by simple iterative improvement algorithms: allowing *worsening search steps*, that is, search steps which achieve no improvement in the given evaluation or objective function, such as in *Simulated Annealing (SA)*, *Tabu Search (TS)* and many *Iterated Local Search (ILS) algorithms* and *Evolutionary Algorithms (EAs)*; dynamically modifying the evaluation function, as exemplified in *Dynamic Local Search (DLS)*; and using adaptive constructive search methods for providing better initial candidate solutions for perturbative search methods, as seen in *GRASP*, *Adaptive Iterated Construction Search (AICS)* and *Ant Colony Optimisation (ACO)*.

Each of these approaches has certain drawbacks. Allowing worsening search steps introduces the need to balance the ability to quickly reach good candidate solutions (as realised by a greedy search strategy) *vs* the ability to effectively escape from local optima and plateaus. Dynamic modifications of the evaluation function can eliminate local optima, but at the same time typically introduces new local optima; in addition, as we will see in Chapter 6, it can be difficult to amortise the overhead cost introduced by the dynamically changing evaluation function by a reduction in the number of search steps required for finding (high-quality) solutions. The use of adaptive constructive search methods for obtaining good initial solutions for subsequent perturbative SLS methods raises a very similar issue; here, the added cost of the construction method needs to be amortised.

Beyond the underlying approach for avoiding the problem of search stagnation due to local optima, the SLS algorithms presented in this chapter share or differ in a number of other fundamental features, such as the combination of simple search strategies into *hybrid methods*, the use of *populations* of candidate solutions and the use of *memory* for guiding the search process. These features form a good basis not only for a classification of SLS methods, but also for understanding their characteristics as well as the role of the underlying approaches (see also Vaessens et al. [1995]).

Our presentation made a prominent distinction between 'simple' and *hybrid SLS methods*, where hybrid methods can be seen as combinations of various 'simple' SLS methods. In some cases, such as ILS and EAs, the components of the hybrid method are various perturbative SLS processes. In other cases, such as GRASP, AICS and ACO, constructive and perturbative search mechanisms are combined. All these hybrid methods can use different types of 'simple' SLS algorithms as their components, including simple iterative improvement methods as well as more complex methods, such as SA, TS or DLS, and a variety of constructive search methods. In this sense, the hybrid SLS methods presented here are higher-order algorithms that require complex procedural or functional parameters, such as a subsidiary SLS procedure, to be specified in order to be applied to a given problem.

It is interesting to note that some of the hybrid algorithms discussed here, including ACO and EAs, originally did not include the use of perturbative local search for improving individual candidate solutions. However, adding such perturbative local search mechanisms has been found to significantly improve the performance of the algorithm in many applications to combinatorial problems.

Two of the SLS methods discussed here, ACO and EAs, can be characterised as *population-based search methods*; these maintain a population of candidate solutions that is manipulated and evaluated during the search process. Most state-of-the-art population-based SLS approaches integrate features from the individual elements of the population in order to guide the search process. In ACO, this integration is realised by the *pheromone trails* which provide the basis for the probabilistic construction process, while in EAs, it is mainly achieved through *recombination*. In contrast, all the 'simple' SLS algorithms discussed in Section 2.2 as well as ILS, GRASP and AICS manipulate only a single candidate solution in each search step. In many of these cases, such as ILS, various population-based extensions are easily conceivable [Hong et al., 1997; Stützle, 1998c].

Integrating features of populations of candidate solutions can be seen as one (rather indirect) mechanism that uses *memory* for guiding the search process towards promising regions of the search space. The weights used in AICS serve exactly the same purpose. A similarly indirect form of memory is represented by the penalties used by DLS; only here, the purpose of the memory is at least as much to guide the search away from the current, locally optimal search position, as to guide it towards better candidate solutions. The prototypical example of an SLS method that strongly exploits an explicit form of memory for directing the search process is Tabu Search.

Many SLS methods were originally inspired by natural phenomena; examples of such methods are SA, ACO and EAs, along with several other SLS methods that we did not cover in detail. Because of the original inspiration, the terminology used for describing these SLS methods is often heavily based on jargon from the corresponding natural phenomenon. Yet, closer study of these computational methods, and in particular of the high-performance algorithms derived from them, often reveals that their performance has little or nothing to do with the aspects and features that are important in the context of the corresponding natural processes. In fact, it can be argued that the most successful nature-inspired SLS methods for combinatorial problem solving are those that have been liberated to a large extent from the context of the phenomenon that originally motivated them, and use the new mechanisms and concepts derived from that original context to effectively guide the search process.

Finally, it should be pointed out that in virtually all of the local search methods discussed in this chapter, the use of random or probabilistic decisions results in significantly improved performance and robustness of these algorithms when solving combinatorial problems in practice. One of the reasons for this lies in the diversification achieved by stochastic methods, which is often crucial for effectively avoiding or overcoming stagnation of the search process. In principle, it would certainly be preferable to altogether obliviate the need for diversification by using strategies that guide the search towards (high-quality) solutions in an efficient and reliable way. But given the inherent hardness of the problems to which SLS methods are typically applied, it is hardly surprising that in practice, such strategies are typically not available, leaving stochastic local search as one of the most attractive solution approaches.

Exercises

2.1 **[*Easy*]** What is the role of 2-exchange steps in the Lin-Kernighan Algorithm?

2.2 **[*Medium*]** Design and describe a variable depth search algorithm for SAT.

2.3 **[*Easy*]** Show that the condition for the independence of a pair of 2-exchange steps to be considered for a complex dynasearch move is necessary to guarantee feasibility of the tour obtained after executing a pair of 2-exchange moves. To do so, consider what happens if we have $i < k < j < l$ for the indices of the 2-exchange moves that delete edges (u_i, u_{i+1}), (u_j, u_{j+1}) and (u_k, u_{k+1}), (u_l, u_{l+1}), respectively.

2.4 **[*Easy*]** Show that Iterative Improvement and Randomised Iterative Improvement can be seen as special cases of Probabilistic Iterative Improvement.

2.5 **[*Easy*]** Explain the impact on the value of the tabu tenure parameter in Simple Tabu Search on the diversification *vs* intensification of the search process.

2.6 **[*Medium*]** Which tabu attributes would you choose when applying Simple Tabu Search to the TSP? Are there different possibilities for deciding when a move is tabu? Characterise the memory requirements for efficiently checking the tabu status of solution components.

2.7 **[*Medium*]** Why is it preferable in Dynamic Local Search to associate penalties with solution components rather than with candidate solutions?

2.8 [*Medium; Implementation*] Implement the Guided Local Search (GLS) algorithm for the TSP described in Example 2.6 (page 84). (You can make use of the 2-opt implementation available at www.sls-book.net; if you do so, think carefully about how to best integrate the edge penalties into the local search procedure.) For the search initialisation, use a tour returned by Uninformed Random Picking.

Run this implementation of GLS on TSPLIB instance pcb3038 (available from TSPLIB [2003]). Perform 100 independent runs of the algorithm with n search steps each, where $n = 3\,038$ is the number of vertices in the given TSP instance. Record the best solution quality reached in each run and report the distribution of these solution quality values (cf. Example 2.1, page 64f.).

For comparison, modify your implementation such that instead of Guided Local Search, it realises a variant of the randomised first improvement algorithm for TSP described in Example 2.1 (page 64f.) that initialises the search by Uninformed Random Picking.

Measure the distribution of solution qualities obtained from 100 independent runs of this algorithm on TSPLIB instance pcb3038, where each run is terminated when a local minimum is encountered. Compare the solution quality distribution thus obtained with that for GLS — what do you observe?

What can you say about the run-time required by both algorithms, using the same termination criteria as in the previous experiments?

2.9 [*Medium*] Show precisely how Memetic Algorithms fit the definition of an SLS algorithm from Chapter 1, Section 1.5.

2.10 [*Medium*] The various SLS methods described in this chapter can be classified according to different criteria, including: (1) the use of a population of solutions, (2) the explicit use of memory (other than just for storing control parameters), (3) the number of different neighbourhood relations used in the search, (4) the modification of the evaluation function during the search and (5) the inspiring source of an algorithm (e.g., by natural phenomena). Classify the SLS methods discussed in this chapter according to these criteria.

*The purpose of models is not to fit the data
but to sharpen the questions.*
—Samuel Karlin, Mathematician & Bioinformatician

3 GENERALISED LOCAL SEARCH MACHINES

In this chapter, we introduce Generalised Local Search Machines (GLSMs), a formal framework for stochastic local search methods. The underlying idea is that most efficient SLS algorithms are obtained by combining simple (pure) search strategies using a control mechanism; in the GLSM model, the control mechanism is essentially realised by a non-deterministic finite state machine. GLSMs provide a uniform framework capable of representing most modern SLS methods in an adequate way; they facilitate representations which clearly separate between search and search control.

After defining the basic GLSM model, we establish the relation between our definition of stochastic local search algorithms and the GLSM model. Next, we discuss several aspects of the model, such as state types, transitions types and structural GLSM types; we also show how various well-known SLS methods can be represented in the GLSM framework. Finally, we address extensions of the basic GLSM model, such as co-operative, learning and evolutionary GLSMs.

3.1 The Basic GLSM Model

Many high-performance SLS algorithms are based on a combination of several simple search strategies, such as Iterative Best Improvement and Random Walk or the subsidiary local search and perturbation procedures in Iterated Local Search. Such algorithms can be seen as operating on two levels: at a lower level, the underlying simple search strategies are executed, while activation of and transitions between different strategies is controlled at a higher level. The main idea underlying the concept of a General Local Search Machine (GLSM) is to explicitly represent the higher-level search control mechanism in the form of a finite state machine.

Finite state machines (FSMs) are one of the most prominent formal models in the computing sciences [Hopcroft et al., 2001; Sipser, 1997]. They can be seen as abstractions of systems characterised by a finite number of states. Starting in a specific state, the current state of an FSM can change as a response to certain events, for example, a signal received from its environment; these changes in system state are called *state transitions*. As one of the simplest control paradigms, FSMs are widely used to model systems, processes and algorithms in many domains, such as hardware design or state-of-the-art computer games.

Intuitively, a *Generalised Local Search Machine (GLSM)* for a given problem Π is an FSM in which each state corresponds to a simple SLS method for Π. The machine starts in an initial state z_0; it then executes one step of the SLS method associated with the current state and selects a new state according to a transition relation Δ in a nondeterministic manner. This is iterated until a termination condition is satisfied; this termination condition typically depends on the search state (e.g., evaluation or objective function value of the current candidate solution), search history (e.g., number of local search steps or state transitions performed), or resource bounds (e.g., total CPU time consumed). For most SLS methods, the termination predicate used typically depends more on the specific application context than on the underlying search strategy. Therefore, for simplicity's sake, a termination condition is not explicitly included in our GLSM model; instead, we consider it as part of the run-time environment. (Note, however, that analogously to standard FSM models, termination conditions could easily be included in the GLSM model in the form of absorbing final states and appropriate state transitions, such that when a final state is reached, the machine halts and the search process is terminated.) Like SLS algorithms, GLSMs can make use of additional memory, for example, to model parameters of the underlying search strategies or to memorise previously encountered candidate solutions.

DEFINITION 3.1   **Generalised Local Search Machine**

A Generalised Local Search Machine (GLSM) is formally defined as a tuple $\mathcal{M} := (Z, z_0, M, m_0, \Delta, \sigma_Z, \sigma_\Delta, \tau_Z, \tau_\Delta)$ where Z is a set of states and $z_0 \in Z$ the initial state. As in Definition 1.10 (page 38f.), M is a set of memory states and $m_0 \in M$ is the initial memory state. $\Delta \subseteq Z \times Z$ is the transition relation for $\mathcal{M}$; σ_Z and σ_Δ are sets of state types and transition types, respectively, while $\tau_Z : Z \mapsto \sigma_Z$ and $\tau_\Delta : \Delta \mapsto \sigma_\Delta$ associate states and transitions with *their corresponding types. We call $\tau_Z(z)$ the type of state z and $\tau_\Delta((z_1, z_2))$ the type of transition (z_1, z_2), respectively.*

Note that in this definition, state types are used to formally represent the (typically simple) SLS methods associated with the GLSM states, and transition types capture the various strategies that are used to switch between the GLSM states. (Transition types are further discussed in the next section, and practically relevant examples of state types are given in Section 3.3.) It is often useful to assume that σ_Z and σ_Δ do not contain any types that are not associated with at least one state or transition of the given machine (i.e., τ_Z, τ_Δ are surjective). In this case, we define the *type of machine* $\mathcal{M}$ as $\tau(\mathcal{M}) := (\sigma_Z, \sigma_\Delta)$. We allow for several states of $\mathcal{M}$ having the same type (i.e., τ_Z need not be injective). However, in many cases there is precisely one state of each state type; in the following, when this is the case we will use the same symbols for denoting states and their respective types, as long as their meaning is clear from the context. Furthermore, for simplicity, we assume that different state types always represent different search strategies.

Note that we do not require that each of the states in Z can be actually reached when starting in state z_0; as we will shortly see, it is generally not trivial to decide this form of reachability. Nevertheless, it is desirable to ensure, whenever possible, that a given GLSM does not contain unreachable states.

EXAMPLE 3.1 A Simple 3-State GLSM

The following GLSM models a hybrid SLS strategy which after initialising the search (state z_0), probabilistically alternates between two search strategies (states z_1 and z_2):

$$\mathcal{M} := (\{z_0, z_1, z_2\}, z_0, \{m_0\}, m_0, \Delta, \sigma_Z, \sigma_\Delta, \tau_Z, \tau_\Delta)$$

where

$$\Delta := \{(z_0, z_1), (z_1, z_2), (z_2, z_1), (z_1, z_1), (z_2, z_2)\}$$
$$\sigma_Z := \{z_0, z_1, z_2\}$$
$$\sigma_\Delta := \{\mathrm{PROB}(p) \mid p \in \{1, p_1, p_2, 1-p_1, 1-p_2\}\}$$
$$\tau_Z(z_i) := z_i, i \in \{1, 2, 3\}$$
$$\tau_\Delta((z_0, z_1)) := \mathrm{PROB}(1)$$
$$\tau_\Delta((z_1, z_2)) := \mathrm{PROB}(p_1)$$
$$\tau_\Delta((z_2, z_1)) := \mathrm{PROB}(p_2)$$
$$\tau_\Delta((z_1, z_1)) := \mathrm{PROB}(1-p_1)$$
$$\tau_\Delta((z_2, z_2)) := \mathrm{PROB}(1-p_2)$$

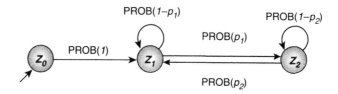

Figure 3.1 Simple 3-state GLSM, representing a hybrid SLS method that, after initialising the search process, probabilistically alternates between two search strategies.

Intuitively, transitions of type $\mathrm{PROB}(p)$ from the current state will be executed with probability p. The generic transition type $\mathrm{PROB}(p)$ formally corresponds to unconditional, probabilistic transitions with an associated transition probability p; it will be presented in more detail in Section 3.2. Note that there is only one memory state, m_0, which indicates that the search process does not make use of memory other than the GLSM state.

This type of GLSM can be used, for example, to model a variant of Randomised Iterative Improvement, in which case the state types z_0, z_1, z_2 may represent the simple SLS methods Uninformed Random Picking, Best Improvement and Uninformed Random Walk (see also Section 3.3).

We will usually specify GLSMs more intuitively using a standard graphic representation for the finite state machine part. In this graphic notation, GLSM states are represented by circles labelled with the respective state types, and state transitions are represented by arrows, labelled with the respective transition types; the initial state is marked by an incoming arrow that does not originate from another state. That arrow may be annotated by the initial memory state, m_0, or a procedure that generates m_0, if the given GLSM uses additional memory, that is, it has a memory space M with more than one element. The graphic representation for the GLSM from Example 3.1 is shown in Figure 3.1.

GLSM Semantics

Having formally defined the structure and components of a Generalised Local Search Machine, we now need to specify the semantics of this formal model, that is, the way in which it works when applied to a specific instance π of a combinatorial problem Π.

Obviously, the operation of any given GLSM $\mathcal{M}$ crucially depends on the search strategies represented by its states and on the nature of the transitions between these states. These, together with the problem instance to be solved as well as the corresponding search space, the solution set, the neighbourhood relations for the SLS methods associated with each state type and the termination predicate of the stochastic local search process to be modelled, form the *run-time environment of* $\mathcal{M}$. The separation between the components included in the GLSM definition and those included in the run-time environment is mainly motivated by the goal to have the GLSM model capture the higher-level search control mechanism, independent of the subsidiary SLS methods associated with the states and independent of the given problem instance.

> **Remark:** The memory space M is the only component of a GLSM that may depend on the given problem instance. However, since in many cases memory states play an important part in the context of the high-level control strategy of a hybrid SLS algorithm, such as Dynamic Local Search or Reactive Tabu Search, it makes sense to include it in the definition of a GLSM rather than in that of the run-time environment. The same argument applies, to a lesser extent, to the search space, since features of the current candidate solution can also be used for search control purposes. In the context of the formal definition of GLSM semantics, the separation of the components into two mathematical objects, the GLSM and its run-time environment, is not important.

For now, we will focus on the part of the search process performed by a GLSM that is independent of the run-time environment, and in particular of the nature of the state and transition types. (Transition types are further discussed in the next section, and practically relevant examples of state types are given in Section 3.3.)

Generally, the operation of a GLSM $\mathcal{M}$ in a given run-time environment can be described as follows. First, the state of $\mathcal{M}$ is set to the initial state z_0, the memory state is set to m_0, and an initial candidate solution is determined. Rather than specifying the initial candidate solution as a part of the GLSM model, we generate it using the search method associated with the initial state z_0, which in the simplest case is Uninformed Random Picking. After initialisation, $\mathcal{M}$ performs GLSM steps, each of which consists of a search step according to the current state of $\mathcal{M}$, followed by a state transition, which may depend on the new candidate solution and memory state of $\mathcal{M}$, and may result in a new memory state. These GLSM steps are iterated until the termination predicate specified in the run-time environment is satisfied.

EXAMPLE 3.2 **Semantics of a Simple 3-State GLSM**

The operation of the simple 3-state GLSM from Example 3.1 (page 115f.) can be described intuitively as follows. For a given problem instance π, the local search process is initialised by setting the machine to its initial state z_0. The memory state is set to m_0, but since it never changes, we ignore it in the following (m_0 is the only memory state). Then, one local search step is performed according to state type z_0 (this step is designed to initialise the subsequent local search process, for example, by randomly generating or selecting a candidate solution), and with a probability of one, the machine switches to state z_1. Now, one search step is performed according to the search strategy associated with state z_1, and with probability p_1, the machine switches to state z_2, otherwise it remains in z_1. The behaviour in state z_2 is very similar: first, a z_2 step is performed, and then with probability p_2, the machine switches back to state z_1, otherwise it remains in z_2.

This results in a local search process which repeatedly and nondeterministically switches between phases of z_1 and z_2 steps. However, only once in each run of the machine, a z_0 step is performed, and that is at the very beginning of the local search process. As previously discussed, a number of different termination criteria can be used for this type of search process.

Note that the operation of $\mathcal{M}$ is uniquely characterised by the state of $\mathcal{M}$, the candidate solution of the search process realised by $\mathcal{M}$, and the memory state of $\mathcal{M}$ for any given point in time, where time is measured in GLSM steps. This information can be captured in the form of two functions, the *actual state trajectory* and *actual search trajectory of $\mathcal{M}$*, which specify the machine state and the candidate solution of $\mathcal{M}$ as well as the memory state over time.

However, due to the inherent stochasticity of GLSMs and the SLS algorithms they model, the outcome of each GLSM step is generally not a single candidate solution, machine state and memory state, but rather a set of probability distributions over machine states, candidate solutions and memory states. Therefore, to completely characterise the behaviour of a given GLSM, two functions are used that define the probability distribution over candidate solutions and machine states as well as memory states over time, the *probabilistic search trajectory* and the *probabilistic state trajectory*. Technically, when additional memory is used, we allow the memory state to be affected by state transitions (this will be further discussed in Section 3.2 and examples are given in Section 3.3), but not by search steps, and consequently we consider it to be part of the state trajectory. This reflects our view of the memory state as a part of the higher-level search control mechanism represented by the GLSM.

IN DEPTH FORMAL DEFINITION OF GLSM SEMANTICS

In the following, we show how the semantics of a GLSM, in terms of its probabilistic and actual search and state transition functions, can be formally defined. In the following, for simplicity's sake, we ignore the termination predicate that is part of the given run-time environment. (Note, however, that in principle the termination predicate can be integrated into the following definitions in a rather straightforward way.)

We assume that the semantics of each state type τ, which are formally part of the run-time environment, are defined in the form of search transition functions $\gamma_\tau : S \times M \mapsto \mathcal{D}(S)$, where S denotes the set of candidate solutions in the search space induced by the given problem instance, M is the given set of memory states and $\mathcal{D}(S)$ represents the set of probability distributions over S. Intuitively, γ_τ determines for each candidate solution in S and memory state in M the resulting candidate solution after one τ-step has been performed; it corresponds to the step function of the search strategy associated with τ and can be defined functionally or procedurally (see also Chapter 1).

Based on the subsidiary search transition functions γ_τ, we now define the direct search transition function $\gamma_S : S \times M \times Z \mapsto \mathcal{D}(S)$ which, for a given search position, memory state and GLSM state, determines the distribution over search positions after one step of the GLSM:

$$\gamma_S(s, m, z)(s') := \gamma_\tau(s, m)(s')$$

where $\tau := \tau_Z(z)$ is the type of state z and $\gamma_\tau(s, m)$ is the distribution over candidate solutions and memory states reached from s, m after one τ-step. (Recall that according to our earlier convention, $\gamma_\tau(s, m)(s')$ denotes the probability of s' under that distribution, and $\gamma_S(s, m, z)(s')$ denotes the probability of s' under distribution $\gamma_S(s, m, z)$.)

Furthermore, we assume that the semantics of the state transitions in the given GLSM are specified in the form of a function $\gamma_Z : S \times M \times Z \mapsto \mathcal{D}(Z \times M)$ that models the direct transitions between states of the given GLSM. (Note that this allows for state transitions to depend on and to affect memory states, as will be discussed in detail later, in Section 3.2.) Formally, this function can be defined on the basis of the transitions from a given state z and the semantics of the respective transition types; for example, given an unconditional, probabilistic transition between two states z_i, z_k, that is, $\tau_\Delta((z_i, z_k)) = \text{PROB}(p)$, applying the direct state transition function γ_Z to a search position s, memory state m and GLSM state z_i, gives a distribution D over GLSM states and memory states such that $D((z_k, m)) = p$, where here and in the following, $D(e)$ denotes the probability of e under distribution D. Note that in this example, the transition does not affect the memory state m.

The direct transition function γ_S and the state transition function γ_Z can be generalised to the case that models the effects of a single GLSM step on a given distribution of candidate solutions, memory states or GLSM states. In the case of γ_S this is modelled by a function Γ_S defined as:

$$\Gamma_S(D_S, D_{ZM})(s') := \sum_{s \in S, m \in M, z \in Z} \gamma_S(s, m, z)(s') \cdot D_S(s) \cdot D_{ZM}(z, m)$$

Note that the probability of candidate solution s' after the step is obtained from the probabilities of going in state z from candidate solution s and memory state m to candidate solution s', weighted by the probability of s, m and z under the given distributions D_S and D_{ZM}.

Analogously, we generalise γ_Z to the function Γ_Z, which models the effects of one GLSM step on a given distribution of GLSM states:

$$\Gamma_Z(D_S, D_{ZM})(z', m') := \sum\nolimits_{s \in S, m \in M, z \in Z} \gamma_Z(s, m, z)(z', m') \cdot D_S(s) \cdot D_{ZM}(z, m)$$

Note that the memory state may be affected only when the GLSM state is updated, but not as a side-effect of a search step. This reflects the view that GLSM states and memory states are conceptually closely related, since both model aspects of the search control mechanism.

Based on Γ_S and Γ_Z, we can now give the following inductive definition of the probabilistic search and state trajectory functions $\gamma_S^* : \mathbb{N} \mapsto \mathcal{D}(S)$ and $\gamma_Z^* : \mathbb{N} \mapsto \mathcal{D}(Z \times M)$:

$$\gamma_S^*(0)(s) := 1/\#S$$
$$\gamma_S^*(t+1)(s) := \Gamma_S(D_S, D_{ZM}),$$
$$\text{where } D_S := \gamma_S^*(t) \text{ and } D_{ZM} := \gamma_Z^*(t)$$

$$\gamma_Z^*(0)(z, m) := \begin{cases} 1 & \text{if } z = z_0 \text{ and } m = m_0 \\ 0 & \text{otherwise} \end{cases}$$
$$\gamma_Z^*(t+1)(z, m) := \Gamma_Z(D_S, D_{ZM}),$$
$$\text{where } D_S := \gamma_S^*(t+1) \text{ and } D_{ZM} := \gamma_Z^*(t)$$

The interlocked inductive definitions of γ_S^* and γ_Z^* reflect the intended operation of a GLSM, where in each step, first a new candidate solution s' is determined based on the current candidate solution s, memory state m, and GLSM state z, and then, a new GLSM state z' and memory state m' are determined based on s', m and z. Note that the choice of the initial distribution of candidate solutions is somewhat arbitrary, since the true starting point of the search trajectory is typically determined by means of the first search step in state z_0 (with initial memory state m_0) in a way that is effectively independent of the current candidate solution.

The actual search and state trajectories are formally defined in a similar manner, in the form of two functions $\delta_S^* : \mathbb{N} \mapsto S$ and $\delta_Z^* : \mathbb{N} \mapsto Z \times M$:

$$\delta_S^*(0) := draw(\gamma_S^*(0))$$
$$\delta_S^*(t+1) := draw(\gamma_S(D_S, D_{ZM}))$$
$$\text{where } D_S := \gamma_S^*(t) \text{ and } D_{ZM} := \gamma_Z^*(t)$$

$$\delta_Z^*(0) := (z_0, m_0)$$
$$\delta_Z^*(t+1) := draw(\gamma_Z(D_S, D_{ZM}))$$
$$\text{where } D_S := \gamma_S^*(t+1) \text{ and } D_{ZM} := \gamma_Z^*(t)$$

In these definitions, the function $draw(D)$ randomly selects an element from the domain of a given probability distribution D such that element e is chosen with probability $D(e)$. Note the similarity between this definition and the definition of the probabilistic search and state trajectories γ_S^* and γ_Z^*; the only difference is that in the case of the actual trajectories, in each GLSM step a single candidate solution and state are randomly chosen according to the respective distributions.

GLSMs as Factored Representations of SLS Strategies

Technically, a GLSM represents the higher-level search control of an SLS strategy, that is, the way in which the initialisation and step function of the SLS method are composed from the respective functions of subsidiary component SLS methods. In this sense, a GLSM is a factored representation of an SLS strategy. Note that the memory used by many SLS strategies, such as Tabu Search or Ant Colony Optimisation, is treated as an explicit part of the high-level search control mechanism that is modelled by a GLSM. In particular, as previously mentioned, in a GLSM formalisation, all modifications of the memory state are performed by means of GLSM state transitions and not in combination with the actual search steps of the underlying component search strategies, which may depend on the memory state, but only affect the current candidate solution. (Technically, the memory modification performed along with a state transition may depend on the current search position and memory state; memory updates that depend on the precise nature of the last search step, such as required in Tabu Search, can be implemented by keeping the previous candidate solution in memory.) The other components of an SLS algorithm, namely the search space, solution set and neighbourhood relation (all of which are induced by the given problem instance), as well as the termination predicate, form part of the run-time environment of a GLSM representation of that algorithm.

When modelling an existing or novel hybrid SLS algorithm by a GLSM, the component search strategies associated with the GLSM states are often derived from existing simple SLS algorithms for the given problem. Theoretically, these component SLS strategies can be complex or hybrid strategies; but according to the primary motivation behind the GLSM model, they should be as pure and simple as possible, in order to achieve a clean separation between simple search strategies and search control.

Typically, the component search strategies have the same search space and solution set, which is part of the GLSM's run-time environment. The respective initialisation functions are either modelled by the initialisation state of the GLSM, or not needed at all, since the respective component search strategy is applied to the result of another component search strategy, in which case the respective initial probability distribution over candidate solutions is implicitly given by the context in which the corresponding GLSM state is activated. The termination predicates of these subsidiary SLS methods, particularly when they are based on aspects of the current candidate solution or memory state, are often reflected in conditional transitions leaving the respective GLSM state.

The simple SLS algorithms modelled by the states of a given GLSM $\mathcal{M}$ can be based on different neighbourhood relations, which become part of the run-time environment of $\mathcal{M}$. It is always possible to define a unified relation that contains the neighbourhood relation for each state type as a subset. However, in the case of the subsidiary SLS methods associated with the states of a given GLSM this can be problematic, since formally, the presence of the initial state often implies that under the unified neighbourhood relation any two candidate solutions are direct neighbours. This is true, for example, for Uninformed Random Picking, which uses a complete neighbourhood (see Chapter 1, Section 1.5). But if the search initialisation is considered separately, as in the formal definition of an SLS algorithm (Definition 1.10, page 38*f.*), unifying the neighbourhood relations of the remaining component search strategies can be useful.

Various component search strategies of a given GLSM may also use different memory spaces. These can always be combined into a single, unified memory space, as required by the formal definition of a GLSM. It may be noted that technically, memory states are not needed in the GLSM model, since the memory states used in any SLS algorithm can always be folded into GLSM states. But this often leads to unnecessarily complex and cumbersome GLSM representations, in which aspects of high-level and low-level search control are not cleanly separated. Although there are cases where the distinction between high-level and low-level search control is somewhat debatable, for most hybrid SLS algorithms the decision which aspects of the search control mechanism should be modelled by GLSM states and state transitions as opposed to memory states is fairly obvious.

3.2 State, Transition and Machine Types

In order to completely specify a GLSM, definitions for the search methods associated with each state type need to be given. Formally, this can be done in the form of search transition functions $\gamma_\tau : S \times M \mapsto \mathcal{D}(S)$ for each state type τ (see also the in-depth section on page 119*ff.*). But as in the case of SLS algorithms in general, it is often clearer and more convenient to define the semantics of GLSM state types in a procedural way, usually in the form of pseudo-code. However, in some cases, more adequate descriptions of complex state types can be obtained by using other formalisms. This is particularly the case for simple SLS strategies whose search steps are based on a multi-stage selection process — these are often amenable to concise decision tree representations. (Examples of such SLS methods include the WalkSAT algorithm family for SAT, cf. Chapter 6, Section 6.3.)

While concrete examples for various state types will be given in Section 3.3 and in subsequent chapters, it is worth discussing some fundamental distinctions between certain state types. One of these concerns the role of the respective states within the general definition of stochastic local search (Definition 1.10, page 38*f.*). Although we are modelling search initialisation and local search steps using the same mechanism, namely GLSM states, there is a clear distinction between the states that realise these two components of an SLS algorithm. An initialising state is usually different from a search step state in that it is left after one corresponding step has been performed. Also, while search step states correspond to moves in a restricted local neigbourhood (like flipping one variable in SAT), a single initial-isation step can typically lead to arbitrary candidate solutions. (As an example, consider Uninformed Random Picking.) Formally, we define an *initialising state type* as a state type τ for which the local search position after one τ-step is inde-pendent of the local search position before the step; states of an initialising type τ are called *initialising states*. Generally, each GLSM will have at least one initial-ising state, which is also its initial state. A GLSM can, however, have more than one initialising state and use these states to implement certain forms of restart strategies.

Furthermore, state types may be distinguished based on whether or not their semantics depends on the given memory state. The former are called *parametric state types*, and states of such types are called *parametric states*. An example for a parametric state can be found in the GLSM representation of Simulated Annealing, where the behaviour of the underlying SLS method depends on the temperature parameter (cf. Section 3.3).

Finally, there are many combinatorial problems with a particular, natural neighbourhood relation N. In these cases, it is often useful to distinguish be-tween *single-step states* and *multi-step states* with respect to that neighbourhood: given a neighbourhood N and a current search position s, one search step in a single-step state always leads to a direct neighbour of s under N, while one search step in a multi-step state may lead to a candidate solution whose distance to s in the neighbourhood graph induced by N is greater than one. For example, most SLS algorithms for SAT use a 1-exchange neighbourhood relation, under which two variable assignments are direct neighbours if, and only if, they differ in ex-actly one variable's value. In this context, a single-step state would flip one vari-able's value in each step, whereas a multi-step state could flip several variables per local search step. Consequently, initialising states are an extreme case of multi-step states, since they can affect the values of all variables at the same time.

Transition Types

The search control mechanism of a GLSM is realised by its state transitions. While the possible transitions between states are specified in the form of a transition relation, the precise conditions under which a transition (z, z') is executed are captured in its type, $\tau_\Delta((z, z'))$; consequently, the definition of transition types forms an important part of GLSM semantics. In the following, we introduce the transition types that provide the basis for the GLSM models of most practically relevant SLS methods. These can be conveniently presented as a hierarchy of increasingly complex and expressive transition types, ranging from simple deterministic transitions to conditional probabilistic transitions.

DET—Unconditional deterministic transitions.   This is the most basic transition type. When a GLSM is in a state z with an outgoing transition (z, z') of type DET, it will invariably switch to state z' after a single GLSM step. This means that unless $z' = z$, only one step of the search strategy corresponding to z is performed before switching to a different state. Furthermore, for each GLSM state there can be at most one outgoing transition of this type, which severely restricts the class of GLSM structures (and hence search control mechanisms) that can be realised when only DET transitions are used. Although the use of unconditional deterministic transitions is somewhat limited, they frequently occur in the GLSM models of practically relevant SLS algorithms, where they are mostly utilised for leaving initialising states.

PROB(p)—Unconditional probabilistic transitions.   A transition of type PROB(p) from a GLSM state z to another state z' takes a GLSM that is in state z directly into state z' with probability p. Clearly, DET transitions are equivalent to a special case of this transition type, namely to PROB(1), and hence do not have to be considered separately in the following. If the set of transitions leaving a state z is given as $\{t_1, \dots, t_n\}$, where the type of transition t_j is PROB(p_j) for every j, the semantics of this state type require $\sum_{j=1}^{n} p_j = 1$ in order to ensure that the selection of the transition from z is based on a proper probability distribution.

Note that without loss of generality, by using PROB(0) transitions we can restrict our attention to fully connected GLSMs, where for each pair of states (z_i, z_k), a transition of type PROB(p_{ik}) is defined. The uniform representation thus obtained facilitates theoretical investigations as well as practical implementations.

It is also interesting to note that for a given GLSM $\mathcal{M}$ any state z that can be reached from the initial state of $\mathcal{M}$ by following transitions of type PROB(p) with $p > 0$ will eventually be reached with arbitrarily high probability in any sufficiently long run of $\mathcal{M}$. Furthermore, in any state z with a PROB(p) self-transition (z, z), the number of GLSM steps before leaving z, and hence the number of consecutive search steps performed according to the search strategy associated with state z, is distributed geometrically with mean and variance $1/p$.

CPROB(C, p) and CDET(C)—Conditional transitions. While until now we have focused on transitions whose execution only depends on the actual state, the following generalisation from PROB(p) introduces context-dependent transitions. A CPROB(C, p) transition from state z to state z' is executed with a probability proportional to p only when a *condition predicate C* is satisfied. If C is not satisfied, all transitions CPROB(C, p) from the current state are *blocked*, that is, they cannot be executed. Note that we do not (and generally cannot) require that the p values of the outgoing CPROB(C) transitions of a given GLSM state sum to one; consequently, these values do not directly correspond to transition probabilities, but only determine the ratios between probabilities within a set of unblocked transitions from the same state. In particular, if in a given situation there is only one unblocked CPROB(C, p) transition with $p > 0$ from the current GLSM state, that transition will be taken with probability one, regardless of the value of p.

Obviously, PROB(p) transitions are equivalent to conditional probabilistic transitions CPROB$(\top, p)$, where $\top$ is the predicate that is always true. Without loss of generality, we can therefore restrict our attention to GLSMs in which all transitions are of type CPROB(C, p). An important special case of conditional transitions is *conditional deterministic transitions*, in particular, transitions of the type CDET$(C) =$ CPROB$(C, 1)$. Conditional deterministic conditions also arise when for a given GLSM state z all but one of its outgoing transitions are blocked at any given time. Note that a deterministic GLSM $\mathcal{M}$ is obtained if all condition predicates for the transitions leaving each state of $\mathcal{M}$ are mutually exclusive. Generally, depending on the nature of the condition predicates used, the decision whether a conditional transition is deterministic or not can be rather difficult. For the same reasons it can be difficult to decide for a given GLSM with conditional probabilistic transitions whether a particular state is reachable from the initial state.

For practical uses of GLSMs with conditional transitions, it is important to ensure that all condition predicates can be evaluated in a sufficiently efficient way (compared to the cost of executing local search steps); ideally, this evaluation should require at most linear (better: constant) time w.r.t. the size of the given problem instance.

We distinguish between two kinds of condition predicates. The first of these captures properties of the current candidate solution and its local neighbourhood; the second kind is based on search control aspects, such as the time that has been spent in the current GLSM state, the overall run-time or the current memory state. Naturally, these two kinds of conditions can also be combined.

$\top$	always true
count(k)	total number of GLSM steps $\geq k$
countm(k)	total number of GLSM steps modulo $k = 0$
scount(k)	number of GLSM steps in current state $\geq k$
scountm(k)	number of GLSM steps in current state modulo $k = 0$
lmin	current candidate solution is a local minimum w.r.t. the given neighbourhood relation
evalf(y)	current evaluation function value $\leq y$
noimpr(k)	incumbent candidate solution has not been improved within the last k steps

Table 3.1 Commonly used simple condition predicates.

Some concrete examples for condition predicates are listed in Table 3.1. Note that all these predicates are based on local information only and can thus be efficiently evaluated during the search process.

Usually, for each condition predicate, a positive as well as a negative (negated) form will be defined. By using propositional connectives, such as '$\wedge$' or '$\vee$', these simple predicates can be combined into compound predicates. However, it is not difficult to see that allowing compound transition predicates does not increase the expressive power of the GLSM model, since every GLSM using compound condition predicates can be reduced to an equivalent GLSM using only simple predicates by introducing additional states and/or transitions.

Transition Actions

None of the transition types introduced above have any effect on the memory state of the given GLSM. While in principle, the condition predicates used in conditional probabilistic (or deterministic) transitions may depend on the memory state, in the absence of a means for effecting changes in memory, GLSMs cannot make any use of memory states. Hence, in order to model SLS algorithms that use additional memory, such as Simulated Annealing, Iterated Local Search or Tabu Search, we introduce the concept of *transition actions*. Transition actions are associated with individual transitions and are executed whenever the GLSM executes the corresponding transition.

Generally, transition actions can be added to each of the transition types defined above, and the semantics of the transition in terms of its effect on the immediate successor state of the GLSM are not affected. If T is a transition type, we let $T : A$ denote the same transition type with associated action A. Formally, a transition action can be seen as a function mapping search positions and memory states into memory states. This accurately reflects the fact that the effect of any transition action is limited to the memory state of the given GLSM; in particular, transition actions cannot be used to modify the current GLSM state or candidate solution. However, transition actions have access to the current search position, and hence they can be used, for example, to memorise the current candidate solution. In practice, the memory used in a given GLSM is often factored into a number of separate attributes or data structures, each of which can be manipulated independently of the others. Hence, when specifying transition actions procedurally, it is often advantageous to represent a single transition action by multiple assignments or procedure calls.

It may also be noted that by assigning special roles to parts of a structured memory space, transition actions can be used for realising input/output functionality in actual GLSM implementations or for communicating information between individual machines in co-operative GLSM models (these are further discussed in Section 3.4). Also, by using a special action NOP ('no operation') that has no effect on the memory state (formally modelled as an identity function on the given memory space), we can obtain uniform GLSMs in which all transitions have associated actions.

Machine Types

The types of the states and transitions form an important part of the complete specification of a given GLSM and determine crucial aspects of the behaviour of the underlying search algorithm. However, it can sometimes be useful to abstract from these types and to focus on the structure of the search control mechanism, as reflected in the states and the transition relation of a GLSM. For example, one may be interested in the difference between a GLSM with five states that are connected sequentially (such that they are visited one after the other in a fixed sequence), and a GLSM with three states that, in principle, allows arbitrary transitions between its states. This motivates the following categorisation of GLSMs into structural classes or machine types.

> **1-state machines:** This is the minimal form of a GLSM. Since every GLSM needs an initialising state in order to generate the initial candidate solution for the local search process, 1-state machines essentially realise iterated sampling processes, such as Uninformed Random Picking. Such extremely simple search algorithms can be useful for analytical purposes, for example, as a reference model when evaluating other types of GLSMs. Nevertheless, their overall practical relevance is rather limited.

1-state+init machines: These machines have one state for search initialisation and one working state. Machines of this type can be further classified as *sequential 1-state+init machines*, which visit the initialisation state z_0 only once, and *alternating 1-state+init machines*, which may visit z_0 multiple times in the course of the search process. The structure of these machine models is shown in Figure 3.2. Many simple SLS methods naturally correspond to sequential 1-state+init GLSMs, while alternating 1-state+init machines are good models for simple SLS algorithms that use a restart strategy.

2-state+init sequential machines: This machine type has three states, one of which is an initialisation state that is only visited once, while the other two are working states. However, once the machine has switched from the first state to the second, it will never switch back (see Figure 3.3, left side);

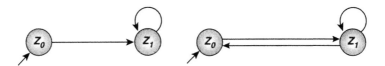

Figure 3.2 Sequential *(left)* and alternating *(right)* 1-state+init GLSM.

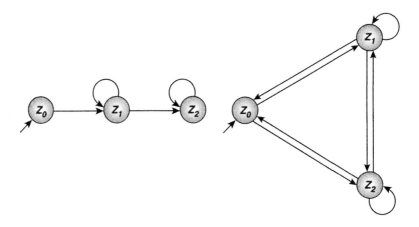

Figure 3.3 Sequential *(left)* and alternating *(right)* 2-state+init GLSM.

analogously, the GLSM switches from the second to the third state only once. Thus, each search trajectory of such a machine can be partitioned into three phases: one initialisation step, a number of steps in the first working state and a number of steps in the second working state.

2-state+init alternating machines: Like a 2-state+init sequential machine, this machine type has one initialisation state and two working states. Here, however, arbitrary transitions between all states are possible (see Figure 3.3, right side). An interesting special case arises when the initial state can only be visited once, while the machine might arbitrarily switch between the two working states. Another case that might be distinguished is a *uni-directional cyclic machine model*, which allows the three states to be visited only in one fixed cyclic order.

Obviously, the categorisation can easily be continued in this manner by successively increasing the number of working states. However, as we will see later, to describe state-of-the-art stochastic local search algorithms, machines with up to three states are often sufficient. We conclude our categorisation with a brief look at two potentially interesting cases of the k-state+init machine types:

k-state+init sequential machines: As a straightforward generalisation of sequential 2-state+init machines, in this machine type we have $k + 1$ states that are visited in a linear order. Consequently, after a machine state has been left, it will never be visited again (see Figure 3.4, top). After initialisation, each search trajectory of this type of GLSM can be partitioned into up to k contiguous segments, each of which consists of a sequence of search steps performed in the same GLSM state.

k-state+init alternating machines: These machines allow arbitrary transitions between the $k + 1$ states and may therefore re-initialise the search process and switch between strategies as often as desired (see Figure 3.4, bottom). Two interesting special cases are the *uni-* and *bi-directional cyclic machine models*, which can switch between states in a cyclic manner. In the former case, the cyclic structure can be traversed only in one direction, while in the latter case the machine can switch from any state to both its neighbouring states (see Figure 3.5).

This categorisation of GLSMs according to their structure provides a very high-level view of the respective search control mechanism, which can be refined in many ways. Nevertheless, as we will see in the following, the abstraction of machine types presented here can be very useful for capturing fundamental differences between various stochastic local search methods.

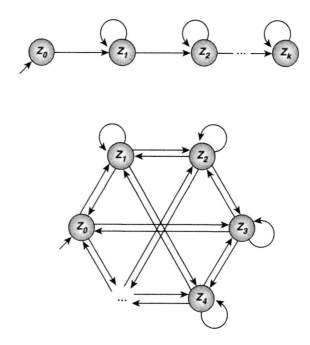

Figure 3.4 Sequential *(top)* and alternating *(bottom)* k-state+init GLSM.

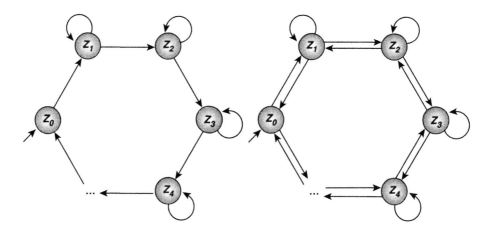

Figure 3.5 Uni-directional *(left)* and bi-directional *(right)* cyclic k-state+init GLSM.

Figure **3.6** GLSM for Uninformed Random Picking.

3.3   Modelling SLS Methods Using GLSMs

Up to this point, we have introduced the GLSM model and discussed various types of GLSM states, transitions and structures. We now demonstrate applications of the model by specifying and discussing GLSM representations for many of the well-known SLS methods described in Chapter 2; this way, important similarities and differences between SLS methods are highlighted. GLSM representations of other SLS methods and algorithms are covered in later chapters and exercises.

Uninformed Random Picking and Random Walk

The simplest possible SLS algorithm is Uninformed Random Picking, as introduced in Chapter 1, Section 1.5. When cast into our definition of a stochastic local search algorithm, the *init* and *step* functions of Uninformed Random Picking are identical and perform a random uniform selection of a candidate solution from the underlying search space. The corresponding GLSM is shown in Figure 3.6. It has only one state of type RP, formally defined by $\tau_{RP}(s, m)(s') = 1/\#S$ for all s', s and m. Since functional state type definitions of more advanced SLS strategies can get rather complex and difficult to understand, it is often preferable to define the step functions for GLSM state types procedurally. Such a definition for the random picking state type is shown in Figure 3.7. Note that, as in previous chapters, in these procedural descriptions we generally only mention those parts of the memory state (if any) that are actually used in the corresponding search mechanism.

The Uninformed Random Walk algorithm (cf. Chapter 1, Section 1.5) requires an additional state type, RW, whose semantics are defined in Figure 3.8. In its simplest form, the search is initialised by random picking, followed by a series of uninformed random walk steps. The corresponding GLSM is shown in Figure 3.9. In practice, many SLS algorithms are extended by a restart mechanism, by which, in the simplest case, after every k search steps (where k is a parameter

procedure *step-RP*(π, s)

    input: *problem instance* $\pi \in \Pi$, *candidate solution* $s \in S(\pi)$

    output: *candidate solution* $s \in S(\pi)$

    $s' := random(S)$;

    return s'

end *step-RP*

Figure 3.7 Procedural specification of GLSM state RP; the function *random*(S) returns an element of S selected randomly according to a uniform distribution over S.

procedure *step-RW*(π, s)

    input: *problem instance* $\pi \in \Pi$, *candidate solution* $s \in S(\pi)$

    output: *candidate solution* $s \in S(\pi)$

    $s' := random(N(s))$;

    return s'

end *step-RW*

Figure 3.8 Procedural specification of GLSM state RW.

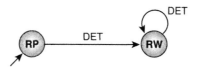

Figure 3.9 GLSM for Uninformed Random Walk.

of the algorithm), the search process is reinitialised. Generally, other conditions can be used for determining when a restart should occur. Figure 3.10 shows the GLSM for Uninformed Random Walk with Random Restart; it is obtained from the GLSM for the basic Uninformed Random Walk algorithm without restart by a simple modification of the state transitions. Note how the GLSM representations for both algorithms indicate the fact that Uninformed Random Walk can already be seen as a (albeit very simple) hybrid SLS algorithm, using two types of search steps, Uninformed Random Picking and Uninformed Random Walk. While using a restart mechanism for Uninformed Random Walk does not appear to be useful other than for illustrative purposes, the analogous extension of Iterative Improvement, covered in the following, leads to significant performance improvements.

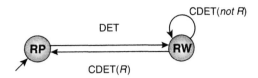

Figure **3.10** GLSM for Uninformed Random Walk with Random Restart; R is the restart predicate, for example, countm(k).

procedure *step-BI*(π, s)
  input: *problem instance* $\pi \in \Pi$, *candidate solution* $s \in S(\pi)$
  output: *candidate solution* $s \in S(\pi)$

  $g^* := \min\{g(s') \mid s' \in N(s)\}$;
  $s' := random(\{s' \in N(s) \mid g(s') = g^*\})$;
  return s'
end *step-BI*

Figure **3.11** Procedural specification of GLSM state BI.

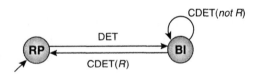

Figure **3.12** GLSM for Iterative Best Improvement with Random Restart; R is the restart predicate, for example, lmin.

Iterative Improvement

The GLSM model for Iterative Improvement (cf. Chapter 1, Section 1.5) is similar to that for Uninformed Random Walk. Again, we use an RP state to model the search initialisation by random picking, but the second state now captures the semantics of iterative improvement search steps. A procedural specification of a GLSM state BI that models best improvement search steps is given in Figure 3.11; Figure 3.12 shows the GLSM for Iterative Best Improvement Search with Random Restart. Note that the random restart mechanism will enable the algorithm to escape from local minima of the evaluation function and can hence

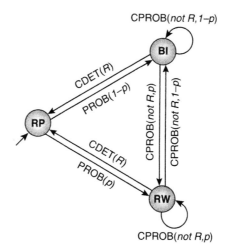

Figure 3.13 GLSM for Randomised Iterative Best Improvement with Random Restart; R is the restart predicate, for example, countm(m).

be expected to improve its performance. Notice that the only difference between this GLSM and the one for Uninformed Random Walk with Random Restart shown in Figure 3.10 lies in the type of one state (BI *vs* RW). This reflects the common structure of the search control mechanism underlying both of these simple SLS algorithms. Similarly, the GLSM models for other variants of iterative improvement search, such as First Improvement or Random Improvement, are obtained by replacing the BI state by a state of an appropriately defined type that reflects the semantics of these different kinds of iterative improvement steps.

Using the RP, RW and BI state types, it is easy to construct a GLSM model for Randomised Iterative Improvement, one of the simplest SLS algorithms (cf. Chapter 2, Section 2.2). The 2-state+init GLSM shown in Figure 3.13 represents the hybrid search mechanism in an explicit and intuitive way. Since the addition of the Uninformed Random Walk state enables the algorithm in principle to escape from local minima, the restart mechanism included in this GLSM is practically not as important as in the previous case of pure Iterative Best Improvement. It may be noted that the same SLS algorithm could be modelled by a 1-state+init GLSM, using a single state for Randomised Iterative Improvement steps. This representation, however, would be substantially inferior to the 2-state+init GLSM introduced above, since the structure of the respective GLSM model does not adequately capture the search control strategy.

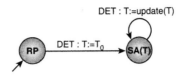

Figure **3.14** GLSM for Simulated Annealing; the initial temperature T_0 and temperature update function *update* implement the annealing schedule.

Simulated Annealing

To model Simulated Annealing by a GLSM, we use a parameterised state type $SA(T)$ to represent the probabilistic iterative improvement strategy that forms the core of Simulated Annealing. This state type can be specified procedurally by the step function introduced in Chapter 2 (cf. Figure 2.7 on page 76). The initialisation and modifications of the temperature parameter T prescribed by the annealing schedule are realised by transition actions.

This leads to the GLSM model shown in Figure 3.14. Note how this representation separates the basic search process (which corresponds to the changes in search position) from modifications of the temperature T, which is a search control parameter. Many variants of Simulated Annealing, including Constant Temperature Simulated Annealing and more complex, hybrid algorithms that combine Simulated Annealing steps with other types of search steps, can be easily represented by similar GLSMs.

Furthermore, transition actions can be used in a similar way for modelling Tabu Search algorithms (cf. Chapter 2, Section 2.2); in this case, a state type representing basic tabu search steps is used along with transition actions that update the tabu status of solution components.

Iterated Local Search

As a hybrid SLS method, Iterated Local Search performs two basic types of search steps in its subsidiary local search and perturbation phases. Obviously, these as well as the search initialisation are modelled by separate GLSM states LS, PS and RP. A slight complication is introduced by the acceptance criterion, which is used in ILS to determine whether or not the search continues from the new candidate solution obtained from the last perturbation and subsequent local search phase. There are various ways of modelling this acceptance mechanism in a GLSM.

procedure *step-AC*(π, s, t)

    input: *problem instance* $\pi \in \Pi$, *candidate solution, s,t* $\in S(\pi)$

    output: *candidate solution s* $\in S(\pi)$

    if $C(\pi, s, t)$ **then**

        return *s*

    else

        return *t*

    end

end *step-AC*

Figure **3.15** Procedural specification of GLSM state AC; this state type uses a candidate solution t, stored earlier in the search process and a selection predicate $C(\pi, s, t)$, which returns $\top$ if s is to be selected as the new search position, and $\bot$ otherwise. A selection predicate that is often used in the context of ILS is $better(\pi, s, t) := (g(s) < g(t))$, where g is the evaluation function for the given problem instance, π.

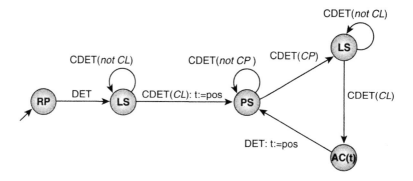

Figure **3.16** GLSM representation of Iterated Local Search; CP and CL are condition predicates that determine the end of the perturbation phase and the local search phase, respectively, and pos denotes the current search position. For many ILS algorithms, $CL := $ lmin.

Figure 3.16 shows a GLSM representation of ILS in which the application of the acceptance criterion is modelled as a separate state, AC (for a procedural definition of AC, see Figure 3.15). Note the use of transition actions for memorising the current candidate solution, which is needed when applying the acceptance criterion in state AC. Furthermore, it may be noted that our GLSM model allows for several perturbation steps to be performed in a row; perturbation mechanisms of this type can be found in various ILS algorithms (for an example, see Chapter 7, page 331).

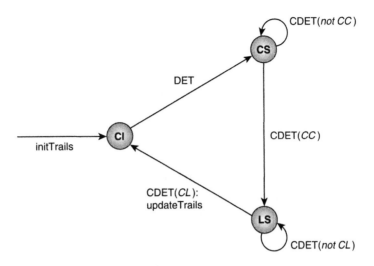

Figure **3.17** GLSM representation of an ACO algorithm, such as Ant System; the condition predicates CC and CL determine the end of the construction and local search phases, respectively. For many ACO algorithms, $CL := \mathsf{lmin}$. Note that some ACO algorithms, such as Ant Colony System, remove pheromone while constructing solutions. This type of pheromone update can be represented by an additional transition action associated with transition (CS, CS).

Ant Colony Optimisation

There are different approaches to representing population-based SLS algorithms as GLSMs, corresponding to different views of the underlying stochastic local search process itself. As briefly discussed in Chapter 2, Section 2.3, one can view populations of individual candidate solutions for the given problem instance π as search positions; under this view, the search space of a population-based SLS algorithm consists of sets of candidate solutions of π. Ant Colony Optimisation, for instance, can then be represented by the GLSM shown in Figure 3.17. State CI initialises the construction search, and state CS performs a single construction step for all ants (cf. Example 2.10, page 98*f.*). The LS state performs a single step of this local search procedure for the entire population of ants. For a typical iterative improvement local search, this means that iterative improvement steps are performed for all ants that have not reached a locally minimal candidate solution of the given problem instance; in this case, usually a condition predicate CL is used that is satisfied when all ants have obtained a locally minimal candidate solution. Initialisation and update of the pheromone trails are modelled using transition actions.

An alternative GLSM model for ACO is based on a view under which the search space consists of probability distributions of candidate solutions for the given problem instance. Note that the probabilistic construction process carried out by each ant induces a probability distribution in which, ideally, higher-quality candidate solutions have higher probability of being constructed. The subsequent local search phase then biases this probability distribution further towards better candidate solutions. Finally, updating the pheromone values modifies the probability distribution underlying the next construction phase. This representation has two disadvantages. Firstly, it does not reflect the fact that ACO effectively samples the probability distribution central to this model in each cycle in order to obtain a set of candidate solutions for the given problem instance; secondly, it does not capture the compact implicit representation of the probability distributions given by the pheromone values. However, this view on ACO is interesting for theoretical reasons, for example, in the context of analysing important theoretical properties of ACO, such as the probability of obtaining a specific solution quality within a given time bound. Also note that the GLSM model corresponding to this view does not require transition actions for manipulating the pheromone trails because these are now essential components of the search position.

Another general approach for modelling population-based SLS algorithms as GLSMs is to represent each member of the population by a separate GLSM. The resulting co-operative GLSM models will be discussed in more detail in the next section.

3.4   Extensions of the Basic GLSM Model

In this section, we discuss various extensions of the basic GLSM model. One of the strengths of the GLSM model lies in the fact that these extensions arise quite naturally and can be easily realised within the basic framework. Some extended GLSM models, in particular co-operative GLSMs, have immediate applications in the form of existing SLS algorithms, while others have not yet been studied in detail.

Co-operative GLSM Models

A natural extension of the basic GLSM model that is particularly suited for representing population-based SLS algorithms, is to apply several GLSMs simultaneously to the same problem instance. We call such extensions *co-operative*

GLSM models, since they capture the idea of solving a given problem instance through the co-operative effort of an ensemble of agents. In the simplest case, such an ensemble consists of a number of identical GLSMs without any communication between the individual machines. The semantics of this *homogeneous co-operative GLSM model* are conceptually equivalent to executing multiple independent runs of an individual GLSM. This model is particularly attractive for parallelisation because it is very easy to implement. It involves virtually no communication overhead (other than making the given problem instance available to all agents and possibly terminating the runs of all individual GLSMs when a solution has been found) and can be almost arbitrarily scaled in principle.

The restrictions of this model can be relaxed in two directions. One is to allow ensembles of different GLSMs. This *heterogeneous co-operative GLSM model* is particularly useful for modelling *algorithm portfolios*, that is, robust combinations of various SLS algorithms, each of which is likely to show superior performance on certain types of instances, when the features of the given problem instances are not known *a priori* [Huberman et al., 1997; Gomes and Selman, 1997b]. Generally, the heterogenous co-operative model has very similar advantages to its homogeneous variant; it is easy to implement and almost free of communication overhead.

Another generalisation is to allow communication between the individual GLSMs of a co-operative model. This is required for explicitly modelling population-based SLS algorithms in which the individual search trajectories are not independent. As an example, consider variants of Ant Colony Optimisation that allow only the ants that obtained the best solution quality in a given iteration to update the pheromone trails (iteration-best pheromone update) [Stützle and Hoos, 2000]; in this case, communication between the ants is required in order to determine the best candidate solution.

In principle, co-operative GLSM models can be extended with various communication schemes, including blackboard mechanisms, synchronous broadcasting and one-to-one message passing in a fixed network topology. There are various ways of formally realising these techniques within the GLSM framework. One approach is to allow transition conditions and transition actions to access a shared memory state, that is, information that is shared between the individual GLSMs. Another option is to use special transition actions for communication (e.g., *send* and *receive*).

Many population-based SLS algorithms can be naturally represented as homogeneous co-operative GLSMs. Most ACO algorithms, for example Ant System [Dorigo et al., 1991; 1996], can be easily modelled in the following way. The basic GLSMs corresponding to the individual ants have the same structure as the GLSM model in Figure 3.17 (page 137); only now, the GLSM states represent the construction and local search steps performed by an individual ant, and the transition action *updateTrails* performs a synchronised pheromone trail update for all individual ants. Note that in this case, the pheromone values are shared information between the individual ants' GLSMs.

Co-operative GLSMs with communication are more difficult to design and to implement than those without communication, since issues such as preventing and detecting deadlocks and starvation situations generally have to be considered. Furthermore, the communication between individual GLSMs usually involves a certain amount of overhead. This overhead has to be amortised by the performance gains that may be realised in terms of speedup when applied to a specific class of problem instances and/or in terms of increased robustness over different problem classes.

Generally, one way of using communication to improve the performance of co-operative GLSMs is to propagate candidate solutions with low evaluation function values (or other attractive properties) within the ensemble such that individual GLSMs that detect a stagnation of their search can pick up these 'hints' and continue their local search from there. (This approach is very similar to the co-operative model described by Clearwater et al. [1991; 1992], which uses a 'blackboard' for communicating hints between agents executing a rather simple search strategy.) This type of co-operative search method can be easily modelled by a homogeneous co-operative GLSM with communication. In such a model, the search effort will be more focused on exploring promising parts of the search space than in a co-operative model without communication.

Another general scheme uses two types of GLSMs, analysts and solvers. Analysts do not attempt to find solutions but rather analyse features of the search space. The solvers try to use this information to improve their search strategy. This architecture is an instance of the heterogeneous co-operative GLSM model with communication. It can be extended in a straightforward way to allow for different types of analysts and solvers, or several independent sub-ensembles of analysts and solvers.

Learning via Dynamic Transition Probabilities

One of the features of the basic GLSM model with probabilistic transitions is the fact that the transition probabilities are static, that is, they are fixed when designing the GLSM. An obvious generalisation, along the lines of learning automata theory [Narendra and Thathachar, 1989], is to let the transition probabilities evolve over time as the GLSM is running. The search control in this model corresponds to a variable-structure learning automaton. The environment in which such a dynamic GLSM is operating is given by the evaluation function induced by an individual problem instance or by a class of evaluation functions induced by a class of instances. In the first case (*single-instance learning*), the idea is to optimise the control strategy on one instance during the local search process. The second case (*multi-instance learning*), is based on the assumption that for a given problem domain (or sub-domain), all instances share certain features to which the search control strategy can be adapted.

The modification of the transition probabilities can either be realised by an external mechanism (external adaption control) or within the GLSM framework by means of specialised transition actions (internal adaption control). In both cases, suitable criteria for transition probability updates have to be defined. Two classes of such criteria are those based on search trajectory information and those based on GLSM statistics. The latter category includes state occupancies and transition frequencies, while the former primarily comprises basic descriptive statistics of the evaluation or objective function value along the search trajectory, possibly in conjunction with discounting of past observations. The approach as outlined here captures only a specific form of parameter learning for a given parameterised class of GLSMs. Conceptually, this can be further extended to allow for dynamic changes of transition types (which is equivalent to parameter learning for a more general transition model, such as conditional probabilistic transitions).

In principle, concepts and methods from learning automata theory can be used for analysing and characterising dynamic GLSMs; basic properties, such as expedience or optimality can easily be defined [Narendra and Thathachar, 1989]. We conjecture, however, that theoretically proving such properties will be extremely difficult, as the theoretical analysis of standard SLS behaviour is already complex and rather limited in its results. Nevertheless, we believe that empirical methodology can provide a sufficient basis for developing and analysing interesting and useful dynamic GLSM models.

Evolutionary GLSM Models

For co-operative GLSMs, another form of learning can be realised by letting the number or type of the individual GLSMs vary over time. The population dynamics of these *evolutionary GLSM models* can be interpreted as a learning mechanism. As for the learning GLSMs described above, we can distinguish between *single-instance* and *multi-instance learning* and base the process for dynamically adapting the population on similar criteria.

In the conceptually simplest case, the evolutionary process only affects the composition of the co-operative ensemble: machines that are performing well will spawn off numerous offspring replacing individuals showing inferior performance. This mechanism can be applied to both, homogeneous and heterogeneous models for single-instance learning. In the former case, the selection is based on trajectory information of the individual machines and achieves a similar effect as described above for certain types of homogeneous co-operative

GLSMs with communication: The search is concentrated on exploring promising parts of the search space. When applied to heterogeneous models, this approach allows the realisation of self-optimising algorithm portfolios, which can be useful for single-instance as well as multi-instance learning.

This concept can be further extended by introducing mutation and possibly recombination operators as known from Evolutionary Algorithms. It is also easily conceivable to combine evolutionary and individual learning, for example, by evolving ensembles of dynamic GLSMs. And finally, one could consider models that additionally allow communication within the ensemble. By combining different extensions we can arrive at very complex and potentially powerful GLSM models; while these are very expressive, in general they will also be extremely difficult to analyse. Nevertheless, their implementation is rather straightforward, and an empirical approach for analysing and optimising their behaviour appears to be viable. We believe that such complex models, which allow for a very flexible and fine-grained search control, will likely be most effective when applied to problem classes with varied and salient structural features (see also Chapter 5). There is little doubt that, to some extent, this is the case for most real-world problem domains.

Continuous GLSM Models

The basic GLSM model and the various extensions discussed up to this point model local search algorithms for solving discrete combinatorial problems. Yet, by using continuous instead of discrete local search strategies for the GLSM state types, the model can be easily and naturally extended to continuous optimisation approaches. Although SLS methods for continuous optimisation problems are beyond the scope of this book, it should be noted that the GLSM model's main feature, the clear distinction between simple search strategies and search control, is as useful an architectural and conceptual principle for continuous optimisation algorithms as it is in the discrete case. Furthermore, the GLSM model is particularly well-suited for modelling algorithms for hybrid problems that involve discrete and continuous solution components, as well as for modelling approaches that combine phases of discrete and continuous search in order to solve continuous or hybrid optimisation problems.

3.5 Further Readings and Related Work

The main idea underlying the GLSM model, namely to adequately represent complex algorithms as a combination of several simple strategies, is one of the fundamental concepts in the computing sciences. Here, we have applied this general idea to SLS algorithms for combinatorial decision and optimisation problems, using suitably extended finite state machines for representing search control mechanisms. The GLSM model is partly inspired by Pnueli's work on hybrid systems (see, e.g., Maler et al. [1992]) and Henzinger's work on hybrid automata; the latter is conceptually related to the GLSM model in that it uses finite state machines to model systems with continuous and discrete components and dynamics [Alur et al., 1993; Henzinger, 1996].

The GLSM definition and semantics are heavily based on well-known concepts from automata theory (for general references, see Harrison [1978] or Rozenberg and Salomaa [1997]). However, when using conditional transitions or transition actions, the GLSM model extends the conventional model of a finite state machine. In its most general form, the GLSM model bears close resemblance to a restricted form of Petri nets [Krishnamurthy, 1989], that uses only one token.

While in principle, any local search algorithm can be represented as a GLSM, this representation does not always offer substantial advantages, particularly when the underlying search control mechanisms are very simple. Note, however, that some of the most successful local search algorithms for various problem classes (such as Novelty$^+$ for SAT [Hoos, 1999a], H-RTS for MAX-SAT [Battiti and Protasi, 1997a] and iterated local search algorithms for TSP [Martin et al., 1991; Johnson, 1990; Johnson and McGeoch, 1997]) rely on rather complex search control mechanisms that are adequately captured by the GLSM model.

It is also worth noting that many existing algorithmic frameworks for local search, such as GenSAT [Gent and Walsh, 1993], can be easily and adequately represented using the GLSM model. These frameworks are generally more specific than the GLSM model and emphasise more details of the respective algorithm families; however, they can be easily realised as generic GLSMs without losing any detail of description. This is achieved by using structured generic state types to capture the more specific aspects of the framework to be modelled.

There exist various languages and programming environments that are specifically designed to facilitate the formulation and implementation of local search algorithms (see, for example, di Gaspero and Schaerf [2002], Fink and Voß [2002], Laburthe and Caseau [1998], Michel and van Hentenryck [1999; 2001] or van Hentenryck and Michel [2003]). Some of these support abstract representations of various aspects of the search process, such as the composition of different neighbourhoods [Hentenryck and Michel, 2003]. For an overview of several such approaches we refer to Voß and Woodruff [2002]. However, the main focus of these languages and environment is usually on simplifying the implementation of SLS algorithms by providing programming language concepts and constructs that directly support the natural and efficient implementation of key components of the respective search mechanism [Michel and Hentenryck, 2000; 2001]. To a large degree, these approaches complement conceptual frameworks such as the

GLSM model, which are primarily designed to represent higher-level aspects of the search control mechanisms underlying complex, hybrid SLS methods.

The various extensions of the basic GLSM model discussed in this chapter are closely related to established work on learning automata [Narendra and Thathachar, 1989], parallel algorithm architectures [Jájá, 1992] and Evolutionary Algorithms [Bäck, 1996]. While most of the proposed extensions have not yet been implemented and empirically evaluated, they appear to be promising, especially when considering existing work on multiple independent tries parallelisation [Shonkwiler, 1993; Gomes et al., 1998], algorithm portfolios [Gomes and Selman, 1997b] and learning local search approaches for solving hard combinatorial problems [Boyan and Moore, 1998; Minton, 1996].

3.6 Summary

Based on the intuition that high-performance SLS algorithms are usually obtained by combining several simple search strategies, in this chapter we introduced the model of *Generalised Local Search Machines (GLSMs)*. This conceptual framework formalises the search control mechanisms underlying most hybrid SLS methods using a finite state machine (FSM) model that associates simple component search strategies with the FSM states.

FSMs belong to the most basic and yet fruitful concepts in computer science; using them to model search control mechanisms offers a number of advantages over other formalisms, such as pushdown automata or rule-based systems. Firstly, FSM-based models are conceptually simple; consequently, they can be implemented easily and efficiently. Secondly, the formalism is expressive enough to allow for the adequate representation of a broad range of SLS algorithms. Finally, there is a huge body of work on FSMs; many results and techniques are in principle directly applicable to GLSMs, which may be of interest in the context of analysing and optimising SLS algorithms.

Of course, formalisms equivalent to the FSM model, such as rule-based descriptions, could be chosen instead. While this might be advantageous in certain contexts (such as reasoning about properties of a given GLSM), we find that the automaton model provides a more intuitive and accessible framework for designing and implementing SLS algorithms whose nature is primarily procedural.

In our experience, the GLSM model facilitates the development and design of new, hybrid SLS algorithms. In this context, both conceptual and implementational aspects play a role: due to the conceptual simplicity of the GLSM model and its clear representational distinction between *search strategies* and *search control*, hybrid combinations of existing SLS methods can be easily formalised and explored. Using a generic GLSM simulator, which is not difficult to implement, new, hybrid SLS methods can be realised and evaluated in a very efficient way.

Based on the formal definition of the GLSM model, the semantics of a GLSM can be specified in a rather straightforward way. Furthermore, there is a close relationship between the GLSM model and the standard definition of an SLS algorithm; in particular, GLSMs provide a factored representation of the step functions underlying complex, hybrid SLS methods that conceptually separates the underlying simple search strategies from the higher-level search control mechanism. Categorisations of *state*, *transition* and *machine types* provide a basis for systematically studying the search control mechanisms underlying many high-performance SLS algorithms.

It may be noted that most SLS methods can be represented by rather simple GLSMs. But the fact that many high-performance SLS algorithms tend to be based on search control mechanisms that correspond to structurally slightly more complex GLSMs suggests that further performance improvements may be achieved through the development and systematic study of complex combinations of simple search strategies, as facilitated by the GLSM model.

As we have shown, the basic GLSM model can be easily extended in various ways. *Co-operative GLSM models* comprise ensembles of GLSMs and capture a wide range of multi-agent approaches to combinatorial problem solving in an adequate way, including simple population-based approaches, algorithm portfolios and co-operative search methods. Various forms of learning can be modelled by GLSMs with dynamically changing transition probabilities or evolutionary GLSM models. Finally, GLSM models can be easily generalised to SLS methods for continuous and hybrid optimisation problems. These extensions are very natural generalisations that not only demonstrate the scope of the general idea but also suggest numerous avenues for further research.

Overall, we believe that the GLSM model is very useful as a unifying framework that facilitates the understanding, development and analysis of stochastic local search algorithms.

Exercises

3.1 **[Easy]** Consider the GLSM from Example 3.1 (page 115*f.*). What is the probability that directly after entering state z_2, a sequence of three successive z_2 steps is performed?

3.2 **[Easy]** Specify the SLS method realised by the GLSM shown at the top of the next page in the form of pseudo-code.

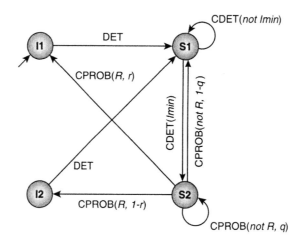

3.3   [*Medium; Implementation*]   Implement a simple GLSM simulator that supports only transitions of type PROB and no transition actions; in each step, instead of performing an actual search step, simply generate a line of output that indicates the type of respective search step.

Use your simulator to perform two independent runs of the GLSM from Example 3.1 (page 115*f.*) with 20 steps each. Show the output of your simulator along with the probability for observing the respective state trajectories.

3.4   [*Medium*]   Explain how any SLS algorithm can be formalised as a 1-state+init GLSM. Why is this formalisation not desirable?

3.5   [*Easy*]   Decide for each of the condition predicates from Table 3.1 (page 126) with the exception of $\top$ to which of the following categories it belongs:

- A: solely based on properties of the current candidate solution

- B: solely based on search control aspects, including memory state

3.6   [*Medium*]   Give a good GLSM representation for the following hybrid SLS algorithm for TSP:

1. Start from a nearest neighbour tour for the given TSP instance.

2. Perform $10\,000$ iterations of Simulated Annealing, using a 3-exchange neighbourhood and a geometric annealing schedule with a starting temperature of 10 and a temperature reduction by a factor of 0.8 every 500 search steps.

3. Perform best improvement steps in the 3-exchange neighbourhood until a local minimum is found.

4. If given target solution quality has not been reached, perform a single, random 4-exchange step and go to Step 3.

It is sufficient to specify the semantics for each state type and transition action in the form of a precise and concise natural language description.

3.7   [*Medium*] Show how a Memetic Algorithm (see Chapter 2, Section 2.4) can be modelled as a GLSM.

3.8   [*Medium*]

(a) Specify a GLSM model for ILS that does not model the acceptance criterion by a separate state.

(b) Discuss advantages and disadvantages of the model from part (a) compared to the representation in Figure 3.16 (page 136).

3.9   [*Hard*]

(a) Give a formal definition of a co-operative GLSM model with a fixed number of individual machines that can communicate with each other via shared memory, building on Definition 3.1 (page 114*f.*).

(b) Describe the semantics of this extended model in an informal, but precise way (cf. Section 3.1).

(c) Use the extended model to give an adequate representation of ACO.

3.10 [*Hard*] Formally define the semantics of a variant of the basic GLSM model in which the memory state can solely be changed during search steps (and not during state transitions). Discuss the advantages or disadvantages of this variant compared to the model specified in Section 3.1.

*There is no higher or lower knowledge,
but one only, flowing out of experimentation.*

—Leonardo da Vinci, Inventor & Artist

4 EMPIRICAL ANALYSIS OF SLS ALGORITHMS

In this chapter, we discuss methods for empirically analysing the performance and behaviour of stochastic local search algorithms. Most of our general considerations and all empirical methods covered in this chapter apply to the broader class of (generalised) Las Vegas algorithms, which contains SLS algorithms as a subclass. After motivating the need for a more adequate empirical methodology and providing some general background on Las Vegas algorithms, we introduce the concept of run-time distributions (RTDs), which forms the basis of the empirical methodology presented in the following. Generally, this RTD-based analysis technique facilitates the evaluation, comparison and improvement of SLS algorithms for decision and optimisation problems; specifically, it can be used for obtaining optimal parameterisations and parallelisations.

4.1 Las Vegas Algorithms

Stochastic Local Search algorithms are typically incomplete when applied to a given instance of a combinatorial decision or optimisation problem; there is no guarantee that an (optimal) solution will eventually be found. However, in the case of a decision problem, if a solution is returned, it is guaranteed to be correct. The same holds for the decision variants of optimisation problems. Another important property of SLS algorithms is the fact that, given a problem instance, the time required for finding a solution (in case a solution is found) is a random variable. These two properties, correctness of the solution computed and run-times characterised by a random variable, define the class of *(generalised) Las Vegas algorithms (LVAs)*.

DEFINITION 4.1 **Las Vegas Algorithm**

An algorithm A for a problem class Π *is a* (generalised) Las Vegas algorithm (LVA) *if, and only if, it has the following properties:*

(1) *If for a given problem instance* $\pi \in \Pi$, *algorithm A terminates returning a solution* s, s *is guaranteed to be a correct solution of* π.

(2) *For each given instance* $\pi \in \Pi$, *the run-time of A applied to* π *is a random variable* $RT_{A,\pi}$.

Remark: The concept of Las Vegas algorithms was introduced by Babai for the (theoretical) study of randomised algorithms [Babai, 1979]. In theoretical computer science, usually two (equivalent) definitions of a Las Vegas algorithm are used. Both of these definitions assume that the algorithm terminates in finite time; they mainly differ in that the first definition always requires the algorithm to return a solution, while the second definition requires that the probability of returning a correct solution is larger or equal than 0.5 [Hromkovič, 2003]. Note that our definition allows for the possibility that, with an arbitrary probability, the algorithm does not return any solution; in this sense, it slightly generalises the established concept of a Las Vegas algorithm. In the following, we will use the term *Las Vegas algorithm* to refer to this slightly more general notion of an LVA.

Here and in the following, we treat the case in which an SLS algorithm does not return a solution as equivalent to the case where the algorithm does not terminate. Under this assumption, any SLS algorithm for a decision problem is a Las Vegas algorithm, as long as the validity of any solution returned by the algorithm is checked. Typically, checking for the correctness of solutions is very efficient compared to the overall run-time of an SLS algorithm, and most SLS algorithms perform such a check before returning any result. (Note that for problems in $\mathcal{NP}$, the correctness of a solution can always be verified in polynomial time.) Based on this argument, in the following we assume that SLS algorithms for decision problems always check correctness before returning a solution.

As an example, it is easy to see that Uninformed Random Picking (as introduced in Section 1.5) is a Las Vegas algorithm. Because a solution is generally never returned without verifying it first (as explained above), condition (1) of the definition is trivially satisfied, and because of the randomised selection process in each search step and/or in the initialisation the time required for finding a solution is obviously a random variable.

In the case of SLS algorithms for optimisation problems, at the first glance, the situation seems to be less clear. Intuitively and practically, unless the optimal value of the objective function is known, it is typically impossible to efficiently verify the optimality of a given candidate solution. However, as noted in Chapter 1, Section 1.1, many optimisation problems include logical conditions that restrict the set of valid solutions. The validity of a solution can be checked efficiently for combinatorial optimisation problems whose associated decision problems are in $\mathcal{NP}$, and SLS algorithms for solving such optimisation problems generally perform such a test before returning a solution. Hence, if only valid solutions are considered correct, SLS algorithms for optimisation problems fit the formal definition of Las Vegas algorithms.

However, SLS algorithms for optimisation problems have the additional property that for fixed run-time, the solution quality achieved by the algorithm, that is, the objective function value of the incumbent candidate solution, is also a random variable.

DEFINITION 4.2 **Optimisation Las Vegas Algorithm**

An algorithm A for an optimisation problem Π' *is a* (generalised) *optimisation Las Vegas algorithm (OLVA) if, and only if, it is a* (generalised) *Las Vegas algorithm, and for each problem instance* $\pi' \in \Pi'$ *the solution quality achieved after any run-time t is a random variable* $SQ(t)$.

Note that for OLVAs, the solution quality achieved within a bounded run-time is a random variable, and the same holds for the run-time required for achieving or exceeding a given solution quality.

Las Vegas algorithms are prominent in various areas of computer science and operations research. A significant part of this impact is due to the successful application of SLS algorithms for solving $\mathcal{NP}$-hard combinatorial problems. However, there are other very successful Las Vegas algorithms that are not based on stochastic local search. In particular, a number of systematic search methods, including some fairly recent variants of the Davis Putnam algorithm for SAT (see also Chapter 6), make use of non-deterministic decisions such as randomised tie-breaking rules and fall into the category of generalised Las Vegas algorithms.

It should be noted that Las Vegas algorithms can be seen as a special case of the larger, and also very prominent, class of *Monte Carlo Algorithms*. Like LVAs,

Monte Carlo algorithms are randomised algorithms with randomly distributed run-times. However, a Monte Carlo algorithm can sometimes return an incorrect answer; in other words, it can generate false positive results (incorrect solutions to the given problem instance) as well as false negative results (missed correct solutions), while for (generalised) Las Vegas algorithms, only false negatives are allowed.

Empirical *vs* Theoretical Analysis

As a result of their inherently non-deterministic nature, the behaviour of Las Vegas algorithms is usually difficult to analyse. For most practically relevant LVAs, in particular for SLS algorithms that perform well in practice, theoretical results are typically hard to obtain, and even in the cases where theoretical results do exist, their practical applicability is often very limited.

The latter situation can arise for different reasons. Firstly, sometimes the theoretical results are obtained under idealised assumptions that do not hold in practical situations. This is, for example, the case for Simulated Annealing, which has been proven to converge towards an optimal solution under certain conditions, one of which is infinitesimally slow cooling in the limit [Hajek, 1988] — which obviously is not practical. Secondly, most complexity results apply to worst-case behaviour, and in the relatively few cases where theoretical average-case results are available, these are often based on instance distributions that are unlikely to be encountered in practice. Finally, theoretical bounds on the run-times of SLS algorithms are typically asymptotic, and do not reflect the actual behaviour accurately enough.

Given this situation, in most cases the analysis of the run-time behaviour of Las Vegas algorithms is based on empirical methodology. In a sense, despite dealing with algorithms that are completely known and easily understood on a step-by-step execution basis, computer scientists are in a sense in the same situation as, for instance, an experimental physicist studying some nondeterministic quantum effect or a microbiologist investigating bacterial growth behaviour. In either case, a complex phenomenon of interest cannot be easily derived from known underlying principles solely based on theoretical means; instead, the classical scientific cycle of observation, hypothesis, prediction, experiment is employed in order to obtain a model that explains the phenomenon. It should be noted that in all empirical sciences, in particular in physics, chemistry and biology, it is largely a collection of these models that constitutes theoretical frameworks, whereas in computer science, theory is almost exclusively derived from mathematical foundations. Historical reasons aside, this difference is largely due to the fact that algorithms are completely specified and mathematically defined at the lowest level. However, in the case of SLS algorithms (and many other complex algorithms or systems), this knowledge is often insufficient to theoretically derive all relevant aspects of their behaviour. In this situation, empirical approaches, based on computational experiments, are often not only the sole way of assessing a given algorithm, but also have the potential to provide insights into practically relevant aspects of algorithmic behaviour that appear to be well beyond the reach of theoretical analysis.

Norms of LVA Behaviour

By definition, Las Vegas algorithms are always correct, while they are not necessarily complete, that is, even if a given problem instance has a solution, a Las Vegas algorithm is generally not guaranteed to find it. Completeness is not only an important theoretical concept for the study of algorithms, but it is often also relevant in practical applications. In the following, we distinguish not only between complete and incomplete Las Vegas algorithms, but also introduce a third category, the so-called *probabilistically approximately complete* LVAs. Intuitively, an LVA is complete, if it can be guaranteed to solve any soluble problem instance in bounded time; it is probabilistically approximately complete (PAC), if it will solve each soluble problem instance with arbitrarily high probability when allowed to run long enough; and it is essentially incomplete, if even arbitrarily long runs cannot be guaranteed to find existing solutions. These concepts can be formalised as follows:

DEFINITION 4.3 **Asymptotic Behaviour of LVAs**

Consider a Las Vegas algorithm A for a problem class Π, and let $P_s(RT_{A,\pi} \leq t)$ denote the probability that A finds a solution for a soluble instance $\pi \in \Pi$ in time less than or equal to t.
A is called

- complete *if, and only if, for each soluble instance $\pi \in \Pi$ there exists some t_{max} such that $P_s(RT_{A,\pi} \leq t_{max}) = 1$;*

- probabilistically approximately complete (PAC) *if, and only if, for each soluble instance $\pi \in \Pi$, $\lim_{t \to \infty} P_s(RT_{A,\pi} \leq t) = 1$;*

- essentially incomplete *if, and only if, it is not PAC, that is, if there exists a soluble instance $\pi \in \Pi$, for which $\lim_{t \to \infty} P_s(RT_{A,\pi} \leq t) < 1$.*

Probabilistic approximate completeness is also refered to as the *PAC property*, and we will often use the term 'approximately complete' to characterise algorithms that are PAC.

Furthermore, we will use the terms *completeness*, *probabilistic approximate completeness* and *essential incompleteness* also with respect to single problem instances or subsets of a problem Π, if the respective properties hold for the corresponding sets of instances instead of Π.

Examples for complete Las Vegas algorithms are randomised systematic search procedures, such as Satz-Rand [Gomes et al., 1998]. Many stochastic local search methods, such as Randomised Iterative Improvement and variants of Simulated Annealing, are PAC, while others, such as basic Iterative Improvement, many variants of Iterated Local Search and most tabu search algorithms are essentially incomplete (see also in-depth section on page 155*ff.*).

Theoretical completeness can be achieved for any SLS algorithm by using a restart mechanism that systematically re-initialises the search such that eventually the entire search space has been visited. However, the time limits for which solutions are guaranteed to be found using this apppproach are typically far too large to be of practical interest. A similar situation arises in many practical situations for search algorithms whose completeness is achieved by different means, such as systematic backtracking.

Essential incompleteness of an SLS algorithm is usually caused by the algorithm's inability to escape from attractive local minima regions of the search space. Any mechanism that guarantees that a search process can eventually escape from arbitrary regions of the search space, given sufficient time, can be used to make an SLS algorithm probabilistically approximately complete. Examples for such mechanisms include random restart, random walk and probabilistic tabu-lists; however, as we will discuss in more detail later (see Section 4.4), not all such mechanisms necessarily lead to performance improvements relevant to practical applications.

For optimisation LVAs, the concepts of completeness, probabilistic approximate completeness and essential incompleteness can be applied to the associated decision problems in a straightforward way, using the following generalisations:

DEFINITION 4.4 **Asymptotic Behaviour of OLVAs**

Consider an optimisation Las Vegas algorithm A' for a problem Π', and let $P_s(RT_{A',\pi'} \leq t, SQ_{A',\pi'} \leq r \cdot q^*(\pi'))$ denote the probability that A' finds a solution of quality $\leq r \cdot q^*(\pi')$ for a soluble instance $\pi' \in \Pi'$ in time $\leq t$, where $q^*(\pi')$ is the optimal solution quality for instance π'.

A′ is called

- *r*-complete *if, and only if, for each soluble instance $\pi' \in \Pi'$ there exists some t_{max} such that $P_s(RT_{A',\pi'} \leq t_{max}, SQ_{A',\pi'} \leq r \cdot q^*(\pi')) = 1$;*

- probabilistically approximately *r*-complete (*r*-PAC) *if, and only if, for each soluble instance $\pi' \in \Pi'$, $\lim_{t \to \infty} P_s(RT_{A',\pi'} \leq t, SQ_{A',\pi'} \leq r \cdot q^*(\pi')) = 1$;*

- essentially *r*-incomplete *if, and only if, it is not approximately r-complete, i.e., if there exists a soluble problem instance $\pi' \in \Pi'$, for which $\lim_{t \to \infty} P_s(RT_{A',\pi'} \leq t, SQ_{A',\pi'} \leq r \cdot q^*(\pi')) < 1$.*

With respect to finding optimal solutions, we use the terms *complete, approximately complete* and *essentially incomplete* synonymously for 1-*complete, approximately* 1-*complete* and *essentially* 1-*incomplete*, where q' is the optimal solution quality for the given problem instance.

IN DEPTH PROBABILISTIC APPROXIMATE COMPLETENESS AND 'CONVERGENCE'

The PAC property states that by running an algorithm sufficiently long, the probability of not finding a (optimal) solution can be made arbitrarily small. Hence, an increase of the run-time typically pays off in the sense that it also increases the chance that the algorithm finds a solution.

As previously stated, several extremely simple SLS algorithms have the PAC property. For example, it is rather straightforward to show that Uninformed Random Picking is PAC. Under some simple conditions, several more complex SLS algorithms can also be proven to have the PAC property. One condition that is sufficient for guaranteeing the PAC property for an SLS algorithm A is the following: there exists $\epsilon > 0$ such that in each search step, the distance to an arbitary, but fixed (optimal) solution s^* is reduced with probability greater than or equal to ϵ, where distance is measured in terms of a minimum length search trajectory of A from its current position to s^*. To see why this condition implies that A is PAC, consider a situation where the distance between the current candidate solution and s^* is equal to l. In that case, we can compute a lower bound on the probability of reaching s^* in exactly l steps as ϵ^l. Since the diameter Δ of the given neighbourhood graph is an upper bound for l, we can give a worst-case estimate for the probability of reaching s^* from an arbitrary candidate solution within Δ steps as ϵ^Δ. Any search trajectory of length $t > \Delta$ can be partitioned into segments of length Δ, for each of which there is an independent probability of at least ϵ^Δ of reaching solution s^*; consequently, the probability that A does not reach s^* within a trajectory of length t can be bounded by

$$(1 - \epsilon^\Delta)^{\lfloor t/\Delta \rfloor}.$$

Since $\epsilon^{\Delta} > 0$, by choosing t sufficiently large, this failure probability can be made arbitrarily small, and consequently, the success probability of A converges to 1 as the run-time approaches infinity, that is, A is PAC.

A proof along these lines can easily be applied to SLS methods such as Randomised Iterative Improvement (see Chapter 2, page 72*ff.*) and — with some additional assumptions on the maximum difference between the evaluation function values of neighbouring candidate solutions — Probabilistic Iterative Improvement (see Chapter 2, page 74*f.*). The PAC property has also been proven for a number of other algorithms, including Simulated Annealing [Geman and Geman, 1984; Hajek, 1988; Lundy and Mess, 1986], specific Ant Colony Optimization algorithms [Gutjahr, 2002; Stützel and Dorigo, 2002], Probabilistic Tabu Search [Faigle and Kern, 1992], deterministic variants of Tabu Search [Glover and Hanafi, 2002; Hanafi, 2001] and Evolutionary Algorithms [Rudolph, 1994].

For many SLS algorithms, properties that are stronger than the PAC property have been proven. (In some sense, the strength of the result depends on the notion of probabilistic convergence proven; see Rohatgi [1976] for the different notions of probabilistic convergence.) In particular, in some cases it can be proven that if s_k is the candidate solution at step k, then $\lim_{k \mapsto \infty} P(s_k \in S) = 1$ (i.e., the probability that the current search position s_k is a (optimal) solution tends to one as the number of iterations approaches infinity). In other words, if run sufficiently long, the probability for the algorithm to visit any non-solution position becomes arbitrarily small. This is exactly the type of result that has been proven, for example, for Simulated Annealing [Hajek, 1988]. This type of convergence can be nicely contrasted with the definition of PAC which implies directly that $\lim_{k \mapsto \infty} P(\hat{s}_k \in S) = 1$, where $\hat{s}_k$ is the incumbent solution after step k. Clearly, the former type of convergence result is stronger in that it implies the PAC property but not vice versa.

However, for practical purposes, this stronger sense of convergence is irrelevant. This is because for decision problems, once a solution is found that satisfies all logical conditions, the search is terminated, and for optimisation problems, the best candidate solution encountered so far is memorised and can be accessed at any time throughout the search. Additionally, these stronger convergence proofs are often based on particular parameter settings of the algorithm that result in an extremely slow convergence of the success probability that is not useful for practically solving problems; this is the case for practically all 'strong' convergence proofs for Simulated Annealing.

The significance of a PAC result is that the respective algorithm is guaranteed to not get permanently trapped in a non-solution area of the search space. What is of real interest in practice, however, is the rate at which the success probability approaches one. While PAC proofs typically give, as a side effect, a lower bound on that rate, this bound is typically rather poor, since it does not adequately capture the heuristic guidance utilised by the algorithm. In fact, in proofs of the PAC property, one typically has to assume a worst-case scenario in which the search heuristic is maximally misleading. However, empirical results indicate that in many cases variants of SLS methods that are PAC perform significantly better in practice than non-PAC variants [Hoos, 1999a; Hoos and Stützle, 2000a; Stützle and Dorigo, 2002], which gives a strong indication that in practice the convergence rate of PAC algorithms is much higher than the theoretical analyses would suggest. In general, proving better bounds on the convergence rate of state-of-the-art SLS algorithms appears to be very challenging, but is doubtlessly an interesting direction of theoretical work on SLS algorithms.

Application Scenarios and Evaluation Criteria

For the empirical analysis of any algorithm it is crucial to use appropriate evaluation criteria. In the case of Las Vegas algorithms, depending on the characteristics of the application context, different evaluation criteria are appropriate. Let us start by considering Las Vegas algorithms for decision problems and classify possible application scenarios in the following way:

Type 1: There are no time limits, that is, we can afford to run the algorithm as long as it needs to find a solution. Basically, this scenario is given whenever the computations are done off line or in a non-realtime environment where it does not really matter how long it takes to find a solution. In this situation we are interested in the expected time required for finding a solution; this can be estimated easily from a number of test runs.

Type 2: There is a hard time limit for finding the solution such that the algorithm has to provide a solution after some given time t_{max}; solutions that are found later are of no use. In real-time applications, such as robotic control or dynamic task scheduling, t_{max} can be very small. In this situation we are not so much interested in the expected time for finding a solution, but in the probability that after the hard deadline t_{max} a solution has been found.

Type 3: The usefulness or utility of a solution depends on the time that was needed to find it. Formally, if utilities are represented as values in $[0, 1]$, we can characterise these scenarios by specifying a utility function $U : \mathbb{R}^+ \mapsto [0, 1]$, where $U(t)$ is the utility of finding a solution at time t. As can be easily seen, application types 1 and 2 are special cases of type 3 which can be characterised by utility functions that are either constant (type 1) or step functions $U(t) := 1$ for $t \leq t_{max}$ and $U(t) := 0$ for $t > t_{max}$ (type 2).

While in the case of no time limits being given (type 1), the mean run-time of a Las Vegas algorithm might suffice to roughly characterise its run-time behaviour, in real-time situations (type 2) this measure is basically meaningless. Type 3 is not only the most general class of application scenario, but these scenarios are also the most realistic. The reason for this is the fact that real-world problem solving usually involves time-constraints that are less strict than the hard deadline given in type 2 scenarios. Instead, at least within a certain interval, the value of a solution gradually decreases over time. In particular, this situation is given when taking into account the costs (in particular, CPU time) of finding a solution.

As an example, consider a situation where hard combinatorial problems have to be solved on line using expensive hardware in a time-sharing mode. Even if the immediate benefit of finding a solution is invariant over time, the costs for performing the computations will diminish the final payoff. Two common ways of modelling this effect are constant or proportional discounting, which use utility functions of the form $U(t) := \max\{u_0 - c \cdot t, 0\}$ and $U(t) := e^{-\lambda \cdot t}$, respectively (see, e.g., [Poole et al., 1998]). Based on the utility function, the weighted solution probability $U(t) \cdot P_s(RT \le t)$ can be used as a performance criterion. If $U(t)$ and $P_s(RT \le t)$ are known, optimal cutoff times t^* that maximise the weighted solution probability can be determined as well as the expected utility for a given time t'. These evaluations and calculations require detailed knowledge of the solution probabilities $P_s(RT \le t)$, potentially for arbitrary run-times t.

In the case of optimisation Las Vegas algorithms, solution quality has to be considered as an additional factor. One might imagine application contexts in which the run-time is basically unconstrained, such as in the type 1 scenarios discussed above, but a certain solution quality needs to be obtained, or situations in which a hard time-limit is given, during which the best possible solution is to be found. Typically, however, one can expect to find more complex tradeoffs between run-time and solution quality. Therefore, the most realistic application scenario for optimisation Las Vegas algorithms is a generalisation of type 3, where the utility of a solution depends on its quality as well as on the time needed to find it. This is modelled by utility functions $U(t, q) : \mathbb{R}^+ \times \mathbb{R}^+ \mapsto [0, 1]$, where $U(t, q)$ is the utility of a solution of quality q found at time t. Analogous to the case of decision LVAs, the probability $P_s(RT \le t, SQ \le q)$ for obtaining a certain solution quality q within a given time t, weighted by the utility $U(t, q)$ can be used as a performance criterion.

4.2 **Run-Time Distributions**

As we have argued in the previous section, it is generally not sufficient to evaluate LVAs based on the expected time required for solving a problem instance or for achieving a given solution quality, or the probability of solving a given instance within a given time. Instead, application scenarios are often characterised by complex utility functions, or Las Vegas algorithms are evaluated without *a priori* knowledge of the application scenario, such that the utility function is unknown, but cannot be assumed to correspond to one of the special cases characterising type 1 or 2 application scenarios. Therefore, LVA evaluations should be based on a detailed knowledge and analysis of the solution probabilities $P_s(RT \le t)$ for decision problems and $P_s(RT \le t, SQ \le q)$ for optimisation problems, respectively. Obviously, these probabilities can be determined from the probability distributions of the random variables characterising the run-time and solution quality of a given LVA.

DEFINITION 4.5 **Run-Time Distribution**

Consider a Las Vegas algorithm A for decision problems class Π, *and let* $P_s(RT_{A,\pi} \le t)$ *denote the probability that A finds a solution for a soluble instance* $\pi \in \Pi$ *in time less than or equal to t. The* run-time distribution (RTD) *of A on* π *is the probability distribution of the random variable* $RT_{A,\pi}$, *which is characterised by the* run-time distribution function $rtd : \mathbb{R}^+ \mapsto [0,1]$ *defined as* $rtd(t) = P_s(RT_{A,\pi} \le t)$.

Similarly, given an optimisation Las Vegas algorithm A' for an optimisation problem Π' *and a soluble problem instance* $\pi' \in \Pi'$, *let* $P_s(RT_{A',\pi'} \le t,$ $SQ_{A',\pi'} \le q)$ *denote the probability that A' applied to* π' *finds a solution of quality less than or equal to q in time less than or equal to t. The* run-time distribution (RTD) *of A' on* π' *is the probability distribution of the bivariate random variable* $(RT_{A',\pi'}, SQ_{A',\pi'})$, *which is characterised by the* run-time distribution function $rtd : \mathbb{R}^+ \times \mathbb{R}^+ \mapsto [0,1]$ *defined as* $rtd(t,q) = P_s(RT_{A',\pi'} \le t, SQ_{A',\pi'} \le q)$.

Since RTDs are completely and uniquely characterised by their distribution functions, we will often use the term 'run-time distribution' or 'RTD' to refer to the corresponding run-time distribution functions.

EXAMPLE 4.1 **RTDs for Decision and Optimisation LVAs**

Figure 4.1 (left) shows a typical run-time distribution for an SLS algorithm applied to an instance of a hard combinatorial decision problem. The RTD is represented by a cumulative probability distribution curve $(t, \widehat{P}_s(RT \le t))$ that has been empirically determined from 1 000 runs of WalkSAT, one of the most prominent SLS algorithms for SAT, on a hard Random 3-SAT instance with 100 variables and 430 clauses (for details on the algorithm and problem class, see Chapter 6); $\widehat{P}_s(RT \le t)$ represents an empirical estimate for the success probability $P_s(RT \le t)$.

Figure 4.1 (right) shows the bivariate RTD for an SLS optimisation algorithm applied to an instance of a hard combinatorial optimisation problem. The plotted surface corresponds to the cumulative probability distribution of an empirically measured RTD, in this case determined from 1 000 runs of an iterated local search algorithm applied to instance pcb442 with 442 vertices from TSPLIB, a benchmark library for the TSP (details on SLS algorithms

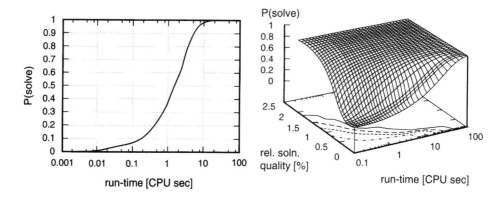

***Figure* 4.1** Typical run-time distributions for SLS algorithms applied to hard combinatorial decision (*left*) and optimisation problems (*right*); for details, see text.

and benchmark problems for the TSP will be discussed in Chapter 8). Note how the contours of the three-dimensional bivariate RTD surface projected into the run-time/solution quality plane reflect the tradeoff between run-time and solution quality: for a given probability level, better solution qualities require longer runs, while vice versa, shorter runs yield lower quality solutions.

The behaviour of a Las Vegas algorithm applied to a given problem instance is completely and uniquely characterised by the corresponding RTD. Given an RTD, other performance measures or evaluation criteria can be easily computed. For decision LVAs, measures such as the mean run-time for finding a solution, its standard deviation, median, quantiles or success probabilities for arbitrary time limits are often used in empirical studies. For optimisation LVAs, popular evaluation criteria include the mean or standard deviation of the solution quality for a given run-time (cutoff time) as well as basic descriptive statistics of the run-time required for obtaining a given solution quality.

Unlike these measures, however, knowledge of the RTD allows the evaluation of Las Vegas algorithms for problems and application scenarios which involve more complex trade-offs. Some of these can be directly represented by a utility function, while others might concern preferences on properties of the RTDs. As an example for the latter case, consider a situation where for a given time-limit t', one SLS algorithm gives a high mean solution quality but a relatively large standard deviation, while another algorithm produces slightly inferior solutions in a more consistent way. RTDs provide a basis for addressing such trade-offs quantitatively and in detail.

Qualified Run-Time Distributions

Multivariate probability distributions, such as the RTDs for optimisation LVAs, are often more difficult to handle than univariate distributions. Therefore, when analysing and characterising the behaviour of optimisation LVAs, instead of working directly with bivariate RTDs, it is often preferable to focus on the (univariate) distributions of the run-time required for reaching a given solution quality threshold.

DEFINITION 4.6 **Qualified Run-Time Distribution**

Let A' be an optimisation Las Vegas algorithm for an optimisation problem Π' and let $\pi' \in \Pi'$ be a soluble problem instance. If $rtd(t, q)$ is the RTD of A' on π', then for any solution quality q', the qualified run-time distribution (QRTD) of A' on π' for q' is defined by the distribution function $qrtd_{q'}(t) := rtd(t, q')$ $= P_s(RT_{A',\pi'} \leq t, SQ_{A',\pi'} \leq q')$.

The qualified RTDs thus defined are marginal distributions of the bivariate RTD; intuitively, they correspond to cross-sections of the two-dimensional RTD graph for fixed solution quality values. Qualified RTDs are useful for characterising the ability of a SLS algorithm for an optimisation problem to solve the associated decision problems (cf. Chapter 1). In practice, they are commonly used for studying an algorithm's ability to find optimal or close-to-optimal solutions (if the optimal solution quality is known) or feasible solutions (in cases where hard constraints are given). Analysing series of qualified RTDs for increasingly tight solution quality thresholds can give a detailed picture of the behaviour of an optimisation LVA.

   An important question arises with respect to the solution quality bounds used when measuring or analysing qualified RTDs. For some problems, benchmark instances with known optimal solutions are available. In this case, bounds expressed as relative deviations from the optimal solution quality are often used; the relative deviation of solution quality q from optimal solution quality q^* is calculated as $q/q^* - 1$; the *relative solution qualities* thus obtained are often expressed in percent. (In cases where $q^* = 0$, sometimes the solution quality is normalised by dividing it by the maximal possible objective function value.) If optimal solutions are not known, one possibility is to evaluate the SLS algorithms w.r.t. the best

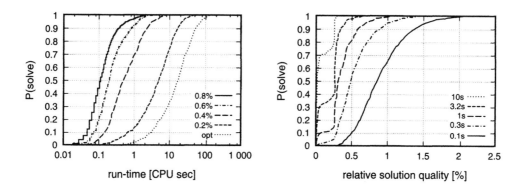

Figure 4.2 *Left:* Qualified RTDs for the bivariate RTD from Figure 4.1. *Right:* SQDs for the same RTD .

known solutions. This method, however, has the potential disadvantage that best known solutions may change. Therefore, it is sometimes preferable to use lower bounds of the optimal solution quality, especially if these are known to be close to the optimum, as is the case for the TSP [Held and Karp, 1970; Johnson and McGeoch, 1997]. Alternatively, there are statistical methods for estimating optimal solution qualities in cases where tight lower bounds are not available [Dannenbring, 1977; Golden and Steward, 1985].

EXAMPLE 4.2 **Qualified Run-Time Distributions**

Figure 4.2 (left) shows a set of qualified RTDs which correspond to marginal distributions of the bivariate empirical RTD from Example 4.1 (page 159*f.*). Note that when tightening the solution quality bound, the qualified RTDs get shifted to the right and appear somewhat steeper in the semi-log plot. This indicates that not only the run-time required for finding higher-quality solutions is higher, but also the relative variability of the run-time (as reflected, for example, in the variation coefficient, that is, the standard deviation of the RTD divided by its mean). The latter observation reflects a rather typical property of SLS algorithms for hard optimisation problems.

Solution Quality Distributions

An orthogonal view of an optimisation LVAs behaviour is given by the distribution of the solution quality for fixed run-time limits.

DEFINITION 4.7 **Solution Quality Distribution**

*Let A' be an optimisation Las Vegas algorithm for an optimisation prob-
lem Π' and let $\pi' \in \Pi'$ be a solvable problem instance. If $rtd(t, q)$ is the
RTD of A' on π', then for any run-time t', the solution quality distribution (SQD) of A' on π' for t' is defined by the distribution function $sqd_{t'}(q)$
$:= rtd(t', q) = P_s(RT_{A', \pi'} \leq t', SQ_{A', \pi'} \leq q)$.*

Like qualified RTDs, solution quality distributions are marginal distributions of
a bivariate RTD. They correspond to cross-sections of the two-dimensional RTD
graph for fixed run-times; in this sense they are orthogonal to qualified RTDs.
SQDs are particularly useful in situations where fixed cutoff times are given (such
as in type 2 application scenarios). Furthermore, they facilitate quantitative and
detailed analyses of the trade-offs between the chance of finding a good solution
fast and the risk of obtaining only low-quality solutions.

Different from run-time, solution quality is inherently bounded from below
by the quality of the optimal solution of the given problem instance. This con-
strains the SQDs of typical SLS algorithms, such that for sufficiently long run-
times, an increase in mean solution quality is often accompanied by a decrease
of solution quality variability. In particular, for a probabilistically approximately
complete algorithm, the SQDs for increasingly large time-limits t' approach a
degenerate probability distribution that has all probability mass concentrated on
the optimal solution quality.

EXAMPLE 4.3 **Solution Quality Distributions**

Figure 4.2 (right) shows a set of SQDs, that is, marginal distributions of the
bivariate empirical RTD from Example 4.1 (page 159*f.*), which offer an orthogonal view to the qualified RTDs from Example 4.2. The SQDs show
clearly that for increasing run-time, the entire probability mass is shifted
towards higher-quality solutions, while the variability in solution quality decreases. It is also interesting to note that the SQDs for large run-times are
multimodal, as can be seen from the fact that they have multiple steep segments which correspond to the peaks in probability density (modes).

An interesting special case arises for iterative improvement algorithms that cannot escape from local minima regions of the given evaluation function. Once
they have encountered such a local minima region, these essentially incomplete

algorithms are unable to obtain any further improvements in solution quality. Consequently, as the run-time is increased towards infinity, the respective SQDs approach a non-degenerate probability distribution. For simple iterative improvement methods that only allow strictly improving steps and always perform such steps when they are possible, this *asymptotic SQD* is reached after finite run-time on any given problem instance. Moreover, in this case asymptotic SQDs can be easily sampled empirically by simply performing multiple runs of the algorithm and recording the quality of the incumbent solution upon termination of each of these runs. Asymptotic SQDs are useful for characterising the performance of simple iterative improvement algorithms (cf. Example 2.1, page 64*f.*).

The information provided by asymptotic SQDs is typically well complemented by the (univariate) run-time distribution that captures the time spent by the algorithm before terminating, independent of the final solution quality reached in this run. To distinguish this type of run-time distribution from the complete bivariate run-time distribution that characterises the behaviour of any optimisation LVA and from the notion of a qualified run-time distribution discussed above, we refer to it as *termination-time distribution (TTD)*. Note that unlike qualified RTDs, TTDs are not marginal distributions of the underlying bivariate RTD.

Asymptotic SQDs are also very useful for characterising the performance of purely constructive search algorithms, such as the Nearest Neighbour Heuristic for the TSP (cf. Chapter 1, Section 1.4), which terminate as soon as a complete candidate solution has been obtained. Unlike (perturbative) iterative improvement algorithms, constructive search algorithms typically terminate after a fixed, instance-dependent number of search steps. Consequently, they typically show much less variability in run-time (or no variability at all), which simplifies comparative performance analyses.

Time-Dependent Summary Statistics

Instead of dealing with a set of SQDs for a series of time limits, researchers (and practitioners) often just look at the development of certain solution quality statistics over time *(SQTs)*. A common example of such an SQT is the function $\overline{SQ}(t)$, which characterises the time-dependent development of the mean solution quality achieved by a given algorithm. It is often preferable to use SQTs that reflect the development of quantiles (e.g., the median) of the underlying SQDs over time, since quantiles are typically statistically more stable than means. Furthermore, SQTs based on SQD quantiles offer the advantage that they can be seen as horizontal sections or contour lines of the underlying bivariate RTD surfaces. Combinations of such SQTs can be very useful for summarising certain aspects of

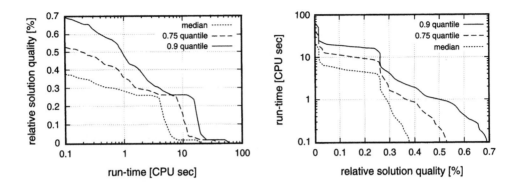

Figure 4.3 *Left:* Development of median solution quality, 0.75 and 0.9 SQD quantiles over time for the same TSP algorithm and problem instance as used in Figure 4.1 (page 160). *Right:* RTQ for the same algorithm and problem instance.

a full SQD series, and hence a complete bivariate RTD; they are particularly well suited for explicitly illustrating trade-offs between run-time and solution quality. Especially individual SQTs, however, offer a fairly limited view of an optimisation Las Vegas algorithm's run-time behaviour in which important details can be easily missed.

EXAMPLE 4.4 **Solution Quality Statistics Over Time**

Figure 4.3 (left) shows the development of median solution quality and its variability over time, obtained from the same empirical data underlying the bivariate RTD from Example 4.1 (page 159*f.*). From this type of evaluation, which is often used in the literature, we can easily see that in the given example the algorithm behaves in a very desirable way: with increasing run-time, the median solution quality as well as the higher SQD quantiles improve substantially and consistently; in this particular example, we can also see that there is a large and rapid improvement in solution quality after 4–20 CPU seconds. The gradual decrease in solution quality variability during the first and final phase of the search is rather typical for the behaviour of high-performance SLS algorithms for hard combinatorial optimisation problems; it indicates that for longer runs the algorithm tends to find better solutions in a more consistent way. Note, however, that interesting properties, such as the fact that in our example the SQDs for large run-times are multimodal, or that the variation in run-time increases when higher-quality solutions need to be obtained, cannot be observed from the SQT data shown here.

It is interesting to note that, while SQTs are commonly used in the literature for evaluating and analysing the behaviour of SLS algorithms for optimisation problems, the orthogonal concept of qualified RTD statistics dependent on solution quality *(RTQs)* does not appear to be used at all. Possibly the reason for this lies in the fact that SQTs are more intuitively related to the run-time behaviour of an optimisation LVA, and that empirical SQTs can be measured more easily (the latter issue will be discussed in more detail in the next section). Nevertheless, RTQs can be useful, for instance, in cases where trade-offs between the mean and the standard deviation of the time required for reaching a certain solution quality q' have to be examined in dependence of q', but where the details offered by a series of qualified RTDs (or the full bivariate RTD) are not of interest.

EXAMPLE 4.5   **Run-Time Statistics Depending on Solution Quality**

Figure 4.3 (right) illustrates several quantiles of the qualified RTDs from Figure 4.2 (page 162) for relative solution quality q in dependence of q. Note the difference to the SQT plots in Figure 4.3 (left), which show SQD statistics as a function of run-time.

Empirically Measuring RTDs

Except for very simple algorithms, such as Uninformed Random Picking, it is typically not possible to analytically determine RTDs for a given Las Vegas algorithm. Hence, the true RTDs characterising a Las Vegas algorithm's behaviour are typically approximated by empirical RTDs. For a given instance π of a decision problem, the empirical RTD of an LVA A can be easily determined by performing k independent runs of A on π and recording for each successful run the time required to find a solution. The empirical run-time distribution is given by the cumulative distribution function associated with these observations. Each run corresponds to drawing a sample from the true RTD of A on π, and clearly, the more runs are performed, the better will the empirical RTD obtained from these samples approximate the true underlying RTD. For algorithms that are known to be either complete or probabilistically approximately complete (PAC), it is often desirable (although not always practical) to terminate each run only after a solution has been found; this way, a complete empirical approximation of A's RTD on π can be obtained. In cases where not all runs are successful, either because the algorithm is essentially incomplete or because some runs were terminated before a solution could be found, a truncated approximation of the true RTD can be obtained from the successful runs. Practically, nearly always a cutoff time is used as a criterion for terminating unsuccessful runs.

More formally, let k be the total number of runs performed with a cutoff time t', and let $k' \leq k$ be the number of successful runs, that is, runs during which a solution was found. Furthermore, let $rt(j)$ denote the run-time for the jth entry in a list of all successful runs, ordered according to increasing run-times. The cumulative empirical RTD is then defined by $\widehat{P}_s(RT \leq t) := \#\{j \mid rt(j) \leq t\}/k$. The ratio $sr := k'/k$ is called the success ratio of A on π with cutoff t'. For algorithms that are known or suspected to be essentially incomplete, the success ratio converges to the asymptotic maximal success probability of A on the given problem instance π, which is formally defined as $p_s^* := \lim_{t \to \infty} P_s(RT_{A,\pi} \leq t)$. For sufficiently high cutoff time, the empirically determined success ratio can give useful approximations of p_s^*.

Unfortunately, in the absence of theoretical knowledge on the success probability or the speed of convergence of the success ratio, the decision whether a given cutoff time is high enough to obtain a reasonable estimate of the success probability needs to be based on educated guessing. In practice, the following criterion is often useful in situations, where a reasonably high number of runs (typically between 100 and 10 000) can be performed: When increasing a given cutoff t' by a factor of τ (where τ is typically between 10 and 100) does not result in an increased success ratio, it is assumed that the asymptotic behaviour of the algorithm is observed and that the observed success ratio is a reasonably good approximation of the asymptotic success probability.

Note that in these situations, as well as in cases where success ratios equal to one cannot be achieved for practical reasons (e.g., due to limited computing resources), certain RTD statistics, in particular all quantiles lower than sr, are still available. Other RTD statistics, particularly the mean time for finding a solution, can be estimated using the following approach: When for cutoff time t', k' out of k runs were successful, the probability for any individual run with cutoff t' to succeed can be estimated by the success ratio $sr := k'/k$. Consequently, for n successive (or parallel) independent runs with cutoff t', the probability that at least one of these runs is successful is $1 - (1 - sr)^n$. Using this result, quantiles higher than sr can be estimated for the variant of the respective algorithm that re-initialises the search after each time interval of length t' (static restart). Furthermore, the expected time for finding a solution can be estimated from the mean time over the successful runs by taking into account the expected number of runs required to find a solution as well as the mean run-time of the failed runs (see also Parkes and Walser [1996]):

$$\widehat{E}(RT') = \widehat{E}(RT_s) + (\frac{1}{sr} - 1) \cdot \widehat{E}(RT_f), \tag{4.1}$$

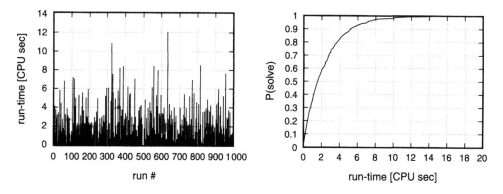

Figure 4.4 Run-time data for WalkSAT/SKC, a prominent SLS algorithm for SAT, applied to a hard Random 3-SAT instance for approx. optimal noise setting, 1 000 tries. *Left:* bar diagram $rt(j)$; *right:* corresponding RTD .

where $\widehat{E}(RT_s) := 1/k' \cdot \sum_{j=1}^{k'} rt(j)$ is the average run-time of a successful run, and $\widehat{E}(RT_f)$ is the average run-time of a failed run. Using a static restart mechanism with a fixed cutoff time t' results in $\widehat{E}(RT_f) := t'$. Note that in this case, $\widehat{E}(RT')$ depends on the cutoff time t'; in fact, RTD information can be used for determining values t' that lead to optimal performance in the sense of minimal expected solution time $\widehat{E}(RT')$ (cf. Section 4.4).

EXAMPLE 4.6 **Raw Run-Time Data *vs* Empirical RTDs**

Figure 4.4 (left) shows the raw data from running WalkSAT/SKC, a prominent SLS algorithm for SAT, on a hard problem instance with 100 variables and 430 clauses; each vertical line represents one run of the algorithm and the height of the lines indicates the CPU time needed for finding a solution. The right side of the same figure shows the corresponding RTD as a cumulative probability distribution curve $(t, \widehat{P}_s(RT \leq t))$. Note that the run-time is extremely variable, which is typical for SLS algorithms for hard combinatorial problems. Clearly, the RTD representation gives a much more informative picture of the run-time behaviour of the algorithm than simple descriptive statistics summarising the data shown on the left side of Figure 4.4, and, as we will see later in this chapter, it also provides the basis for more sophisticated analyses of algorithmic behaviour. (The graphs shown in Figure 4.4 are based on the same data used in Example 4.1 on page 159*f.*)

For empirically approximating the bivariate RTDs of an optimisation LVA A' on a given problem instance π', a slightly different approach is used. During each run of A', whenever the incumbent solution (i.e., the best candidate solution found during this run) is improved, the quality of the improved incumbent solution and the time at which the improvement was achieved is recorded in a *solution quality trace*. The empirical RTD is derived from the solution quality traces obtained over multiple (independent) runs of A' on π'. Formally, let k be the number of runs performed and let $sq(t, j)$ denote the quality of the best solution found in run j until time t. Then the cumulative empirical run-time distribution of A' on π' is defined by $\widehat{P}_s(RT \leq t', SQ \leq q') := \#\{j \mid sq(t', j) \leq q'\}/k$. Qualified RTDs and SQDs as well as SQT and RTQ data and, where appropriate, asymptotic SQDs and TTDs can also be easily derived from the solution quality traces. With regard to the use of cutoff times and their impact on the completeness of the empirical RTDs, considerations very similar to those discussed for the case of decision problems apply.

CPU Time *vs* Operation Counts

Up to this point, and consistent with a large part of the empirical analyses of algorithmic performance in the literature, we have used CPU time for measuring and reporting the run-time of algorithms. Obviously, a CPU time measurement is always based on a concrete implementation and run-time environment (i.e., machine and operating system). However, it is often more appropriate, especially in the context of comparative studies of algorithmic performance, to measure run-time in a way that allows one to abstract from these factors and that facilitates comparisons of empirical results across various platforms. This can be done using *operation counts*, which reflect the number of operations that are considered to contribute significantly towards an algorithm's performance, and *cost models*, which relate the cost (typically in terms of run-time per execution) of these operations relative to each other or absolute in terms of CPU time for a given implementation and run-time environment [Ahuja and Orlin, 1996].

Generally, using operation counts and an associated cost model rather than CPU time measurements as the basis for empirical studies often gives a clearer and more detailed picture of algorithmic performance. This approach is especially useful for comparative studies involving various algorithms or different variants of one algorithm. Furthermore, it allows one to explicitly address trade-offs in the design of SLS algorithms, such as complexity *vs* efficacy of different types of local search steps. To make a clear distinction between run-time measurements corresponding to actual CPU times and abstract run-times measured in operation counts, we refer to the latter as *run-lengths*. Similarly, we refer to RTDs obtained

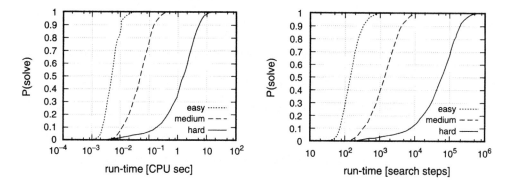

Figure 4.5 RTDs *(left)* and RLD *(right)* for WalkSAT/SKC, a prominent SLS algorithm for SAT, applied to three Uniform Random 3-SAT instances of varying difficulty, based on 1 000 runs per instance (using an approx. optimal noise parameter setting).

from run-times measured in terms of operation counts as *run-length distributions* or *RLDs*.

For SLS algorithms, a commonly used operation count is the number of local search steps. In the case of pure SLS methods, such as Iterative Improvement, there is only one type of local search step, and while the cost or time complexity of such a step typically depends on the size and other properties of the given problem instance, in many cases it is constant or close to constant within and between runs of the algorithm on the same instance. In this situation, measuring run-time in terms of local search steps as elementary operations is often the method of choice; furthermore, run-times measured in terms of CPU time and run-lengths based on local search steps as basic operations are related to each other by scaling with a constant factor.

EXAMPLE 4.7   **RTDs vs RLDs**

Figure 4.5 shows RTD and RLD data for the same experiments (solving three Uniform Random 3-SAT instances with 100 variables and 430 clauses each using WalkSAT/SKC, a prominent SLS algorithm for SAT). The operations counted for obtaining RLDs are local search steps; in the case of WalkSAT/SKC, each local search step corresponds to flipping the truth value assigned to one propositional variable. Note that, when comparing the RTDs and the corresponding RLDs in a semi-log plot, both distributions always have the same shape. This reflects the fact that the CPU time per step is roughly constant. However, closer examination of the RTD and RLD data reveals that the CPU time per step differs between the three instances; the reason for this is the fact that the hard problem was solved on a faster machine than the medium and easy instances. In this example, the CPU time per search step is 0.027ms for the hard instance, and 0.035ms for the medium and easy instances; the time required for search initialisation is 0.8ms for the hard instance and 1ms for the medium and easy instances. These differences result solely from the difference in CPU speed between the two machines used for running the respective experiments.

In the case of hybrid SLS algorithms characterised by GLSM models with multiple frequently used states, such as Iterated Local Search (cf. Chapter 2, Section 2.3 and Chapter 3, Section 3.3), the search steps for each state of the GLSM model may have significantly different execution costs (i.e., run-time per step) and, consequently, they should be counted separately. By weighting these different operation counts relative to each other, using an appropriate cost model, it is typically possible to aggregate them into run-lengths or RLDs. Alternatively, or in situations where the cost of local search steps can vary significantly within a run of the algorithm or between runs on the same instance, it may be necessary to use finer-grained elementary operations, such as the number of evaluations of the underlying objective function, or the number of updates of internal data structures used for implementing the algorithm's step function.

4.3   RTD-Based Analysis of LVA Behaviour

After having introduced RTDs (and related concepts) in the previous section, we now show how these can be used for analysing and characterising the behaviour and relative performance of Las Vegas algorithms. We will start with the quantitative analysis of LVA behaviour based on single RTDs; next, we will show how this technique can be generalised to cover sets and distributions of problem instances. We will then explain how RTDs can be used for the comparative analysis of several algorithms before returning to individual algorithms, for which we discuss advanced analysis techniques, including the empirical analysis of asymptotic behaviour and stagnation.

Basic Quantitative Analysis based on Single RTDs

When analysing or comparing the behaviour of Las Vegas Algorithms, the empirical RTD (or RLD) data can be used in different ways. In many cases, graphic representations of empirical RTDs provide a good starting point. As an example,

Figures 4.6 and 4.7 show the RTD for the hard problem instance from Figure 4.5 (page 170) in three different views. Compared to standard representations, semi-log plots (as shown on the right side of Figure 4.7) give a better view of the distribution over its entire range; this is especially relevant for RTDs of SLS algorithms, which often show an extreme variability in run-time. Also, when using semi-log plots to compare RTDs, uniform performance differences characterised by a constant factor can be easily detected, as they correspond to simple shifts along the horizontal axis (for an example, see Figure 4.5, page 170). On the other hand, log-log plots of an RTD or its associated failure rate decay function,

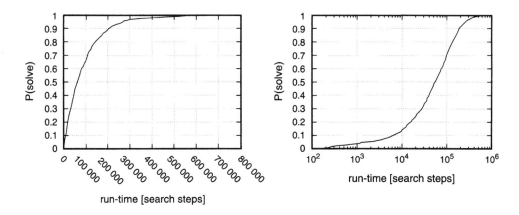

Figure **4.6** *Left:* RLD for WalkSAT/SKC, a prominent SLS algorithm for SAT, on a hard Random 3-SAT instance for approx. optimal noise parameter setting. *Right:* Semi-log plot of the same RLD.

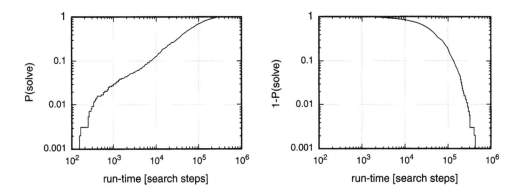

Figure **4.7** Log-log plot of the same RLD as in Figure 4.6 *(left)* and log-log plot of the corresponding failure probability over time *(right)*.

$1 - rtd(t)$, are often very useful for examining the behaviour of a given Las Vegas algorithm for extremely short or extremely long run-times (cf. Figure 4.7).

While graphical representations of RTDs are well-suited for investigating and describing the qualitative behaviour of Las Vegas Algorithms, quantitative analyses are usually based on summarising the RTD data with basic descriptive statistics. For our example, some of the most common standard descriptive statistics, such as the empirical mean, standard deviation, minimum, maximum and some quantiles, are reported in Table 4.1. Note again the huge variability of the data, as indicated by the large standard deviation and quantile ratios. The latter, like the *variation coefficient*, $vc := stddev/mean$, have the advantage of being invariant to multiplication of the data by a constant, which – as we will see later – is often advantageous when comparing RTDs.

In the case of optimisation LVAs, analogous considerations apply to graphical representations and standard descriptive statistics of qualified RTDs for various solution quality bounds. Similarly, different graphical representations and summary statistics can be used for analysing and characterising empirical SQDs for various run-time bounds or time-dependent statistics of solution quality; this approach is more commonly followed in the literature, but not always preferable over studying qualified RTDs.

Generally, it should be noted that for directly obtaining sufficiently stable estimates for summary statistics, the same number of test-runs have to be performed as for measuring reasonably accurate empirical RTDs. Thus, measuring RTDs does not cause a computational overhead in data acquisition when compared to measuring only a few simple summary statistics, such as averages and empirical standard deviations. At the same time, arbitrary quantiles and other descriptive statistics can be easily calculated from the RTD data. Furthermore, in the case of optimisation LVAs, bivariate RTDs, qualified RTDs, SQDs and SQTs can all be easily determined from the same solution quality traces without significant overhead in computation time. Because qualified RTDs, SQDs and

mean	57 606.23	median	38 911
min	107	$q_{0.25}$; $q_{0.1}$	16 762; 5 332
max	443 496	$q_{0.75}$; $q_{0.9}$	80 709; 137 863
stddev	58 953.60	$q_{0.75}/q_{0.25}$	4.81
vc	1.02	$q_{0.9}/q_{0.1}$	25.86

Table 4.1 Basic descriptive statistics for the RLD shown in Figures 4.6 and 4.7; q_x denotes the x-quantile; the variation coefficient $vc := stddev/mean$ and the quantile ratios q_x/q_{1-x} are measures for the relative variability of the run-length data.

SQTs merely present different views on the same underlying bivariate RTD, and since similar considerations apply to all of these, in the following discussion of empirical methodology we will often just explicitly mention RTDs.

Because of the high variability in run-time over multiple runs on the same problem instance that is typical for many SLS algorithms, empirical estimates of mean run-time can be rather unstable, even when obtained from relatively large numbers of successful runs. This potential problem can be alleviated by using quantiles and quantile ratios instead of means and standard deviations for summarising RTD data with simple descriptive statistics.

Basic Quantitative Analysis for Ensembles of Instances

In many applications, the behaviour of a given algorithm needs to be tested on a set of problem instances. In principle, the same method as described above for single instances can be applied — RTDs are measured for each instance, and the corresponding sets of graphs and/or associated descriptive statistics are reported.

Often, LVA behaviour is analysed for a set of fairly similar instances (such as instances of the same type, but different size, or instances from the same random instance distribution). In this case, the RTDs will often have similar shapes (particularly as seen in a semi-log plot) or share prominent qualitative properties, such as being uni- or bi-modal, or having a very prominent right tail. A simple example can be seen in Figure 4.8 (left side), where very similarly shaped RTDs are obtained when applying the same SLS algorithm for SAT (WalkSAT/SKC) to three randomly generated instances from the same instance

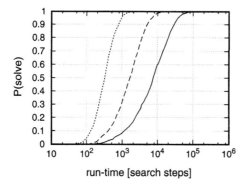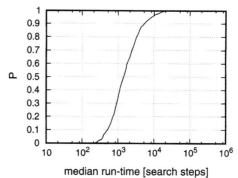

Figure 4.8 *Left:* RLDs for WalkSAT/SKC (using an approx. optimal noise parameter setting), a prominent SLS algorithm for SAT, applied to three hard Random 3-SAT instances. *Right:* Distribution of median local search cost for the same algorithm across a set of 1 000 Uniform Random 3-SAT instances.

distribution (Uniform Random 3-SAT with 100 variables and 430 clauses). In such cases, a representative or typical instance can be selected for presentation or further analysis, while the analogous data for the other instances are only briefly summarised. It is very important, however, to not naïvely assume properties of or similarities between RTDs based on a few selected examples only, but to carefully test such assumptions by manual or automated analysis of all or sufficiently many RTDs. In Section 4.4, we will demonstrate how in certain cases, the latter can be done in an elegant and informative way by using functional approximations of RTDs and statistical goodness-of-fit tests.

For bigger sets of instances, such as the sets obtained from sampling random distributions of problem instances, it becomes important to characterise the performance of a given algorithm on individual instances as well as across the entire ensemble. Often (but not always!) when analysing the behaviour of reasonably optimised, probabilistically approximately complete SLS algorithms in such situations, there is a fairly simple scaling relationship between the RTDs for individual problem instances: Given two instances and a desired probability of finding a solution, the ratio of the run-times required for achieving this solution probability for the two instances is roughly constant. This is equivalent to the observation that in a semi-log plot, the two corresponding RTDs essentially differ only by a shift along the time axis. If this is the case, the performance of the given algorithm across the ensemble can be summarised by one RTD for an arbitrarily chosen instance from the ensemble and the distribution of the mean (or any quantile) of the individual RTDs across the ensemble. The latter type of distribution intuitively captures the cost for solving instances across the set; in the past it has often been referred to as 'hardness distribution' – however, it should be noted that without further knowledge, the underlying notion of hardness is entirely relative to the algorithm used rather than intrinsic to the problem instance, and hence this type of distribution is technically more appropriately termed a *search cost distribution (SCD)*. An example for such a SCD, here for an SLS algorithm for SAT (WalkSAT/SKC) applied to a set of 1 000 Uniform Random 3-SAT instances with 100 variables and 430 clauses each, is shown in Figure 4.8 (right side).

In reality, the simple multiplicative scaling relationship between any two instances of a given ensemble will hardly ever hold exactly. Hence, depending on the degree and nature of variation between the RTDs for the given ensemble, it is often reasonable and appropriate to report cost distributions along with a small set of RTDs that have been carefully selected from the ensemble such that they representatively illustrate the variation of the RTDs across the sets. Sometimes, distributions (or statistics) of other basic descriptive RTD statistics across the ensemble of instance—for example, a distribution of variation coefficients or quantile ratios—can be useful for obtaining a more detailed picture of the algorithm's behaviour on the given ensemble. It can also be very informative to investigate the correlation between various features of the RTD across the ensemble; specifically, the correlation between the median (or mean) and some measure of variation can be very interesting for understanding LVA behaviour.

Finally, it should be mentioned that when dealing with sets of instances that have been obtained by systematically varying some parameter, such as problem size, it is natural and obvious to study characteristics and properties of the corresponding RTDs (or the cost distributions) in dependence of this parameter. Otherwise, similar considerations as discussed above for ensembles of instances apply. Again, choosing an appropriate graphical representation, such as a semi-log plot for the functional dependence of mean run-time on problem size, is often the key for easily detecting interesting behaviour (e.g., exponential scaling).

IN DEPTH BENCHMARK SETS

The selection of benchmark instances is an important factor in the empirical analysis of an algorithm's behaviour, and the use of inadequate benchmark sets can lead to questionable results and misleading conclusions. The criteria for benchmark selection depend significantly on the problem domain under consideration, on the hypotheses and goals of the empirical study, and on the algorithms being analysed. There are, however, some general issues and principles which will be discussed in the following.

Typically, benchmark sets should mainly consist of problem instances that are intrinsically hard or difficult to solve for a broad range of algorithms. While easy instances can be sometimes useful for illustrating or investigating properties of specific algorithms (for example polynomially solvable instances that are hard for certain, otherwise high-performing algorithms), they should not be used as general benchmark problems, as this can easily lead to heavily biased evaluations and assessments of the usefulness of specific algorithms. Similar considerations apply to problem size; small problem instances can sometimes lead to atypical SLS behaviour that does not generalise to larger problem sizes. To avoid such problems and to facilitate studies on the scaling of SLS performance it is generally advisable to include problem instances of different sizes into benchmark sets.

Furthermore, benchmark sets should contain a diverse collection of problem instances. An algorithm's behaviour can substantially depend on specific features of problem instances, and in many cases at least some of these features are not known *a priori*. Using a benchmark set comprising a diverse range of problem instances reduces the risk of incorrectly generalising from behaviour or performance results that only apply to a very limited class of problem instances.

We distinguish three types of benchmark instances: instances obtained from real-world applications, artificially crafted problem instances and randomly generated instances. Some combinatorial problems have no real-world applications; where real-world problem instances are available, however, they often provide the most realistic test-bed for algorithms of potential practical interest. Artificially crafted problem instances can be very useful for studying specific properties or features of an algorithm; they are also often used in situations where real-world instances are not available or unsuitable for a specific study (e.g., because they are too large, too difficult to solve, or only very few real-world instances are available). Random problem instance generators have been developed and widely used in many domains, including SAT and TSP. These generators effectively sample from distributions of problem instances with controlled syntactic properties, such as instance size or expected number of solutions. They offer the advantage that large test-sets can be generated easily, which facilitates the application of statistical tests. However, basing the evaluation of an algorithm on randomly generated problem instances only carries the risk of obtaining results that are misleading or meaningless w.r.t. to practical applications.

Ideally, benchmark sets used for empirical studies should comprise instances of all three types. In some cases, it can also be beneficial to additionally use suitable encoded problem instances from other domains. The performance of SAT algorithms, for example, is often evaluated on SAT-encoded instances from domains such as graph colouring, planning or circuit verification (see, e.g., Hoos and Stützle [2000a]). In these cases, it is often important to ensure that the respective encoding schemes do not produce undesirable features that, for instance, may render the resulting instances abnormally difficult for the algorithm(s) under consideration.

In principle, artificially crafted and randomly generated problem instances can offer the advantage of carefully controlled properties; in reality, however, the behaviour of SLS algorithms is often affected by problem features that are not well understood or difficult to control. (This issue will be further discussed in Chapter 5.) Randomly generated instance sets often show a large variation w.r.t. their non-controlled features, leading to the kind of diversity in the benchmark sets that we have advocated above. On the other hand, this variation often also causes extreme differences in difficulty for instances within the same sample of problem instances (see, e.g., Hoos [1998], Hoos and Stützle [1999]). This can easily lead to substantial differences in difficulty (as well as other properties) between test-sets sampled from the same instance distribution. As a consequence, comparative analyses should always evaluate all algorithms on identical test-sets.

To facilitate the reproducibility of empirical analyses and the comparability of results between studies, it is important to use established benchmark sets and to make newly created test-sets available to other researchers. In this context, public benchmark libraries play an important role. Such libraries exist for many domains; widely known examples include TSPLIB (containing a variety of TSP and TSP-related instances), SATLIB (which includes a collection of benchmark instances for SAT), ORLIB (comprising test instances for a variety of problems from Operations Research), TPTP (a collection of problem instances for theorem provers) and CSPLIB (a benchmark library for constraints). Good benchmark libraries are regularly updated with new, challenging problems. Using severely outdated or static benchmark libraries for empirical studies gives rise to various, well-known pitfalls [Hooker, 1994; 1996] and should therefore be avoided as much as possible. Furthermore, good benchmark libraries will provide descriptions and explanations of all problem instances offered, ideally accompanied by references to the relevant literature. Generally, a good understanding of all benchmark instances used in the context of an empirical study, regardless of their source, is often crucial for interpreting the results correctly and conclusively.

Comparing Algorithms Based on RTDs

Empirical investigations of algorithmic behaviour are frequently performed in the context of comparative studies, often with the explicit or implicit goal to establish the superiority of a new algorithm over existing techniques. In this situation, given two Las Vegas algorithms for a decision problem, one would empirically show that one of them consistently gives a higher solution probability than the other. Likewise, for an optimisation problem, the same applies for a specific (e.g., the optimal) solution quality or for a range of solution qualities. Formally, this can be captured by the concept of probabilistic domination, defined in the following way:

DEFINITION 4.8 **Probabilistic Domination**

Let $\pi \in \Pi$ an instance of a decision problem Π, and let A and B be two Las Vegas algorithms for Π. A probabilistically dominates B on π if, and only if, $\forall t : P_s(RT_{A,\pi} \leq t) \geq P_s(RT_{B,\pi} \leq t)$ and $\exists t : P_s(RT_{A,\pi} \leq t) > P_s(RT_{B,\pi} \leq t)$.

Similarly, for an instance $\pi' \in \Pi'$ of an optimisation problem Π' and optimisation LVAs A' and B' for Π', A' probabilistically dominates B' on π' for solution quality less than or equal to q if, and only if, $\forall t : P_s(RT_{A',\pi'} \leq t, SQ_{A',\pi'} \leq q) \geq P_s(RT_{B',\pi'} \leq t, SQ_{B',\pi'} \leq q)$ and $\exists t : P_s(RT_{B',\pi'} \leq t, SQ_{B',\pi'} \leq q) > P_s(RT_{B',\pi'} \leq t, SQ_{B',\pi'} \leq q)$.

A' probabilistically dominates B' on π' if, and only if, A' probabilistically dominates B' on π' for arbitrary solution quality bounds q.

Remark: A probabilistic domination relation holds between two Las Vegas algorithms on a given problem instance if, and only if, their respective (qualified) RTDs do not cross each other. This provides a simple method for graphically checking probabilistic domination between two LVAs on individual problem instances.

In practice, performance comparisons between Las Vegas algorithms are complicated by the fact that even for a single problem instance, a probabilistic domination does not always hold. This situation is characterised by the occurrence of cross-overs between the corresponding RTDs, indicating that which of the two algorithms performs better, that is, obtains higher solution probabilities (for a given solution quality bound), depends on the time the algorithm is allowed to run.

Statistical tests can be used to assess the significance of performance differences. In the simplest case, the Mann-Whitney U-test (or, equivalently, the Wilcoxon rank sum test) can be applied [Sheskin, 2000]; this test determines whether the medians of two samples are equal, hence a rejection indicates significant performance differences. This test can also be used to determine whether the median solution qualities achieved by two SLS optimisation algorithms are identical. (The widely used t-test generally fulfils a similar purpose, but requires the assumption that the given samples are normally distributed with identical variance; since this assumption is often violated in the context of the empirical analysis of SLS behaviour, in many cases, the t-test is not applicable.) The more specific hypothesis whether the theoretical RTDs (or SQDs) of two algorithms are identical can be tested using the Kolmogorov-Smirnov test for two independent samples [Sheskin, 2000].

One important question when assessing the statistical significance of performance differences observed between algorithms is that of sample size: How many runs should be performed for measuring the respective empirical RTDs? Generally, the precision of statistical tests, that is, their ability to correctly distinguish situations in which the given null hypothesis is correct from those where it is incorrect, crucially depends on sample size. Table 4.2 shows the performance differences between two given RTDs that can be detected by the Mann-Whitney U-test for standard significance levels and power values in dependence of sample size. (Note that the significance level and power value indicate the maximum probabilities that the test incorrectly rejects or accepts the null hypothesis that the medians of the given RTDs are equal, respectively.)

sign. level 0.05, power 0.95		*sign. level 0.01, power 0.99*	
sample size	m_1/m_2	*sample size*	m_1/m_2
3 010	1.1	5 565	1.1
1 000	1.18	1 000	1.24
122	1.5	225	1.5
100	1.6	100	1.8
32	2	58	2
10	3	10	3.9

Table **4.2** Performance differences detectable by the Mann-Whitney U-test for various sample sizes (runs per RTD); m_1/m_2 denotes the ratio between the medians of the two given RTDs. (The values in this table have been obtained using a standard procedure based on adjusting the statistical power of the two-sample t-test to the Mann-Whitney U-test using a worst-case Pitman asymptotic relative efficiency (ARE) value of 0.864.)

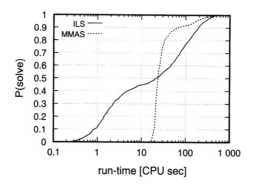

Figure 4.9 Qualified RTDs for two SLS algorithms for the TSP that, applied to a standard benchmark instance, are required to find a solution of optimal quality. The two RTDs cross over between 20 and 30 CPU seconds.

EXAMPLE 4.8   **Comparative RTD Analysis**

Figure 4.9 shows the qualified RTDs for two SLS algorithms for the TSP, $\mathcal{MAX}$–$\mathcal{MIN}$ Ant System ($\mathcal{MMAS}$) and Iterated Local Search (ILS) under the requirement of finding a solution of optimal quality for TSPLIB instance lin318 with 318 vertices, each RTD is based on 1 000 runs of the respective algorithm. Although the Mann-Whitney U-test rejects the null hypothesis that the medians of the two RTDs are equal at a significance level $\alpha = 0.05$ (the p-value is $6.4 \cdot 10^{-5}$), taking into consideration the sample size of 1 000 runs per RTD, the difference between the medians is slightly too small to be considered significant at a power of 0.8. On the other hand, the significance of the obvious differences between the two distributions is confirmed by the Kolmogorov-Smirnov test, which rejects the null hypothesis that the observed run-times for the two algorithms stem from the same distribution at a significance level of $\alpha = 0.05$ (the p-value is $\leq 2.2 \cdot 10^{-16}$).

Clearly, there is no probabilistic domination between the two algorithms. The qualified RTD curves cross over at one specific point between 20 and 30 CPU seconds, and ILS gives a higher solution probability than $\mathcal{MMAS}$ for shorter runs, whereas $\mathcal{MMAS}$ is more effective for longer runs. Both algorithms eventually find optimal solutions in all runs and hence do not show any evidence for essentially incomplete behaviour on this problem instance. Interestingly, it appears that $\mathcal{MMAS}$ has practically no chance of finding an optimal solution in less than 10 CPU seconds, while ILS finds optimal solutions with a small probability after only 0.2 CPU seconds. (This salient difference in performance is partly explained by the fact that population-based algorithms such as $\mathcal{MMAS}$ typically incur a certain overhead from maintaining multiple candidate solutions.)

Comparative Analysis for Ensembles of Instances

As previously mentioned, empirical analyses of LVA behaviour are mostly performed on ensembles of problem instances. For comparative analyses, in principle this can done by comparing the respective RTDs on each individual problem instance. Ideally, when dealing with two algorithms A and B, one would hope to observe probabilistic domination of A by B (or vice versa) on every instance of the ensemble. In practice, probabilistic domination does not always hold for all instances, and even where it holds, it may not be consistent across a given set of instances. Hence, an instance-based analysis of probabilistic domination (based on RTDs) can be used to partition a given problem ensemble into three subsets: (i) those on which A probabilistically dominates B, (ii) those on which B probabilistically dominates A and (iii) those for which probabilistic domination is not observed, that is, for which A's and B's RTDs cross each other. The relative sizes of these partitions give a rather realistic and detailed picture of the algorithms' relative performance on the given set of instances.

Statistical tests can be used to assess the significance of performance differences between two algorithms applied to the same ensemble of instances. These tests are applied to performance measures, such as mean run-time or an RTD quantile, for each algorithm on any problem instance in the given ensemble; hence, they do not capture qualitative differences in performance, particularly as given in cases where there is no probabilistic domination of one algorithm over the other. The binomial sign test as well as the Wilcoxon matched pairs signed-rank test measure whether the median of the paired differences is statistically significantly different from zero, indicating that one algorithm performs better than the other [Sheskin, 2000]. The Wilcoxon test is more sensitive, but requires the assumption that the distribution of the paired differences is symmetric. It may be noted that the widely used t-test for two dependent samples requires assumptions on the normality and homogeneity of variance of the underlying distributions of search cost over the given test-set; this test should not be used for comparing the performance of SLS algorithms, where these assumptions are typically not satisfied.

Particularly for large instance ensembles, it is often useful to refine this analysis by looking at particular performance measures, such as the median run-time, and to study the correlation between A and B w.r.t. these. For qualitative analyses of such correlations, scatter plots can be used in which each instance is represented by one point in the plot, whose coordinates correspond to the performance measure for A and B applied to that instance. Quantitatively, the correlation can be summarised using the empirical correlation coefficient. When the nature of an observed performance correlation seems to be regular (e.g., a roughly linear trend in the scatter plot), a simple regression analysis can be used to model the corresponding relationship in the algorithms' performance.

To test whether the correlation between the performance of two algorithms is significant, non-parametric tests like Spearman's rank order test or Kendall's tau test can be employed [Sheskin, 2000]. These tests determine whether there is a significant *monotonic* relationship in the performance data. They are preferable over tests based on the Pearson product-moment correlation coefficient, which require the assumption that the two random variables underlying the performance data stem from a bivariate normal distribution.

EXAMPLE 4.9 **Comparative Analysis on Instance Ensembles**

Figure 4.10 shows the correlation between the performance of an ILS algorithm and an ACO algorithm for TSP applied to a set of 100 randomly generated Euclidean TSP instances (the algorithms and problem class are described in Chapter 8). The ILS algorithm has a lower median run-time than the ACO algorithm for 66 of the 100 problem instances; this performance difference is statistically significant, because the Wilcoxon matched pairs signed-rank test rejects the null hypothesis that the performance of the two

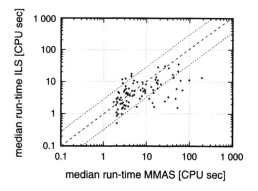

Figure **4.10** Correlation between median run-time required by $\mathcal{MMAS}$ *vs* ILS for finding the optimal solutions to instances of a set comprising 100 TSP instances with 300 vertices each; each median was measured from 10 runs per algorithm. The band between the two dashed grey lines indicates performance differences that, based on the sample size of the underlying RTDs, cannot be assumed to be statistically significant.

algorithms is equal at a significance level of $\alpha = 0.05$ (the p-value is $7 \cdot 10^{-5}$). (It may be noted that based on the sample size of 10 runs per instance that was used for the RTDs underlying each median value, performance differences of less than a factor of three can not be assumed to be statistically significant, which follows from a power analysis of the Mann-Whitney U-test that is used for assessing such performance differences, when using $\alpha = 0.05$ and a power of 0.95.)

The median run-times required for finding optimal solutions show a significant correlation (the correlation coefficient is equal to 0.39 and Spearman's rank order test rejects the null hypothesis that the performance for the two algorithms is uncorrelated at significance level $\alpha = 0.05$; the p-value is $9 \cdot 10^{-11}$), which indicates that instances that are difficult for one algorithm tend to also be difficult for the other. This suggests that similar features are responsible for rendering instances from this class of TSP instances difficult for both SLS algorithms, a hypothesis that can be investigated further through additional empirical analysis (cf. Chapter 5).

Peak Performance *vs* Robustness

Most state-of-the-art SLS algorithms have parameters (such as the noise parameter in Randomised Iterative Improvement, or the mutation and crossover rates in Evolutionary Algorithms) that need to be set manually; often, these parameter settings have a very significant impact on the respective algorithm's performance. The existence of such parameters complicates the empirical investigation of LVA behaviour significantly. This is particularly the case for comparative studies, where 'unfair parameter tuning', that is, the use of unevenly optimised parameter settings, can bring about extremely misleading results. Many comparative empirical studies of algorithms in the literature use peak performance w.r.t. parameter settings as the measure for comparing parameterised algorithms. This can be justified by viewing peak performance as a measure of potential performance; more formally, it can be seen as a tight upper bound on performance over algorithm parameterisations.

For peak performance analyses, it is important to determine optimal or close to optimal parameterisations of the respective algorithms. Since differently parameterised versions of the same algorithm can be viewed as distinct algorithms, the RTD-based approach described above can be applied. For continuous parameters, such as the noise parameter mentioned before, a series of such experiments can be used to obtain approximations of optimal values. Peak performance analysis can be very complex, especially when multiple parameters are involved whose effects are typically not independent from each other, or when dealing with complex parameters, such as the temperature schedule for Simulated Annealing, for which the domain of possible settings are extremely large and complex. In such cases, it can be infeasible to obtain reasonable approximations of optimal parameter settings; in the context of comparative studies, this situation should then be clearly acknowledged and approximately the same effort should be spent in tuning the parameter settings for every algorithm participating in a direct comparison. An alternative to hand-tuning is the use of automated parameter tuning approaches that are based on techniques from experimental design [Xu et al., 1998; Coy et al., 2001; Birattari et al., 2002].

In practice, optimal parameter settings are often not known *a priori*; furthermore, optimal parameter settings for a given algorithm can differ considerably between problem instances or instance classes. Therefore, robustness of an SLS algorithm w.r.t. suboptimal parameter settings is an important issue. This notion of robustness can be defined as the variation in an algorithm's RTD (or some of its basic descriptive statistics) caused by specific deviations from an optimal parameter setting. It should be noted that typically, such robustness measures can be easily derived from the same data that have been collected for determining optimal parameter settings.

A more general notion of robustness of an LVA's behaviour additionally covers other types of performance variation, such as the variation in run-time for a fixed problem instance and a given algorithm (which is captured in the corresponding RTD) as well as performance variations over different problem instances or domains. In all these cases, using RTDs rather than just basic descriptive statistics often gives a much clearer picture of more complex dependencies and effects, such as qualitative changes in algorithmic behaviour which are reflected in the shape of the RTD s. More advanced empirical studies should attempt to relate variation in LVA behaviour over different problem instances or domains to specific features of these instances or domains; such features can be of entirely syntactic nature (e.g., instance size), or they can reflect deeper, semantic properties. In this context, for SLS algorithms, features of the corresponding search spaces, such as density and distribution of solutions, are particularly relevant and often studied; this approach will be further discussed in Chapter 5.

4.4 Characterising and Improving LVA Behaviour

Up to this point, our discussion of the RTD-based empirical methodology has been focused on analysing specific quantitative and qualitative aspects of algorithmic behaviour as reflected in RTDs. In this section, we first discuss more advanced aspects of empirical RTD analysis. This includes the analysis of asymptotic and stagnation behaviour, as well as the use of functional approximations for mathematically characterising entire RTDs. Then, we discuss how a more detailed and sophisticated analysis of RTDs can facilitate improvements in the performance and run-time behaviour of a given Las Vegas algorithm.

Asymptotic Behaviour and Stagnation

In Section 4.1, we defined various norms of LVA behaviour. It is easy to see that all three norms of behaviour — completeness, probabilistic approximate completeness (PAC property) and essential incompleteness — correspond to properties of the given algorithm's theoretical RTDs. For complete algorithms, the theoretical cumulative RTDs will reach one after a bounded time (where the bound depends on instance size). Empirically, for a given time bound, this property can be falsified by finding a problem instance on which at least one run of the algorithm did not produce a solution within the respective time bound. However, it should be clear that a completeness hypothesis can never be verified experimentally, since the instances for which a given bound does not hold might be very rare, and the probability for producing longer runs might be extremely small.

SLS algorithms for combinatorial problems are often incomplete, or in the case of complete SLS algorithms, the time bounds are typically too high to be of any practical relevance. There are, however, in many cases empirically observable and practically significant differences between essentially incomplete and PAC algorithms [Hoos, 1999a]. Interestingly, neither property can be empirically verified or falsified. For an essentially incomplete algorithm, there exists a problem instance for which the probability of not finding a solution in an arbitrarily long run is greater than zero. Since only finite runs can be observed in practice, arbitrarily long unsuccessful runs could hypothetically always become successful after the horizon of observation. On the other hand, even if unsuccessful runs are never observed, there is always a possibility that the failure probability is just too small compared to the number of runs performed, or the instances on which true failure can occur are not represented in the ensemble of instances tested. However, empirical run-time distributions can provide evidence for (rather than proof of) essential incompleteness or PAC behaviour and hence provide the basis for hypotheses which, in some cases, can then be proven by theoretical analyses. Such evidence primarily takes the form of an apparent limiting success probability that is asymptotically approached by a given empirical RTD .

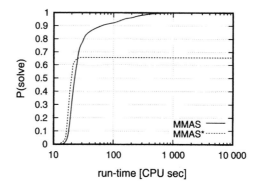

run-time [CPU sec]

Figure **4.11** Qualified RTDs for two SLS algorithms for the TSP that are required to find an optimal solution of a well-known benchmark instance; $\mathcal{MMAS}$ is provably PAC, whereas $\mathcal{MMAS}^*$ is an essentially incomplete variant of the same algorithm (see text for details). Each RTD is based on 1 000 independent runs of the respective algorithm.

EXAMPLE 4.10 **Asymptotic Behaviour in Empirical RTDs**

Figure 4.11 shows the qualified RTDs for two variants of an ACO algorithm required to find an optimal solution for TSPLIB instance lin318 with 318 vertices. The RTD for $\mathcal{MMAS}^*$ shows severe stagnation behaviour; after 26 CPU seconds, the probability for finding a solution does not increase any further, and up to 10 000 CPU seconds, not a single additional solution is found. This provides strong evidence (but no proof) that $\mathcal{MMAS}^*$ is essentially incomplete. Conversely, all 1 000 runs of $\mathcal{MMAS}$ were successful and the underlying RTD appears to asymptotically approach one, suggesting that $\mathcal{MMAS}$ is probabilistically approximately complete. In fact, $\mathcal{MMAS}$, a slight extension of $\mathcal{MMAS}^*$, is provably PAC, while $\mathcal{MMAS}^*$ is essentially incomplete. The two algorithms differ only in the key feature that renders $\mathcal{MMAS}$ PAC [Stützle and Dorigo, 2002] (details on $\mathcal{MMAS}$ can be found in Chapter 8, Section 8.4).

In practice, true asymptotic behaviour (such as probabilistic approximate completeness) is less relevant than the rate at which the failure probability of a given LVA decreases over time. Intuitively, a drop in this rate indicates a stagnation in the algorithm's progress towards finding solutions of the given problem instance. Here, we adopt a slightly different view of stagnation, which turns out to be consistent with the intuition described before. This view is based on the fact that in many cases, the probability of obtaining a solution of a given problem instance by using a particular Las Vegas algorithm can be increased by restarting the algorithm after a fixed amount of time (the so-called cutoff time) rather than letting it run longer and longer. Whether or not such a static restart strategy yields the desired improvement depends entirely on the respective RTD, and it is easy to see that only for RTDs identical to exponential distributions (up to discretisation effects), static restart does not result in any performance loss or improvement [Hoos and Stützle, 1999].

Exponential RTDs are characterised by a constant rate of decay in their right tail, which corresponds to the failure probability, a measure of the probability that the given algorithm fails to find an existing solution of a given problem instance within a given amount of time. When augmenting any LVA with a static restart mechanism, the resulting algorithm will show RTDs with exponentially decaying right tails. Based on this observation, efficiency and stagnation can be measured by comparing the decay rate of the failure probability at time t, denoted $\lambda(t)$, with the tail decay rate obtained when using static restarts with cutoff t, denoted $\lambda^*(t)$. This leads to the following definition:

DEFINITION 4.9 **LVA Efficiency and Stagnation**

Let A be a Las Vegas algorithm for a given combinatorial problem Π, and let $rtd_{A,\pi}(t)$ be the cumulative run-time distribution function of A applied to a problem instance $\pi \in \Pi$.

Then we define $\lambda_{A,\pi}(t) := -d/dt[\ln(1 - rtd_{A,\pi})](t) = 1/(1 - rtd_{A,\pi}(t)) \cdot d/dt[rtd_{A,\pi}](t)$, where $d/dt[f]$ denotes the first derivative of a function f in t. Furthermore, we define $\lambda^*_{A,\pi}(t) := -\ln(1 - rtd_{A,\pi}(t))/t$.

The efficiency of A on π at time t is then defined as $eff_{A,\pi}(t) := \lambda_{A,\pi}(t)/ \lambda^*_{A,\pi}(t)$. Similarly, the stagnation ratio of A on π at time t is defined as $stagr_{A,\pi}(t) := 1/eff_{A,\pi}(t)$, and the stagnation of A on π at time t is given by $stag_{A,\pi}(t) := \ln(stagr_{A,\pi}(t))$.

Finally, we define the minimal efficiency of A on π as $eff_{A,\pi} := \inf\{eff_{A,\pi}(t) \mid t > 0\}$ and the minimal efficiency of A on a problem class Π as $eff_{A,\Pi} := \inf\{eff_{A,\pi} \mid \pi \in \Pi\}$. The maximum stagnation ratio and maximum stagnation on problem instances and problem classes are defined analogously.

Remark: For empirical RTDs, the decay rates $\lambda_{A,\pi}(t)$ are approximated using standard techniques for numerical differentiation of discrete data such that artifacts due to discretisation effects are avoided as much as possible.

It is easy to see that according to the definition, for any essentially incomplete algorithm A there are problem instances on which the minimal efficiency of

A is zero. Constant minimal efficiency of one is observed if, and only if, the corresponding RTD is an exponential distribution. LVA efficiency greater than one indicates that restarting the algorithm rather then letting it run longer would result in a performance loss; this situation is often encountered for SLS algorithms during the initial search phase.

It should be clear that our measure of LVA efficiency is a relative measure; hence, the fact that a given algorithm has high minimal efficiency does *not* imply that this algorithm cannot be further improved. As a simple example, consider Uninformed Random Picking as introduced in Chapter 1, Section 1.5; this primitive search algorithm has efficiency one for arbitrary problem instances and run-times, yet there are many other SLS algorithms which perform significantly better than Uninformed Random Picking, some of which have a smaller minimal efficiency. Hence, LVA efficiency as defined above cannot be used to determine the optimality of a given Las Vegas algorithm's behaviour in an absolute way. Instead, it provides a quantitative measure for relative changes in efficiency of a given LVA over the course of its run-time. (However, the definition can easily be extended such that an absolute performance measure is obtained; this is done by using the restart decay rate λ^* over a set of algorithms instead of $\lambda^*_{A,\pi}$ in the definition of LVA efficiency.)

Functional Characterisation of LVA Behaviour

Obviously, any empirical RTD, as obtained by running a Las Vegas algorithm on a given problem instance, can be completely characterised by a function — a step function that can be derived from the empirical RTD data in a straightforward way. Typically, if an empirical RTD is a reasonably precise approximation of the true RTD (i.e., if the number of runs underlying the empirical RTD is sufficiently high), this step function is rather regular and can be approximated well using much simpler mathematical functions.

Such approximations are useful for summarising the observed algorithmic behaviour as reflected in the raw empirical RTD data. But more importantly, they can provide the basis for modelling the observed behaviour mathematically, which is often a key step in gaining deeper insights into an algorithm's behaviour. It should be noted that this general approach is commonly used in other empirical disciplines and can be considered one of the fundamental techniques in science.

In the case of empirical RTDs, approximations with parameterised families of continuous probability functions known from statistics, such as exponential or normal distributions, are particularly useful. Given an empirical RTD and a parameterised family of cumulative probability functions, good approximations can be found using standard model fitting techniques, such as the Marquart-Levenberg algorithm [Marquardt, 1963] or the expectation maximisation (EM) algorithm [Dempster et al., 1977]. The quality of the approximation thus obtained can be assessed using standard statistical goodness-of-fit tests, such as the well-known χ^2-test or the Kolmogorov-Smirnov test [Sheskin, 2000]. Both of these tests are used to decide if a sample comes from a population with a specific distribution. While the Kolmogorov-Smirnov test is restricted to continuous distributions, the χ^2 goodness-of-fit test can also be applied to discrete distributions.

EXAMPLE 4.11 **Functional Approximation of Empirical RTDs**

Looking at the empirical RLD of WalkSAT/SKC applied to a hard Uniform Random 3-SAT instance shown in Figure 4.6 (page 172), one might notice that the RLD graph resembles that of an exponential distribution. This leads to the hypothesis that on the given problem instance, the algorithm's behaviour can be characterised by an exponential RLD. To test this hypothesis, we first fit the RLD data with a cumulative exponential distribution function of the form $ed[m](x) := 1 - e^{x/m}$, using the Marquart-Levenberg algorithm (as realised in C. Gramme's Gnufit software) to determine the optimal value for the parameter m. This approximation is shown in Figure 4.12 (left side).

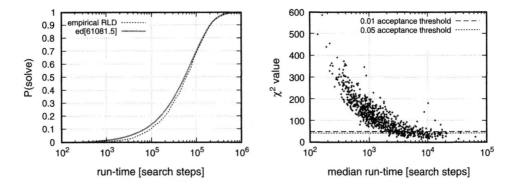

Figure **4.12** *Left:* Best-fit approximation of the RLD from Figure 4.6 (page 172) by an exponential distribution; this approximation passes the χ^2 goodness-of-fit test at significance level $\alpha = 0.05$. *Right:* Correlation between median run-length and χ^2 values from testing RLDs of individual instances versus a best-fit exponential distribution for a test-set of 1 000 hard Random 3-SAT instances; the horizontal lines indicate the acceptance thresholds for the 0.01 and 0.05 acceptance levels of the χ^2-test.

Then, we applied the χ^2 goodness-of-fit test to examine the hypothesis whether the resulting exponential distribution is identical to the theoretical RTD underlying the empirically observed run-lengths. In the given example, the resulting χ^2 value of 26.24 indicates that our distribution hypothesis passed the test at a standard significance level $\alpha = 0.05$.

It is worth noting that, since Las Vegas algorithms (like all algorithms) are of an inherently discrete nature, their true (theoretical) RTDs are always step functions. However, there are good reasons for the use of continuous probability functions for approximation. For increasing problem sizes, these step functions will become arbitrarily detailed — an effect which, especially for computationally hard problems, such as SAT or TSP, becomes relevant even for relatively modest and certainly realistically solvable problem sizes. Furthermore, abstracting from the discrete nature of RTDs often facilitates a more uniform characterisation that is mathematically easier to handle. However, for 'very easy' problem instances, that is, instances that can be solved by a given algorithm in tens or hundreds of basic operations or CPU cycles, the discrete nature of the respective true RTDs can manifest itself — an effect which needs to be taken into account when fitting parameterised functions to such data and testing the statistical significance of the resulting approximations.

Functional Characterisation for Instance Ensembles

Like the previous RTD-based analytical approaches, the functional characterisation of LVA behaviour can be extended from single problem instances to ensembles of instances in a rather straightforward way. For small instance sets, it is generally feasible to perform the approximation and goodness-of-fit test for each instance as described above; for larger ensembles, it becomes necessary to automate this procedure, and to analyse and summarise its results in an appropriate way. Overall, similar considerations apply as described in the previous section.

Using this approach, hypotheses on the behaviour of a given LVA on classes or distributions of problem instances can be tested. Hypotheses on an LVA's behaviour on infinite or extremely large sets of instances, such as the set of all SAT instances with a given number of clauses and variables, cannot be proven by this method; however, it allows one to falsify such hypotheses or to collect arbitrary amounts of evidence for their validity.

EXAMPLE 4.12 **Functional RTD Approximation for Instance Ensembles**

A simple generalisation from the result presented in the previous example results in the hypothesis that for an entire class of SAT instances WalkSAT/SKC's behaviour can be characterised by exponential run-time distributions. Here, we test this hypothesis for a set of 1 000 Uniform Random 3-SAT instances with 100 variables and 430 clauses. By fitting the RLD data for the individual instances with exponential distributions and calculating the χ^2 values as outlined above, we obtained the result shown on the right side of Figure 4.12 (page 189), which shows the median values of the RLDs plotted against the corresponding χ^2 values: Although, for most instances, the distribution hypothesis is rejected, we observe a clear correlation between the solution cost of the instances and the χ^2 values, and for almost all of the hardest instances, the distribution hypothesis passes the test. Thus, although our original generalised hypothesis could not be confirmed, the results suggest an interesting modification of this hypothesis. (Further analysis of the easier instances, for which the RLDs could not be well approximated by exponential distributions, shows that there is a systematic deviation in the left tail of the RLDs, while the right tail matches that of an exponential distribution; details on this result can be found in Hoos and Stützle [1999; 2000a].)

This functional characterisation approach can also be used for analysing and modelling the dependency of LVA behaviour on algorithmic parameters or properties of problem instances (in particular, problem size). Furthermore, it facilitates comparative studies of the behaviour of two or more LVA algorithms. In all of these cases, reasonably simple, parameterised models of the algorithms' run-time behaviour provide a better basis for the respective analysis than the basic properties and statistics of RTDs discussed before. For example, when studying the scaling of an algorithm's run-time behaviour with problem size, knowledge of good parameterised functional approximations of the RTDs reduces the investigation to an analysis of the impact of problem size on the model parameters (e.g., the median of an exponential distribution).

As we will see in the following, such characterisations can also have direct consequences for important issues such as parallelisation or optimal parameterisation of Las Vegas algorithms. At the same time, they can suggest novel interpretations of LVA behaviour and thus facilitate an improved understanding of these algorithms.

Optimal Cutoff Times for Static Restarts

A detailed analysis of an algorithm's RTDs, particularly with respect to asymptotic behaviour and stagnation, can often suggest ways of improving the performance of the algorithm. Arguably the simplest way to overcome stagnation of an SLS algorithm is to restart the search after a fixed amount of time (cutoff time). Generally, based on our definition of search efficiency and stagnation, it is easy to decide whether such a *static restart strategy* can improve the performance of a Las Vegas algorithm A for a given problem instance π. If for all run-times t, the efficiency of A on π at time t, $\mathit{eff}_{A,\pi}(t)$, is larger than one, restart with any cutoff-time t will lead to performance loss. Intuitively, this is the case when with increasing t, the probability of finding a solution within a given time interval increases, which is reflected in an cumulative RTD graph that is steeper at t than the exponential distribution $ed[m]$ for which $ed[m](t) = rtd_{A,\pi}(t)$ (an example for such an RTD is shown in Figure 4.13). Furthermore, if, and only if, $\mathit{eff}_{A,\pi}(t) = 1$ for all t, restart at any time t will not change the success probability for any time t'; as mentioned in Section 4.3, this condition is satisfied if, and only if, the RTD of A on π is an exponential distribution. Finally, if there exists a run-time t' such that $\mathit{eff}_{A,\pi}(t) \leq 1$ for all $t > t'$, then restarting the algorithm at time t will lead to an increased solution probability for some run-time $t'' > t$. This is equivalent to the condition that from t' on the cumulative RTD graph of A on π is less steep for any time $t > t'$ than the exponential distribution $ed[m]$ for which $ed[m](t) = rtd_{A,\pi}(t)$.

In the case where random restart is effective for some cutoff-time t', an optimal cutoff time t_{opt} can intuitively be identified by finding the 'left-most' exponential distribution, $ed[m^*]$, that touches the RTD graph of A on π, and

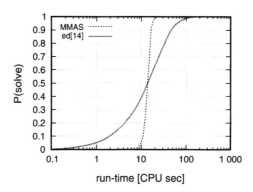

Figure **4.13** Qualified RTD of an ACO algorithm for TSP ($\mathcal{MMAS}$) on TSPLIB instance lin318 with 318 vertices, based on 1 000 independent runs, and exponential distribution with identical median. The fact that this RTD is consistently steeper than an exponential indicates that restart with any fixed cutoff time will lead to performance loss.

the minimal t for which $ed[m^*](t) = rtd_{A,\pi}(t)$. Formally, this is achieved using the following definitions:

$$m^* \quad := \quad \min\{m \mid \exists t > 0 : ed[m](t) = rtd_{A,\pi}(t)\} \tag{4.2}$$

$$t_{opt} \quad := \quad \min\{t \mid t > 0 \wedge ed[m^*](t) = rtd_{A,\pi}(t)\} \tag{4.3}$$

where $rtd_{A,\pi}(t)$ is the theoretical run-time distribution of A on π, and A is incomplete, that is, $P_s(RT \leq t) < 1$ for any finite run-time t (note that A may still be probabilistically approximately complete).

Generally, there are two special cases to be considered when solving these two equations. Firstly, we might not be able to determine m^* because the set over which we minimise in the first equation has no minimum. In this case, if the infimum of the set is zero, it can be shown that the optimal cutoff time is either equal to zero, or it is equal to $+\infty$ (depending on the behaviour of t_{opt} as m^* approaches zero). Secondly, if m^* as defined by the first equation exists, it might still not be possible to determine t_{opt}, because the set in the second equation does not have a minimum. In this case, there are arbitrarily small times t for which $ed[m^*](t) = rtd_{A,\pi}(t)$, that is, the two curves are identical on some interval $[0, t']$, and the optimal cutoff time is equal to zero. In practice, optimal cutoff times of zero will hardly occur, since they could only arise if A would solve π with probability larger than zero for infinitesimally small run-times.

Equations 4.2 and 4.3 apply to theoretical as well as to empirical RTDs. In the latter case, however, it is sufficient to consider only run-times t in Equations 4.2 and 4.3 that have been observed in one of the runs underlying the empirical RTD. There is one caveat with this method: cases in which the optimal cutoff time determined from Equation 4.3 is equal to one of the longest run-times underlying the given empirical RTD should be treated with caution. The reason for this lies in the fact that the high quantiles of empirical RTDs, which correspond to the longest runs, are often rather statistically unstable. Still, using cutoffs based on such extreme run-times may be justified if there is evidence that the algorithm shows stagnation behaviour.

In the case of SLS algorithms for optimisation problems, optimal cutoff times are determined from qualified RTDs. Clearly, such optimal cutoff times depend on the solution quality bound. In many cases, tighter solution quality bounds (i.e., bounds that are closer to the optimal solution quality) lead to higher optimal cutoff times; yet, for weak solution quality bounds, restart with any cutoff time typically leads to performance loss.

EXAMPLE 4.13 **Determining Optimal Cutoff Times for Static Restarts**

Figure 4.14 shows the empirical qualified RTD of a simple ILS algorithm for the TSP for finding optimal solutions to TSPLIB instance pcb442 with

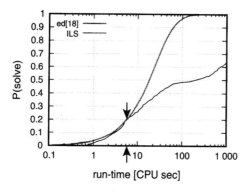 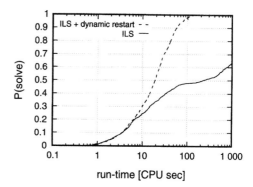

Figure **4.14** Qualified RTD for an ILS algorithm required to find optimal solutions for TSPLIB instance pcb442; note the stagnation behaviour apparent from the RTD graph. *Left:* Optimal cutoff time for static restarts, t_{opt}, and corresponding exponential distribution $ed[m^*]$. *Right:* Effect of dynamic restart strategy. (Details are given in the text.)

$n = 442$ vertices. The algorithm was run 1 000 times on a Pentium 700MHz machine with 512MB RAM, and unsuccessful runs were terminated after 1 000 CPU seconds. This qualified RTD shows strong stagnation behaviour; note that this behaviour could not have been observed when limiting the maximal run-time of the algorithm to less than 5 CPU seconds. Figure 4.14 shows the optimal cutoff time for static restarts, t_{opt}, and the corresponding exponential distribution $ed[m^*]$, determined according to Equations 4.2 and 4.3. The same exponential distribution characterises the shape of the RTD for the algorithm using static restarts with cutoff time t_{opt}.

Dynamic Restarts and Other Diversification Strategies

One drawback of using a static restart strategy lies in the fact that optimal cutoff times typically vary considerably between problem instances. Therefore, it would be preferable to re-initialises the search process not after a fixed cutoff time, but depending on search progress. A simple example of such a *dynamic restart strategy* is based on the time that has passed since the current incumbent candidate solution was found; if this time interval exceeds a threshold θ, a restart is performed. (In this scheme, incumbent candidate solutions are not carried over restarts of the search.) The time threshold θ is typically measured in search steps; θ corresponds to the minimal time interval between restarts and is often defined depending on syntactic properties of the given problem instance, in particular, instance size.

EXAMPLE 4.14 **Improving SLS Behaviour Using Dynamic Restarts**

> Figure 4.14 (right) shows the effect of the simple dynamic restart strategy described above on the ILS algorithm and TSP instance from Example 4.13. Here, for a TSP instance with n vertices, $\theta := n$ is used as the minimal time-interval between restarts. Interestingly, the RTD of ILS with this dynamic restart mechanism is basically identical to the RTD of ILS with static restart for the optimal cutoff-time determined in the previous example. This indicates that the particular dynamic restart mechanism used here is very effective in overcoming the stagnation behaviour of the ILS algorithm without restart.

Restarting an SLS algorithm from a new initial solution is typically a rather time-consuming operation. Firstly, a certain *setup time* is required for generating a new candidate solution from which the search is started and for initialising the data structures used by the search process accordingly. This setup time is often substantially higher than the time required for performing a search step. Secondly, after initialising the search process, SLS algorithms almost always require a certain number of search steps to reach regions of the underlying search space in which there is a non-negligible chance of finding a solution. These effects are reflected in extremely low success probabilities in the extreme left tail of the respective RTDs. Furthermore, they typically increase strongly with instance size, rendering search restarts a costly operation.

These disadvantages can be avoided by using diversification techniques that are less drastic than restarts in order to overcome stagnation behaviour. One such technique called *fitness-distance diversification* has been used to enhance the ILS algorithm for the TSP mentioned in Example 4.13; the resulting algorithm shows substantially better performance than the variant using dynamic restarts from Example 4.14. (Details on this enhanced ILS algorithm can be found in Chapter 8, page 396.)

Another diversification technique that also has the theoretical advantage of rendering the respective SLS algorithm probabilistically approximately complete (PAC), is the so-called *random walk extension* [Hoos, 1999a]. In terms of the GLSM models of the respective SLS algorithms, the random walk extension consists of adding a random walk state in such a way that throughout the search, arbitrarily long sequences of random walk steps can be performed with some (small) probability. This technique has been used to obtain state-of-the-art SLS algorithms for SAT, such as Novelty$^+$ (for details, see Chapter 6). Generally, effective techniques for overcoming search stagnation are important components of advanced SLS methods, and improvements in these techniques can be expected to play a major role in designing future generations of SLS algorithms.

Multiple Independent Runs Parallelisation

Las Vegas algorithms lend themselves to a straightforward parallelisation approach by performing independent runs of the same algorithm in parallel. From the discussion in the previous sections we know that if an SLS algorithm has an exponentially distributed RTD, such a strategy is particularly effective. Based on a well-known result from the statistical literature [Rohatgi, 1976], if for a given algorithm the probability of finding a solution in t time units is exponentially distributed with median m, then the probability of finding a solution in at least one of p independent runs of time t each is exponentially distributed with median m/p. Consequently, if we run such an algorithm once for time t, we obtain exactly the same success probability as when running the algorithm p times for time t/p. By executing these p independent runs in parallel on p processors, an optimal *parallelisation speedup* $S_p := RT_1/RT_p = p$ is achieved, where $RT_1 = t$ is the sequential run-time and $RT_p = t/p$ is the parallel computation time, using p processors. This theoretical result holds for arbitrary numbers of processors.

In practice, SLS algorithms do not have perfectly exponential RTDs; as explained previously, there are typical deviations in the left tail which reflect the setup time and initial search phase. Therefore, when the number of processors is high enough that each of the parallel runs becomes very short, the parallelisation speedup will generally be less than optimal. Given an empirical RTD, the parallelisation speedup S_p for reaching a certain success probability p_s can be calculated as follows. RT_1, the sequential run-time required for reaching a solution probability p_s, can be directly determined from the given RTD; technically, $RT_1 := \min\{t' \mid \widehat{P_s}(RT \leq t') \geq p_s\}$. Then the parallel time required for reaching the same solution probability by performing multiple independent runs on p processors is given by

$$RT_p := \min\{t' \mid \widehat{P_s}(RT \leq t') \geq 1 - (1 - p_s)^{1/p}\} \qquad (4.4)$$

Using this equation, the minimal number of processors required for achieving the desired success probability within a maximal accumulated parallel run-time t_{max} can be easily determined. (The accumulated parallel run-time is the total run-time over all processors.) It is interesting to note that for higher success probabilities, the maximal number of processors for which optimal parallelisation can be achieved is typically also higher.

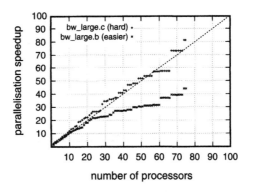

Figure **4.15** Speedup achieved by multiple independent runs parallelisation of a high–performing SLS algorithm for SAT applied to two SAT-encoded instances of a hard planning problem. The diagonal line indicates optimal parallelisation speedup. Note that for the easier instance, the parallelisation speedup is increasingly suboptimal for more than 10 processors. (For details, see text.)

EXAMPLE 4.15 **Speedup Through Independent Parallel Runs**

Figure 4.15 shows the parallelisation speedup S_p as a function of the number of processors (computed using Equation 4.4) for a high-performance SLS algorithm for SAT (Novelty) applied to two well-known benchmark instances for SAT, the SAT-encoded planning problems bw_large.b and bw_large.c. The underlying empirical RTDs (determined using instance-specific optimal noise parameter settings of Novelty) are based on 250 successful runs each, and all points of the speedup curves are based on no fewer than ten runs. A desired success probability of $p_s = 0.95$ was used for determining the sequential and parallel run-times.

Instance bw_large.c is much harder than bw_large.b, and allows approx. optimal speedup for more than 70 processors; the underlying RTD is almost perfectly approximated by an exponential distribution. For the easier instance, the parallelisation speedup becomes suboptimal for more than 10 processors; this is due to the larger relative impact of the setup time and initial search phase on overall run-time.

Generally, using multiple independent runs is an attractive model of parallel processing, since it involves basically no communication overhead and can be easily implemented for almost any parallel hardware and programming environment, from networks of standard workstations to specialised multiple instruction / multiple data (MIMD) machines with thousands of processors. The resulting parallel SLS algorithms are precisely captured by the homogeneous co-operative GLSM model without communication introduced in Chapter 3. They are of particular interest in the context of SLS applications to time-critical tasks (such as robot control or on-line scheduling), as well as to the distributed solving of very large and hard problem instances.

4.5 Further Readings and Related Work

The term *Las Vegas algorithm* was originally introduced by Babai [1979]. Although the concept is widely known, the literature on Las Vegas algorithms is relatively sparse. Luby, Sinclair and Zuckerman have studied optimal strategies for selecting cutoff times [Luby et al., 1993]; closely related theoretical work on the parallelisation of Las Vegas algorithms has been published by Luby and Ertel [1994]. The application scenarios for Las Vegas algorithms and norms of LVA behaviour covered here have been introduced by Hoos and Stützle [1998].

Run-time distributions have been occasionally observed in the literature for a number of years [Taillard, 1991; Battiti and Tecchiolli, 1992; Taillard, 1994; ten Eikelder et al., 1996]. Their use, however, has been typically restricted to purely descriptive purposes or to obtaining hints on the speedup achievable by performing independent parallel runs of a given sequential algorithm [Battiti and Tecchiolli, 1992; Taillard, 1991]. Taillard specifies general conditions under which super-optimal speedups can be achieved through multiple independent tries parallelisation [Taillard, 1994]. The use of RTDs at the core of an empirical methodology for studying SLS algorithms was first proposed by Hoos and Stützle [1998]. Since then, RTD-based methods have been used for the empirical study of a broad range of SLS algorithms for numerous combinatorial problems [Aiex et al., 2002; Hoos and Stützle, 2000a; Hoos and Boutilier, 2000; Stützle and Hoos, 2001; Stützle, 1999; Tulpan et al., 2003].

There is some related work on the use of search cost distributions over instance ensembles for the empirical analysis of complete search algorithms. Kwan showed that for different types of random CSP instances, the search cost distributions for several complete algorithms cannot be characterised by normal distributions [Kwan, 1996]. Frost, Rish and Vila use continuous probability distributions for approximating the run-time behaviour of complete algorithms applied to randomly generated Random 3-SAT and binary CSPs from the phase transition region [Frost et al., 1997]. In Rish and Frost [1997], this approach is extended to search cost distributions for unsolvable problems from the over-constrained region.

Gomes and Selman studied run-time distributions of backtracking algorithms based on the Brelaz heuristic for solving instances of the Quasigroup Completion

Problem, a special type of CSP, in the context of algorithm portfolios design [Gomes and Selman, 1997a]. Interestingly, the corresponding RTDs for the randomised systematic search algorithms they studied can (at least in some cases) be approximated by 'heavy-tailed' distributions, a fact which can be exploited for improving the performance of these algorithms by using a static restart mechanism [Gomes et al., 1997]. Similar results have been obtained for randomised complete algorithms for SAT; at the time, the resulting algorithms showed state-of-the-art performance on many types of SAT instances [Gomes et al., 1998]. Interestingly, the RTDs for some of the most widely known and best-performing SLS algorithms for SAT appear to be well approximated by exponential distributions [Hoos, 1998; Hoos and Stützle, 1999; Hoos, 1999a] or mixtures of exponentials [Hoos, 2002b]. To our best knowledge, heavy-tailed RTDs have generally not been observed for any SLS algorithm.

A number of specific techniques have proven to be useful in the context of certain types of experimental analyses. Estimates for optimal solution qualities for combinatorial optimisation problems can be obtained using techniques based on insights from mathematical statistics [Dannenbring, 1977; Golden and Steward, 1985]. Using solution quality distributions, interesting results have been obtained regarding the behaviour of SLS algorithms as instance size increases [Schreiber and Martin, 1999]. Techniques from experimental design were shown to be helpful in deriving automated (or semi-automated) procedures for tuning algorithmic parameters [Xu et al., 1998; Coy et al., 2001; Birattari et al., 2002].

Various general aspects of empirical algorithms research are covered in a number of publications. There have been several early attempts to provide guidelines for the experimental investigation of algorithms for combinatorial optimisation problems and to establish reporting procedures that improve the reproducibility of empirical results [Crowder et al., 1979; Jackson et al., 1990]. Guidelines on how to report results that are more specific to heuristic methods, including SLS algorithms, are given in Barr et al. [1995].

Hooker advocates a scientific approach to experimental studies in operations research and artificial intelligence [Hooker, 1994]; this approach is based on the formulation and careful experimental investigation of hypotheses about algorithm properties and behaviour. General guidelines for the experimental analysis of algorithms are also given by McGeoch and Moret [McGeoch, 1996; McGeoch and Moret, 1999; Moret, 2002]. A recent article by Johnson provides an extensive collection of guidelines and potential pitfalls in experimental algorithms research, including some very practical advice on the topic [Johnson, 2002]. Gent et al. give a similar, but more limited, overview of potential problems in the experimental analysis of algorithms [Gent et al., 1997].

Statistical methods are at the core of any empirical approach to investigate the behaviour and the performance of SLS algorithms. Cohen's book on empirical methods in artificial intelligence is becoming a standard text and reference book for the presentation and application of statistical methods not only in AI but also in other fields of computer science [Cohen, 1995]. For an additional introduction to statistical methods we also recommend the book by Papoulis [1991]. The handbook by Sheskin [2000] is an excellent guide to statistical tests and their proper application; a more specialised introduction to non-parametric statistics can be found in Conover [1999] and Siegel et al. [1988]. Furthermore, for general techniques of experimental design and the analysis of experimental data we refer to the work of Dean and Voss [2000] and Montgomery [2000].

4.6 Summary

Empirical methods play a crucial role in analysing the performance and behaviour of SLS algorithms, and appropriate techniques are required for conducting empirical analyses competently and correctly. In this chapter, we motivated why *run-time distributions (RTDs)* provide a good basis for empiricially analysing the behaviour of SLS algorithms and more generally, members of the broader class of *(generalised) Las Vegas algorithms*. We discussed the asymptotic behaviour of Las Vegas algorithms and introduced three application scenarios with different requirements for empirical performance analyses. We then introduced formally the concepts of *run-time distributions (RTDs)*, *qualified run-time distributions (QRTDs)* and *solution quality distributions (SQDs)*, as well as *time-dependent solution quality statistics (SQTs)* and *solution quality dependent run-time statistics (RTQs)*. Empirical RTDs can be easily obtained from the same data required for stable estimates of mean run-times or time-dependent solution quality. We presented and discussed RTD-based methods for the empirical analysis of individual LVAs as well as for the comparative analysis of LVAs, on single problem instances and instance ensembles. We also contrasted *peak-performance* and *robustness analysis* and argued that the latter is important to capture dependencies of an algorithm's performance on parameter settings, problem instances or instance size. The measures of *efficiency* and *stagnation* are derived from a given RTD and characterise an algorithm's performance over time; intuitively, these measures indicate how much an algorithm's performance can be improved by a *static restart mechanism*.

Functional approximations of RTDs with known probability distributions can be used to summarise and mathematically model the behaviour of Las Vegas algorithms. The regularities of LVA behaviour captured by such functional characterisations can facilitate performance analysis, for example, by suggesting simplified experimental designs in which only the parameter values of a functionally characterised family of RTDs are analysed instead of the entire distributions. Applied to SLS algorithms, this approach can also reveal fundamental properties of the algorithm and provide deeper insights into its behaviour.

Results from the empirical analysis of an SLS algorithm can provide significant leverage for further improvement of its performance. We gave an overview of various approaches to achieving such improvements, including *static and dynamic restart mechanisms, adaptive diversification, random walk extension* and *parallelisation based on multiple independent tries*.

Overall, the importance of empirical analyses in the development and application of SLS algorithms can hardly be overestimated. We believe that the methods and techniques presented in this chapter provide a solid basis for sound and thorough empirical studies on SLS algorithms and thus facilitate the development of better algorithms and an improved understanding of their characteristics and behaviour.

Exercises

4.1   [*Easy*] Give three examples for (generalised) Las Vegas algorithms and identify all stochastic elements in these.

4.2   [*Easy*] Describe a concrete application domain where the utility of a solution to a given problem instance changes over time.

4.3   [*Medium*] Prove that Uninformed Random Picking (see Chapter 1, Section 1.5) has the PAC property.

4.4   [*Easy*] Explain the difference between a run-time distribution (RTD), a solution quality distribution (SQD) and a search cost distribution (SCD).

4.5   [*Medium*] In order to investigate the behaviour of an SLS algorithm for a combinatorial optimisation problem on a given problem instance, solution quality traces over m independent runs are recorded. In each of these runs, the known optimal solution quality for the given instance is reached. Explain how qualified run-time distributions (RTDs) for various solution quality bounds and solution quality distributions (SQDs) for various run-time bounds can be obtained from these solution quality traces.

4.6   [*Easy; Hands-On*] Study the behaviour of a simple iterated local search algorithm for the TSP (available from `www.sls-book.net`) on TSPLIB instance lin318 (available from TSPLIB [Reinelt, 2003]). In particular, report and compare the solution quality-distributions (SQDs) for increasingly high run-time bounds. (The provably optimal solution quality for this instance is 42 029.) Describe how the SQDs change with the run-time bounds and explain the reasons underlying this phenomenon.

4.7   [**Medium**] You are comparing the performance of two SLS algorithms A and B for a combinatorial decision problem. Applied to a well-known benchmark instance, these algorithms were found to exhibit the RTDs shown below.

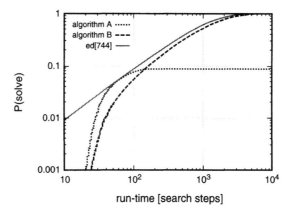

What do you learn from these RTDs? Which further experiments do you suggest to decide which algorithm is superior?

4.8   [**Medium**] Explain why it is desirable to mathematically model observed RTDs using functional approximations. Do the approximations have to be perfect to be useful?

4.9   [**Medium**] What happens if the equations used for determining optimal restart times according to Equations 4.2 and 4.3 (page 193) are applied to a complete Las Vegas algorithm?

4.10 [**Hard**] Outline an empirical approach to run and decide a competition on the best SLS algorithm for the TSP. Discuss all relevant aspects of the competition (selection of problem instances, performance measures, experimental protocol) and justify your approach.

5 SEARCH SPACE STRUCTURE AND SLS PERFORMANCE

The performance of SLS algorithms crucially depends on structural aspects of the spaces being searched. Studying the nature of this dependency can significantly improve our understanding of SLS behaviour and facilitate the further improvement and successful application of SLS methods.

In this chapter, we introduce various aspects of search space structure and discuss their impact on SLS performance. These include fundamental properties of a given search space and neighbourhood graph, such as size, connectivity, diameter and solution density, as well as global and local properties of the search landscapes encountered by SLS algorithms, such as the number and distribution of local minima, fitness distance correlation, measures of ruggedness, and detailed information on the plateau and basin structure of the given space.

Some of these search space features can be determined analytically, but most have to be measured empirically, often involving rather complex search methods. We exemplify the type of results obtainable from such analyses of search space features and their impact on SLS performance for our standard example problems, SAT and TSP.

5.1 Fundamental Search Space Properties

The search process carried out by any SLS algorithm when applied to a given problem instance π can be seen as a walk on the neighbourhood graph associated with π, $G_N(\pi)$. Recall from Chapter 1, Section 1.5 that $G_N(\pi) := (S(\pi), N(\pi))$, where $S(\pi)$ is the search space of π, that is, the set of all candidate solutions, and $N(\pi)$ is the given neighbourhood relation. Obviously, the properties of the search space and the corresponding neighbourhood graph have an impact on the behaviour and performance of SLS algorithms.

> **Remark:** For simplicity and generality, in this chapter we mainly use the term *search position* (or short: *position*) to refer to candidate solutions; other equivalent terms that are often found in the literature are *state* or *configuration*.

Search Space Size and Diameter

It is intuitively clear that the order of the neighbourhood graph, that is, the size of the search space in terms of the number of candidate solutions it comprises, plays an important role: Generally, finding any of a fixed number of (optimal) solutions becomes harder as the size of the search space increases. For example, we would expect that a SAT instance with 20 variables is substantially easier to solve than one with 200, considering that in the former case, the search space comprises only $2^{20} = 1\,048\,576$ variable assignments, compared to $2^{200} \approx 1.61 \cdot 10^{60}$ in the latter case. This correlation between search space size and search cost certainly exists for the simplest SLS methods, Uninformed Random Picking and Uninformed Random Walk, and it typically also holds for more powerful SLS strategies. However, search complexity is not always correlated with search space size, as other factors, which will be discussed in the following, can have an impact on SLS behaviour that is substantial enough to completely dominate the effect of search space size. This has been shown, for example, for certain types of SAT instances and algorithms [Hoos, 1999b].

Another property of the neighbourhood graph that is intuitively related to search cost is the *diameter*, $diam(G_N(\pi))$, which is defined as the maximal distance between any two vertices s, s' in $G_N(\pi)$ in terms of the number of edges that have to be traversed to reach s from s' or vice versa. For example, in the case of a SAT instance with n variables under the 1-flip neighbourhood, the diameter of the neighbourhood graph is n. The neighbourhood graphs underlying typical SLS algorithms are usually connected; that is, there is a path between any two vertices; consequently, the diameter of such neighbourhood graphs is finite. All else being equal, graphs with larger diameters are intuitively harder to search than compact graphs characterised by a small diameter. For Uninformed Random Walk, this is clearly the case; since in almost all SLS algorithms the search steps predominantly correspond to traversing single edges in the underlying neighbourhood graph, a similar correlation can be expected.

There is a relationship between the size of the search space, the diameter of the neighbourhood graph and the local neighbourhood size (where the latter corresponds to the vertex degree of the neighbourhood graph): Intuitively, for fixed search space size, the larger the neighbourhood size, the smaller the diameter of $G_N(\pi)$. Neighbourhood graphs are typically regular, that is, each candidate solution has the same number of direct neighbours. In this case, a bound on this relationship can be formally derived from the well-known Moore bound as $diam(G_N(\pi)) \geq \log_{d-1}(m - 2 \cdot m/d)$, where $d := |N(s)|$ is the neighbourhood size and $m := |S(\pi)|$ denotes the size of the given search space (see, e.g., Biggs [1993]). However, in most cases this bound is rather weak, and a more precise determination of the diameter is desirable.

EXAMPLE 5.I **Fundamental Search Space Properties for the TSP**

> For the symmetric TSP, when candidate solutions are represented as permutations of the vertices, a problem instance with n vertices has a search space of size $(n-1)!/2$. (Note that for the symmetric TSP, the direction in which a tour is traversed does not matter.) It is easy to see that the neighbourhood size is $\binom{n}{2} = n \cdot (n-1)/2$ and $\binom{n}{3} = n \cdot (n-1) \cdot (n-2)/6$ for the 2- and 3-exchange neighbourhoods, respectively. Unfortunately, the exact diameter of the neighbourhood graphs is unknown. For the 2-exchange neighbourhood, an upper bound of $n - 1$ can be given [Stadler and Schnabl, 1992], while an immediate lower bound is $n/2$. For the 3-exchange neighbourhood, the same upper bound applies, but the lower bound is now $n/3$. In both cases, the true diameter is more likely to be close to the respective lower bounds.

As illustrated in the example, in some cases the diameter of the search space cannot be determined exactly. A similar problem often arises in the computation of the distance between candidate solutions s and s', which is usually defined to be the length of the shortest path between s and s' in the given neighbourhood graph. If the exact distance cannot be computed easily, surrogate distances measures are often used that give an approximation of the true distance. For example, in the case of the TSP, the distance $d(s, s')$ between candidate solutions s and s' is usually measured in terms of the number of edges that are part of s but not s'; more precisely, $d(s, s') := n - k$, where n is the number of vertices in the given graph and k is the number of edges contained in both s and s' (this distance metric is called *bond distance*). The bond distance is tightly correlated with the true distance in the standard 2- and 3-exchange neighbourhoods and is used in practically all studies of the search landscape for the TSP [Boese, 1996; Mühlenbein, 1991; Merz and Freisleben, 2001; Stützle and Hoos, 2000]. In general, when using surrogate distance measures in the context of analysing SLS behaviour, it is important to consider properties of candidate solutions and search steps of the given algorithm in an appropriate way. For example, the bond distance for the TSP takes into account that the crucial features of a candidate tour are based on the relative order of its vertices, rather than their absolute position in a list representation of the respective cyclic path.

Number and Density of Solutions

Another factor that has a rather obvious impact on search cost is the *number of (optimal) solutions* of a given problem instance: For fixed search space size, the more (optimal) solutions there are, the shorter is the expected time to find one of these using Uninformed Random Picking; again, it is reasonable to expect a qualitatively similar negative correlation between the number of (optimal) solutions and search cost for more complex SLS algorithms. In the case of Uninformed Random Picking, it is easy to verify that for a problem instance π with search space $S(\pi)$ and k (optimal) solutions, the expected number of search steps required for finding a (optimal) solution is $\#S(\pi)/k$.

In this simple example, the expected search cost is inversely proportional to $k/\#S(\pi)$, the *density of (optimal) solutions*. Different from the number of solutions, solution density can be meaningfully compared between problem instances that differ in size. Because solution density values are typically very small, it is often more convenient to report them as $-\log_{10}(k/\#S(\pi))$.

As illustrated in the following example, for SLS algorithms more powerful than Uninformed Random Picking, the search cost can be strongly correlated with solution density, but the relationship is usually not precisely an inverse proportionality.

EXAMPLE 5.2 **Solution Density vs Search Cost for GWSAT**

The impact of solution density on search cost has been demonstrated clearly for Uniform Random 3-SAT, a well-known class of benchmark instances

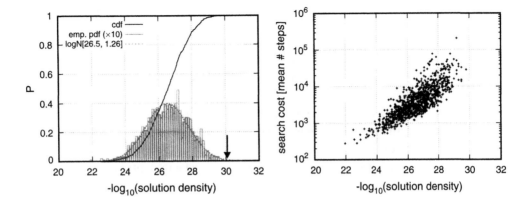

Figure 5.1 *Left:* Distribution of solution density over a set of hard Uniform Random-3-SAT instances with 100 variables and 430 clauses; the corresponding empirical probability density function (histogram) is well approximated by a log-normal distribution (dashed grey curve); the arrow indicates the solution density value of an instance with a single solution. *Right:* Correlation between solution density and search cost for GWSAT. (For details, see text.)

for SAT (see also Chapter 6, Section 6.1) [Clark et al., 1996; Hoos, 1999b]. The left side of Figure 5.1 shows the distribution of the solution density over a set of hard Uniform Random 3-SAT instances with 100 variables and 430 clauses (as determined by a suitably modified systematic search algorithm). Clearly, there is a large variability in the solution density over this test-set. Furthermore, as shown in the figure, the distribution of the solution density, and hence the distribution of the number of solutions, is well approximated by a log-normal distribution. This is an effect of the random process used for generating these problem instances, during which each point in the search space (i.e., each variable assignment) is uniformly affected by each independently generated clause. (It may be noted that the empirical distribution of the logarithms of the solution density values does not pass the Shapiro-Wilk normality test at the 0.05 significance level; however, as can be easily seen from a quantile-quantile plot, the reason for this solely lies in a discretisation effect for extremely small solution densities.)

As can be seen from the right side of Figure 5.1, there is a strong correlation between the solution density and the search cost of GWSAT, a well-known randomised best-improvement algorithm for SAT (see also Section 6.2, page 269*f.*). Here, the search cost for a given instance represents the mean number of search steps required for finding a solution, using GWSAT with an approximately optimal noise parameter setting and no restart. A more detailed analysis reveals that there is a strong linear dependency between the negative logarithm of the solution density, $-\log_{10}(sd)$, and the logarithm of the search cost, $\log_{10}(sc)$ (the Pearson correlation coefficient is 0.83); this indicates a polynomial relationship between sd and sc. Notably, there is a substantial variation in search cost, especially for small numbers of solutions, which indicates that factors other than solution density have a significant impact on search cost. Analogous results have been obtained for sets of randomly generated instances of graph colouring and constraint satisfaction problems [Clark et al., 1996; Hoos, 1999b].

For typical combinatorial problems, the density of (optimal) solutions is very low. For example, while it has been shown for critically constrained Uniform Random 3-SAT that the average number of solutions scales exponentially with the number of variables, n [Monasson and Zecchina, 1996; 1997], closer examination reveals that the solution density drops exponentially with n. In many cases, structured instances of combinatorial decision problems have unique or very few solutions (e.g., this is the case for a number of well-known SAT-encoded instances of blocksworld planning problems first used by Kautz and Selman [1996]); the same applies to many types of combinatorial optimisation problems.

For very small problem instances, solution density can be determined by exhaustive enumeration or estimated using simple sampling techniques. In some cases, in particular for SAT, the number of solutions for moderately large instances can be determined using suitably modified variants of systematic search algorithms (see, e.g., Birnbaum and Lozinskii [1999]; Bayardo Jr. and Pehoushek [2000]). Furthermore, there are situations in which (expected) solution densities can be analytically derived; this is the case, for instance, for the previously mentioned class of Uniform Random 3-SAT [Monasson and Zecchina, 1996; 1997].

In most cases, however, determining or estimating solution densities is a difficult problem, for which no general solution methods are available. One method that is fairly regularly used in practice for sampling the set of (optimal) solutions of a given problem instance is based on performing multiple runs of a high-performance SLS algorithm. Typically, this method leads to very biased samples, and it is often infeasible to perform sufficiently many runs to obtain a reasonably good estimate of the solution density. Additionally, in the case of optimisation problems, it requires knowledge of the optimal solution quality, which in many cases is not available.

Distribution of Solutions

For problem instances with multiple (optimal) solutions, and particularly in the case of non-vanishing solution densities, the distribution of the solutions across the search space can have an impact on search cost. This can be illustrated by considering the two extreme cases of evenly distributed and tightly clustered solutions. If solutions are evenly distributed within the search space, the expected number of steps for reaching a solution using Uninformed Random Walk is very similar across the entire search space. If all solutions are tightly clustered in a relatively small region of the space, the search cost varies considerably depending on the starting point of the walk. Intuitively, more efficient SLS algorithms should exhibit similar behaviour.

One method for measuring how evenly solutions are distributed within a given search space is based on the pairwise distances within a sample $\hat{S}$ of solutions that has been obtained, for example, by running a high-performance stochastic algorithm multiple times. (Ideally, the entire set of solutions should be used; but as discussed above, this is typically infeasible except in the case of small problem instances.) We now consider the distributions $D(s)$ of distances $d(s, s')$ between any solution $s \in \hat{S}$ and all other solutions $s' \in \hat{S}$ and note that large variations between the $D(s)$ indicate an uneven solution distribution. For this approach to work well it is important to use an unbiased sample of the solution set; unfortunately, obtaining such unbiased samples is often hard or impossible.

In many cases, decision problems such as SAT have a large number of solutions that occur in the form of tightly connected regions. These *solution plateaus* will be discussed in more detail in Section 5.5.

5.2 Search Landscapes and Local Minima

In all but the simplest SLS algorithms, the search process is guided by an evaluation function $g(\pi)$ which, in the case of optimisation problems, is often identical with the given objective function $f(\pi)$. To capture the impact of the evaluation function in combination with search space properties and the neighbourhood relation, we introduce the well-known concept of a *search landscape*.

DEFINITION 5.1 **Search Landscape**

Given an SLS algorithm A and a problem instance π with associated search space $S(\pi)$ and neighbourhood relation $N(\pi)$, as well as an evaluation *function* $g(\pi) : S \mapsto \mathbb{R}$, *the* search landscape *(or short* landscape*) of* π, $L(\pi)$, *is defined as* $L(\pi) := (S(\pi), N(\pi), g(\pi))$.

For a given search landscape $L = (S, N, g)$, we will refer to the evaluation function value $g(s)$ of a given position s also as the *level of s*. Furthermore, for convenience, we will sometimes refer to the fundamental properties of S and $G_N = (S, N)$ discussed in Section 5.1 as properties of L; and following common usage, we will occasionally use the term *search space structure* to refer to the structure of L.

Remark: Formal definitions that differ from the one given here can be found in the literature; the differences are, however, superficial only and do not affect the common underlying concept. Frequently, the term *fitness landscape* is used to refer to the same concept. This term dates back to some of the earliest studies of search landscapes, which were performed in the context of evolutionary theory [Wright, 1932], and has since been used in the study of the factors underlying the behaviour of Evolutionary Algorithms [Kallel et al., 2001].

Basic Landscape Properties

Both global and local properties of search landscapes can have an impact on the behaviour and performance of SLS algorithms. The following properties provide the basis for a useful classification of landscapes:

DEFINITION 5.2 **Invertible, Locally Invertible and Non-Neutral Landscapes**

Let $L := (S, N, g)$ be a search landscape. L is:

- invertible *(or* non-degenerate*), if no two positions in S have the same level, that is, $\forall s, s' \in S : [s \neq s' \Rightarrow g(s) \neq g(s')]$. Landscapes that do not have this property are also called* degenerate.

- locally invertible, *if any local neighbourhood in L is invertible, that is, $\forall r \in S : [\forall s, s' \in N(r) \cup \{r\} : [s \neq s' \Rightarrow g(s) \neq g(s')]]$.*

- non-neutral, *if neighbouring positions in L always have different levels, that is, $\forall s \in S : [\forall s' \in N(s) : [s \neq s' \Rightarrow g(s) \neq g(s')]]$. Landscapes in* *which neighbouring positions may have the same level are also called neutral.*

Obviously, every invertible landscape is locally invertible, and every locally invertible landscape is non-neutral; however, there are non-neutral landscapes that are not locally invertible and locally invertible landscapes that are not invertible [Flamm et al., 2002].

Although exceptions exist, the search landscapes encountered by SLS algorithms for combinatorial problems are often degenerate. For example, when using the number of violated clauses as an evaluation function in the case of SAT, the landscapes are typically degenerate and neutral; in fact, as we will discuss in more detail in Section 5.5, these landscapes are characterised by large and numerous plateaus. On the other hand, there are many classes of combinatorial optimisation problems that give rise to locally invertible landscapes; for example, the landscapes for Euclidean TSP instances obtained from a set of randomly placed vertices under the standard 2-edge-exchange neighbourhood can be expected to be locally invertible.

Note that problem instances with invertible landscapes always have unique (optimal) solutions. The converse, however, does not hold: single solution instances can have degenerate landscapes. Locally invertible landscapes have the interesting property that when performing an iterative best improvement search, the end point of the respective search trajectory is uniquely determined by its starting point [Flamm et al., 2002].

Position Types and Position Type Distributions

While basic properties, such as invertibility or neutrality can be useful for characterising landscapes at a global level, the detailed analysis of search landscapes is often based on local features. The following definition provides a natural classification of search positions according to the topology of their local neighbourhood.

DEFINITION 5.3 **Position Types**

Let $L := (S, N, g)$ be a search landscape. For a position $s \in S$, we define the following functions, which determine the number of upwards, sidewards and downwards steps from s to its direct neighbours:

$$upw(s) \quad := \quad \#\{s' \in N(s) \mid g(s') > g(s)\}$$
$$sidew(s) \quad := \quad \#\{s' \in N(s) \mid g(s') = g(s)\}$$
$$downw(s) \quad := \quad \#\{s' \in N(s) \mid g(s') < g(s)\}$$

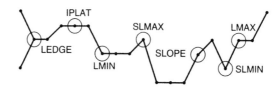

Figure 5.2 Examples for the various types of search positions.

Based on these functions, we define the following position types:

$$
\begin{aligned}
SLMIN(s) &:\Leftrightarrow & downw(s) = sidew(s) = 0 \\
LMIN(s) &:\Leftrightarrow & downw(s) = 0 \wedge sidew(s) > 0 \wedge upw(s) > 0 \\
IPLAT(s) &:\Leftrightarrow & downw(s) = upw(s) = 0 \\
LEDGE(s) &:\Leftrightarrow & downw(s) > 0 \wedge sidew(s) > 0 \wedge upw(s) > 0 \\
SLOPE(s) &:\Leftrightarrow & downw(s) > 0 \wedge sidew(s) = 0 \wedge upw(s) > 0 \\
LMAX(s) &:\Leftrightarrow & downw(s) > 0 \wedge sidew(s) > 0 \wedge upw(s) = 0 \\
SLMAX(s) &:\Leftrightarrow & sidew(s) = upw(s) = 0
\end{aligned}
$$

The positions defined by these predicates are called strict local minima (SLMIN), local minima (LMIN), interior plateau (IPLAT), ledge (LEDGE), slope (SLOPE), local maxima (LMAX) *and* strict local maxima (SLMAX) positions. *For an illustration of the various position types, see Figure 5.2.*

Note that for any landscape L, the classes of search positions induced by these predicates form a complete partition of S, that is, every search space position falls into exactly one of these types. Note also that these types can be weakly ordered according to the restrictiveness of their defining predicates when assuming that defining equalities are more restrictive than inequalities; in this respect, SLMIN, SLMAX and IPLAT are most restricted, followed by LMIN, LMAX and SLOPE, while LEDGE is least restricted. This ordering can be further refined based on the observation that equalities on the number of sideward steps are less restrictive than those on the number of upward or downward steps; consequently, type SLOPE is less constrained than LMIN and LMAX, while IPLAT is more restrictive than SLMIN and SLMAX. For random landscapes, we would therefore expect a distribution of the position types according to this ordering, that is, the more constrained a position type is, the more rarely it should occur.

The relative abundance of position types encountered in a given search landscape can give interesting insights into SLS behaviour; it can be summarised in the form of *position type distributions* that specify for each position type T the fraction of type T positions within the given search space. For small search spaces, position type distributions can be precisely determined by exhaustive enumeration. In many cases, however, the search spaces to be analysed are too large for this approach, and sampling methods have to be applied. Unbiased random sampling often suffers from the problem that positions of a type that are particularly relevant for SLS behaviour, such as LMIN, SLMIN or IPLAT positions, can be extremely rare. (Note that most effective SLS algorithms mainly search a small subspace of high-quality candidate solutions, which can be structurally very different from the rest of the search space.) One way of overcoming this problem is to sample position types along SLS trajectories [Hoos, 1998]. The position type distributions thus obtained characterise the regions of the search space seen by the respective SLS algorithm and are hence likely to contain position type information related to the algorithm's behaviour and performance. The following example illustrates the methods used for determining position type distributions as well as the results obtained from this type of search space analysis.

EXAMPLE 5.3 **Position Type Distributions for SAT**

In the context of an analysis of the search space features of various types of SAT instances, the following results on position type distributions have been obtained [Hoos, 1998]: Table 5.1 shows the complete distributions of

Instance	avg sc	SLMIN	LMIN	IPLAT
uf20-91/easy	13.05	0%	0.11%	0%
uf20-91/medium	83.25	< 0.01%	0.13%	0%
uf20-91/hard	563.94	< 0.01%	0.16%	0%

Instance	SLOPE	LEDGE	LMAX	SLMAX
uf20-91/easy	0.59%	99.27%	0.04%	< 0.01%
uf20-91/medium	0.31%	99.40%	0.06%	< 0.01%
uf20-91/hard	0.56%	99.23%	0.05%	< 0.01%

Table **5.1** Distribution of position types for critically constrained Uniform Random 3-SAT instances with low, intermediate and high search cost for GWSAT, a high-performance SLS algorithm for SAT based on exhaustive enumeration of the search space. (See text for details.)

position types for three small Uniform Random 3-SAT instances from the solubility phase transition region (see also Chapter 6, Section 6.1); the fractions of positions for each type were obtained by exhaustive enumeration of the entire search space. The instances were selected based on the search cost for GWSAT, a high-performance randomised iterative improvement algorithm for SAT (see also Chapter 6, Section 6.2), measured as the mean number of search steps required by GWSAT to find a solution, using close to optimal parameter settings and estimated from 100 successful runs per instance; the search cost values are indicated as 'avg sc' in the table. Entries reading '$< 0.01\%$' indicate that the corresponding values are in the open interval $(0\%, 0.01\%)$.

The results are consistent with the ordering based on the restrictiveness of the position types discussed above: LEDGE positions are predominant, followed by SLOPE, LMIN and LMAX positions; SLMIN and SLMAX positions occur very rarely, and no IPLAT positions were found for any of the instances analysed here or later. This suggests that the search spaces of Uniform Random 3-SAT instances show some structural properties similar to entirely random search spaces; but while random search spaces can be expected to contain equal numbers of LMIN and LMAX as well as of SLMIN and SLMAX positions, this is apparently not the case for the search spaces of Uniform Random 3-SAT instances, where LMIN positions occur more frequently than LMAX positions. This is most probably an effect of the CNF generation mechanism for Uniform Random 3-SAT instances; since each added three-literal CNF clause 'lifts' the evaluation function for one eighth of the search positions by one, while the remaining positions remain unaffected, local maxima are more likely to be eliminated when more and more clauses are added.

Our results also suggest that for Uniform Random 3-SAT instances from the solubility phase transition, the hardness of instances for SLS algorithms such as GWSAT might be correlated with the number of LMIN positions, which is consistent with the intuition that local minima impede local search (cf. Yokoo [1997]). It is interesting to note that there are no IPLAT positions, and consequently, randomised iterative improvement algorithms, such as GWSAT, can always either perform an improving or a worsening step, but are never forced to search the interior of large plateaus.

Table 5.2 shows position type distributions for Uniform Random 3-SAT instances that are too large for exhaustive enumeration; since for these instances, random sampling yields almost exclusively LEDGE (and, very seldomly, SLOPE) positions, we sampled the position distributions along trajectories of GWSAT. For obtaining these samples, we used 100 runs of the

Instance	avg sc	SLMIN	LMIN	IPLAT
uf50-218/medium	615.25	0%	47.29%	0%
uf100-430/medium	3 410.45	0%	43.89%	0%
uf150-645/medium	10 231.89	0%	41.95%	0%

Instance	SLOPE	LEDGE	LMAX	SLMAX
uf50-218/medium	< 0.01%	52.71%	0%	0%
uf100-430/medium	0%	56.11%	0%	0%
uf150-645/medium	0%	58.05%	0%	0%

Table 5.2 Distribution of position types for critically constrained Uniform Random-3-SAT instances of different size, sampled along GWSAT trajectories. (See text for details.)

algorithm with a maximum of 1 000 steps each. Any of these short runs during which a solution was found were terminated at that point in order to prevent solution positions from being over-represented in the sample. (GWSAT will, once it has found a solution, return to that solution over and over again with a very high probability, unless the search is restarted.) Therefore, the actual sample sizes vary; however, we made sure that each sample contained at least 50 000 search positions. It may be noted that the differences in search cost between the instances shown in Table 5.2 can be entirely attributed to differences in solution density.

The results of this analysis clearly reflect the fact that GWSAT is strongly attracted to local minima positions. They also suggest that neither SLMIN, nor IPLAT or SLOPE positions play a significant role in SLS behaviour on Uniform Random 3-SAT instances, which is consistent with our earlier observation for small instances, that these position types are extremely rare or do not occur at all. Similar results have been obtained for other types of SAT instances (cf. Hoos [1998]); IPLAT positions were never encountered and SLOPE as well as SLMIN positions occur only occasionally. Instead, particularly when sampling along GWSAT trajectories, the position type distributions are dominated by LEDGE positions, which mainly represent the gradient descent and plateau escape behaviour of GWSAT, and LMIN positions, which characterise plateau search phases of GWSAT's search. Somewhat surprisingly, our results regarding the ratio of LEDGE and LMIN positions along GWSAT's trajectories, when applied to instances from various problem domains, do not give clear evidence for the plateau search phase having substantially larger or smaller impact on SLS performance than the gradient descent and plateau escape phases.

Clearly, in non-neutral landscapes, all local minima are strict [Flamm et al., 2002]. Hence, the results described in Example 5.3 illustrate the previously mentioned fact that SAT instances typically have neutral search landscapes; furthermore, the observed position type distributions show that this neutrality is manifested for the vast majority of all search positions. This latter observation exemplifies how, in general, position type distributions can be used to elaborate and quantify basic landscape properties such as degeneracy, local non-invertibility and neutrality.

Number, Density and Distribution of Local Minima

Local minima are amongst the most relevant landscape features in terms of their impact on SLS behaviour. (As always, we assume that any given SLS algorithm attempts to minimise its respective evaluation function.) Clearly, in the case of landscapes that have no local minima other than the global minima (which are technically speaking also local minima), even simple iterative improvement algorithms should find (optimal) solutions relatively easily. The hardness of solving typical combinatorial optimisation problems can therefore be attributed to the fact that the respective search landscapes typically contain a large number of local minima that are not global optima.

These observations suggest that the number of local minima should be positively correlated with the hardness of the respective problem instance for SLS algorithms. For degenerate landscapes, we define the *number of local minima* as the total number of LMIN and SLMIN positions; this reflects the intuition that positions of these types have a detrimental effect on SLS behaviour, since they do not allow improving search steps. (In Sections 5.5 and 5.6, we will discuss the larger plateau and basin structures that are not captured by this definition.) Analogous to the case of the number of solutions, rather than the number of local minima, m, the respective *local minima density*, defined as $m/\#S$, should be considered; again, given the typically very small values of this measure, it is often more convenient to use $-\log_{10}(m/\#S)$. In Example 5.3 (Table 5.1), we saw some evidence supporting the correlation between the local minima density and search cost (all three SAT instances shown there have the same search space size).

Local minima density can typically not be measured directly (except for extremely small problem instances), and no general estimation methods are available. However, in some cases, local minima density can be analytically determined or estimated. For example, in completely random landscapes over a neighbourhood graph with $\#S$ vertices and vertex degree D, the expected local minima density is known to be $\#S/(D+1)$ [Kauffmann et al., 1988; Stadler, 1995]. In some cases, local minima density can be estimated based on auto-correlation length, a measure of landscape ruggedness that will be discussed in Section 5.4 [Stadler and Schnabl, 1992; Stadler, 1995]; based on this method, it has been shown for a number of combinatorial problems and neighbourhood relations, including the TSP with the 2-vertex-exchange neighbourhood, that the number of local minima increases exponentially with problem size n (for TSP, n is the number of vertices in the given graph); at the same time, in most cases, the local minima density decreases exponentially with n (see [Stadler, 1995]).

It is intuitively clear that for a given local minima density, the distribution of local minima within a given landscape may have an impact on SLS behaviour: While for a uniform distribution, the localised density of local minima would be the same across the entire landscape, highly non-uniform distributions are associated with a large variability in localised local minima density.

A commonly used method for studying the local minima distribution is based on the pairwise distances between a set of local minima sampled by means of SLS algorithms. For sampling the local minima, typically relatively simple SLS methods, such as iterative improvement algorithms, are used, although in principle, more complex and effective methods can be equally well applied. It may be noted that the SLS method used in this context will typically introduce a bias into the sampling process; in order to obtain a meaningful sample, it is important to ensure that the sampling algorithm (i) is well diversified and (ii) reaches reasonably high-quality local minima. The first of these criteria is typically achieved by means of randomisation, while the second calls for the application of an SLS algorithm with reasonably high performance.

Given a sample of local minima, either the empirical distribution of the distances between the respective positions in the underlying search landscape is calculated, or the empirical distribution between all local minima positions and the closest (optimal) solution. In either case, the range of observed distances and the modes of the distance distributions reflect important properties of the distribution of local minima across the landscape. For example, a multi-modal distance distribution indicates the concentration of local minima in a number of clusters, where the lower modes correspond to the intra-cluster distances, while the higher modes represent the inter-cluster distances. A more detailed analysis of the clustering of local minima can be performed using standard statistical techniques for cluster analysis [Everitt et al., 2001]. As illustrated in the following example, this method can be used for showing that for many

Euclidean TSP instances, all local minima occur in a relatively small region of the search space.

EXAMPLE 5.4 **Distribution of Local Minima for the TSP**

In order to analyse the distribution of local minima within the search landscapes for various TSP instances (all taken from TSPLIB), a set M of distinct local minima was obtained from 1 000 runs of a 3-opt first-improvement local search algorithm, where each run started from a different random initial solution. The 3-opt algorithm uses standard speed-up techniques based on fixed-radius nearest neighbour searches with candidate lists of the 40 nearest neighbours (see Chapter 8, Section 8.2 for details on these speed-up techniques).

The pairwise distances $d(s, s')$ between any two locally minimal candidate solutions s and s' in M were measured in terms of the bond distance, that is, $d(s, s')$ is equal to the number of edges that occur in s but not in s'. Furthermore, for all elements of M, we determined the minimal bond distance to a set of optimal tours; for any given problem instance, this set was obtained by selecting the best candidate solutions generated in 100 independent runs of a high-performance SLS algorithm for the TSP. (It may be noted that using this method, for all instances considered here, except for pr1002, multiple optimal solutions were found; in all cases, the provably optimal solution quality is known.)

The same experiment was performed for an ILS algorithm that uses the previously described 3-opt algorithm as its subsidiary local search procedure and is capable of reaching substantially higher-quality local minima. This ILS algorithm uses a single random double-bridge move for perturbation and an acceptance criterion that always selects the better of two given candidate solutions. In this experiment, it was run 200 times on each given TSP instance in order to obtain a set of sample local minima; each run was terminated after n iterations, where n is the number of vertices in the given graph.

The results of these experiments are shown in Table 5.3 and Figure 5.3. Generally, the average distance between the local minima is very small compared to the maximal possible bond distance of n between any two candidate solutions. This indicates that all local minima are concentrated in a relatively small region of the respective search landscape. Furthermore, the average and maximum distances between local minima and between locally minimal and globally optimal tours are rather similar; however, the average distance between local minima is slightly larger than the average distance from a local minimum to its (purportedly) closest global optimum. These observations

Instance	n	avg sq [%]	avg d_{lmin}	min d_{opt}	avg d_{opt}	max d_{opt}
			Results for 3-opt			
rat783	783	3.45	197.8	147	185.90	231
pr1002	1 002	3.58	242.0	141	208.6	273
pcb1173	1 173	4.81	274.6	198	246.0	293
			Results for Iterated Local Search			
rat783	783	0.92	142.2	67	123.1	166
pr1002	1 002	0.85	177.2	67	143.2	207
pcb1173	1 173	1.05	177.4	58	151.8	208

Table 5.3 Experimental results on the distribution and solution quality of local minima for three well-known TSP instances from TSPLIB. The data on local minima were collected from 1 000 runs of a simple 3-opt algorithm (upper part of the table) and from 200 runs of a more powerful ILS algorithm (lower part of the table); n denotes the number of vertices in the given graph, *avg sq* is the average relative solution quality of the local minima (measured as percentage deviation from the optimal solution quality of the given instance), *avg d_{lmin}* is the average distance between the local optima, and *min d_{opt}*, *avg d_{opt}* and *max d_{opt}* denote the minimum, average and maximum distance between a local minimum and the (purportedly) closest globally optimal solution. (For details, see text.)

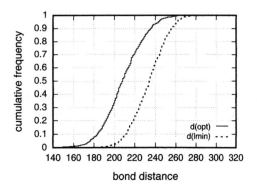

Figure 5.3 Distributions of the distance between local minima and the respective (purportedly) closest global optimum (left curve) and of the average distance between local minima (right curve); the underlying set of local minima has been obtained from 1 000 runs of a simple 3-opt algorithm on TSPLIB instance pr1002. (For further details, see text.)

suggest that optimal solutions are located centrally within the region of high local minima density.

Further evidence for this hypothesis comes from the fact that the higher-quality local minima found by the ILS algorithm tend to be significantly closer to each other and to the closest (purportedly) global optimum than observed for the lower-quality local minima found by 3-opt. This indicates that higher-quality local minima tend to be concentrated in smaller regions of the given search space. As a result, in the case of the TSP, local minima are clearly not uniformly distributed within the search landscape. (We will see more evidence for this hypothesis in the next section, in the context of fitness-distance analysis for the TSP.)

The results illustrated in this example, that is, concentration of the local minima in a small region of the landscape and central position of optimal solutions within that region, have been shown to hold for many types of TSP instances, including randomly generated and structured instances, as well as for other combinatorial optimisation problems, such as the Graph Bi-Partitioning Problem [Boese et al., 1994; Merz and Freisleben, 2000b]. However, for other problems, including certain types of instances of the Quadratic Assignment Problem (see also Chapter 10, Section 10.2), the local minima for commonly used neighbourhood relations are distributed across the entire landscape, and the average distance between local minima is close to the diameter of the underlying search graph [Merz and Freisleben, 2000a; Stützle, 1999].

In certain types of highly neutral landscapes, such as the ones typically encountered for SAT instances, local minima positions are clustered in the form of large plateau regions. While this phenomenon is clearly an important aspect of the distribution of local minima within the given landscape, depending on the size and distribution of these plateaus within the given landscape, it may be hard to detect from an empirically sampled local minima distance distribution. (The properties of such plateaus and their impact on SLS behaviour will be discussed in more detail in Section 5.5.)

5.3 Fitness-Distance Correlation

The evaluation function value is the primary guidance mechanism that is used by SLS algorithms in their search for a (globally optimal) solution to a given problem instance. For this guidance to be effective, ideally, the better the candidate solutions are rated, the closer they should be to an optimal solution. *Fitness-distance analysis (FDA)* aims to evaluate the nature of the relationship between the solution quality and the distance between solutions within a given search landscape. This relationship can be summarised using a simple measure of the correlation between the quality of solutions in terms of their evaluation function value and their distance to the closest globally optimal solution [Jones and Forrest, 1995].

DEFINITION 5.4 **Fitness-Distance Correlation Coefficient**

Given a candidate solution $s \in S$, let $g(s)$ be the evaluation function value of s, and let $d(s)$ be the distance of s to the closest global optimum. Given fitness-distance pairs $(g(s), d(s))$ for all $s \in S$, the fitness-distance correlation coefficient (FDC coefficient) is defined as

$$\rho_{fdc}(g, d) := \frac{Cov(g, d)}{\sigma(g) \cdot \sigma(d)} = \frac{\langle g(s) \cdot d(s) \rangle - \langle g(s) \rangle \cdot \langle d(s) \rangle}{\sqrt{\langle g^2(s) \rangle - \langle g(s) \rangle^2} \sqrt{\langle d^2(s) \rangle - \langle d(s) \rangle^2}}, \quad (5.1)$$

where $Cov(g, d)$ denotes the covariance of fitness-distance pairs $(g(s), d(s))$ over all $s \in S$; $\sigma(g)$ and $\sigma(d)$ are the respective standard deviations of the evaluation function and the distance values for all $s \in S$; and $\langle g(s) \rangle$, $\langle g^2(s) \rangle$, $\langle g(s) \cdot d(s) \rangle$ denote the averages of $g(s)$, $g^2(s)$ and $g(s) \cdot d(s)$, respectively, over all candidate solutions $s \in S$.

Remark: The term 'fitness-distance correlation' was originally introduced in the literature on evolutionary algorithms, where — motivated by the notion of evolutionary fitness of an organism — the term 'fitness' is often used to refer to evaluation or objective function values. In the context of minimisation problems, the use of the term 'fitness' is somewhat counterintuitive; but given the common usage of the terms 'fitness' and 'fitness-distance correlation' in the literature, we will use the same terminology in this section.

By definition, we have that $-1 \leq \rho_{fdc}(g, d) \leq 1$. The extreme values are taken in case of a perfect *linear* correlation between fitness and distance. For minimisation problems, a large positive value of ρ_{fdc} indicates that the lower the evaluation function value, the closer the respective positions are, on average, to a globally optimal solution. A value close to zero indicates that the evaluation function does not provide much guidance towards globally optimal solutions, while for negative correlations, the evaluation function is actually misleading.

Empirical Evaluation of the Fitness-Distance Relationship

The computation of the exact FDC coefficient would require to compute averages over all solutions in the search space. Obviously, this is infeasible even for relatively small instances of combinatorial problems. Therefore, in practice, the FDC coefficient is computed based on a sample of candidate solutions. Given a sample of m candidate solutions $\{s_1, \ldots, s_m\}$ with an associated set of fitness-distance pairs $\{(g_1, d_1), \ldots, (g_m, d_m)\}$, the estimate r_{fdc} of ρ_{fdc} is computed as

$$r_{fdc} := \frac{\widehat{Cov}(g, d)}{\widehat{\sigma}(g) \cdot \widehat{\sigma}(d)}, \tag{5.2}$$

where

$$\widehat{Cov}(g, d) := \frac{1}{m - 1} \sum_{i=1}^{m} (g_i - \bar{g})(d_i - \bar{d}), \tag{5.3}$$

$$\widehat{\sigma}(g) := \sqrt{\frac{1}{m - 1} \sum_{i=1}^{m} (g_i - \bar{g})^2}, \qquad \widehat{\sigma}(d) := \sqrt{\frac{1}{m - 1} \sum_{i=1}^{m} (d_i - \bar{d})^2}, \tag{5.4}$$

and $\bar{g}$, $\bar{d}$ are the averages over the sets $G := \{g_1, \ldots, g_m\}$ and $D := \{d_1, \ldots, d_m\}$, respectively; $\widehat{Cov}(g, d)$ is the sample covariance of the pairs (g_i, d_i), while $\widehat{\sigma}(g)$ and $\widehat{\sigma}(d)$ are the sample standard deviations of G and D, respectively.

The FDC coefficient can be estimated using random samples of candidate solutions [Jones and Forrest, 1995]. However, in the context of powerful SLS algorithms, it is typically more interesting to focus the analysis towards locally optimal solutions. The main reason for this lies in the fact that efficient SLS methods are highly biased towards sampling good candidate solutions, and several high-performance SLS methods, such as Iterated Local Search or Memetic Algorithms, can be seen as searching the space of local optima. Consequently, in many studies, fitness-distance analysis and in particular the computation of the FDC coefficient is based on samples of local optima [Boese, 1996; Merz and Freisleben, 1999; 2000b; Reeves, 1999; Stützle and Hoos, 2000]. Such samples are typically obtained from multiple runs of an iterative improvement algorithm that uses Uninformed Random Picking as its initialisation procedure. Occasionally, fitness distance analysis is also based on candidate solutions obtained from short runs of a high-performance SLS algorithm, which allows the sampling of higher-quality candidate solutions than those typically obtained from iterative improvement algorithms [Boese, 1996; Stützle, 1999].

The same data that are used for estimating FDC coefficients can be graphically displayed in the form of *fitness-distance plots*. These are scatter plots in which every fitness-distance pair (g, d) is represented by a point with x-coordinate d and y-coordinate g; evaluation function values g are sometimes measured as percentage deviations from a (purportedly) global optimum. Because the FDC coefficient only measures the linear correlation between distances and evaluation function values, nonlinear relationships that are easily visible in an FDC plot may sometimes be overlooked in an analysis that is solely based on the FDC coefficient. The following example illustrates the use of fitness-distance plots.

EXAMPLE 5.5 **Fitness-Distance Analysis for the TSP**

Using the same 3-opt algorithm as described in Example 5.4 (page 218*ff.*), we sampled 2 500 local optima for TSPLIB instance rat783. The fitness-distance plot for the respective local minima is shown on the left side of Figure 5.4. Clearly, there is a significant fitness-distance correlation; the corresponding FDC coefficient of 0.68 confirms this observation. Note that the maximum possible distance between candidate solutions for this TSP instance is $n = 783$. Hence, the distances between local optima and the respective closest global optima are generally relatively small, which confirms the result regarding the distribution of local minima illustrated in Example 5.4.

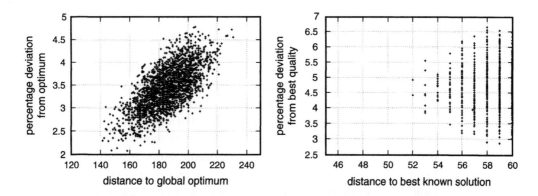

Figure **5.4** Fitness-distance plots for TSP instance rat783, based on 2 500 local optima obtained from a 3-opt algorithm (left side) and for QAP instance tai60a, based on 1 000 local optima obtained from a 2-opt algorithm for the QAP (right side). The maximum possible distance is 783 for the TSP instance and 60 for the QAP instance.

However, not all combinatorial problems give rise to search landscapes with significant fitness-distance correlations. The fitness-distance plot shown on the right side of Figure 5.4 was obtained from 1 000 runs of a 2-opt algorithm for the Quadratic Assignment Problem (QAP), applied to `tai60a`, a well-known benchmark instance from QAPLIB. (The QAP will be discussed in more detail in Chapter 10, Section 10.2.) Clearly, there is no significant fitness-correlation ($r_{fdc} = 0.03$); furthermore, it may be noted that here, different from our previous observations for TSP instances, the distances between the local optima and the (unique) best known solution is generally close to the maximum possible distance of 60. (Since provably optimal solutions for this instance are not known, the best known solution has been used instead.) This indicates that for QAP instances such as the one used in this experiment, the local minima are much more uniformly distributed across the search space than for the TSP. These observations suggest that SLS algorithms for the QAP may require much stronger diversification mechanisms than SLS algorithms for the TSP — a hypothesis that is well supported by more direct evidence.

Applications and Limitations of Fitness-Distance Analysis

The existence of a strong correlation between the evaluation function value and the distance to the closest global optimum, as observed for many types of TSP instances, is often referred to as a *big valley structure* of the underlying landscape [Boese, 1996]; intuitively, in this structure a (optimal) solution is surrounded by a large number of local minima with evaluation function values that deteriorate with increasing distance from the (optimal) solution. In the case of maximisation problems, the analogous phenomenon is typically referred to as the *massif central*. Further evidence for a big valley structure can be obtained from results on the correlation between the solution quality and the average distance between one element and any other element of a given set of locally optimal candidate solutions [Mühlenbein et al., 1988; Mühlenbein, 1991; Kauffman, 1993; Boese et al., 1994]. An example of such a correlation is illustrated graphically in Figure 5.5; the correlation between solution quality and average distance values indicates that high-quality local minima tend to be located centrally within the region containing all local minima represented by this sample.

The presence of a big valley structure and high FDC has an impact on the design and behaviour of SLS algorithms. In the case of search landscapes with a marked big valley structure, the use of strong intensification mechanisms usually

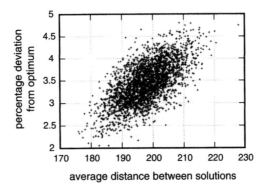

Figure 5.5 Average distance among a sample of 2 500 local minima sampled using a 3-opt algorithm versus the percentage deviation from the optimum solution quality for TSP instance `rat783`.

leads to better performing SLS algorithms. For example, in Iterated Local Search this can be achieved by a greedy acceptance criterion that always accepts the better of two given candidate solutions and thus helps to focus the search around the best candidate solution found so far. Also, for landscapes with high FDC, the use of restart strategies can be rather detrimental, and better performance can typically be achieved by using weaker forms of search diversification or perturbation.

Fitness-distance analysis can also be used to compare the effectiveness of neighbourhood relations used in different SLS algorithms. Neighbourhoods that lead to landscapes with higher FDC coefficients typically provide the basis for more efficient SLS algorithms. This has been shown in the case of the TSP [Boese, 1996; Merz and Freisleben, 2001]; similar results have been obtained for the Graph Bi-Partitioning Problem [Merz and Freisleben, 2000b].

FDC is widely used to analyse the difficulty of problems and problem instances based on the idea that, all else being equal, high fitness-distance correlations should render problem instances easier to solve for SLS algorithms, since the respective search landscapes provide more global guidance. However, fitness-distance analysis (FDA) has a number of shortcomings and limitations.

Firstly, FDA is based on the knowledge of (optimal) solutions, and it can typically only be used as an *a posteriori* method for assessing the hardness of problem instances or the performance of SLS algorithms. However, this is not a strong drawback if FDA results for individual problem instances can be generalised to larger instance classes; current experience suggests that this is often the case. Furthermore, even if used as an *a posteriori* method, FDA can play an important role in the context of explaining and understanding the behaviour of SLS algorithms. In many cases, the observations and insights gained from FDA can be exploited to develop new and better performing algorithms [Boese et al., 1994; Stützle, 1998c; Finger et al., 2002].

Secondly, for many instances of optimisation problems, optimal solutions cannot be readily computed. In these cases, sometimes the best known solutions are used as the basis of a fitness-distance analysis (cf. Example 5.5). This appears to be reasonable, since in many cases, the goal of an SLS application is to reach high-quality, but not necessarily optimal, solutions; however, proceeding in this way may lead to erroneous conclusions if these best known solutions are rather distant to true global optima.

Finally, it has been shown in the literature that fitness-distance analysis can lead to counterintuitive results if the FDC coefficient is used as a basis for classifying problem instances according to their hardness for SLS algorithms. For example, Altenberg [1997a] constructed a function that, based on FDC results, should be difficult for a given evolutionary algorithm, but experimentally turns out to be very easy. However, closer examination revealed that when using a distance measure that reflects the effective neighbourhood used by the evolutionary algorithm rather than plain Hamming distance, a high FDC coefficient is obtained, which correctly characterises this problem instance as easy for the given algorithm. A similar case arises for the so-called ridge function $R(s)$, which is defined as

$$
R(s) := \begin{cases} n + 1 + \#_1 s, & \text{if } \exists i \in \{0, 1, \ldots, n\} : s = 1^i 0^{n-i}; \\ n - \#_1 s, & \text{otherwise}; \end{cases} \tag{5.5}
$$

where any candidate solution s is a string of zeros and ones, n is the length of s and $\#_1 s$ denotes the number of ones in s [Quick et al., 1998]. While global optima of this function are easily found by an iterative improvement algorithm, the FDC averaged over all 2^n candidate solutions is very small, which erroneously suggests that the function would be difficult to optimise.

For further criticism of fitness-distance analysis we refer to Naudts and Kallel [2000]. However, despite of its known shortcomings, fitness-distance analysis has proven to be a valuable tool for the analysis of search landscapes and has yielded significant insights, particularly into the behaviour of hybrid SLS algorithms.

5.4   Ruggedness

One of the features that strongly influences the performance of SLS algorithms is the *ruggedness* of the search landscape. Ruggedness describes the degree of

Figure 5.6 Example of a smooth (left side) and a rugged (right side) search landscape.

variability between the levels of neighbouring positions. Intuitively, landscape ruggedness is related to the number of local optima: with a high density of distinct local optima are typically very rugged, while smooth landscapes can be expected to have fewer local optima (see Figure 5.6). Consequently, more rugged landscapes are intuitively harder to search for SLS algorithms. Conversely, smoother landscapes allow iterative improvement algorithms to perform a larger number of steps before hitting a local optimum, and hence to explore larger parts of the search space, which typically results in higher-quality candidate solutions.

Landscape Correlation Functions

There have been several attempts to formally define the concept of landscape ruggedness. One possibility is to measure the correlation between the levels of positions at a fixed distance i in the given landscape [Weinberger, 1990].

DEFINITION 5.5 **Search Landscape Correlation Function**

Given a search landscape $L := (S, N, g)$, *the* search landscape correlation function *of* L, $\rho(i)$, *is defined as*

$$\rho(i) := \frac{\langle g(s) \cdot g(s') \rangle_{d(s,s')=i} - \langle g(s) \rangle^2}{\langle g^2(s) \rangle - \langle g(s) \rangle^2} \tag{5.6}$$

where $\langle g(s) \rangle$ *and* $\langle g^2(s) \rangle$ *denote the average of* $g(s)$ *and* $g^2(s)$, *respectively, measured over all positions in the given space* S, *and* $\langle g(s) \cdot g(s') \rangle_{d(s,s')=i}$ *is the average over* $g(s) \cdot g(s')$ *for all pairs of solutions* $s, s' \in S$ *for which* $d(s, s') = i$.

The correlation structure of a given landscape is characterised by the values $\rho(i)$ for different distances i. The most important correlation is the first-order correlation, $\rho(1)$, which captures the statistical dependency between the level of a position and its direct neighbours. Values of $\rho(1)$ that are close to one indicate that neighbouring positions tend to have very similar levels; intuitively, this corresponds to a smooth landscape. (Note that a high correlation value $\rho(1)$ implies that positions whose level is far below the average over the entire landscape can be expected to have direct neighbours at similarly low levels.) In contrast, a $\rho(1)$ value close to zero indicates that there is basically no correlation between the levels of neighbouring positions. In this case, the evaluation function values of any candidate solution is essentially independent of the evaluation function value of its neighbours, which gives rise to very rugged landscapes and renders the problem very hard for local search.

Intuitively, a high correlation value $\rho(1)$ occurs, for example, in the TSP: A neighbouring solution under the standard 2-exchange neighbourhood (see Chapter 1, page 42*f.*) differs only in two of the n possible edges; hence, $n-2$ edge weights remain the same and we expect the cost of two neighbouring tours to be rather similar. Zero correlation results, for example, if for each candidate solution an independently generated random value is assigned, that is, in the case of completely random landscapes.

In principle, when given a particular instance of some combinatorial problem, the landscape ruggedness could be computed exactly by measuring the evaluation function values of all candidate solutions and the distances between every pair of solutions. In practice, this is not feasible because of the enormous size of the given search spaces. However, for some problems, the quantities that are used in the definition of the landscape correlation function can be determined analytically from the instance data [Angel and Zissimopoulos, 2000; 2001].

In cases where the landscape correlation function cannot be determined analytically, estimation methods have to be used. One fairly common approach is to measure the correlations between neighbouring solutions by means of a (uninformed) random walk (cf. Chapter 1, page 45) [Weinberger, 1990]. Starting from a randomly selected initial candidate solution, a random walk of m steps can be used to generate a sequence of evaluation function values $(g_1, \ldots, g_m)$, based on which the *(empirical) autocorrelation function* of the walk can be determined as:

$$r(i) := \frac{1/(m-i) \cdot \sum_{k=1}^{m-i}(g_k - \bar{g}) \cdot (g_{k+i} - \bar{g})}{1/m \cdot \sum_{k=1}^{m}(g_k - \bar{g})^2} \tag{5.7}$$

where $\bar{g} := 1/m \cdot \sum_{k=1}^{m} g_k$ is the average of the observed evaluation function values. This autocorrelation function gives statistical information about the strength of the correlation between two points that are i steps apart in the random walk.

Note that if the empirical autocorrelation function of a random walk is to correctly summarise the information on the correlation structure of the search landscape, one necessary assumption is that the landscape is *isotropic*; this basically means that the starting point of the random walk has no influence on the statistical information obtained from its trajectory. In this situation, any random walk is representative of the entire search landscape.

For many problems, the empirical autocorrelation functions of random walks show an exponential decay of the form $r(i) = e^{-i/l}$, where the parameter l is called the *(empirical) correlation length* of the search landscape. Based on this observation, as long as $r(1) \neq 0$, we can define the *(empirical) correlation length* as

$$l := \frac{1}{\ln(|r(1)|)}. \tag{5.8}$$

Like $r(1)$, the correlation length summarises the ruggedness of the given landscape: the larger the correlation length, the smoother is the search landscape. Correlation length typically depends on instance size [Stadler, 1995]; therefore, it is often scaled in dependence of the diameter of the neighbourhood graph and defined as $l' := l/diam(G_N(\pi))$.

A measure similar to the correlation length, called *autocorrelation coefficient*, was introduced by Angel and Zissimopoulos [Angel and Zissimopoulos, 1998]. It is defined as $\xi := 1/(1 - \rho(1))$; intuitively, large values of ξ correspond to smooth landscapes.

Random Landscapes

Landscape correlation functions can also be computed for random distributions of problem instances. For example, some well-known random instance distributions for symmetric TSP instances are obtained by assigning random weights to the edges of a given graph, while others are defined by randomly placing points in the Euclidean plane (see Chapter 8, Section 8.1 for more details). In such cases, the evaluation function value of each position can be seen as a random variable, and instead of the quantities of Equation 5.6 (page 227), which refer to the landscape of one particular problem instance, one can compute the expected values of the corresponding random variables. For an introduction to the resulting theory we refer to Stadler [1996]. Examples of how this can be done have been given by Stadler and Schnabl [1992] for the TSP and in Angel and Zissimopoulos [1998], as well as Stadler and Happel [1992] for the Graph Bi-Partitioning Problem.

In this context, a particularly important role is played by the so-called $AR(1)$ *landscapes*, where all correlations are completely determined by the correlations between directly neighbouring positions. Formally, a landscape is an AR(1) landscape if, and only if, it is *isotropic*, *Gaussian* and *Markovian*. A random landscape is called *isotropic* if all candidate solutions have the same expected evaluation function value and if for any pairs of candidate solutions (s, t) and (u, v) with $d(s, t) = d(u, v)$ the expectations $\mathbf{E}[g(s) \cdot g(t)]$ and $\mathbf{E}[g(u) \cdot g(v)]$ are equal. A random landscape is *Gaussian* if the evaluation function value for each individual search position follows a Gaussian (normal) distribution. For example, it can be argued that the evaluation function values of random TSP instances are normally distributed. The reason for this is that the length of a tour can be seen as the sum of n random variables corresponding to the distance between each pair of neighbouring vertices in the tour drawn from some (possibly unknown) distribution with a fixed mean μ and a fixed variance σ^2; by the central limit theorem (see, e.g., Rohatgi [1976]), for reasonably large instance sizes n, the resulting tour lenghts will be normally distributed. Analogous arguments can be applied to many other combinatorial problems. Finally, a random landscape is *Markovian* if the evaluation function value of any position statistically only depends on the evaluation function values of all its neighbours. It is easy to see that (at least in the limit), a random walk in an AR(1) landscape is an AR(1) (first-order autoregressive) stochastic process [Box and Jenkins, 1970], which leads to an autocorrelation function of the form

$$\rho(i) = (\rho(1))^i = e^{-i/\lambda} \tag{5.9}$$

where λ is the correlation length.

Among the AR(1) landscapes we find random instance distributions of the TSP, the Graph Bi-Partitioning Problem and several others combinatorial problems [Stadler, 1995]. Prototypical examples for random AR(1) landscapes are the *NK*-landscapes (for details, see the in-depth section on page 231*ff*.). The *NK*-landscape model is important in the context of *epistasis*, a concept originating from genetics (where it refers to the masking of the effects of a set of genes by another set of genes), which is related to landscape ruggedness. In the literature on evolutionary algorithms, the term epistasis has been used to refer to the interaction between the solution components of a given combinatorial problem in terms of their effect on the evaluation function. Intuitively, in cases where the effects of the solution components on the evaluation function are mutually independent, the corresponding optimisation problem is easy to solve, since suitable values for each solution component can be determined independently of the others. But with increasing interactions among solution components, the problem of finding optimal values of the given evaluation function becomes more difficult.

Several measures for epistasis have been proposed and used in the literature [Davidor, 1991; Fonlupt et al., 1998]. However, some concerns have been raised regarding these epistasis measures [Naudts and Kallel, 2000; Rochet et al., 1997], and their use is largely limited to research on the behaviour of evolutionary algorithms.

IN DEPTH *NK*-LANDSCAPES

NK-landscapes are a statistical model of search landscapes [Kauffman, 1993]. Given n binary variables, $s_1, \ldots, s_n$, the evaluation function value of a candidate solution $s := (s_1, \ldots, s_n)$ is defined as

$$g(s) := \frac{1}{n} \sum_{i=1}^{n} g_i(s_i, s_{i_1}, \ldots, s_{i_k}), \qquad (5.10)$$

that is, $g(s)$ is the average over contributions of the n individual solution components. The contribution of an invidual solution component s_i depends on the value of s_i as well as on the values of k other variables. It is specified in the form of a function $g_i : \{0,1\}^{k+1} \mapsto [0,1]$, whose value for each possible assignment of the $k+1$ input variables is a random number, which is typically chosen according to a uniform distribution over the interval $[0,1]$. In implementations of *NK*-landscape generators, the possible values for g_i can be determined efficiently by using a lookup table that contains the function values for all 2^{k+1} possible inputs (see Altenberg [1997b] for a discussion of how to avoid the explicit storage of all function values). Note that for $k > 0$, changing the value of a single variable generators, s_i will generally affect the evaluation function contributions of all other variables that depend on s_i.

The k variables that form the context of the evaluation function contribution of a solution component s_i can be chosen according to different models. The two models originally investigated by Kaufmann are the *random neighbourhood* model, where the k variables are chosen randomly according to a uniform distribution among the $n-1$ variables other than s_i, and the *adjacent neighbourhood* model, in which the k variables are chosen that are closest to s_i in a total ordering $s_1, s_2, \ldots, s_n$. (The adjacent neighbourhood model uses periodic boundaries, that is, variable s_1 is a direct neighbour of variable s_n). While no significant differences between the two models were found in terms of global properties of the respective families of search landscapes, such as mean number of local optima or autocorrelation length [Weinberger, 1991; Kauffman, 1993], they differ substantially with respect to computational complexity. While the optimisation problem for general *NK*-landscapes is $\mathcal{NP}$-hard, in the case of the adjacent neighbourhood model, it can be solved by a dynamic programming algorithm in time $O(n \cdot 2^k)$ [Weinberger, 1996; Wright et al., 2000]. For the random neighbourhood model, on the other hand, only the special case of $k = 1$ is polynomially solvable [Wright et al., 2000].

NK-landscapes were designed as a prototype of tunably rugged search landscapes. The properties of the resulting landscapes are influenced by the two parameters n and k. The parameter k corresponds to the order of the interaction between variables; low values of k indicate a low interaction, while high values of k indicate a strong interaction between the variables. At the extremes, $k = 0$ corresponds to landscapes in which the evaluation function contribution of each variable is independent of all other variables; for $k = n - 1$, on the other hand, the contribution of every variable depends on the values of all other variables, which leads to completely random landscapes, in which the evaluation function value of any candidate solution is statistically independent of its neighbours.

When considering the 1-exchange neighbourhood typically used by SLS algorithms for NK-landscapes, under which the value of exactly one variable is flipped in each search step, interesting differences arise for the two extreme cases $k = 0$ and $k = n - 1$.

For $k = 0$, all variables can be optimised independently of each other. Consequently, any search position that is not globally optimal has a direct neighbour with a better evaluation function value, and every local optimum is also a global optimum. Note that as a result of using random numbers as the values of the component contributions g_i, these NK-landscapes are expected to have a unique optimal solution. When performing Iterative Improvement on such a landscape, the Hamming distance to the optimal solution is reduced by one in each step; consequently, when initialising the search at a randomly chosen position, the global optimum is reached in $n/2$ search steps on average.

In the case of $k = n - 1$, the landscapes are completely random. In this case, it can be shown that the probability for an arbitrarily chosen candidate solution to be a local optimum is $1/(n + 1)$, which leads to an expected number of $2^n/(n + 1)$ local optima. Furthermore, the expected number of iterative improvement steps to reach a local optimum scales approximately as $\ln(n - 1)$.

By varying the parameter k, interpolations between these two extreme cases can be obtained. Estimates for the number of local optima and the length of iterative improvement trajectories for large k have been derived by Weinberger [1991]. The empirical autocorrelation function and the correlation length can be approximated by $r(i) \approx (1 - (k + 1)/n)^i$ and $l \approx n/(k + 1)$, respectively [Fontana et al., 1993]. Furthermore, the autocorrelation functions $\rho(i)$ for various types of NK-landscapes have been analytically derived [Fontana et al., 1993; Weinberger, 1996]. It has also been shown experimentally that while for low k, NK-landscapes have significant fitness-distance correlations [Kauffman, 1993], with increasing k, the FDC coefficient quickly approaches zero [Merz, 2000].

Landscape Ruggedness and Local Optima

As stated at the beginning of this section, ruggedness is closely related to the number of local minima in a given landscape; in fact, it has been proposed to call a family of landscapes rugged if the number of local optima increases exponentially with search space size [Palmer, 1991]. This suggests that the number or density of local minima may be a good measure for landscape ruggedness. Since it is typically infeasible to determine the exact number of local minima for realistically sized problem instances, estimation methods have to be used instead.

In the context of the random field approach to the analysis of search landscapes, the relationship between landscape ruggedness and the number of local optima is made more precise by the *correlation length conjecture*. This conjecture establishes a tight relationship between the number of local minima and the correlation length of a given isotropic elementary landscape [Stadler and Schnabl,

1992]. (For the definition of an elementary landscape, see, for example, Stadler [2002a].) More precisely, the correlation length conjecture can be stated as follows. Given an isotropic elementary landscape L, let λ be the correlation length of L (as determined from a random walk on L), and let $r(\lambda)$ denote the average distance in L that can be reached from a given initial position by a random walk of length λ. Furthermore, let $B(r(\lambda))$ be the number of search positions in a region of radius $r(\lambda)$. Then, it is conjectured that L has approximately $\#S/B(r(\lambda))$ local optima. The conjecture is based on the observation that for typical elementary landscapes, the correlation length λ directly determines the expected size of correlated regions in L, and in the absence of other distinctive features, each correlated region can be expected to have only a very small number of local optima (see also García-Pelayo and Stadler [1997]).

The correlation length conjecture has been tested on several combinatorial optimisation problems including the TSP, the Graph Bi-Partitioning Problem [Krakhofer and Stadler, 1996] and p-spin models [Stadler and Krakhofer, 1996]; these studies have yielded empirical evidence in favour of the conjecture. However, the correlation length conjecture cannot be expected to hold when the underlying isotropy assumption is not satisfied. For example, a study of discrete XY-Hamiltonian spin glasses has shown that, as the degree of anisotropy is raised, the predictions obtained from the correlation length hypothesis become increasingly inaccurate [García-Pelayo and Stadler, 1997].

Landscape Ruggedness and Algorithm Behaviour

Intuitively, landscape ruggedness has an impact on the behaviour of specific SLS algorithms and the general effectiveness of stochastic local search for solving a given problem instance. Therefore, at the first glance, there should be some hope that measures of ruggedness, such as autocorrelation or scaled correlation length, could help to explain the variation in search cost between different problem instances. However, for several combinatorial problems, including the TSP and the Graph Bi-Partitioning Problem, it has been shown (usually with the help of random landscape theory) that expected autocorrelation and correlation length values only depend on instance size, but not on particular instance features [Stadler and Happel, 1992; Stadler and Schnabl, 1992; Angel and Zissimopoulos, 1998]. Considering the substantial variation in difficulty between instances of the same size, even when the instances stem from the same random instance distribution, this casts some doubt on the suitability of ruggedness measures for explaining or predicting the hardness of individual problem instances.

These doubts are further confirmed by empirical results. For example, it has been shown that in the case of SAT, the correlation length is independent of the ratio of clauses to variables, which is known to critically determine average instance hardness [Rana, 1999]. In a similar vein, Kinnear Jr. [1994] reported in a study of genetic programming algorithms that he was unable to observe any correlation between the landscape autocorrelation and the difficulty of problem instances, where difficulty was measured in terms of a difficulty index derived from run-length distributions. (In the same study, the average number of steps required by an iterative improvement algorithm to reach a local optimum was found to be a better predictor for difficulty.)

While measures of landscape ruggedness are often insufficient for distinguishing between the hardness of individual problem instances, they can occasionally be useful for analysing differences between various neighbourhood relations for a given problem, for studying the impact of parameter settings of an SLS algorithm on its behaviour or for classifying the relative difficulty of combinatorial problems at an abstract level. In the following, we briefly exemplify each of these applications.

Most properties of a search landscape are strongly determined by the choice of the underlying neighbourhood relation. In particular, varying the neighbourhood relation of a given search landscape can have substantial effects on its ruggedness. This gives rise to the general idea of using measures of ruggedness for selecting one of several possible neighbourhood relations; intuitively, a neighbourhood relation that minimises landscape ruggedness should provide a better basis for SLS methods that can solve the given problem effectively.

EXAMPLE 5.6 **Neighbourhoods and Ruggedness for the Symmetric TSP**

Stadler and Schnabl computed the correlation length of several neighbourhoods for the symmetric and the asymmetric TSP [Stadler and Schnabl, 1992]. In particular, they focused on the transposition neighbourhood, under which two tours are direct neighbours if, and only if, they differ by the positions of two vertices, and the standard 2-exchange neighbourhood (see Figure 1.6, page 44). They have shown that the transposition neighbourhood leads to a correlation length of $n/4$, while the 2-exchange neighbourhood leads to twice as large a correlation length of $n/2$. This analysis indicates that the 2-exchange neighbourhood results in smoother landscapes, which should be beneficial for stochastic local search. This is confirmed by the empirical observation that SLS algorithms based on the 2-exchange neighbourhood usually perform significantly better than algorithms based on the transposition neighbourhood [Reinelt, 1994].

Care has to be taken when comparing neighbourhoods of different size. For example, a well known result by Stadler and Schnabl predicts that the landscape ruggedness obtained for the k-exchange neighbourhood for the symmetric TSP is well approximated by an AR(1) process with a correlation length of $\lambda = n/k + O(1)$ [Stadler and Schnabl, 1992]. In fact, this estimate is rather independent of the particular distance matrix, and it was shown to hold for Euclidean TSP instances (independent of the dimensionality of the space), as well as for TSP instances with random distance matrices. This result suggests that for larger k, smaller correlation lengths — indicative of more rugged landscapes — are obtained. However, the increased ruggedness needs to be contrasted with the smaller diameter of the search space and the empirical results on the solution quality achieved by iterative improvement algorithms based on the respective neighbourhoods.

A second use of ruggedness measures arises in the context of finding good settings for parameters of an SLS algorithm that can influence the shape of the search landscape. Probably the most noteworthy example for this approach was given by Angel and Zissimopoulos [1998], who examined the landscape correlation coefficient for the Graph Bi-Partitioning Problem in dependence of a parameter α that determines the penalisation of infeasible candidate solution in a simulated annealing algorithm for this problem [Johnson et al., 1989]. Interestingly, they found that the landscape correlation coefficient depends on α, and that when using a value α^* that maximises the correlation coefficient, the simulated annealing algorithm they studied achieved superior performance [Angel and Zissimopoulos, 1998].

Finally, measures of landscape ruggedness have been used as the basis for high-level classifications of the difficulty of combinatorial optimisation problems. One such result led to a ranking of combinatorial problems, including the Travelling Salesman Problem (under different neighbourhood structures), the Quadratic Assignment Problem (see also Chapter 10, Section 10.2), the Weighted Independent Set Problem, the Graph Colouring Problem (see also Chapter 10, Section 10.1), the Maximum Cut Problem and the Low Autocorrelation Bit String Problem, based on the observation of a rough correspondence between the known autocorrelation coefficients and the general difficulty of these problems reported in the literature [Angel and Zissimopoulos, 2000].

5.5 Plateaus

So far, our discussion of search space features has been focused on global features of search landscapes, such as local minima density or the fitness distance correlation coefficient, and on features of local neighbourhoods, such as position types. Now, we shift our attention to landscape structures and features that are encountered at an intermediate scale—in particular, to the plateau regions that are characteristic for the neutral landscapes obtained for many types of combinatorial problems, including SAT.

Plateaus and Their Basic Properties

Before discussing plateau properties, we define the concept of a region and its border. Intuitively, a region is a connected set of search positions, and its border is formed by those positions that have neighbours outside the region. Formally, these concepts can be defined as follows.

DEFINITION 5.6 **Region, Border**

> *Let $L := (S, N, g)$ be a search landscape and $G_N := (S, N)$ the corresponding neighbourhood graph. A* region *in L is a set $R \subseteq S$ of search positions that induces a connected subgraph of G_N, that is, for each pair of positions $s', s'' \in R$, a connecting path $s' = s_0, s_1, \ldots, s_k = s''$ exists, for which all s_i are in R and all s_i, s_{i+1} are direct neighbours w.r.t. N.*
>
> *The* border *of a region R is defined as the set of positions within R that have at least one neighbour outside of R, that is, $border(R) := \{s \in R \mid \exists s' \in S - R : N(s, s')\}$.*

Plateau regions are simply regions of positions that are all at the same level; *plateaus* are plateau regions that cannot be further extended by expanding their border. This is captured by the following definition.

DEFINITION 5.7 **Plateau Region, Plateau**

> *Let $L := (S, N, g)$ be a search landscape. A region R in L is a* plateau region, *if, and only if, all positions $s' \in R$ have the same evaluation function value, that is, $\exists l \in \mathbb{N} : \forall s' \in R : g(s') = l$; in this case, $level(R) := l$ is called the* level *of plateau region R.*
>
> *A* plateau *in L is a maximal plateau region, that is, a plateau region P for which no position in $border(P)$ has any neighbour $s'' \in S - P$ with $g(s'') = level(P)$.*

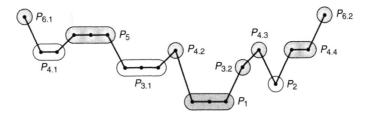

Figure 5.7 Plateaus in a simple search landscape; note the solution plateau P_1 and the various degenerate (single position) plateaus, such as P_2 and $P_{3.2}$.

Finally, we refer to plateaus that consist entirely of solutions as solution plateaus, *while all other plateaus are called* non-solution plateaus.

According to this definition, every search position is contained in exactly one plateau; hence the set of plateaus forms a partition of the search space (see Figure 5.7). It may be noted that in non-neutral landscapes, all plateaus are degenerate in the sense that they consist of a single search position only. Plateaus of size one also occur in most neutral landscapes, such as the landscapes encountered for SAT instances.

Two important properties of plateaus are size and diameter. The *size* of a plateau P is simply the number of positions in P, that is, $size(P) := \#P$. The *diameter* of P is defined as the diameter of the corresponding subgraph of G_N (the neighbourhood graph of the given landscape), that is, the maximal distance between any two positions $s, s' \in P$ (see also Section 5.1, page 204). It should be noted that the diameter of a plateau P in a given landscape L can be larger than the diameter of the search graph of L, G_N because intuitively, P can be 'folded' into G_N.

Another important property of a plateau P is its *width*, defined as the minimal length of a path between any position in P and any position not in P. Plateau width has an impact on the efficacy of plateau escape mechanisms: Intuitively, plateaus of lower width are easier to escape from. Interestingly, for many classes of SAT instances, as previously indicated by the lack of internal plateau positions (see Example 5.3, page 213*ff.*), all plateaus appear to be of width one (see also Hoos [1998]).

Finally, the *plateau branching factor* of a position s is defined as the fraction of direct neighbours of s that are in the same plateau as s. (When restricted to LMIN positions, the plateau branching factor is also called *local minima branching factor* [Hoos, 1998; 1999b].) Average plateau branching factors close to one indicate highly branched plateaus, while low branching factors are characteristic for weakly branched plateaus. Note that the extreme cases of branching factors of zero and one correspond to degenerate plateaus that consist of a single SLMIN, SLMAX or SLOPE position, and internal plateau positions, respectively. The effectivity of exploration and escape mechanisms can be affected by plateau branching; in particular, simple escape mechanisms, such as Uninformed Random Walk, can be expected to be less effective for escaping from more highly branched plateaus.

For small problem instances, basic plateau properties can be determined by exhaustive plateau exploration. For this purpose, various standard search algorithms can be used. Depth-first search (see, e.g., Russel and Norvig [2003]) has the advantage that for a given plateau P, it requires only $O(diam(P))$ space for storing the search frontier. Depending on the plateau topology and the starting point, breadth-first search (see, e.g., Russel and Norvig [2003]) may require substantially more space, but can be adapted more easily for partial plateau exploration by abandoning certain search positions from the frontier. More advanced techniques for partial plateau exploration include hybrid search techniques that alternate between phases of breadth-first search with bounded depth and random walk, as well as other types of stochastic sampling methods. Partial plateau exploration can yield useful lower bounds on basic plateau characteristics, such as size and diameter.

Typically, rather than measuring the properties of a single plateau, one is interested in the plateau properties that are characteristic for a given landscape or family of landscapes. One approach for investigating such characteristic plateau properties is to determine a sample of search positions, for example, from the endpoints of SLS trajectories or by using dedicated statistic sampling techniques, and to start complete or partial plateau exploration from these endpoints.

Exits and Exit Distributions

Even extensive plateaus do not necessarily impede search progress. As an example for this, imagine a plateau in which every position has a neighbour at a lower level (such a position is called an *exit*) — clearly, even a simple iterative improvement algorithm would not be adversely affected by such a plateau. On the other hand, plateaus that have few or no exits may cause substantial problems for SLS algorithms. The concept of exits and the related notion of open *vs* closed plateaus are captured in the following definition.

DEFINITION 5.8 **Exits, Open and Closed Plateaus**

Let $L := (S, N, g)$ be a search landscape and P a plateau in L. An exit of P is a position s on the border of P that has a neighbour at a lower level than P, that is, s is an exit of P if, and only if, $s \in border(P)$ and $\exists s' \in S - P : [N(s, s') \wedge g(s') < level(P)]$; in this context, such a position s' is called a target of exit s, and (s, s') is called an exit-target pair. We use $Ex(P)$ to denote the set of exits of plateau P.

A plateau P is called open if, and only if, it has exits, (i.e., $Ex(P) \neq \emptyset$), and closed otherwise, (i.e., $Ex(P) = \emptyset$).

Note that the definition of a closed plateau includes strict local minima as a special case. Simple iterative improvement methods can escape from open, but not from closed plateaus if only non-worsening search steps are allowed. Clearly, for open plateaus, the number, or more precisely, the density of exits and their distribution within the plateau has an impact on SLS behaviour: Intuitively, plateaus with a high exit density and uniform exit distribution should be less detrimental for SLS efficiency than plateaus with a small number of exits that are all clustered together. The *exit distance distribution*, that is, the distribution of pairwise distances between exits on a given plateau, captures both exit density and exit distribution; in particular, the average exit distance provides a measure for the expected number of steps required for finding an exit from any point of a given plateau.

In cases where plateaus can be explored exhaustively, the number of exits and their location within the plateau can be recorded in a straightforward way. From this information, the exit density is directly calculated as $\#Ex(P)/\#P$. The exit distribution can be determined using any algorithm for finding shortest paths in graphs, such as Dijkstra's algorithm [Dijkstra, 1959]. In fact, this shortest path determination can be built into the search algorithm that is used for exploring the plateau. In principle, similar techniques can be used for estimating exit densities and exit distributions (or their characteristics, such as the mean exit distance) of plateaus that are too large for exhaustive exploration; in this case, the information underlying the estimate is obtained from a partial plateau search or sampling method.

Plateau Connectivity and Plateau Connection Graphs

The notion of plateaus and their connections via exits provides a convenient basis for obtaining an abstract, yet relatively detailed view of a neutral landscape.

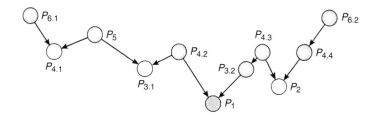

Figure 5.8 Plateau connection graph for the simple search landscape from Figure 5.7 (page 237); the edge-weights are not explicitly indicated and are all equal to one.

Intuitively, this is achieved by collapsing all positions that belong to the same plateau into a single 'macro position' (or 'macro state'). These form the vertices of a so-called *plateau connection graph (PCG)*, whose edges indicate the existence of exits that directly connect the respective plateaus. (For a simple example, see Figure 5.8.)

DEFINITION 5.9   **Plateau Connection Graph**

The plateau connection graph (PCG) *of a landscape L is a directed graph $G := (V, E)$ whose vertices are the plateaus in L and in which there is an edge $e := (P, P') \in E$ if, and only if, P has an exit with a target in P', that is, $\exists s \in Ex(P), s' \in P' : [N(s, s') \wedge g(s') < g(s)]$.*

It is often useful to refine the notion of a PCG by assigning weights to the edges of the graph that indicate how strongly two given plateaus are connected or how likely a particular SLS algorithm is to reach one from the other. One way of defining such a *weighted plateau connection graph* is to define the weight of an edge between plateaus P and P' as $w((P, P')) := \#ETP(P, P')/\#ETP(P)$, where $ETP(P)$ is the set of all exit-target pairs (s, s') for which $s \in Ex(P)$, and $ETP(P, P') := \{(s, s') \in ETP(P) \mid s' \in P'\}$. In this case, $w((P, P'))$ intuitively corresponds to the probability of reaching P' from P under the simplifying assumption that all exit/target pairs (s, s') are used equally likely.

Determining complete plateau connection graphs typically requires exhaustive enumeration of the entire search space. In cases where this is feasible, the PCG can be constructed during the enumeration process based on information on the sets $ETP(P, P')$. Note that by exploring the plateaus one at a time, the PCG can be built vertex by vertex, and the space requirement of the overall algorithm is dominated by the space needed for storing the frontier of the search process used for plateau exploration. When exhaustive enumeration of the search space is infeasible, it is often useful to construct certain subgraphs of the full PCG by exhaustively searching a subset of the plateaus, for example, plateaus below a certain level or plateaus that a given SLS algorithm is likely to visit. The latter can be determined by exploring plateaus starting from positions that have been sampled from sets of SLS trajectories. In cases where any exhaustive plateau exploration is infeasible, parts of a given PCG can still be approximated based on partial plateau exploration techniques; this approach typically does not yield proper subgraphs of the PCG, since some of the exits between two partially explored plateaus may not have been found during the partial plateau search.

EXAMPLE 5.7 **Plateau Connection Graphs for SAT**

Figures 5.9 and 5.10 show the partial weighted plateau connection graphs for two critically constrained Uniform Random 3-SAT instances; the first of

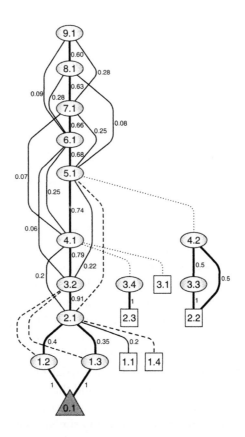

Figure **5.9** Partial plateau connection graph of an easy satisfiable Uniform Random 3-SAT instance with 20 variables and 91 clauses. (For further explanations, see text.)

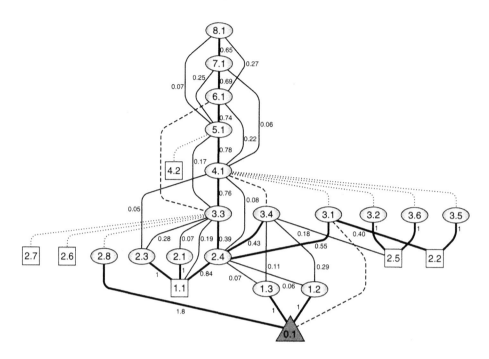

Figure 5.10 Partial plateau connection graph of a hard satisfiable Uniform Random 3-SAT instance with 20 variables and 91 clauses. (For further explanations, see text.)

these is relatively easy for high-performance SLS algorithms for SAT, such as GWSAT (see Chapter 6, Section 6.2), while the other is relatively hard.

In these illustrations, open plateaus are shown as elliptical nodes, closed non-solution plateaus as rectangular nodes and solution plateaus as triangular nodes. The y-coordinates of the nodes as well as the node labels indicate the respective plateau level; more precisely, a node label of the form $l.i$ is used for plateau number i at level l. The edge-weights of the PCG, which have been determined as previously explained (cf. page 240), are indicated by edge labels and line style; in particular, dashed lines indicate edges of weight < 0.05 and dotted lines indicate some of the edges of weight < 0.01. For both instances, there is only a single plateau per level with exit density < 1 at level 5 and above. (Plateaus with exit density 1 are essentially irrelevant in the context of SLS behaviour, because they pose no challenges for the search process.) These partial PCGs have been obtained by complete exploration of all plateaus and their respective exits.

As can be seen from their respective (partial) PCGs, both the easy and the hard instance have a very similar number of closed plateaus and a single solution plateau. However, for the easy instance, the plateaus are connected in such a way that following the maximal weight path leads directly to the solution plateau, while for the hard instance, the maximum weight path leads to closed non-solution plateau 1.1. Intuitively, the easy instance has a very attractive solution that can be easily reached from almost everywhere in the search space, while the hard instance has a very attractive non-solution region, a 'trap', from which it is quite hard to reach a solution. There is some indication that such differences in plateau connectivity are typically underlying the observed differences in hardness (for SLS algorithms) between otherwise very similar SAT instances [Hoos, 2002b].

As illustrated in Example 5.7, certain features of the plateau connection graph have a significant impact on SLS behaviour; for example, the occurrence of closed non-solution plateaus, especially when they are located at the bottom of a system of plateaus that all feed predominantly into that same 'sink', can be expected to impede search progress and may cause search stagnation. Structures such as these 'deep basins' or 'traps' can occur in non-neutral as well as in neutral landscapes. As illustrated in the example, they are important for understanding SLS behaviour.

5.6   **Barriers and Basins**

Not all strict local minima or closed plateaus are equally hard to escape from. One factor that is intuitively connected with the difficulty of achieving an improvement in the given evaluation function g from a strict local minimum or closed plateau is the difference in g that needs to be overcome in order to reach a position at a lower level. The following definition formalises this idea for arbitrary positions in a landscape (cf. Hajek [1988], Flamm et al. [2002]).

DEFINITION 5.10 **Mutual Accessibility, Barrier Level, Depth**

Given a landscape $L := (S, N, g)$, two positions $s, s' \in S$ are mutually accessible at level l if, and only if, there is a path in the neighbourhood graph $G_N := (S, N)$ that connects s and s', visiting only positions t with $g(t) \le l$.

The barrier level between positions s, s', denoted $bl(s, s')$ is the lowest level l at which s and s' are mutually accessible. The barrier height between s and s', denoted $bh(s, s')$, is defined as $bh(s, s') := bl(s, s') - g(s)$.

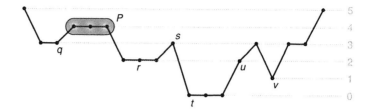

Figure 5.11 Some advanced landscape features. (See text for discussion.)

Finally, we define the depth of a position $s \in S$ as the minimal barrier height $bh(s, s')$, where s' is any position with $g(s') < g(s)$; the depth of s is defined as ∞ if s is a global minimum of L.

EXAMPLE 5.8   **Mutual Accessibility, Barrier Level and Depth**

Consider the simple search landscape in Figure 5.11. Positions r and t are mutually accessible at level 3, but not at level 2; level 3 is also the barrier level between r and t. The barrier height between r and t is 1, while the barrier height between r and q is 2. Note that barrier level is symmetric, while barrier height is not: $bl(u, v) = bl(v, u) = 3$, but $bh(u, v) = 1 \neq 2 = bh(v, u)$. The depth of q, u and v is 1, 0 and 2, respectively.

For many types of SLS algorithms that can escape from strict local minima or closed plateaus, such as Probabilistic Iterative Improvement, the probability of moving from one search position to another within a given number of search steps is negatively correlated with the barrier height between the two positions. The concepts of barrier height and depth of local minima play an important role in the theory of such algorithms; this applies particularly to Simulated Annealing, where local minima depth has been analytically linked to the convergence properties of certain types of cooling schedules [Hajek, 1988].

Basins and Saddles

Closely related to the notion of local minima depth and barrier height is the notion of a basin, which intuitively describes a region of positions at or below a given level. Formally, basins can be defined as follows:

DEFINITION 5.11 **Basin**

Given a landscape $L := (S, N, g)$ and a position $s \in S$, the basin of s at level *l is the set of all search positions s' such that $g(s') \leq l$ and s, s' are mutually accessible at level l; the* basin of s at level $g(s)$ *is also referred to simply as the* basin of s.

A basin of s below level l *is a maximal, connected subset of the basin of s at level l that consists solely of positions s' for which $g(s') < l$, or, equivalently, B is a basin of s below level l if, and only if, there is some t in the basin of s at level l such that $B = \{s' \in S \mid bh(s, s') = 0 \wedge bl(s', t) < l\}$. The basin of s below level $g(s)$ is referred to as the* basin below s.

For example, in Figure 5.11, the basin of u at level 2 contains t and the two other positions at level 0 along with u itself; the basin of u at level 3, however, contains 12 positions, including r, s, t and v. Similarly, the basin of r contains just r and its two neighbours, while the basin of r at level 3 is identical to the basin of u at level 3, as well as to the basin of s.

Note that depending on the given landscape and the given level l, there can be more than one basin at level l. To move between two such basins, a search trajectory intuitively has to reach their barrier level and cross a *saddle*, that is, a position or plateau at the barrier level. The following definition captures this notion of a saddle (see also Flamm et al. [2002]):

DEFINITION 5.12 **Saddle Point, Saddle Plateau**

Given a landscape $L := (S, N, g)$ and positions $r, s, t \in S$, s is a saddle point *between r and t if, and only if, there is a walk w from r to t in $G_N := (S, N)$ such that*

(i)   *w visits s;*

(ii) *$g(s) > \max\{g(r), g(t)\}$;*

(iii) *$g(s)$ is the maximum level reached by any position on w;*

(iv) *$g(s)$ is the barrier level between r and t;*

(v)   *w never enters the same basin below s more than once, that is, for any position u visited by w, the positions in any basin below u visited by w form a connected subpath of w.*

A plateau P in L is called a saddle plateau *between r and t if, and only if, it consists entirely of saddle points between r and t.*

For example, in Figure 5.11, position s is a saddle point between r and t, and P is a saddle plateau between q and r. Note that according to our definition of saddle points, if one position in a given plateau P is a saddle point, all other positions in P are also saddle points, and hence P is a saddle plateau. Condition (ii) of this definition ensures that in neutral landscapes, saddles (i.e., saddle points or saddle plateaus) represent only connections between true basins and not, for example, the direct connections between two plateaus via an exit from one to the other. Condition (v) rules out walks that climb to the saddle level $g(s)$ of a basin, but then drop below that level again without leaving the basin, before reaching and crossing the true saddle; in cases where multiple basins are connected by saddles at the same level, it also prevents each of these saddles from being considered a saddle between two arbitrary basins.

Note that according to our definition, saddle points can have any position type except SLMIN. Furthermore, all saddle plateaus are open plateaus with exits to at least two different (lower-level) plateaus that are not connected at any lower level of the plateau connection graph of the given landscape.

While the concepts of basins and saddles as defined here can be very useful for characterising search landscapes and understanding SLS behaviour, they do not always precisely capture the constraints on the trajectories of a given SLS algorithm. In particular, there is no guarantee that all positions in the basin below a position s can actually be visited by a given SLS algorithm starting from s; as an example, consider Iterative Best Improvement in a non-degenerate landscape for the case where the basin below s contains more than one strict local minimum. Similarly, even if two basins are only separated by a relatively low barrier, a given SLS algorithm may be unable or unlikely to cross the respective saddle.

Remark: In other literature, sometimes basins are referred to as 'cups' [Hajek, 1988] or 'cycles' [Flamm et al., 2002]. A related, but slightly different definition of saddle points has been proposed by Flamm et al. [2002].

Basin Trees and Basin Partition Trees

It is relatively easy to see that for any search landscape $L := (S, N, g)$, given two positions $s, s' \in S$, the basins of s and s', B and B', are either disjoint, or one contains the other (the latter includes the case $B = B'$). Hence, the basins in L form a hierarchy. In this context, the basins that reach just below any adjacent saddle play a special role; these so-called barrier-level basins give rise to the notion of a *basin tree*. A basin tree represents the hierarchical relationship between the barrier-level basins of L; its leaves correspond to the closed plateaus of L (including strict local minima), while the internal nodes are closely related to the barriers between them. The following definition formalises these concepts (see also the work of Flamm et al. [2000; 2002]).

DEFINITION 5.13 Barrier-Level Basin, Basin Tree

Let $L := (S, N, g)$ be a landscape. For each position $s \in S$, consider the set $BL(s)$ of all barrier levels between s and any other position $s' \in S$ for which $bl(s, s') > \max\{g(s), g(s')\}$. Let $BLB(s)$ denote the set of all basins of s below any level in $BL(s)$; the elements of $BLB(s)$ are referred to as barrier-level basins *of s.*

For barrier-level basins $B, B' \in BLB(s)$ we say that B' is directly contained in B if, and only if, $B' \subset B$ and there is no $B'' \in BLB(s)$ with $B' \subset B'' \subset B$.

The basin tree *(BT) of L is an edge-weighted tree $BT := (V, E, w)$ whose vertices represent the barrier-level basins of the positions in S, that is, $V := \bigcup \{BLB(s) \mid s \in S\} \cup \{S\}$; the edges in E connect any $B \in V$ with the barrier-level basins that are directly contained in B (the children of B in T), that is, $E := \{(B, B') \mid B, B' \in V \text{ and } B' \text{ is directly contained in } B\}$; and edge-weights are defined as $w((B, B')) := l - l'$, where l and l' are the minimal levels of any point in B and B' that occurs in none of the children of B and B', respectively.*

An example of a basin tree for a very simple landscape is shown in Figure 5.12. Notice that many positions of a given landscape L will be represented by more than one vertex of the respective basin tree. However, there is an easy way of obtaining a closely related (isomorphic) tree structure called the basin partition tree, in which every position is represented by exactly one vertex.

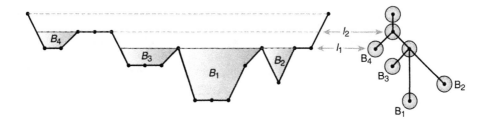

Figure **5.12** Barrier-level basins and basin tree for the simple search landscape from Figure 5.7 (page 237); the edge-weights are not indicated and correspond to the vertical distances between the nodes in the tree.

DEFINITION 5.14 **Basin Partition Tree**

> *The* basin partition tree (BPT) of L, $\overline{T} := (\overline{V}, E, w)$, *is obtained from the* basin tree of L, $T = (V, E, w)$, *by replacing each vertex $B \in V$ that has children $B_1, B_2, \ldots, B_k$ by a new vertex $\overline{B}$ that represents only those positions of B that are not contained in any of its children, i.e., $\overline{B} := B - \bigcup\{B_1, B_2, \ldots, B_k\}$ and $\overline{V} := \{\overline{B} \mid B \in V\}$.*

The notion of a basin partition tree can be intuitively illustrated with the following metaphor [Flamm et al., 2000]. Imagine the search landscape being flooded with water such that at some point in time, all positions at a given level l and below are submerged. Clearly, at any time, there is a distinct number of separate, water-filled basins (this number may be one for sufficiently high water levels). Now, we consider all critical water levels at which the 'land bridges' between two or more basins just become submerged. Below any such level l, there are two or more distinct basins that, as the water raises above l, are merged into one bigger basin. Under this view, the vertices in the basin partition tree T of the given landscape consist of all positions in a basin between one such critical level and the next higher critical level. Furthermore, any two vertices v', v'' of T have the same parent v if, and only if, there is a critical level l for which, as the water reaches l, the distinct basins corresponding to v' and v'' become connected. For degenerate landscapes, this can involve the simultaneous flooding of multiple landbridges (i.e., saddles) between v' and v''.

Note that, as desired, the position sets represented by the vertices of a basin partition tree form a complete partition of the respective landscape L, i.e., every position in L is represented by exactly one vertex of the tree. (See also Figure 5.13.) The same property holds for plateau connection graphs, and it is not hard to see that the vertices of the BPT of a given landscape L can be obtained by

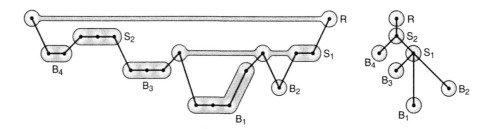

Figure 5.13 Basin partition tree for the landscape from Figure 5.7 (page 237); the edge-weights are not indicated and correspond to the vertical distances between the nodes in the tree.

merging sets of vertices in the PCG of L. Hence, basin partition trees can be seen as abstractions of plateau connection graphs which summarise the connectivity between closed plateaus.

5.7 Further Readings and Related Work

The topic of search space analysis has been receiving a considerable amount of general interest in recent years, as witnessed by a steadily growing number of studies in this area. Search landscapes of combinatorial problems are studied in such diverse fields as theoretical biology [Wright, 1932; Stadler, 2002a] and chemistry [Mezey, 1987], physics [Frauenfelder et al., 1997; Kirkpatrick and Toulouse, 1985], evolutionary computation [Merz and Freisleben, 1999; Kallel et al., 2001; Reeves and Rowe, 2003], operations research [Reeves, 1999; Boese et al., 1994] and artificial intelligence [Yokoo, 1997; Hoos, 1998; Watson et al., 2003].

Some of the simplest approaches to search space analysis focus on the number of solutions and its link to search cost [Clark et al., 1996; Hoos, 1999b]. However, typically the number of solutions cannot be determined exactly for large instance sizes; furthermore, it is often insufficient to explain even fairly drastic differences in SLS behaviour. Somewhat more detailed analyses consider the relative frequency of occurrence for different types of search positions [Frank et al., 1997; Hoos, 1999b] or the distribution of the local minima over a given search landscape [Kirkpatrick and Toulouse, 1985; Mühlenbein et al., 1988; Kauffman, 1993; Boese, 1996]. Typically, these analyses are based on samples of the given search space. While the number of local minima is usually also difficult to determine exactly, some estimation methods exist [Bray and Moore, 1980; Garnier and Kallel, 2002; Stadler and Schnabl, 1992; Tanaka and Edwards, 1980].

Fitness-distance analysis has received a significant amount of attention in the literature. While precursors of the concept had been proposed and used before [Kirkpatrick and Toulouse, 1985; Mühlenbein et al., 1988; Boese et al., 1994], the fitness-distance correlation coefficient was first defined by Jones and Forrest [1995]. Since then, a number of problems have been analysed using this technique, including the Travelling Salesman Problem [Stützle and Hoos, 2000; Merz and Freisleben, 2001], the Flow Shop Scheduling Problem [Reeves, 1999; Watson et al., 2003], the Quadratic Assignment Problem [Merz and Freisleben, 2000a; Stützle, 1999; Stützle and Hoos, 2000], the Graph Bi-Partitioning Problem [Merz and Freisleben, 2000b], the Set Covering Problem [Finger et al., 2002], the Linear Ordering Problem [Schiavinotto and Stützle, 2003] and many more. However, as illustrated in the work of Naudts and Kallel [2000], there are some pitfalls in solely relying on this measure.

Similarly widely used is the analysis of landscape ruggedness. The ruggedness of search landscapes has been empirically measured for a number of problems, using random walks [Weinberger, 1990] or trajectories of iterative improvement algorithms [Kinnear Jr., 1994]. In addition, there is a substantial amount of theoretical work, much of which is based on the theory of random landscapes [Stadler, 1995]. In particular, there exist strong theoretical results linking spectral landscape theory [Stadler, 2002b] and ruggedness measures such as autocorrelation functions [Stadler, 1996]. For more general overviews of mainly theoretical developments, we refer to the work of Reidys and Stadler [2002] and Stadler [2002a].

There has been some work on more detailed aspects of search space structure. Plateaus and neutrality in search landscapes have been studied in the context of problems from artificial intelligence [Frank et al., 1997; Hoos, 1998] as well as from theoretical biology and chemistry [Huynen et al., 1996; Reidys and Stadler, 2001]. Similarly, there are several theoretical and empirical studies of barriers, basins and related concepts [Ferreira et al., 2000; Flamm et al., 2000; 2002; Hordijk et al., 2003; Stadler and Flamm, 2003]. Plateau connection graphs and basin partition trees have been developed in the context of recent work by Hoos, and are described here for the first time. So far, while these more advanced approaches to landscape analysis hold much promise in the context of understanding the behaviour and performance of SLS algorithms, they are still largely unexplored.

5.8 Summary

Search space features and properties have an important impact on the behaviour and performance of SLS algorithms. In this chapter, we introduced and discussed a wide range of measures and techniques that can be used for analysing various aspects of search space structure. We covered a number of fundamental properties, such as *search space size* and *neighbourhood size*, the *diameter of the neighbourhood graph*, and the *number, density and distribution of (optimal) solutions*.

The concept of a *search landscape* captures the set of candidate solutions, the neighbourhood relation and the evaluation function used in an SLS algorithm, but abstracts from details of the actual search process. Search landscapes can be classified into various *landscape types*, which have important implications on the behaviour of certain SLS algorithms. To capture local features, search positions, (i.e., candidate solutions) can also be classified into different *position types* according to their local neighbourhood in the given landscape. As we have illustrated, the analysis of *position type distributions* can yield important information about a given search landscape.

Among the different types of positions, *local minima* are particularly relevant, since they tend to have a detrimental effect on SLS performance. Landscape features such as the *number, density* and *distribution of local minima* positions play an important role in analysing the hardness of problem instances for given SLS methods and for understanding SLS behaviour.

Fitness-distance analysis (FDA) is an important and widely used method for analysing and characterising search landscapes. FDA captures the correlation between the evaluation function value of search positions and their distance to the closest (optimal) solution. This correlation can be summarised in the *fitness distance correlation (FDC) coefficient* or studied in more detail by using *fitness-distance plots*. We introduced methods for empirically determining FDC coefficients and discussed various applications and limitations of fitness-distance analysis.

Another important property that is intuitively related to problem hardness and SLS behaviour is *landscape ruggedness*. Intuitively, for rugged landscapes, the evaluation function value of a search position is only weakly correlated with its direct neighbours. This intuition is captured by the concept of *landscape correlation functions*. In practice, correlation functions are often approximated using the empirical autocorrelation functions of random walks, which can be summarised by means of *correlation length*, a widely used measure for landscape ruggedness. The theory of random landscapes provides a mathematical framework for the analysis of landscape ruggedness. In fact, many random search landscapes can be shown to be $AR(1)$ *landscapes*, in which case the correlation structure is fully defined by the correlation between neighbouring candidate solutions or, analogously, by the correlation length. There is an interesting and intuitive relationship between ruggedness and local minima density. In particular, under certain circumstances, the number and density of local minima can be estimated based on the *correlation length* of a given landscape; the latter can sometimes be determined analytically or estimated empirically with relatively low computational cost. Measures of landscape ruggedness have been widely used for analysing or predicting the hardness of problem instances. They can also be useful for assessing the relative merits of different neighbourhood relations as the basis for SLS algorithms. However, as in the case of FDC, the usefulness of typical measures of ruggedness for these applications is limited in various ways.

Finally, we described various approaches for a more detailed analysis of search landscapes. In many cases, the search landscapes encountered by SLS algorithms for combinatorial decision or optimisation problems contain large *plateaus*. Features such as the *size* and *diameter* of plateaus can have a substantial impact on the behaviour of SLS methods. We distinguished between two fundamentally different types of plateaus, *open* and *closed plateaus*, depending on the existence of *exits* to lower levels. The density and distribution of exits for the plateaus of a given landscape can have substantial effects on SLS behaviour and performance. *Plateau connection graphs* capture the way in which plateaus are connected within a given landscape and are often extremely useful for understanding the hardness of problem instances and the behaviour of SLS algorithms.

The concepts of *basins*, *barrier levels* and *saddles*, as well as the related concept of *local minimum depth* provide further means for the detailed analysis and characterisation of search landscapes. They form the basis for the notions of *basin partition trees*, which can be seen as abstractions of plateau connection graphs; like those, they provide high-level, yet detailed characterisations of landscape structure.

The analysis of the spaces and landscapes searched by SLS algorithms is crucial for understanding SLS behaviour and performance, and in many cases provides key insights that can be used for improving existing SLS algorithms. Many relatively well established types of search space analyses are computationally expensive and suffer from various limitations; they can also be rather difficult to perform and care needs to be taken to correctly interpret the results. Nevertheless, techniques such as fitness-distance or autocorrelation analysis can yield useful insights. More advanced methods, such as the ones based on measuring plateau connection graphs or basin partition trees are computationally very expensive, since they require enumeration of large parts of the search space. But at the same time, this type of search space analysis facilitates a much deeper understanding of search space structure and is hence likely to become increasingly important and prominent in the context of analysing and explaining the behaviour and performance of SLS algorithms.

Exercises

5.1 [**Easy**] Explain why it is possible that for a family of instances of a given combinatorial problem, the number of solutions increases exponentially, while the solution density decreases exponentially, as instance size increases.

5.2 [**Easy**] Prove that the expected number of search steps required by Uninformed Random Picking for finding a (optimal) solution for a problem instance π with search space $S(\pi)$ and k (optimal) solutions is $\#S(\pi)/k$.

5.3 [**Medium**] Prove that the neighbourhood graph of a SAT instance under the 2-flip neighbourhood in which neighbouring assignments differ in the truth value of exactly two variables is disconnected. Which conclusion can you draw from this fact?

5.4   [*Easy*]   Give a (simple) example for a landscape that is non-neutral, but not locally invertible.

5.5   [*Easy*]   Give a simple argument that intuitively explains why SAT landscapes based on the standard evaluation function, which measures the number of clauses violated under a given assignment, are usually degenerate.

5.6   [*Medium*]   Are there non-neutral search landscapes in which a gradient walk (i.e., a trajectory of Iterative Best Improvement) from a given point is not uniquely defined? Give an example of such a landscape or prove that no such landscape exists.

5.7   [*Easy*]   Give an example for a landscape that has no local minimum other than the global optimum and is yet very hard to search for any standard SLS method.

5.8   [*Medium*]

    (a)   Show a (fictitious) fitness distance plot that indicates an FDC close to zero.

    (b)   Explain why in this situation random restarts can still be detrimental to the performance of a given SLS algorithm.

5.9   [*Medium; Hands-On*]   Perform a fitness-distance analysis for Novelty$^+$, a high-performance SLS algorithm for SAT (available from www.sls-book.net), on SATLIB instance bw_large.a (available from SATLIB [Hoos and Stützle 2003a]; this formula has exactly one model) based on the best candidate solutions from 1 000 runs, each of which is terminated after n steps, where n is the number of variables in the given problem instance. Measure and report the FDC coefficient and show a fitness-distance plot; interpret the results of your analysis.

5.10   [*Medium*]   What can you say about the plateau connection graphs of non-neutral landscapes?

part II

Applications

6 PROPOSITIONAL SATISFIABILITY AND CONSTRAINT SATISFACTION

The Satisfiability Problem in Propositional Logic (SAT) is a conceptually simple combinatorial decision problem that plays a prominent role in complexity theory and artificial intelligence. To date, stochastic local search methods are among the most powerful and successful methods for solving large and hard instances of SAT. In this chapter, we first give a general introduction to SAT and motivate its relevance to various areas and applications. Next, we give an overview of some of the most prominent and best-performing classes of SLS algorithms for SAT, covering algorithms of the GSAT and WalkSAT architectures as well as dynamic local search algorithms. We discuss important properties of these algorithms — such as the PAC property — and outline their empirical performance and behaviour.

Constraint Satisfaction Problems (CSPs) can be seen as a generalisation of SAT; they form an important class of combinatorial problems in artificial intelligence. In the second part of this chapter, we introduce various types of CSPs and give an overview of prominent SLS approaches to solving these problems. These approaches include encoding CSP instances into SAT and solving the encoded instances using SAT algorithms, various generalisations of SLS algorithms for SAT and native CSP algorithms.

6.1 The Satisfiability Problem

As motivated and formally defined in Chapter 1 (page 17*ff.*), the Satisfiability Problem in Propositional Logic (SAT) is to decide for a given propositional formula F, whether there exists an assignment of truth values to the variables in F under which F evaluates to true; such satisfying assignments are called models of F and form the solutions of the respective instance of SAT. When applying SLS algorithms to SAT, we are typically more interested in solving the search variant of SAT (i.e., in finding models of a given formula) rather than the decision variant. It should be noted that typical SLS algorithms for SAT (including all SAT algorithms covered in this chapter) are incomplete and hence cannot determine with certainty that a given formula is unsatisfiable (i.e., that it has no models).

CNF Representations and Transformations

Most algorithms for SAT, including all state-of-the-art SLS algorithms, are restricted to formulae in conjunctive normal form (CNF), that is, to formulae that are conjunctions over disjunctions of literals. Since any propositional formula can be transformed into a logically equivalent CNF formula, in principle this restriction does not limit the class of SAT instances that can be solved by such algorithms. The naïve method of transforming a non-CNF formula into CNF (using the distributive laws of propositional logic to resolve nestings of '∧' and '∨' that are not allowed in CNF) can lead to an exponential growth in the length of the formula. There is, however, an alternative CNF transformation, which avoids this effect at the cost of introducing a number of additional propositional variables that scales linearly with the size of the given formula in the worst case [Poole, 1984]. When representing problems from other domains as SAT instances, in many cases relatively natural and concise CNF formulations can be found directly without using general CNF transformation methods. In particular, this is the case for many classes of CSPs, and we will discuss approaches for encoding CSP instances as SAT in Section 6.5.

Alternative Formulations of SAT

Alternative representations of SAT for CNF formulae are used in various contexts, specifically, when techniques for solving more general problems are applied to SAT. As we will discuss in some more detail in Section 6.5, SAT can be seen as a special case of the more general finite discrete CSP.

Another prominent representation encodes the truth values $\bot$ (false) and $\top$ (true) as integers 0 and 1, and propositional variables as integer variables with domain $\{0, 1\}$. Negated literals $\neg x$ are then encoded as $I(\neg x) := 1 - x$, while positive literals remain unchanged, that is, $I(x) := x$. Finally, the encoding of a CNF clause $c_i = l_1 \lor l_2 \lor l_3 \lor \ldots \lor l_{k(i)}$ is given by $I(c_i) := I(l_1) + I(l_2) + \ldots + I(l_{k(i)})$, and an entire CNF formula $F = c_1 \wedge c_2 \wedge \ldots \wedge c_m$ is encoded as $I(F) := I(c_1) \cdot I(c_2) \cdot \ldots \cdot I(c_m)$. Then, a truth assignment a satisfies c_i if, and only if, the corresponding 0-1 assignment satisfies the inequality $I(c_i) \geq 1$, and the CNF formula F is satisfied under a if, and only if, $I(F) \geq 1$.

Based on this representation, SAT can be seen as a special case of a discrete constrained optimisation problem: Let $u_i(F, a) := 1$ if clause c_i is unsatisfied under assignment a and $u_i(F, a) := 0$ otherwise; furthermore, let $U(F, a) := \sum_{i=1}^{m} u_i(F, a)$. Then any model of F corresponds to a solution of $a^* \in \mathrm{argmin}_{a \in \{0,1\}^n} U(F, a)$ subject to $\forall i \in \{1, 2, \ldots, m\} : u_i(F, a) = 0$. This type of constrained optimisation problem is a particular case of the *0-1 Integer Linear Programming (ILP)* or *Boolean Programming Problem*.

Using these representations, SAT instances can, in principle, be solved using more general CSP or ILP algorithms. In practice, however, this approach has not been able to achieve sufficiently high performance to provide a viable alternative to native SAT solvers, such as the SLS algorithms presented in this chapter (see, e.g., Mitchell and Levesque [1996], Battiti and Protasi [1998], Schuurmans et al. [2001]). However, a number of SAT algorithms, particularly some of the dynamic local search methods presented in Section 6.4, are inspired by more general CSP or constrained optimisation solving techniques. Furthermore, successful SLS algorithms for SAT have been extended to more general classes of CSPs and ILPs, resulting in competitive solvers for these problems (some of these generalised SLS algorithms will be discussed in Section 6.6). Finally, it may be noted that the ILP formulation of SAT can be easily generalised to weighted MAX-SAT, a closely related optimisation problem, for which in some cases more general ILP methods perform much better than for SAT [Resende et al., 1997].

Polynomial Simplification of CNF Formulae

One of the advantages of the native, logical formulation of SAT is that propositional formulae in general, and CNF formulae in particular, can often be substantially simplified using computationally cheap reduction techniques. Such reductions have been shown to be crucial in solving various types of SAT instances more effectively; as preprocessing techniques, they can be used for simplifying the input to any SAT algorithm for CNF formulae.

One of the simplest reductions is the elimination of duplicate literals and clauses from a given CNF formula. Obviously, this can be performed in time $O(n)$, where n is the size of the formula, and results in a logically equivalent CNF formula. Similarly, all clauses that contain a variable and its negation, and are hence trivially satisfied (tautological clauses), can be detected and eliminated in linear time. A slightly more interesting reduction is the elimination of *subsumed clauses*. A clause $c = l_1 \vee l_2 \vee \ldots \vee l_k$ is subsumed by another clause $c' = l'_1 \vee l'_2 \vee \ldots \vee l'_j$ if, and only if, every literal in c' also occurs in c, that is, $\{l'_1, l'_2, \ldots, l'_j\} \subseteq \{l_1, l_2, \ldots, l_k\}$. Detection and elimination of all subsumed clauses can be performed efficiently and leads to a logically equivalent formula. Another linear time reduction is the elimination of clauses containing *pure literals*, that is, variables that occur either only negated or only unnegated in the given formula. Setting such a variable to false or true, respectively, does not change the satisfiability of the formula; hence, all clauses containing such variables can be removed.

One of the most important reduction techniques is based on the unit resolution method: If a CNF formula contains a *unit clause*, that is, a clause consisting of only a single literal, this clause and all clauses containing the same literal can be removed (this is a special case of the subsumption reduction), and all remaining occurrences of the corresponding variable (i.e., the complementary literal) can be removed — this can be seen as a special case of the general resolution rule (see, e.g., Russel and Norvig [2003]). Performing unit resolution for all unit clauses in the original CNF formula leads to a logically equivalent CNF formula; we also refer to this transformation as a single pass of *unit propagation*. It may be noted that unit resolution can lead to empty clauses, rendering the resulting formula trivially unsatisfiable, or eliminate all clauses, leaving an empty CNF formula, which is trivially satisfiable. Furthermore, unit resolution can produce new unit clauses and hence make further unit resolution steps possible. Repeated application of unit resolution eventually leads to a formula without any unit clauses. We refer to this reduction as *complete unit propagation*; it can be performed in time $O(n)$ and forms a crucial component of basically any systematic search algorithm for SAT. Unit propagation alone is sufficient for deciding the satisfiability of Horn formulae, that is, CNF formulae in which every clause contains at most one unnegated variable [Dowling and Gallier, 1984], in linear time w.r.t. to the size of the given formula. It also forms the basis of a linear-time algorithm for solving SAT for 2-CNF formulae [del Val, 2000].

Unit propagation provides the basis for two other efficient and practically useful simplification techniques, unary and binary failed literal reduction. The key idea behind *unary failed literal reduction* is the following: If setting a variable x occurring in the given formula F to true makes F unsatisfiable, then adding the unit clause $c := \neg x$ to F yields a logically equivalent formula F'. Since F' contains at least one unit clause, c, it can be simplified using unit propagation, which can result in a substantially smaller formula. Whether setting x to true renders F unsatisfiable is determined by adding a unit clause $c := x$ to F, and by checking whether subsequent application of unit propagation produces an empty clause. Complete unary failed literal reduction consists of performing this operation for each variable occurring in the given formula and has complexity $O(n^2)$. *Binary failed literal reduction* works analogously but checks whether simultaneously adding any two unary binary clauses, $c_1 := x$ and $c_2 := y$ and applying unit propagation leads to a trivially unsatisfiable formula. If this is the case, the binary clause $c := \neg x \vee \neg y$ is added to F, which potentially leads to further simplifications. Binary failed literal reduction has time complexity $O(n^3)$; it is a fairly expensive operation, but sometimes leads to substantial reductions in the overall time required for solving a given SAT instance (see, e.g., Brafman and Hoos [1999]).

Randomly Generated SAT Instances

Many empirical studies of SAT algorithms have made use of randomly generated CNF formulae. Various such classes of SAT instances have been proposed and studied in the literature; in most cases, they are obtained by means of a random instance generator that samples SAT instances from an underlying probability distribution over CNF formulae. The probabilistic generation process is typically controlled by various parameters, which mostly determine syntactic properties of the generated formulae, such as the number of variables and clauses, in a deterministic or probabilistic way.

One of the earliest and most widely studied classes of randomly generated SAT instances is based on the *random clause length model* (also called *fixed density model*): Given a number of variables, n, and clauses, m, the clauses are constructed independently from each other by including each of the $2n$ literals with fixed probability p [Franco and Paull, 1983]. A variant of this model was used in Goldberg's empirical study on the average case time complexity of the Davis Putnam algorithm [Goldberg, 1979]. Theoretical and empirical results show that this family of instance distributions is mostly easy to solve on average using rather simple deterministic algorithms [Cook and Mitchell, 1997; Franco and Swaminathan, 1997]. As a consequence, the random clause length model is no longer widely used for evaluating the performance of SAT algorithms. Similar considerations apply to other distributions of SAT instances, such as the instances obtained from the AIM instance generator [Asahiro et al., 1996], which can be solved in polynomial time by binary failed literal reduction [Hoos and Stützle, 2000a].

To date, the most prominent class of randomly generated SAT instances that is used extensively for evaluating the performance of SAT algorithms is based on the so-called *fixed clause length model* and known as *Uniform Random k-SAT* [Franco and Paull, 1983; Mitchell et al., 1992]. For a given number of variables, n, number of clauses, m, and clause length k, Uniform Random k-SAT instances are obtained as follows. To generate a clause, k literals are chosen independently and uniformly at random from the set of $2 \cdot n$ possible literals (the n propositional variables and their negations). Clauses are not included into the problem instance if they contain multiple copies of the same literal, or if they are tautological, that is, they contain a variable and its negation. Using this mechanism, clauses are generated and added to the formula until it contains m clauses overall.

Random k-SAT Hardness and Solubility Phase Transition

One particularly interesting property of Uniform Random k-SAT is the occurrence of a phase transition phenomenon, that is, a rapid change in solubility that can be observed when systematically increasing (or decreasing) the number m of clauses for a fixed number of variables n [Mitchell et al., 1992; Kirkpatrick and Selman, 1994]. More precisely, for small m, almost all formulae are underconstrained and therefore satisfiable; when crossing some critical value m^*, the probability of generating a satisfiable instance drops sharply to almost zero. Beyond m^*, almost all instances are overconstrained and thus unsatisfiable. (For an illustration, see Figure 6.1.) For Uniform Random 3-SAT, it has been empirically shown that this phase transition occurs approximately at $m^* = 4.26n$ for large n; for smaller n,

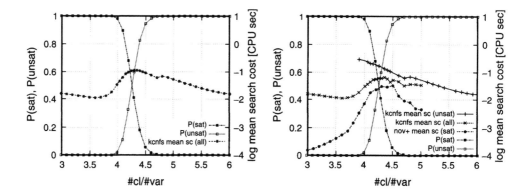

Figure 6.1 The Uniform Random 3-SAT solubility phase transition, illustrated here for formulae with $n = 200$ variables. *Left:* Empirically measured probability of obtaining satisfiable *vs* unsatisfiable instances and mean search cost (measured in terms of CPU time required for solving a given instance) for kcnfs, a state-of-the-art systematic search algorithm for this problem class. *Right:* Mean search cost (*sc*) for kcnfs on unsatisfiable *vs* satisfiable instances for the same test-sets as in the left figure, and mean search cost (measured in terms of mean CPU time for solving a given instance) of Novelty$^+$, a high-performance SLS algorithm. (All logarithms are base 10.)

the critical clauses/variable ratio m^*/n is slightly higher [Mitchell et al., 1992; Crawford and Auton, 1996]. For fixed k, the transition becomes increasingly sharp as n grows; furthermore, the critical value m^* increases with k [Kirkpatrick and Selman, 1994].

Empirical analyses have shown that problem instances from the phase transition region of Uniform Random 3-SAT tend to be particularly hard for both systematic SAT solvers [Cheeseman et al., 1991; Crawford and Auton, 1996] and SLS algorithms [Yokoo, 1997]. Figure 6.1 illustrates this effect for kcnfs [Dubois and Dequen, 2001], a state-of-the-art systematic search algorithm for this problem class and Novelty$^+$ [Hoos, 1999a], a high-performance SLS algorithm for SAT (see also Section 6.3, page 276*ff.*). Striving to evaluate their algorithms on hard problem instances, many researchers are using test-sets sampled from the phase transition region of Uniform Random 3-SAT. Particularly in the context of empirical studies including incomplete SAT algorithms, these test-sets are separated into satisfiable and unsatisfiable instances using state-of-the-art complete SAT solvers [Hoos and Stützle, 2000b]. Although similar results hold for Uniform Random k-SAT with $k > 3$, test-sets from these instance distributions are rarely used.

SAT-Encodings of Other Combinatorial Problems

Since SAT is an $\mathcal{NP}$-complete problem, any other problem in $\mathcal{NP}$ can be encoded into SAT in polynomial time and space. SAT-encoded instances of various combinatorial problems play an important role in evaluating and characterising the performance of SAT algorithms; these combinatorial problems stem from various domains, including mathematical logic, artificial intelligence and VLSI engineering.

Finite, discrete constraint satisfaction problems (CSPs) can be seen as a generalisation of SAT in which variables can have domains other than truth values, and constraints between the values assigned to individual variables can be different from the ones captured by CNF clauses. CSPs are also often a natural intermediate stage in encoding other combinatorial problems into SAT. CSP instances can be encoded into SAT in various ways; CSPs and their encodings into SAT will be further discussed in Section 6.5. It has been shown that certain types of randomly generated CSPs can be solved similarly efficiently by applying current SAT algorithms to SAT-encoded instances as by using state-of-the-art CSP algorithms [Hoos, 1999c; 1999b] (see also Section 6.6).

Other prominent examples of SAT-encoded instances of combinatorial problems include graph colouring, various types of planning and scheduling problems, Boolean function learning, inductive inference, cryptographic key search, and

n-Queens [Gu et al., 1997; Hoos and Stützle, 2000c]. For some of these, particularly in the case of SAT-encoded STRIPS planning problems from the 'blocks world' and 'logistics' domains, applying SAT solvers and reduction techniques to suitably encoded problem instances has been shown to achieve performance levels that are competitive with state-of-the-art algorithms for the respective original problem [Kautz and Selman, 1996]. Key factors underlying such results are the conceptual simplicity of SAT, which facilitates the design and efficient implementation of algorithms, and the large amount of knowledge on techniques for solving SAT and their specific properties. Furthermore, using suitable SAT encodings and reduction techniques is of crucial importance for solving the resulting SAT problems efficiently.

Interestingly, the size of the SAT encodings is not always indicative of the difficulty of solving them. Particularly, it has been shown for various problem domains that compact SAT encodings that result in instances with small search spaces can be much more difficult to solve than sparser encodings that produce instances with substantially larger search spaces [Ernst et al., 1997; Hoos, 1998; 1999b].

Some Practical Applications of SAT

Despite its conceptual simplicity and abstract nature, the SAT problem has various practical applications. Some of the most prominent industrially relevant SAT applications stem from hardware design and verification, in particular, from the verification of reactive systems, such as microprocessor components. In an approach called bounded model checking (BMC), a system and a specification of its formal properties can be encoded into a propositional formula, whose models correspond to bugs, that is, situations in which the behaviour of the system violates its specifications [Biere et al., 1999a; 1999b]. Similar to SAT encodings of planning problems that require the plan length to be bounded, in BMC, the size of the bug, that is, the number of states of the system involved in the bug, is limited by a constant. It may be noted that for proving that a given system does not have any bugs below a certain size, a complete SAT solver is required. Incomplete SAT solvers, such as the SLS algorithms for SAT covered in the following sections, can be used, however, to find bugs efficiently.

Symbolic model checking methods, such as BMC, are increasingly gaining industrial acceptance, because compared to traditional, simulation-based validation techniques, they detect a wider range of bugs, including subtle error conditions. Many traditional formal verification techniques use binary decision diagrams (BDDs) [Bryant, 1986] for representing propositional formulae. By using CNF encodings and standard SAT algorithms in a BMC approach, it is often possible to find bugs faster, and to find bugs of minimal size; the latter is important, since small bugs are typically easier to understand for a human system tester or designer. Furthermore, BDD-based approaches often require extremely large amounts of memory as well as specialised techniques for finding models of the given propositional formula, while the CNF representations are typically more concise and can be solved using standard SAT algorithms [Biere et al., 1999a]. (It should be noted, however, that BDD representations facilitate solving problems beyond SAT, such as finding all solutions of a given formula.)

Another application area in which SAT encodings and solvers have been successfully used for solving real-world problems is asynchronous circuit design [Vanbekbergen et al., 1992; Gu and Puri, 1995]. In one prominent approach to asynchronous circuit synthesis, the circuits are specified using signal transition graphs (STGs). One of the core problems is then to assign a distinguishable binary code to every circuit state. This Complete State Coding (CSC) Problem can be modelled as a SAT problem, but the size and hardness of the formulae thus obtained limits the practical applicability of SAT algorithms for solving CSC instances. However, by partitioning the STG into smaller components and using a SAT algorithm to solve the corresponding CSC subproblems, substantial performance improvements can be obtained for industrial asynchronous circuit design benchmarks [Gu and Puri, 1995].

Finally, SAT algorithms have been recently used for solving real-world sports scheduling problems [Zhang, 2002]. Specifically, the problem of finding fair schedules for college conference basketball tournaments can be encoded into SAT. This encoding is based on a decomposition of the problem into three phases, each of which deals with different constraints of the overall scheduling problem. Using a standard SAT algorithm for solving the SAT instances for the three phases, real-world college conference basketball scheduling problems were solved substantially more efficiently than by previous, specialised techniques, and more balanced schedules were obtained than the ones that are currently used for these tournaments [Zhang, 2002].

Generalisations and Related Problems

Many generalisations of the Propositional Satisfiability Problem have been proposed and studied in the literature. As mentioned above, the Constraint Satisfaction Problem (CSP) can be seen as a generalisation of SAT. Multi-Valued SAT [Béjar and Manyà, 1999; Frisch and Peugniez, 2001] and Pseudo-Boolean CSP [Abramson et al., 1996; Connolly, 1992; Walser, 1997; Løkketangen, 2002] are two special cases of the CSP that are closely related to SAT. *Multi-Valued SAT (MV-SAT)* allows variables whose domains are arbitrary finite sets of values and uses logical constraints similar to CNF clauses. *Pseudo-Boolean CSPs* use binary variables with domain $\{0, 1\}$, but allow more general constraints. Both, MV-SAT and Pseudo-Boolean CSP, as well as general finite discrete CSPs will be further discussed in Section 6.5.

The optimisation variant of SAT, in which the objective is to maximise the number of satisfied clauses, of a given CNF formula rather than completely satisfying every clause, is called *(unweighted) MAX-SAT*. In a further generalisation called *weighted MAX-SAT*, weights (usually positive integer or real numbers) are associated with the clauses of a given CNF formula, and the objective is to find a variable assignment that maximises the total weight of the satisfied clauses. As one of the conceptually simplest combinatorial optimisation problems, and because of its close relation to SAT, MAX-SAT plays an important role in the development and evaluation of search algorithms for hard combinatorial problems. In general, the best known methods for solving MAX-SAT problems are SLS algorithms. MAX-SAT problems and SLS algorithms for MAX-SAT will be discussed in more detail in Chapter 7.

Another interesting generalisation of SAT is *Dynamic SAT (DynSAT)* [Hoos and O'Neill, 2000]; intuitively, in DynSAT, a given CNF formula changes over time and a solution consists of a sequence of models such that at any time, the current CNF formula is satisfied by the current model. Equivalently, DynSAT can be defined in such a way that each problem instance consists of a conventional CNF formula some of whose variables are fixed to specific truth values at certain times. SLS algorithms for SAT can be generalised to DynSAT in a straight-forward way and appear to be well-suited for solving these problems.

Let us mention three other prominent problems that are closely related to SAT. In the *Propositional Validity Problem (VAL)*, the objective is to decide whether a given propositional formula F is valid, that is, whether F is satisfied under all of its variable assignments [Russel and Norvig, 2003]. VAL and SAT are dual problems in the sense that any formula F is valid if, and only if, $\neg F$ is unsatisfiable. Hence, any complete algorithm for SAT can be used for deciding VAL and vice versa. VAL is an important problem in theorem proving and has applications in artificial intelligence and other areas of computer science.

The *Satisfiability Problem for Quantified Boolean Formulae (QSAT)* can be seen as a generalisation of both SAT and VAL. A *quantified Boolean formula (QBF)* is a propositional formula in which all variables are quantified existentially ($\exists$) or universally ($\forall$). A QBF of the form $\exists x : F$ is satisfiable if, and only if, either assigning $x := \top$ or $x := \bot$ makes F satisfiable, and a QBF of the form $\forall x : F$ is satisfiable if, and only if, both $x := \top$ and $x := \bot$ render F satisfiable (see, e.g., Cadoli et al. [2002] or Rintanen [1999b]). Many important problems in artificial intelligence can be mapped directly into QSAT, including conditional planning, abduction and non-monotonic reasoning [Rintanen, 1999a; 1999b]. QSAT also plays a prominent role in complexity theory, where it is prototypical and complete for the problems in the polynomial hierarchy.

Finally, *#SAT* is a variant of SAT in which, given a propositional formula F, the objective is to determine the number of models of F (counting variant) or to decide whether F has at least a given number of models (decision variant) [Roth, 1996; Bailey et al., 2001]. This problem has important applications to approximate reasoning problems in artificial intelligence; it is also of substantial theoretical interest, as the counting variant of #SAT is the prototypical complete problem for the complexity class #$\mathcal{P}$, and the decision variant is a prototypical complete problem for the probabilistic complexity class $\mathcal{PP}$ [Papadimitriou, 1994].

6.2 The GSAT Architecture

The *GSAT* algorithm [Selman et al., 1992] was one of the first SLS algorithms for SAT; it had a very significant impact on the development of a broad range of SAT solvers, including most of the current state-of-the-art SLS algorithms for SAT. Like all SAT algorithms covered in this chapter, GSAT is based on a 1-exchange neighbourhood in the space of all complete truth value assignments of the given formula; under this 'one-flip neighbourhood', two variable assignments are neighbours if, and only if, they differ in the truth assignment of exactly one variable. Furthermore, GSAT uses an evaluation function $g(F, a)$ that maps each variable assignment a to the number of clauses of the given formula F unsatisfied under a. Note that the models of F are exactly the assignments with evaluation function value zero. GSAT and most of its variants are iterative improvement methods that flip the truth value of one variable in each search step. The selection of the variable to be flipped is typically based on the *score* of a variable x under the current assignment a, which is defined as $g(F, a) - g(F, a')$, where a' is the assignment obtained from a by flipping the truth value of x. Algorithms of the GSAT architecture differ primarily in their underlying variable selection method. In the following, we describe some of the most widely known and best-performing GSAT algorithms.

```
procedure GSAT (F, maxTries, maxSteps)
    input: CNF formula F, positive integers maxTries and maxSteps
    output: model of F or 'no solution found'
    for try := 1 to maxTries do
        a := randomly chosen assignment of the variables in formula F;
        for step := 1 to maxSteps do
            if a satisfies F then return a end
            x := randomly selected variable flipping that minimizes
                    the number of unsatisfied clauses;
            a := a with x flipped;
        end
    end
    return 'no solution found'
end GSAT
```

Figure 6.2 The basic GSAT algorithm; all random selections are according to a uniform probability distribution over the underlying sets.

Basic GSAT

The core of the basic GSAT algorithm [Selman et al., 1992] consists of a simple best-improvement search procedure: Starting from a randomly chosen variable assignment, in each local search step, one of the variables with maximal score, that is, a variable that results in a maximal decrease in the number of unsatisfied clauses, is flipped. If there are several variables with maximal score, one of them is randomly selected according to a uniform distribution. The iterative best-improvement search underlying GSAT gets easily stuck in local minima of the evaluation function. Therefore, GSAT uses a simple static restart mechanism that re-initialises the search at a randomly chosen assignment every *maxFlips* flips. The search is terminated when a model of the given formula F has been found, or after *maxTries* sequences (also called 'tries') of *maxFlips* variable flips each have been performed without finding a model of F (see Figure 6.2).

Straightforward implementations of GSAT are rather inefficient, since in each step the scores of all variables have to be calculated from scratch. The key to efficiently implementing GSAT is to compute the complete set of scores only once at the beginning of each try, and then after each flip to update only the scores of those variables that were possibly affected by the flipped variable. Details on these implementation issues for GSAT and related algorithms are discussed in the in-depth section on page 271.

For any fixed number of restarts, GSAT is essentially incomplete [Hoos, 1998; 1999a], and severe stagnation behaviour is observed on most SAT instances.

Still, when it was introduced, GSAT outperformed the best systematic search algorithms for SAT available at that time. To date, basic GSAT's performance is substantially weaker than that of any of the other algorithms described in the following, and the algorithm is mainly of historical interest.

GSAT with Random Walk (GWSAT)

Basic GSAT can be significantly improved by extending the underlying search strategy into a randomised best-improvement method (see Chapter 2, page 72 *ff.*). This is achieved by introducing an additional type of local search step, so-called *conflict-directed random walk steps*. In this type of random walk step, first a currently unsatisfied clause c is selected uniformly at random. Then, one of the variables appearing in c is randomly selected and flipped, thus effectively forcing c to become satisfied. A simple SLS algorithm that initialises the search by randomly picking an assignment (like basic GSAT) and then performs a sequence of these conflict-directed random walk steps has been proven to solve 2-SAT in quadratic expected time [Papadimitriou, 1991]; this result inspired the use of this type of random walk to extend basic GSAT.

The basic idea of *GWSAT* is to decide at each local search step with a fixed probability wp (called *walk probability* or *noise setting*) whether to do a standard GSAT step or a variant of a conflict-directed random walk step, in which a variable is flipped that has been selected uniformly at random from the set of all variables occurring in currently unsatisfied clauses. Note that the variables that can be flipped in this latter type of random walk step are exactly the same as for the conflict-directed random walk steps described above, only the probabilistic bias may differ, depending on the number and length of clauses in which a given variable appears. For any $wp > 0$, this algorithm allows arbitrarily long sequences of random walk steps; this implies that from arbitrary assignments, a model (if existent) can be reached with a positive, bounded probability [Hoos, 1999a]. In particular, this allows the algorithm to escape from any local minima region of the underlying search space. Hence, the probability that GWSAT (without random restart) applied to a satisfiable formula finds a solution converges to one as the run-time approaches infinity, that is, GWSAT is probabilistically approximately complete (PAC). Like all GSAT algorithms, GWSAT uses the same static restart mechanism as basic GSAT.

Generally, GWSAT achieves substantially better performance than basic GSAT. It has been shown that when using sufficiently high noise settings (the precise value varies between problem instances), GWSAT does not suffer from stagnation behaviour. Furthermore, for hard SAT instances, it typically shows exponential RTDs [Hoos, 1998; Hoos and Stützle, 1999]; hence, static restart strategies are ineffective, and optimal speedup can be obtained by multiple independent runs parallelisation (see Chapter 4, Section 4.4). For low noise settings, stagnation behaviour is frequently observed; recently, there has been evidence that the corresponding RTDs can be characterised by mixtures of exponential distributions [Hoos, 2002b].

GSAT with Tabu Search (GSAT/Tabu)

The best-improvement search underlying basic GSAT can be easily extended into a simple tabu search strategy. *GSAT/Tabu* is obtained from basic GSAT by associating a tabu status with propositional variables of the given formula [Mazure et al., 1997; Steinmann et al., 1997]. In GSAT/Tabu, after a variable x has been flipped, it cannot be flipped back within the next tt steps, where the tabu tenure, tt, is a parameter of the algorithm. In each search step, the variable to be flipped is selected as in basic GSAT, except that the choice is restricted to variables that are currently not tabu. Upon search initialisation, the tabu status of all variables is cleared. Efficient implementations of GSAT/Tabu store for each variable x the search step number t_x when it was last flipped. When initialising the search, all the t_x are set to $-tt$; subsequently, every time a variable x is flipped, t_x is set to the current search step number t since the last initialisation of the search process. A variable x is tabu if, and only if, $t - t_x \leq tt$.

Unlike in the case of GWSAT, it is not clear whether GSAT/Tabu with fixed cutoff parameter *maxSteps* has the PAC property. Intuitively, for low tt, the algorithm may not be able to escape from extensive local minima regions without using restart, while for high tt settings, all the routes to a solution may be cut off, because too many variables are tabu. In practice, for very short tabu tenure, GSAT/Tabu often shows severe stagnation behaviour (the tt value for which this occurs depends on the given problem instance). For sufficiently high tabu tenure settings, GSAT/Tabu does not suffer from stagnation behaviour, and for hard problem instances, it shows exponential RTDs. As with GWSAT's noise parameter, very high settings of tt, although not causing stagnation behaviour, uniformly decrease GSAT/Tabu's performance.

Using instance-specific optimised tabu tenure settings for GSAT/Tabu and similarly optimised noise settings for GWSAT, GSAT/Tabu typically performs significantly better than GWSAT, particularly when applied to large and structured SAT instances [Hoos and Stützle, 2000a]. (There are, however, a few exceptional cases where GSAT/Tabu performs substantially worse than GWSAT, including well-known SAT-encoded instances of logistics planning problems.) Analogous to basic GSAT, GSAT/Tabu can be extended with a random walk mechanism; limited experimentation suggests that typically this hybrid algorithm does not perform better than GSAT/Tabu [Steinmann et al., 1997]. Overall, besides the dynamic local search algorithms covered in Section 6.4, GSAT/Tabu is one of the best-performing variants of GSAT known to date. (see Example 6.1 on page 280.)

HSAT and HWSAT

The intuition behind *HSAT* [Gent and Walsh, 1993b] is based on the observation that in basic GSAT, some variables might never get flipped although they are frequently eligible to be chosen. This can cause stagnation behaviour, since one of these variables may have to be flipped to allow the search to make further progress. Therefore, when in a search step there are several variables with identical score, HSAT always selects the least recently flipped variable, that is, the variable that was flipped longest ago. Only shortly after search initialisation, when there are still variables that have not been flipped, HSAT performs the same random tie breaking between variables with identical score as plain GSAT. Apart from this difference in the variable selection mechanism, HSAT is identical to basic GSAT.

Although HSAT was found to show superior performance over basic GSAT [Gent and Walsh, 1993b], it is clear that it is even more likely to get stuck in local minima from which it cannot escape, since the history-based tie-breaking rule effectively restricts the search trajectories when compared to GSAT. To counteract this problem, HSAT can be extended with the same random walk mechanism as used in GWSAT. The resulting variant is called *HWSAT* [Gent and Walsh, 1995]; like GWSAT, HWSAT has the PAC property. Generally, HWSAT shows improved peak performance over GWSAT. Compared to GSAT/Tabu, HWSAT's performance appears to be somewhat better on hard Uniform Random 3-SAT instances and certain types of structured SAT problems, and significantly worse in many other cases [Hoos and Stützle, 2000a].

IN DEPTH EFFICIENTLY IMPLEMENTING GSAT

The key to implementing GSAT algorithms efficiently lies in caching and updating the variable scores that form the basis for selecting the variable to be flipped in each search step. Typically, not all variable scores change after each search step; this suggests that rather than recomputing all variable scores in each step, it should be more efficient to compute all scores when the search is initialised, but to subsequently only update the scores affected by a variable that has been flipped. The following definition will help to explain the precise mechanism for incrementally updating the scores and to analyse its time complexity.

DEFINITION 6.1   **Variable and Clause Dependencies**

Given a CNF formula F and two variables x, x' appearing in F, x' is dependent on x (and vice versa) if, and only if, there is a clause in which both x and x' appear.
Furthermore, we define the set of variables dependent on x *as*

$$V_{dep}(F, x) := \{x' \in Var(F) \mid x\}' \text{ is dependent on } x$$

A clause c of F is dependent on x, *if, and only if, x appears in c, and the* set of clauses dependent on x *is defined as*

$$C_{dep}(F, x) := \{c \text{ is a clause of } F \mid c\} \text{ is dependent on } x$$

A clause c is critically satisfied *by a variable x under assignment a if, and only if, x appears in c, c is satisfied under a, and flipping the value of x makes c unsatisfied. Finally, a variable x' is* critically dependent *on a variable x under assignment a, if, and only if, there is a clause c that is dependent on x and x', and flipping x results in the clause to change its satisfaction status from (i) satisfied to unsatisfied or vice versa, or (ii) satisfied to critically satisfied (by x') or vice versa.*

After flipping a variable x, only clauses dependent on x can change their satisfaction status; hence, in order to update the evaluation function value (i.e., the number of unsatisfied clauses), only the clauses in $C_{dep}(x, F)$ need to be considered. According to the definition of a variable's score, the score of x just changes its sign as a consequence of flipping x. For all other variables $x' \neq x$, the score of x' remains unchanged if x' is not dependent on x, that is, if $x' \notin V_{dep}(F, x)$. Hence, after flipping x, only the scores of the variables in $V_{dep}(F, x)$ need to be updated. In fact, among those, only the scores of variables that critically depend on x can actually change.

For a given formula F with n variables, m clauses, and a clause length (number of literals per clause) bounded from above by $CL(n)$, the time complexity of computing all variable scores is $O(m \cdot CL(n))$. This is achieved by going through all clauses, checking their satisfaction status, and increasing or decreasing the scores of the variables appearing in a clause c, depending on whether c is currently unsatisfied, or whether it is critically satisfied by a given variable. At the end of this process, the evaluation function value, a list of all unsatisfied clauses, and all variable scores have been computed.

After each search step, all variable scores that are affected by the respective flip can be updated in time $O(CD(n) \cdot CL(n))$, where $CD(n)$ is an upper bound on the cardinality of the sets $C_{dep}(F, x)$. This is achieved by going through all clauses that are dependent on the flipped variable, x, and updating the scores of the variables occurring in these, depending on the (critical) satisfaction status of the respective clause before and after the flip of x. In order to perform this operation efficiently, for each variable x, a list is kept of the clauses that are dependent on x; these lists are built when parsing the input formula. For each variable, we furthermore store its current truth value and score, and for each clause, we store its (critical) satisfaction status under the current assignment.

For Uniform Random k-SAT formulae with constant clauses/variable ratio, the average number of dependent clauses for each variable is constant. Therefore, independent of instance size, this implementation of GSAT achieves a time complexity of $\Theta(1)$ for each search step, compared to $\Theta(n^2)$ for a naïve implementation in which all variable scores are computed before every variable flip. For SAT-encoded instances of other combinatorial problems, there are typically more extensive variable dependencies, leading to a somewhat reduced, but still substantial performance advantage of the efficient implementation described above.

The efficient mechanism for caching and updating variable scores described here is also used in Selman and Kautz's publicly available reference implementation of GSAT. Very similar techniques can be used for efficiently implementing other SLS algorithms, such as Galinier and Hao's tabu search algorithm for the CSP, which is outlined in Section 6.6. Interestingly, for the WalkSAT algorithms described in the following, a more straight-forward implementation, which does not use the previously described caching and incremental updating scheme, achieves slightly better performance.

6.3   The WalkSAT Architecture

The WalkSAT architecture is based on ideas first published by Selman, Kautz and Cohen [1994] and was later formally defined as an algorithmic framework by McAllester, Selman and Kautz [1997]. WalkSAT can be seen as an extension of the conflict-directed random walk method that is also used in Papadimitriou's algorithm [1991] and GWSAT. It is based on a 2-stage variable selection process focused on the variables occurring in currently unsatisfied clauses. For each local search step, in a first stage, a clause c that is unsatisfied under the current assignment is selected uniformly at random. In a second stage, one of the variables appearing in c is then flipped to obtain the new assignment. Thus, while the GSAT architecture is characterised by a static neighbourhood relation between assignments with Hamming distance one, using this two-stage procedure, WalkSAT algorithms are effectively based on a dynamically determined subset of the GSAT neighbourhood relation. As a consequence of this substantially reduced effective neighbourhood size, WalkSAT algorithms can be implemented efficiently without caching and incrementally updating variable scores and still achieve substantially lower CPU times per search step than efficient GSAT implementations [Hoos, 1998; Hoos and Stützle, 2000a]. All WalkSAT algorithms considered here use the same random search initialisation and static random restart as GSAT. A pseudo-code representation of the WalkSAT architecture is shown in Figure 6.3.

```
procedure WalkSAT (F, maxTries, maxSteps, slc)
    input: CNF formula F, positive integers maxTries and maxSteps,
        heuristic function slc
    output: model of F or 'no solution found'

    for try := 1 to maxTries do
        a := randomly chosen assignment of the variables in formula F;
        for step := 1 to maxSteps do
            if a satisfies F then return a end
            c := randomly selected clause unsatisfied under a;
            x := variable selected from c according to heuristic function slc;
            a := a with x flipped;
        end
    end
    return 'no solution found'
end WalkSAT
```

Figure 6.3 The WalkSAT algorithm family. All random selections are according to a uniform probability distribution over the underlying sets; WalkSAT algorithms differ in the variable selection heuristic *slc*.

WalkSAT/SKC

The first WalkSAT algorithm, *WalkSAT/SKC,* originally introduced in a paper by Selman, Kautz and Cohen [1994], differs in one important aspect from most of the other SLS algorithms for SAT: The scoring function $score_b(x)$ used by Walk-SAT/SKC counts the number of currently satisfied clauses that will be broken — that is: become unsatisfied — by flipping a given variable x. Using this scoring function, the following variable selection scheme is applied: If there is a variable with $score_b(x) = 0$ in the clause selected in stage 1, that is, if c can be satisfied without breaking another clause, this variable is flipped (*zero damage step*). If more than one such variable exists in c, one of them is selected uniformly at random and flipped. If no such variable exists, with a certain probability *1-p*, the variable with minimal $score_b$ value is selected (*greedy step;* ties are broken uniformly at random); in the remaining cases, that is, with probability p (the so-called *noise setting*), one of the variables from c is selected uniformly at random (*random walk step*).

Conceptually as well as historically, WalkSAT/SKC is closely related to GWSAT. However, there are a number of significant differences between both algorithms, which in combination account for the generally superior performance of WalkSAT/SKC. Both algorithms use closely related types of random walk steps; but WalkSAT/SKC applies them only under the condition that there is no variable with $score_b(x) = 0$. In GWSAT, on the other hand, random walk steps are performed in an unconditional probabilistic way. From this point of view, WalkSAT/SKC is greedier, since random walk steps, which usually increase the number of unsatisfied clauses, are only performed when every variable occurring in the selected clause would break some clauses when flipped. Yet, in a greedy step, due to its two-stage variable selection scheme, WalkSAT/SKC chooses from a significantly reduced set of neighbours and can therefore be considered less greedy than GWSAT. Finally, because of the different scoring function, in some sense, GWSAT shows a greedier behaviour than WalkSAT/SKC: In a best-improvement step, GWSAT may prefer a variable that breaks some clauses, but compensates for this by fixing other clauses, whilst in the same situation, WalkSAT/SKC would select a variable that may lead to a smaller reduction in the total number of unsatisfied clauses, but breaks fewer currently satisfied clauses.

It has been proven that WalkSAT/SKC with fixed *maxTries* parameter has the PAC property when applied to 2-SAT [Culberson et al., 2000], but it is not known whether the algorithm is PAC in the general case. Note that, differently from GWSAT, it is not clear whether WalkSAT/SKC can perform arbitrarily long sequences of random walk steps, since random walk steps are only possible when the selected clause does not allow any zero damage steps. In practice, however, WalkSAT/SKC does not appear to suffer from any stagnation behaviour when using sufficiently high (instance-specific) noise settings, in which case its run-time behaviour is characterised by exponential RTDs [Hoos, 1998; Hoos and Stützle, 1999; 2000a]. Like in the case of GWSAT, stagnation behaviour is frequently observed for low noise settings, and there is some evidence that the corresponding RTDs can be characterised by mixtures of exponential distributions [Hoos, 2002b].

Generally, when using (instance-specific) optimised noise settings, Walk-SAT/SKC probabilistically dominates GWSAT in terms of the number of variable flips required for finding a model to a given formula, but it does not always reach the performance of HWSAT or GSAT/Tabu. When comparing CPU time, however, WalkSAT/SKC typically outperforms all GSAT variants presented in Section 6.2. (See also Example 6.1 on page 280.)

WalkSAT with Tabu Search (WalkSAT/Tabu)

Analogously to GSAT/Tabu, there is also an extension to WalkSAT/SKC that uses a simple tabu search mechanism. *WalkSAT/Tabu* [McAllester et al., 1997] uses the same two-stage selection mechanism and the same scoring function $score_b$

as WalkSAT/SKC and additionally enforces a tabu tenure of tt steps for each flipped variable. (To implement this tabu mechanism efficiently, the same approach is used as described in Section 6.2 for GSAT/Tabu.) If the selected clause c does not allow a zero damage step, of all the variables occurring in c that are not tabu, WalkSAT/Tabu picks the one with the highest $score_b$ value; when there are several variables with the same maximal score, one of them is selected uniformly at random. It may happen, however, that all variables appearing in c are tabu, in which case no variable is flipped (a so-called *null-flip*).

WalkSAT/Tabu with fixed *maxTries* parameter has been shown to be essentially incomplete [Hoos, 1998; 1999a]. Although this is mainly caused by null-flips, it is not clear whether replacing null-flips by random walk steps, for instance, would be sufficient for obtaining the PAC property. In practice, when using sufficiently high (instance-specific) tabu tenure settings, WalkSAT/Tabu's run-time behaviour is characterised by exponential RTDs; but there are cases (particularly for structured SAT instances) in which extreme stagnation behaviour is observed. Typically, however, WalkSAT/Tabu performs significantly better than WalkSAT/SKC, and there are structured SAT instances (e.g., large SAT-encoded blocks world planning problems), for which WalkSAT/Tabu appears to achieve better performance than any other SLS algorithm currently known [Hoos and Stützle, 2000a].

Novelty and Novelty$^+$

Novelty [McAllester et al., 1997] is a WalkSAT algorithm that uses a history-based variable selection mechanism similar to HSAT. In Novelty, the number of local search steps that have been performed since a variable was last flipped is taken into consideration; this value is called the variable's *age*. An important difference of Novelty compared to WalkSAT/SKC and WalkSAT/Tabu is that it uses the same scoring function as GSAT.

In Novelty, after an unsatisfied clause has been chosen, the variable to be flipped is selected as follows. If the variable with the highest score does not have minimal age among the variables within the same clause, it is always selected. Otherwise, it is only selected with a probability of *1-p*, where p is a parameter called the *noise setting*. In the remaining cases, the variable with the next lower score is selected (see also Figure 6.4). When sorting the variables according to their scores, ties are broken according to decreasing age. (If there are several variables with identical score and age, the reference implementation by Kautz and Selman always chooses the one appearing first in the selected clause.)

Note that for $p > 0$, the age-based variable selection of Novelty probabilistically prevents flipping the same variable over and over again; at the same time,

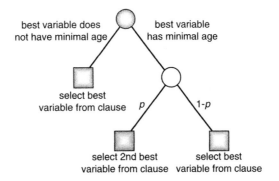

Figure **6.4** Decision tree representation of Novelty's mechanism for selecting a variable to be flipped within a given clause. Deterministic and probabilistic choices are represented by shaded and white circles, respectively; edges are labelled with the respective conditions and probabilities. Shaded boxes indicate variable decision actions.

flips can be immediately reversed with a certain probability if no better choice is available. Generally, the Novelty algorithm is significantly greedier than Walk-SAT/SKC, since always one of the two most improving variables from a clause is selected, where WalkSAT/SKC may select any variable if no improvement can be achieved without breaking other clauses. Also, Novelty is more deterministic than WalkSAT/SKC and GWSAT, since its probabilistic decisions are more limited in their scope and take place under more restrictive conditions. For example, different from WalkSAT/SKC, the Novelty strategy for variable selection within a clause is deterministic for both $p = 0$ and $p = 1$.

On the one hand, this typically leads to a significantly improved performance of Novelty when compared to WalkSAT/SKC. On the other hand, because of this property, it can be shown that, for fixed *maxTries* setting, Novelty is essentially incomplete [Hoos, 1998], because selecting only among the best two variables in a given clause can lead to situations where the algorithm gets stuck in local minima of the objective function. This situation has been observed for a number of commonly used benchmark instances, where it severely compromises Novelty's performance [Hoos and Stützle, 2000a].

By extending Novelty with conflict-directed random walk analogously to GWSAT, the essential incompleteness as well as the empirically observed stagnation behaviour can be overcome. The *Novelty*$^+$ algorithm [Hoos, 1998; 1999a] selects the variable to be flipped according to the standard Novelty mechanism with probability $1 - wp$, and performs a random walk step, as defined above for GWSAT, in the remaining cases. A GLSM model of the resulting algorithm is shown in Figure 6.5.

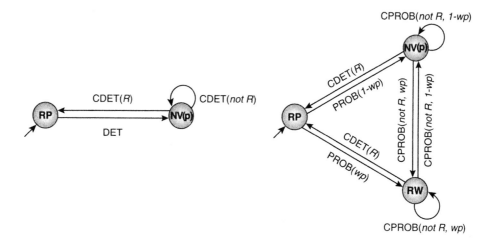

Figure **6.5** GLSM models for Novelty (left side) and Novelty$^+$ (right side); the restart predicate R is equal to countm(m), GLSM state RP initialises the search at a randomly selected variable assignment, NV(p) performs a Novelty step (with noise setting p), and RW performs a random walk step (see text for details).

Novelty$^+$ is provably PAC for $wp > 0$ and shows exponential RTDs for sufficiently high (instance-specific) settings of the primary noise parameter, p. In practice, small walk probabilities, wp, are generally sufficient to prevent the extreme stagnation behaviour that is occasionally observed for Novelty and to achieve substantially superior performance compared to Novelty. In fact, a setting of $wp := 0.01$ seems to result in uniformly good performance [Hoos, 1999a], and the algorithm's performance appears to be much more robust w.r.t. to the wp parameter than w.r.t. to the primary noise setting, p. In cases where Novelty does not suffer from stagnation behaviour, Novelty$^+$'s performance for $wp := 0.01$ is typically almost identical to Novelty's. Overall, Novelty$^+$ is one of the best-performing WalkSAT algorithms currently known and one of the best SLS algorithms for SAT available to date [Hoos and Stützle, 2000a; Hutter et al., 2002].

R-Novelty and R-Novelty$^+$

R-Novelty [McAllester et al., 1997] is a variant of Novelty that is based on the intuition that, when deciding between the best and second best variable (using the same scoring function as for Novelty), the actual difference of the respective scores should be taken into account. The exact mechanism for choosing a variable from the selected clause can be seen from the decision tree representation shown in Figure 6.6. Note that the R-Novelty heuristic is quite complex — as reported by McAllester et al. [1997], it was discovered by systematically testing a large number of WalkSAT variants.

R-Novelty's variable selection strategy is even more deterministic than Novelty's; in particular, it is completely deterministic for any $p \in \{0, 0.5, 1\}$. Since the pure R-Novelty algorithm gets too easily stuck in local minima, an extremely simple diversification mechanism is used: Every 100 steps, a variable is randomly chosen from the selected clause and flipped. As shown in [Hoos, 1998; 1999a], this loop breaking strategy is generally not sufficient for effectively escaping from local minima and leaves R-Novelty essentially incomplete (for fixed *maxTries*); as in the case of Novelty, severe stagnation behaviour is observed in practice for some SAT instances [Hoos and Stützle, 2000a]. R-Novelty's performance is often, but not always, superior to Novelty's.

Replacing the original diversification mechanism in R-Novelty with a random walk mechanism exactly analogous to the one used in Novelty$^+$ leads to the *R-Novelty$^+$* algorithm [Hoos, 1998; 1999a]. Like Novelty$^+$, R-Novelty$^+$ is provably PAC for $wp > 0$ and shows exponential RTDs for sufficiently high (instance-specific) noise settings. Again, a small walk probability of $wp := 0.01$ appears to be generally sufficient for avoiding stagnation behaviour and for robustly achieving good performance in practice. R-Novelty$^+$'s performance for instances on which R-Novelty does not suffer from stagnation behaviour is very similar to R-Novelty's. There is some indication that R-Novelty and R-Novelty$^+$ do not reach the performance of Novelty$^+$ on several classes of structured SAT

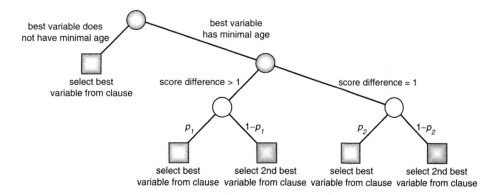

Figure 6.6 Decision tree representation of the mechanism used by R-Novelty for selecting a variable to be flipped in a given clause c; 'score difference' refers to the difference in score between the best and the second best variable in c; $p_1 := \min\{2 - 2p, 1\}$, $p_2 := \max\{1 - 2p, 0\}$.

instances, including SAT-encoded hard graph colouring and planning problems [Hoos and Stützle, 2000a].

EXAMPLE 6.1 **Performance Results for SLS Algorithms for SAT**

To illustrate the performance differences between various GSAT and Walk-SAT algorithms, we empirically analysed their performance on a number of well-known benchmark instances for SAT. All performance results reported in the following are based on at least 100 runs per problem instance, conducted on a PC with a 2GHz Xeon CPU, 512KB cache, and 4GB RAM, running Red Hat Linux 2.4.20-18.9. All algorithms were run with optimised parameters (noise and tabu tenure) and without restart.

The left side of Figure 6.7 shows the run-time distributions of GSAT with Random Walk (GWSAT), GSAT/Tabu, WalkSAT/SKC and Novelty$^+$, determined from 500 runs of the algorithm on a SAT-encoded, hard graph colouring instance with 100 vertices and 239 edges from SATLIB [Hoos and Stützle, 2000a]. The observed performance differences, which are consistent across all percentiles of the RTDs, are typical for many types of SAT instances. Furthermore, all four RTDs can be well approximated by exponential distributions, which again is characteristic for these algorithms when using sufficiently high noise and tabu tenure settings [Hoos and Stützle, 2000a]. It may

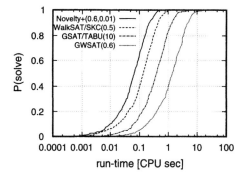
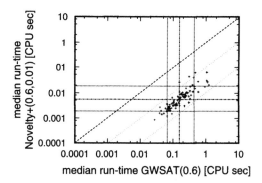

Figure 6.7 Left: Run-time distributions for various GSAT and WalkSAT algorithms on a SAT-encoded graph colouring instance. *Right:* correlation of median search cost between GSAT with Random Walk (GWSAT) and Novelty$^+$ on a set of randomly generated, SAT-encoded graph colouring problems; the horizontal and vertical lines indicate the median, $q_{0.1}$ and $q_{0.9}$ of the search cost for the respective algorithm across the self-test; the diagonal lines indicate equal, 1/10th and 1/100th CPU time of Novelty$^+$ compared to GWSAT. (For further details, see text.)

be noted that basic GSAT, when run on the same problem instance, could not find a solution in 500 runs of 10 CPU seconds each.

The right side of Figure 6.7 illustrates the correlation of search cost between GWSAT and Novelty[+] across a set of 100 instances from the same distribution of randomly generated, hard graph colouring instances as the previously studied instance. Each data point in the graph represents the median CPU time required by GWSAT *vs* Novelty[+] on a single problem instance, determined from an RTD based on 100 runs. Horizontal and vertical lines indicate the median as well as the $q_{0.1}$ and $q_{0.9}$ percentiles of the distribution of search cost for the two algorithms, respectively, across the entire test-set. As can be clearly seen from this correlation plot, Novelty[+] performs substantially better than GWSAT across the entire test-set. Furthermore, the hardness of the problem instances for both algorithms is highly correlated, indicating that both algorithms are affected by the same features of the respective instances (in this case, the solution density, which varies substantially across the test-set; see also Chapter 5, Section 5.1). Similar results hold for all GSAT and WalkSAT algorithms discussed in this chapter.

Table 6.1 summarises performance results for several GSAT and Walk-SAT algorithms on a test-set comprising a hard instance from the solubility phase transition of Uniform Random 3-SAT, as well as SAT-encoded instances of graph colouring, Boolean function learning, and planning problems. These SAT instances range in size from 75 variables and 298 clauses (par8-5-c) to 3 016 variables and 50 457 clauses (bw_large.c). The results

Problem Instance	GSAT/Tabu	WalkSAT/SKC	WalkSAT/ Tabu	Novelty[+]
uf200/hard	45.4 $(9.30 \cdot 10^6)$	1.04 $(0.85 \cdot 10^6)$	4.61 $(3.70 \cdot 10^6)$	**0.82 $(0.61 \cdot 10^6)$**
flat100/hard	0.39 (73 594)	0.15 (192 788)	0.18 (229 496)	**0.06(80 315)**
par8-5-c	0.22 (45 027)	0.010 (13 345)	0.006 (8 388)	**0.003(3 341)**
logistics.d	—	0.58 (398 277)	0.49 (332 494)	**0.21(113 664)**
bw_large.a	0.09 (6 977)	0.02 (13 505)	0.01 (7 563)	**0.01(6 780)**
bw_large.c	11.9 $(1.01 \cdot 10^6)$	23.6 $(9.76 \cdot 10^6)$	**5.6 $(2.00 \cdot 10^6)$**	11.4 $(4.36 \cdot 10^6)$

Table 6.1 Performance of various WalkSAT algorithms on selected benchmark instances for SAT; the table entries are median run-times obtained from RTDs based on 100 or more runs per instance, reported in CPU seconds (search steps). All algorithms solved any given problem instance in every run, with the exception of GSAT/Tabu, which did not solve logistics.d in 10 runs of 10 CPU seconds each for any of a number of tabu tenure settings tested, and which solved bw_large.c in only 247 of 250 runs.

illustrate the excellent performance of Novelty[+] and WalkSAT/Tabu compared to other GSAT and WalkSAT algorithms. Note also that GSAT/Tabu often performs better than the WalkSAT algorithms in terms of search steps required for solving a given problem instance; yet, as previously explained, this rarely results in faster run-times, since search steps of WalkSAT algorithms can be implemented more efficiently than those of GSAT algorithms. More detailed results on the performance and behaviour of GSAT and Walk-SAT algorithms can be found in Hoos and Stützle [2000a].

WalkSAT with Adaptive Noise

The noise parameter, p, which is common to all WalkSAT algorithms discussed here with the exception of WalkSAT/Tabu (where the tabu tenure tt plays a similar role), has a major impact on the performance and run-time behaviour of the respective algorithm. For low noise settings, stagnation behaviour is typically observed, and as a consequence, using an appropriate *maxSteps* setting for the static restart mechanism becomes crucial for obtaining good performance [Hoos and Stützle, 2000a]. For sufficiently high noise settings, however, the *maxSteps* setting has typically little or no impact on the behaviour of the algorithm [Parkes and Walser, 1996; Hoos and Stützle, 1999], since the corresponding RTDs are closely approximated by exponential distributions. (There are exceptions to this general observation, including instances on which essentially incomplete WalkSAT variants show extreme stagnation behaviour, as well as the irregular instances recently described by Hoos [2002b].) Fortunately, for many of the most prominent and best-performing WalkSAT algorithms, including WalkSAT/SKC, WalkSAT/Tabu, Novelty[+] and R-Novelty[+], the noise settings required for reaching peak performance are generally high enough that the cutoff parameter, *maxSteps*, does not affect performance unless it is chosen too low, in which case performance is degraded. This leaves the noise setting, p, to be optimised in order to achieve maximal performance of these WalkSAT algorithms.

Unfortunately, finding the optimal noise setting is typically rather difficult. Because optimal noise settings appear to differ considerably depending on the given problem instance, this task often requires experience and substantial experimentation with various noise values [Hoos and Stützle, 2000a]. It has been shown that even relatively minor deviations from the optimal noise setting can lead to a substantial increase in the expected time for solving a given instance; and to make matters worse, the sensitivity of WalkSAT's performance w.r.t. the noise setting seems to increase with the size and hardness of the problem instance to be solved [Hoos, 2002a]. This complicates the use of WalkSAT for solving SAT instances as well as the evaluation, and hence the development, of new WalkSAT algorithms.

The key idea behind *Adaptive WalkSAT* [Hoos, 2002a] is to use high noise values only when they are needed to escape from stagnation situations in which the search procedure appears to make no further progress towards finding a solution. This idea is closely related to the motivation behind Reactive Tabu Search [Battiti and Tecchiolli, 1994]. More precisely, Adaptive WalkSAT dynamically adjusts the noise setting p, and hence the probability for performing greedy steps, based on search progress, as reflected in the time elapsed since the last improvement in the evaluation function has been achieved. At the beginning of the search process, the search is maximally greedy ($p := 0$). This will typically lead to a series of rapid improvements in the evaluation function value, followed by stagnation (unless a solution to the given problem instance is found). In this situation, the noise value is increased. If this increase is not sufficient to escape from the stagnation situation, that is, if it does not lead to an improvement in evaluation function value within a certain number of steps, the noise setting is further increased. Eventually, the noise setting should be high enough for the search process to overcome the stagnation situation, at which point the noise can be gradually decreased until the next stagnation situation is detected or a solution to the given problem instance is found.

As an indicator for search stagnation, Adaptive WalkSAT uses a predicate that is true if, and only if, no improvement in evaluation function value has been observed over the last $\theta \cdot m$ search steps, where m is the number of clauses of the given problem instance, and θ is a parameter. Every increase in the noise setting is realised as $p := p + (1 - p) \cdot \phi$. The decrements are defined as $p := p - p \cdot \phi/2$, where p is the noise level, and ϕ is an additional parameter. The asymmetry between increases and decreases in the noise setting is motivated by the fact that detecting search stagnation is computationally more expensive than detecting search progress, and by the observation that it is advantageous to approximate optimal noise levels from above rather than from below [Hoos, 2002a]. After the noise setting has been increased or decreased, the current evaluation function value is stored and becomes the basis for measuring improvement, and hence for detecting search stagnation. As a consequence, between increases in noise level, there is always a phase during which the trajectory is monitored for search progress without further increasing the noise. No such delay is enforced between successive decreases in noise level.

It may be noted that the behaviour of the adaptive noise mechanism is controlled by two parameters, θ and ϕ. While one might assume that this merely replaces the problem of tuning one parameter, p, with the potentially more difficult problem of tuning these new parameters, it appears that the performance of Adaptive WalkSAT is much more robust w.r.t. to the settings of θ and ϕ, than WalkSAT is w.r.t. to the noise setting. Using fixed settings of $\theta := 1/6$ and $\phi := 0.2$ for Adaptive Novelty$^+$ generally seems to result in similar performance as observed for Novelty$^+$ with approximately optimal, instance-specific noise settings; in some cases, Adaptive Novelty$^+$ achieves significantly better performance than Novelty$^+$ with approximately optimal static noise [Hoos, 2002a], which makes Adaptive Novelty$^+$ one of the best-performing and most robust SLS algorithms for SAT currently available.

6.4   Dynamic Local Search Algorithms for SAT

The first application of Dynamic Local Search to SAT was proposed around the same time as GWSAT. Since then, a number of DLS algorithms for SAT have been developed, the most recent of which achieve better performance than the best GSAT and WalkSAT variants for many types of SAT instances and can therefore be seen as the best performing SLS algorithms for SAT currently known.

Most DLS algorithms for SAT are based on variants of GSAT as their underlying local search procedure. The solution components that are being selectively penalised are the clauses of the given formula; in the following, we denote the penalty associated with clause c by $clp(c)$. (Here and in the following we assume — without loss of generality — that all clauses of a given CNF formula are pairwise different.) Consistent with the general outline for DLS algorithms from Section 2.2 (page 82*ff.*), typically a modified evaluation function of the form

$$g'(F, a) := g(F, a) + \sum_{c \in CU(F,a)} clp(c)$$

is used within the local search procedure, where $CU(F, a)$ is the set of all clauses in F that are unsatisfied under assignment a. Many DLS algorithms for SAT use the notion of clause weights $clw(c)$ instead of clause penalties, where

$$clw(c) := clp(c) + 1$$

and

$$g'(F, a) := \sum_{c \in CU(F,a)} clw(c).$$

For $g(F, a) := \#CU(F, a)$, the standard evaluation function used by most SLS algorithms for SAT, both definitions of $g'(F, a)$ are equivalent. The major differences between DLS algorithms for SAT are in the details of the local search procedure and in the scheme used for updating the clause penalties or weights.

Most DLS algorithms for SAT perform excellently in terms of the number of variable flips required for finding a model of a given formula. However, the time complexity and frequency of the weight updates is typically rather high, which makes it difficult for DLS algorithms to reach or exceed the time performance of the best-performing WalkSAT variants. Unfortunately, the run-time behaviour of DLS algorithms for SAT has not been as thoroughly investigated as that of GSAT and WalkSAT algorithms. In particular, little is know about these algorithms in terms of their asymptotic run-time behaviour, search stagnation and RTD characterisations.

GSAT with Clause Weights

This early DLS algorithm for SAT is based on the observation that when applied to certain types of structured SAT instances, basic GSAT often finds the same set of clauses unsatisfied at the end of a run [Selman and Kautz, 1993]. In this GSAT variant, weights are associated with each clause. These weights are initially set to one; before each restart, the weights of all currently unsatisfied clauses are increased by $\delta := 1$. The underlying local search procedure is a variant of basic GSAT that uses the modified evaluation function $g'(F, a)$ introduced above. It may be noted that for sufficiently high *maxSteps* settings, this local search procedure will terminate in or very close to a local minima region of the underlying search space. Different from the other DLS methods discussed in this section, *GSAT with Clause Weights* begins each local search phase from a randomly selected variable assignment. (A further extension, called 'Averaging In', uses a modified search initialisation that introduces a bias towards the best candidate solutions reached in previous local search phases [Selman and Kautz, 1993].)

GSAT with Clause Weights performs substantially better than basic GSAT on various classes of structured SAT instances, including SAT-encoded graph colouring problems; there is also some indication that by using the same clause weighting mechanism with GWSAT, further performance improvements can be achieved [Selman and Kautz, 1993]. Today, since its performance is not competitive with any of the more recent DLS algorithms for SAT presented in the following, GSAT with Clause Weights is mainly of historical interest.

Several variants of GSAT with Clause Weights have been studied by Cha and Iwama [1995]. In particular, they proposed and tested a variant that — like the Breakout Method [Morris, 1993], an earlier DLS algorithm for the CSP — performs weight updates whenever a local minimum of the modified evaluation function is encountered and, in its basic form, does not perform restarts. This algorithm appears to perform substantially better than GSAT and GWSAT when applied to a class of randomly generated SAT instances that have only a single model [Asahiro et al., 1996]. (These instances, however, are not intrinsically hard, because they can be solved by polynomial simplifications, and hence they are only of limited use as benchmark problems [Hoos and Stützle, 2000a].) There is no evidence that this variant performs better than the original GSAT with Clause Weights algorithm.

Cha and Iwama also investigated slight variations of the weight update scheme, as well as combinations of their basic algorithm with static restarts and a simple tabu search strategy that, different from GSAT/Tabu or WalkSAT/Tabu, associates a tabu status with the most recently visited variable assignments, rather than with recently flipped variables [Cha and Iwama, 1995]. From their limited empirical results it appears that none of these variations achieves significant performance improvements over their previously described, basic variant of GSAT with Clause Weights.

Methods Using Rapid Weight Adjustments

Frank introduced several variants of GSAT with Clause Weights that perform weight updates after each local search step [Frank, 1996; 1997]. The underlying idea is that GSAT should benefit from discovering which clauses are most difficult to satisfy relative to recent assignments. The most basic of these variants, called *WGSAT*, uses the same weight initialisation and update procedure as GSAT with Clause Weights, but performs only a single GSAT step before updating the clause weights. On hard Random 3-SAT instances, WGSAT achieves a significantly improved performance over HSAT (and hence, basic GSAT) when measuring run-time in terms of variable flips required for finding a solution [Frank, 1996; 1997]. When comparing CPU times however, it appears that due to the computational overhead caused by the frequent weight updates, WGSAT's performance cannot reach that of HSAT or GWSAT.

A modification of this algorithm, called *UGSAT*, uses a best-improvement local search strategy, but restricts the neighbourhood considered in each search step to the set of variables appearing in currently unsatisfied clauses [Frank, 1996]. (Note that this is the same effective neighbourhood as used in the random walk steps of GWSAT.) While this leads to considerable speedups for naïve implementations of the underlying local search procedure, the difference for efficient implementations is likely to be insufficient to render UGSAT competitive with HSAT or GWSAT.

Another variant of WGSAT implements a uniform decay of clause weights over time. The underlying idea is that the relative importance of clauses w.r.t.

their satisfaction status can change during the search, and hence a mechanism is needed that focuses the weighted search on the most recently unsatisfied clauses. In *WGSAT with Decay*, this idea is implemented by uniformly decaying all clause weights in each weight update phase before the weights of the currently unsatisfied clauses are increased; this decay is performed according to the formula $clw(c) := \rho \cdot clw(c)$, where the decay rate ρ (with $0 < \rho < 1$) is a parameter of the algorithm [Frank, 1997]. Empirical results suggest that on larger instances from the phase transition region of Uniform Random 3-SAT, using this decay mechanism slightly improves the performance of WGSAT when measured in terms of variable flips required for finding a model; this improvement, however, appears to be insufficient to amortise the added time complexity of the frequent weight update steps. Nevertheless, as we will see later in this section, similar mechanisms for focusing the search on recently unsatisfied clauses play a crucial role in state-of-the-art DLS algorithms for SAT.

Guided Local Search (GLS)

This DLS algorithm has been applied to a number of combinatorial problems [Voudouris, 1997; Voudouris and Tsang, 1999]. *GLS for SAT (GLSSAT)* [Mills and Tsang, 1999a; 2000] is based on a local search algorithm that, similar to HSAT, Novelty and R-Novelty, implements a bias towards flipping variables whose respective values have not been changed recently. More precisely, in each local search step, from the set of all variables that, when flipped, would lead to a strict decrease in the total penalty of unsatisfied clauses, the one whose last flip has occurred least recently is flipped. If no such strictly improving variable exists, the same selection is made from the set of all variables that, when flipped, do not cause an increase in the evaluation function value. The subsidiary local search procedure terminates when a satisfying assignment is found, or after a fixed number *smax* of consecutive non-improving flips has been made.

Before the actual search begins, GLSSAT performs a complete pass of unit propagation in order to simplify the given formula. Then, all clause penalties are initialised to zero, and the search starts from a variable assignment that is chosen uniformly at random.

After each local search phase, the penalties of all clauses with maximal utilities are incremented by $\delta := 1$, where the utility of a clause c under assignment a is defined as $util(a, c) := 1/(1 + clp(c))$ if clause c is unsatisfied under x, and zero otherwise. Note that this corresponds to incrementing the smallest clause penalties occurring in currently unsatisfied clauses. An important extension of GLSSAT uses an additional mechanism for bounding the range of the clause penalties: If after updating the clause penalties, the maximum penalty exceeds a given threshold, *pmax*, all clause penalties are uniformly decayed by multiplying them with a factor *pdecay*. This clause penalty decay mechanism has a substantial impact on the performance of GLSSAT and significantly improves the algorithm's efficacy in solving large and hard structured instances. A similar modification of GLSSAT, called *GLSSAT2*, was used in another study [Mills and Tsang, 2000]; in this variant, all clause penalties are multiplied by a factor *pdecay* := 0.8 after every 200 penalty updates.

GLSSAT achieves better performance than WalkSAT/SKC on some widely used benchmark instances when measuring run-time in terms of variable flips, but in many cases WalkSAT/SKC is superior in terms of CPU time [Mills and Tsang, 2000]. There are some hard structured SAT instances, however, for which GLSSAT2 appears to perform significantly better than WalkSAT/SKC. Indirect evidence suggests that GLSSAT is generally outperformed by the most recent DLS algorithms for SAT, such as ESG and SAPS (these are described later in this section).

The Discrete Lagrangian Method (DLM)

The basic DLM algorithm for SAT [Shang and Wah, 1998] is motivated by the theory of Lagrange multipliers for continuous optimisation. *Basic DLM* is a DLS algorithm based on GSAT/Tabu with clause weights as its underlying local search procedure; in each search step, it flips a non-tabu variables that maximises the decrease in the total weight of all unsatisfied clauses. This subsidiary local search is terminated when an assignment is reached for which the number of neighbouring assignments with larger or equal evaluation function value exceeds a given threshold θ_1. After each local search phase, the penalties for all unsatisfied clauses are increased by $\delta^+ := 1$; additionally, in order to bound the range of the clause penalties, all penalties are reduced by $\delta^- := 1$ after every θ_2 local search phase. Before the actual search begins, DLM simplifies the given formula by performing a complete pass of unit propagation. As usual, all clause penalties are initialised to zero, and the search process starts from a variable assignment that is chosen uniformly at random.

This basic DLM algorithm has been extended in various ways. *DLM-99-SAT* [Wu and Wah, 1999] uses an additional mechanism for escaping more effectively from local minima of the evaluation function. The idea behind this mechanism is to identify clauses that are frequently unsatisfied in local minima, and to additionally increase their penalties. This is achieved by means of temporary clause penalties t_i, which are initialised at zero and increased by $\delta_w := 1$ for all unsatisfied clauses whenever a local minimum is encountered. After each regular clause penalty update, if the ratio between the maximal t_i and average t_i over all clauses exceeds a threshold θ_3, the regular penalty of the clause with the largest t_i is increased by $\delta_s := 1$. (In another variant, only the t_i of currently unsatisfied clauses are considered when computing the ratio and determining the clause penalty that receives the additional increase.)

A different extension of DLM, called *DLM-2000-SAT*, uses a long-term memory mechanism for preventing the search process from getting stuck repeatedly in certain attractive non-solution areas of the search space. This is implemented by using a list of previously visited assignments and by adding an additional distance penalty to the evaluation function for assignments that are close to the elements of this list. More precisely, during the search process, every w_s variable flips, the current variable assignment is added to a fixed-length queue. Using the assignments a_j in this queue, a distance term for a given variable assignment a is computed as $d := \sum_j \min\{\theta_t, hd(a, a_j)\}$, where $hd(a, a_j)$ is the Hamming distance (i.e., the number of variables assigned different values) between assignments a and a_j. The evaluation function used in the subsidiary local search procedure is then extended to $g'(F, a) := g(F, a) + \sum_{c \in CU(F,a)} clw(c) - d$, where $CU(F, a)$ denotes the number of clauses in F unsatisfied under a. Note that by using a bound $\theta_t \ll n$ on the distance contribution from each assignment a_j, the impact of this mechanism on the search process is fairly localised.

DLM-99-SAT shows substantially better performance than the basic DLM algorithm, particularly on large and structured SAT instances. DLM-2000-SAT, the most recent DLM variant, typically seems to perform better than DLM-99-SAT as well as WalkSAT/SKC. For a considerable time, this dynamic local search algorithm was one of the best known SLS algorithms for SAT.

The Exponentiated Subgradient Algorithm (ESG)

The Exponentiated Subgradient (ESG) algorithm [Schuurmans et al., 2001] is motivated by subgradient optimisation, a well-known method for minimising Lagrangian functions, which is often used for generating good lower bounds in branch and bound techniques or as a heuristic in local search algorithms.

ESG starts its search from a randomly selected variable assignment after initialising all clause weights to one. As its underlying local search procedure, ESG uses a best improvement search method that can be seen as a simple variant of GSAT. In each local search step, the variable to be flipped is selected uniformly at random from the set of all variables that appear in currently unsatisfied clauses and whose flip leads to a maximal reduction in the total weight of unsatisfied clauses. When reaching a local minimum position (i.e., an assignment in which flipping any variable that appears in an unsatisfied clause would not lead to a decrease in the total weight of unsatisfied clauses), with probability η, the search is continued by flipping a variable that is uniformly chosen at random from the set of all variables appearing in unsatisfied clauses; otherwise, the local search phase is terminated.

After each local search phase, the clause weights are updated. This involves two stages: First, the weights of all clauses are multiplied by a factor depending on their satisfaction status; weights of satisfied clauses are multiplied by α_{sat}, weights of unsatisfied clauses by α_{unsat} (scaling stage). Then, all clause weights are smoothed using the formula $clw(c) := clw(c) \cdot \rho + (1-\rho) \cdot \overline{w}$ (smoothing stage), where $\overline{w}$ is the average of all clause weights after scaling, and the parameter ρ has a fixed value between zero and one. The algorithm terminates when a satisfying assignment for F has been found or when a user-specified maximal number of iterations have been completed.

In a straight-forward implementation of ESG, the weight update steps are computationally much more expensive than the weighted search steps, whose cost is determined by the underlying basic local search procedure. Each weight update step requires accessing all clause weights, while a weighted search step only needs to access the weights of the critical clauses, that is, clauses that can change their satisfaction status when a variable appearing in a currently unsatisfied clause is flipped. (The complexity of all other operations is dominated by these operations.) Typically, for the major part of the search, only few clauses are unsatisfied; hence, only a small subset of the clauses is critical, rendering the weighted search steps computationally much cheaper than weight updates.

If weight updates would typically occur very infrequently compared to weighted search steps, the relatively high complexity of the weight update steps might not have a significant effect on the performance of the algorithm. However, experimental evidence indicates that the fraction of weighting steps performed by ESG is quite high; it ranges from around 7% for SAT encodings of large flat graph colouring problems to more than 40% for SAT-encoded all-interval-series problems [Hutter et al., 2002].

Efficient implementations of ESG therefore critically depend on additional techniques in order to achieve the competitive performance results reported by Schuurmans et al. [2001]. The most recent publicly available ESG-SAT implementation by Southey and Schuurmans (Version 1.4), for instance, uses $\alpha_{sat} := 1$ (which avoids the effort of scaling satisfied clauses), replaces $\overline{w}$ by 1 in the smoothing step, and utilises a 'lazy' weight update technique which updates clause weights only when they are needed.

When measuring run-time in terms of search steps, ESG typically performs substantially better than state-of-the-art WalkSAT variants, such as Novelty$^+$. In terms of CPU-time, however, even the optimised ESG-SAT implementation by Southey and Schuurmans does not always reach the performance of Novelty$^+$. Compared to DLM-2000-SAT, ESG-SAT typically requires fewer steps for finding a model of a given formula, but in terms of CPU-time, both algorithms show very similar performance [Schuurmans et al., 2001; Hutter et al., 2002].

Scaling and Probabilistic Smoothing (SAPS)

The SAPS algorithm [Hutter et al., 2002] can be seen as a variant of ESG that uses a modified weight update scheme, in which the scaling stage is restricted to the weights of currently unsatisfied clauses, and smoothing is only performed with a certain probability p_{smooth}. Note that restricting the scaling operation to the weights of unsatisfied clauses ($\alpha_{sat} := 1$) does not affect the variable selection in the weighted search phase, since rescaling all clause weights by a constant factor does not affect the variable selection mechanism. (Southey and Schuurmans' efficient ESG implementation also makes use of this fact.) This reduces the complexity of the scaling step from $\Theta(\#C(F))$ to $\Theta(\#CU(F, a))$, where $C(F)$ is the set of clauses in the given CNF formula F and $CU(F, a)$ is the set of clauses in F that are unsatisfied under assignment a.

After a short initial search phase, typically only a few clauses remain unsatisfied such that $\#CU(F, a)$ becomes rather small compared to $\#C(F)$; this effect seems to be more pronounced for larger SAT instances with many clauses. The smoothing step, however, has complexity $\Theta(\#C(F))$, and now dominates the complexity of the weight update. Therefore, by applying the expensive smoothing operation only occasionally, the time complexity of the weight update procedure can be substantially reduced. It has been shown experimentally that this does not have a detrimental effect on the performance of the algorithm in terms of the number of weighted search steps required for solving a given instance [Hutter et al., 2002]. By having the weight update procedure perform smoothing of all clause weights (using the same formula as shown in the description of ESG above) only with a probability $p_{smooth} \ll 1$, the time complexity of a weight update is reduced to $\Theta(p_{smooth} \cdot \#C(F) + \#CU(F, a))$ compared to $\Theta(\#C(F) + \#CU(F, a))$ for ESG. As a result, the discounted cost of smoothing no longer dominates the algorithm's run-time. Performing the smoothing probabilistically rather than deterministically after a fixed number of steps (like the occasional clause weight reduction in DLM) also has the theoretical advantage of preventing the algorithm from getting trapped in the same kind of cyclic behaviour that renders R-Novelty essentially incomplete. (In practice, SAPS has been found to consistently perform well for small p_{smooth} values of about 0.05.)

The SAPS algorithm as described here does not require additional implementation tricks other than the standard mechanism for efficiently accessing critical clauses, which is used in all efficient implementations of SLS algorithms for SAT. Compared to ESG, SAPS typically requires a similar number of variable flips for

procedure *UpdateClauseWeights* $(F, a; \alpha, \rho, p_{smooth})$

    input: *propositional formula F, variable assignment a;*

        scaling factor α, smoothing factor ρ, smoothing probability p_{smooth}

    $C := \{c \mid c \text{ is a clause of } F\};$

    $U := \{c \in C \mid c \text{ is unsatisfied under } a\};$

    for each $c \in U$ **do**

        $clw(c) := clw(c) \cdot \alpha;$

    end

    with probability p_{smooth} **do**

        for each $c \in C$ **do**

            $clw(c) := clw(c) \cdot \rho + (1 - \rho) \cdot \overline{w};$

        end

    end

end *UpdateClauseWeights*

Figure 6.8 The SAPS weight update procedure; $\overline{w}$ is the average over all clause weights.

finding a model of a given formula, but in terms of time performance it is significantly superior to ESG, DLM-2000-SAT, and the best known WalkSAT variants [Hutter et al., 2002]. However, there are some cases (in particular, hard and large SAT-encoded graph colouring instances), for which SAPS does not reach the performance of Novelty$^+$.

A reactive variant of SAPS, *RSAPS,* automatically adjusts the smoothing probability p_{smooth} during the search, using a mechanism that is very similar to the one underlying Adaptive WalkSAT. RSAPS sometimes achieves significantly better performance than SAPS [Hutter et al., 2002]; however, different from Adaptive WalkSAT, RSAPS still has other parameters, in particular ρ, that need to be tuned manually in order to achieve optimal performance.

6.5 Constraint Satisfaction Problems

An instance of the *Constraint Satisfaction Problem (CSP)* is defined by a set of variables, a set of possible values (or *domain*) for each variable, and a set of constraining conditions (*constraints*) involving one or more of the variables. The *Constraint Satisfaction Problem* is to decide for a given CSP instance whether all variables can be assigned values from their respective domains such that all constraints are simultaneously satisfied. Depending on whether the variable domains are discrete or continuous, finite or infinite, different types of CSP instances and respective subclasses of the CSP can be distinguished. Here, we restrict our attention to the *finite discrete CSP*, a widely studied type of constraint satisfaction problem with many practical applications.

DEFINITION 6.2 **Finite Discrete CSP**

A CSP instance is a triple $P := (V, \mathcal{D}, \mathcal{C})$, where $V := \{x_1, \ldots, x_n\}$ is a finite set of n variables, $\mathcal{D}$ is a function that maps each variable x_i to the set D_i of possible values it can take (D_i is called the domain *of x_i), and $\mathcal{C} := \{C_1, \ldots, C_m\}$ is a finite set of* constraints. *Each constraint C_j is a relation over an ordered set $Var(C_j)$ of variables from V, that is, for $Var(C_j) := (y_1, \ldots, y_k)$, $C_j \subseteq \mathcal{D}(y_1) \times \cdots \times \mathcal{D}(y_k)$. The elements of the set C_j are referred to as* satisfying *tuples of C_j, and k is called the* arity *of the constraint C_j. A CSP instance P is called n-ary, if, and only if, the arity of all constraints in P have arity at most n; in particular,* binary *CSP instances have only constraints of arity at most two.*

P is a finite discrete *CSP instance if, and only if, all variables in P have discrete and finite domains.*

A variable assignment *of P is a mapping $a : V \mapsto \bigcup_{i=1}^{n} D_i$ that assigns to each variable $x \in V$ a value from its domain $\mathcal{D}(x)$. Let $Assign(P)$ denote the set of all possible variable assignments for P; then a variable assignment $a \in Assign(P)$ is a* solution *of P if, and only if, it simultaneously satisfies all constraints in $\mathcal{C}$, that is, for all $C_j \in \mathcal{C}$ with, say, $Var(C_j) = (y_1, \ldots, y_k)$ the assignment a maps $y_1, \ldots, y_k$ to values $v_1, \ldots, v_k$ such that $(v_1, \ldots, v_k) \in C_j$.*

CSP instances for which at least one solution exists are also called consistent, *while instances that do not have any solutions are called* inconsistent.

The finite discrete *CSP is the problem of deciding whether a given finite discrete CSP instance P is consistent.*

Remark: In many cases, the constraint relations involved in CSP instances can be represented more compactly by using standard mathematical relations, such as '=', '$\neq$', '<', '$\leq$', '>', '$\geq$'. In other cases, a more compact representation of a given constraint C_j is obtained by explicitly listing the complement of the set of satisfying tuples, that is, the set of unsatisfying tuples of C_j.

EXAMPLE 6.2 **The Canadian Flag Problem**

Let us consider the problem of colouring the Canadian flag by assigning colours red (r) and white (w) to the four fields L, C, R, M in such a way that

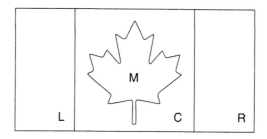

Figure 6.9 A simple CSP instance: the problem of colouring the Canadian flag (see text for details).

any two neighbouring fields are coloured differently (see Figure 6.9). This problem can be formulated as a binary CSP instance as follows:

$$V := \{L, C, R, M\}$$
$$\mathcal{D}(L) := \mathcal{D}(C) := \mathcal{D}(R) := \mathcal{D}(M) := \{r, w\}$$
$$\mathcal{C} := \{C_1, C_2, C_3\}$$
$$\text{with} \quad \textit{Var}(C_1) := (L, C)$$
$$\textit{Var}(C_2) := (C, M)$$
$$\textit{Var}(C_3) := (C, R)$$
$$\text{and} \quad C_1 := C_2 := C_3 := \{(r, w), (w, r)\}$$

There are two solutions to this CSP instance; one assigns red to M, L, R and white to C, while the other assigns white to M, L, R and red to C. By adding a fourth, unary constraint to $\mathcal{C}$ that forces M to be coloured red, the instance can be modified such that only the solution corresponding to the correct colouring of the Canadian flag remains.

This simple CSP instance is an example of a Map Colouring Problem, which in turn can be seen as a special case of the Graph Colouring Problem (GCP). GCP is an important subclass of the CSP in which the objective is to colour the vertices of a given graph in such a way that two vertices connected by an edge are never assigned the same colour. The GCP is covered in more detail in Chapter 10, Section 10.1.

Like SAT, finite discrete CSP is an $\mathcal{NP}$-complete combinatorial problem. This can be proven rather easily based on the following close relationship between SAT and finite discrete CSP. Any instance of SAT (for CNF formulae) can be seen as a finite discrete CSP instance where all the domains contain only the truth values $\top$, $\bot$ and each constraint contains exactly all the satisfying assignments of one particular clause of the given CNF formula F. Vice versa, as we will show in the next section, any finite discrete CSP instance can be directly transformed into a SAT instance rather efficiently.

Encoding CSP Instances into SAT

CSP instances can be encoded into SAT in a rather straight-forward way. The basic idea is to use propositional variables to represent the assignment of values to single CSP variables and clauses to express the constraint relations [de Kleer, 1989]. For the sake of simplicity, we assume in the following, without loss of generality, that the domains of all variables are equal to $\mathbb{Z}_k := \{0, 1, \dots, k-1\}$, where k is an arbitrary positive integer. Furthermore, we use $\sigma(C_j)$ to denote the arity of a constraint C_j, that is, the number of variables involved in C_j.

Given a finite discrete CSP instance $P := (V, \mathcal{D}, \mathcal{C})$ with $V := \{x_1, \dots, x_n\}$, $\mathcal{D}(x) := \mathbb{Z}_k$ for all $x \in V$, and $\mathcal{C} := \{C_1, \dots, C_m\}$, a very natural SAT encoding is based on propositional variables $c_{i,v}$ that, if assigned the value $\top$, represent the assignment $x_i := v$, where $v \in \mathcal{D}(x_i)$. P can then be represented by a CNF formula comprising the following sets of clauses:

(1) $\quad \neg c_{i,v_1} \vee \neg c_{i,v_2}$ $\qquad\qquad (1 \le i \le n; v_1, v_2 \in \mathbb{Z}_k; v_1 < v_2)$

(2) $\quad c_{i,0} \vee c_{i,1} \vee \dots \vee c_{i,k-1}$ $\qquad (1 \le i \le n)$

(3) $\quad \neg c_{i_1,v_1} \vee \neg c_{i_2,v_2} \vee \dots \vee \neg c_{i_s,v_s}$ $\quad (x_{i_1} := v_1; x_{i_2} := v_2; \dots; x_{i_s} := v_s)$
$\qquad\qquad\qquad\qquad\qquad\qquad\qquad$ violates some constraint $C_j \in \mathcal{C}$
$\qquad\qquad\qquad\qquad\qquad\qquad\qquad$ with $\sigma(C_j) := s$

Intuitively, these clause sets ensure that each constraint variable is assigned exactly one value from its domain (sets 1 and 2) and any solution is compatible with all constraints (set 3). The number of propositional variables required for encoding a given CSP instance is linear in the number of constraint variables and their domain sizes, while the number of clauses is at least linear in the number of constraint variables and depends critically on the domain sizes and the arity of the constraints. This encoding is frequently used in the context of translations of combinatorial problems into SAT that use CSP as an intermediate domain (cf. Section 6.1). It is known as the *sparse encoding*, because it generates relative large SAT instances whose models have only a small fraction of the propositional variables set to $\top$. (In the literature, this encoding has also been referred to as the *unary transform* or *direct encoding*.)

By using an alternative SAT encoding of CSP instances, the number of propositional variables required for encoding a given CSP instance can be significantly reduced compared to the sparse encoding. The *compact encoding* (in the literature also referred to as the *binary transform* or *log encoding*) is based on the idea of representing the value v assigned to any constraint variable x_i by a group of $\lceil \log_2 k \rceil$ propositional variables $c_{i,j}$, using a binary encoding of v [Iwama and Miyazaki, 1994; Hoos, 1999c; 1999b]. This leads to a CNF formula with $n \cdot \lceil \log_2 k \rceil$ propositional variables; particularly for large domain sizes k, this can be a substantial reduction compared to the $n \cdot k$ propositional variables required by the sparse encoding. The number of clauses, however, is similar for both encodings, since in either case the same number of clauses is needed for representing the constraint relations (these clauses typically dominate the overall number of clauses). Although the SAT instances generated by the compact encoding have search spaces that are substantially smaller than those obtained from the sparse encoding, and consequently, substantially higher solution densities, they often appear to be much harder to solve using standard SLS algorithms for SAT [Hoos, 1999c; 1999b; Frisch and Peugniez, 2001]. (There is, however, some evidence that for relatively small, structured CSP instances, the SAT instances obtained from the compact encoding can sometimes be solved as efficiently as those obtained from the sparse encoding [Prestwich, 2003].)

There are several other methods for encoding CSP instances into SAT. One of these is the *multivalued encoding*, a variant of the sparse encoding that does not include the binary clauses preventing two values being simultaneously assigned to the same CSP variable (set 1 above); it produces SAT instances that have higher solution densities than those obtained by the sparse encoding and appear to be easier to solve for high-performance SLS algorithms for SAT [Prestwich, 2003]. Even higher solution densities can be achieved by *weakened encodings*, which can be seen as a generalisation of the multivalued encoding. Although recent empirical results suggest that at least one weakened encoding, the so-called *reduced encoding*, can lead to excellent SLS performance, further studies are required to clarify the benefits of weakened encodings [Prestwich, 2003]. Finally, the *support encoding* is similar to the sparse encoding, but rather than ruling out the unsatisfying tuples of the given constraints (set 3 above), it directly captures the satisfying tuples in the form of so-called support clauses [Kasif, 1990; Gent, 2002].

CSP Simplification and Local Consistency Techniques

Similar to the case of SAT, native CSP instances can often be substantially reduced in size and complexity by applying polynomial-time simplification methods. Also known as *local consistency techniques*, these methods are transformations that are applied to (local) subsets of a given CSP instance P [Mackworth, 1977;

Debruyne and Bessière, 2001]. Local consistency techniques can reduce the effective domains of CSP variables by eliminating values that cannot occur in any solution.

One of the most prominent simplification techniques for the CSP is the enforcement of *arc consistency*. A given CSP instance P is made arc consistent w.r.t. to one of its constraints, C, by removing any value v from the domain of any variable x involved in C if, and only if, there exists no CSP assignment that satisfies C, that is, no tuple $t \in C$ for which x has value v. A CSP instance P is arc consistent if, and only if, it is arc consistent w.r.t. all of its constraints. For binary CSP instances with e constraints and maximal domain size k, the best known algorithms for enforcing arc consistency have a time complexity $O(ek^2)$ and space complexity $O(ek)$ [Bessière et al., 1999]. (It may be noted that enforcing arc consistency on a given CSP instance is equivalent to applying unit propagation to its support encoding [Gent, 2002].) A number of further local consistency techniques has been described by Debruyne and Bessière [2001].

Combined with backtracking mechanisms, simplification methods, such as forward checking or enforcing arc consistency, play a crucial role in systematic search algorithms for the CSP [Haralick and Elliot, 1980; Grant and Smith, 1996]. They can also be used as preprocessing techniques before applying SLS-based, incomplete CSP solvers. The high computational cost of enforcing higher levels of local consistency, such as path consistency, are often not amortised by the reduced run-times of CSP solvers that are subsequently applied to the resulting CSP instances. One method for improving this situation is to apply the corresponding local consistency methods to heuristically selected parts of a given CSP instance only [Kask and Dechter, 1995].

Prominent Benchmark Instances for the CSP

There are numerous types of CSP instances that have been commonly used in the literature on the CSP. Many studies have focused on a particular class of randomly generated CSP instances with binary constraint relations, which we call *Uniform Random Binary CSP*. Besides the number of CSP variables and the size of the variable domains, this problem distribution is characterised by two parameters, the *constraint graph density* α and the *constraint tightness* β; α specifies the probability that a constraint relation exists between an arbitrary pair of CSP variables, and β is the expected fraction of value pairs that satisfy a given constraint relation. For this class of CSP instances, a solubility phase transition phenomenon with an associated peak in hardness, similar to the one for Uniform Random 3-SAT, has been observed [Smith, 1994], and test-sets of hard instances can be obtained for specific combinations of α and β values.

Another widely used class of CSP instances stems from the *Graph Colouring Problem* (see Example 6.2), which can be seen as a special case of the finite discrete CSP in which all variables have the same domain, and all constraint relations are binary, allowing a pair of values (x, y) if, and only if, $x \neq y$ (inequality constraint). The Graph Colouring Problem and specific instance classes are discussed in more detail in Chapter 10, Section 10.1. Graph colouring instances with three colours are amongst the most commonly used benchmark instances for the CSP.

A prominent special case of the Graph Colouring Problem is the *Quasigroup Completion Problem (QCP)*, which is derived from the following *Quasigroup Problem* or *Latin Square Problem*: Given an $n \times n$ quadratic grid and n colours, the objective is to assign a colour to each grid cell in such a way that every row and column contains all n colours. In the QCP, the objective is to decide whether a partial solution of the Quasigroup Problem, that is, an incomplete assignment of colours to the given grid such that no two cells in the same row or column have the same colour, can be extended into a complete solution by assigning colours to the remaining cells. In the CSP formulation, the pre-assigned cells can be easily represented by unary constraints. The QCP is known to be $\mathcal{NP}$-complete [Colbourn, 1984], and a phase-transition phenomenon with an associated peak in hardness has been observed [Gomes and Selman, 1997b]. The QCP has a variety of important applications in areas, such as dynamic wavelength routing in fibre optic networks and the design of statistical experiments [Kumar et al., 1999; Laywine and Mullen, 1998].

The *n-Queens Problem* is another prominent CSP; it can be seen as a generalisation of the problem of placing eight queens on a chessboard such that no queen is threatened by any of the other queens. This is achieved by distributing the queens in such a way that no row, column, or diagonal has more than a single queen on it. The 8-Queens Problem can be represented as a CSP instance with 8 variables and 28 binary constraints. In the *n*-Queens Problem, the objective is to place n queens on an $n \times n$ board subject to analogous constraints.

Most of the work on CSP has been focused on binary CSP. One of the reasons for this is that any non-binary CSP instance can be transformed into a binary CSP instance in a rather straight-forward way [Dechter and Pearl, 1989; Rossi et al., 1990; Bacchus et al., 2002]. Another reason lies in the fact that algorithms restricted to binary CSP instances are typically easier to implement than general CSP solvers.

There are numerous other classes of CSP instances, including CSP encodings of the real-world problems mentioned in Section 6.1. Some application-relevant problems include frequency assignment in radio networks, scheduling problems and vehicle routing. A description of many of the different types of constraint satisfaction problems can be found at CSPLIB, a benchmark library for constraints [Gent et al., 2003].

6.6 **SLS Algorithms for CSPs**

Because of the close relationship between CSP and SAT, the respective SLS algorithms for solving these problems are quite similar; historically, there has been significant cross-fertilisation between both domains in terms of SLS algorithm design and development. We distinguish three types of SLS techniques for solving CSPs: SLS algorithms for SAT applied to SAT-encoded CSP instances; generalisations of SLS algorithms for SAT; and native SLS algorithms for CSPs. In the following, we will discuss each of these approaches in more detail and present some of the most prominent and best performing SLS algorithms for CSPs.

The 'Encode and Solve as SAT' Approach

In principle, any CSP instance P can be solved by encoding it into SAT and subsequently applying standard SAT solvers to determine the satisfiability of the resulting CNF formula F. If P is soluble, its solutions can be determined from the models of F. By using any of the SAT encodings of CSPs discussed in Section 6.5, encoding CSP instances as well as decoding their solutions are efficient processes, and the resulting SAT instances are typically not prohibitively large compared to the original CSP instances.

The main advantage of this approach lies in the fact that it allows the use of highly optimised and efficiently implemented 'off-the-shelf' SAT solvers. Besides the SLS algorithms described earlier in this chapter, this includes high-performance systematic SAT solvers and other state-of-the-art SAT algorithms (see Section 6.7 for references).

Furthermore, standard polynomial preprocessing techniques for SAT can be used to simplify SAT-encoded CSP instances prior to applying a SAT solver. CSP preprocessing techniques, such as efficiently computable forms of k-consistency, can be applied before encoding a CSP instance into SAT. However, one potentially major disadvantage of the 'encode and solve as SAT' approach may arise from the inability of standard SAT algorithms to exploit the structure present in given CSP instances.

There is some indication that by using the sparse encoding and high-performing SAT algorithms, such as Novelty or Novelty$^+$, competitive performance compared to state-of-the-art SLS algorithms for the CSP, such as Galinier and Hao's Tabu Search algorithm (which will be described later in this section), can be obtained for Uniform Random Binary CSP instances [Hoos, 1998; 1999b]. Similar results appear to hold for graph colouring instances, but there is some evidence that native CSP algorithms might achieve superior performance for random instances with large variable domains [Frisch and Peugniez, 2001]. Interestingly, when using the compact encoding, SLS algorithms for SAT show substantially weaker performance; this performance difference appears to be caused by aspects of the search space structure induced by the respective encodings; in particular, it has been shown that applied to the same CSP instances, the compact encoding generates search spaces with substantially higher local minima branching factors (see Chapter 5) than the sparse encoding [Hoos, 1998; 1999b].

Clearly, the 'encode and solve as SAT' approach is not limited to the CSP, but can in principle be applied to any $\mathcal{NP}$-complete problem. For the CSP, this approach is particularly attractive, because as a result of the close relationship between SAT and CSP, the encoding of CSP instances into SAT is conceptually simple and very efficiently computable in practice. Whether or not it can achieve competitive performance compared to the best native CSP algorithms, particularly when applied to structured CSP instances, is somewhat unclear at the present time and needs to be further investigated.

Pseudo-Boolean CSP and WSAT($\mathcal{PB}$)

An alternative to the 'encode and solve as SAT' approach discussed in the previous section is to extend high performing SLS algorithms for SAT to more general subclasses of the CSP. One such generalisation of SAT is obtained by maintaining the restriction to Boolean variables, while allowing constraints that are more expressive than CNF clauses. In the *Pseudo-Boolean CSP*, also known as the *(Linear) Pseudo-Boolean Programming*, all variables have Boolean values represented by integers zero and one, and the constraints between variables x_i are of the form

$$\sum_{i=1}^{n} a_{ij} \cdot x_i \geq b_j,$$

where the a_{ij} as well as b_j are rational numbers. Note that analogous constraints that use any of the relations '$\leq$', '$<$', '$>$', '$=$' and '$\neq$' instead of '$\geq$' can be represented using '$\geq$' constraints only. Pseudo-Boolean constraints are more expressive than CNF clauses because any clause can be expressed by a single pseudo-Boolean constraint, but there are pseudo-Boolean constraints that cannot be captured by a single CNF clause. From an operations research point of view, Pseudo-Boolean CSP can be seen as a special case of 0-1 Integer Linear Programming [Nemhauser and Wolsey, 1988].

EXAMPLE 6.3 **Pseudo-Boolean Constraints**

As an example for a Pseudo-Boolean constraint between three variables y_1, y_2, y_3 with domain $\{0, 1\}$, consider the inequality $y_1 + y_2 - y_3 \geq 0$. This constraint is equivalent to $y_1 + y_2 + (1 - y_3) \geq 1$, and hence to the CNF clause $x_1 \vee x_2 \vee \neg x_3$.

The following constraint limits the number of variables that are assigned the value one to a maximum of k:

$$(-y_1) + \ldots + (-y_n) \geq -k$$

Note that in order to express this constraint by a CNF formula, $\binom{n}{k}$ clauses of size $k + 1$ each are required; these encode the condition that for every possible subset of $k + 1$ variables, at least one variable needs to be assigned the value $\perp$.

There are several SLS algorithms for Pseudo-Boolean CSP [Abramson et al., 1996; Connolly, 1992; Walser, 1997; Løkketangen, 2002]. Among these, Walser's WSAT($\mathcal{PB}$) algorithm is of particular interest, since it is based on a direct generalisation of the WalkSAT architecture to Pseudo-Boolean CSP. The WSAT($\mathcal{PB}$) algorithm follows the WalkSAT framework as presented in Section 6.3, but uses a generalised evaluation function and variable selection mechanism. The evaluation function is based on the notion of the *net integer distance* of a constraint from being satisfied. More precisely, for each constraint C, let $d(a, C)$ denote the integer difference between the right-hand side of the inequality C and the value of the left-hand side under assignment a if C is unsatisfied, or zero otherwise; the evaluation function value of assignment a is then defined as the sum of the $d(a, C)$ values over all constraints unsatisfied in a. As in the SAT case, an assignment that satisfies all constraints has an evaluation function value of zero.

Based on this evaluation function, WSAT($\mathcal{PB}$) uses a modified version of the WalkSAT variable selection strategy to determine the variable to be flipped in each search step. First, a constraint is uniformly selected at random from the set of all currently unsatisfied constraints. Then, a variable involved in C is selected according to the following criteria: If flipping any of the variables involved in C leads to a decrease in the evaluation function, the variable that leads to the largest such decrease is selected; if there are several such variables, the one that was flipped least recently is chosen. Otherwise, with a small probability wp, the variable that has been flipped least recently is selected; in the remaining cases, the variable whose flip would cause a minimal increase in the evaluation function is chosen; again, ties are broken in favour of the least recently flipped variable. (At the beginning of the search, ties may arise between variables that have not been flipped yet; such ties are broken uniformly at random.) Additionally, WSAT($\mathcal{PB}$)

uses a simple tabu mechanism, which excludes all variables that have been flipped within the previous tt search steps from the selection process.

Different from conventional WalkSAT, WSAT($\mathcal{PB}$) supports a biased random initialisation of the search process, in which each variable is independently set to zero with probability p_z and to one otherwise; however, experimental results suggest that using a biased initialisation (i.e., $p_z \neq 0.5$) generally does not lead to performance improvements [Walser, 1997].

Furthermore, the WSAT($\mathcal{PB}$) algorithm, as presented by Walser [1997], can also be used to solve an optimisation variant of Pseudo-Boolean CSP in which a subset of the constraints is considered to be 'soft' constraints, and the objective is to find variable assignments that satisfy all conventional ('hard') constraints, while minimising the number of unsatisfied soft constraints. This problem can be seen as a special case of MAX-CSP, the optimisation variant of CSP, which is discussed in more detail in Chapter 7, where we also describe the mechanism used by WSAT($\mathcal{PB}$) to handle soft constraints (see Chapter 7, page 348).

Many practically relevant problems can be formulated quite easily and naturally as Pseudo-Boolean CSP instances. Applied to encodings of the Progressive Party Problem [Smith et al., 1996] and radar surveillance problems (the latter include soft constraints), WSAT($\mathcal{PB}$) has been shown to achieve significantly improved performance over state-of-the-art commercial integer programming and constraint programming packages as well as compared to results from the literature [Walser, 1997].

WalkSAT Algorithms for Many-Valued SAT

Another interesting subclass of CSPs is the class of non-Boolean or many-valued satisfiability problems [Béjar and Manyà, 1999; Frisch and Peugniez, 2001]. In *non-Boolean SAT (NB-SAT)*, each variable can take values from some finite domain D, which may contain more than two values [Frisch and Peugniez, 2001]. Formally, a non-Boolean literal is of the form z/v or $\neg z/v$, where z is a variable and v a value from the domain of z. The value of z/v under the (non-Boolean) variable assignment a is *true* if, and only if, z is set to v in a, and *false* otherwise; the value of $\neg z/v$ under a is obtained by negating the value of z/v under a. Analogously to conventional SAT, non-Boolean SAT is the problem to decide for a given non-Boolean CNF formula, that is, for a conjunction over disjunctions of non-Boolean literals, whether or not it has a satisfying (non-Boolean) assignment. Obviously, any conventional CNF formula can be represented by a non-Boolean CNF formula with the same number of clauses and variables. When encoding NB-SAT instances into SAT, however, a significantly higher number of variables and CNF clauses as used in the non-Boolean formula may be required.

Because NB-SAT instances have the same clause structure as conventional SAT instances, SAT algorithms such as WalkSAT can be generalised to non-Boolean SAT in a rather straightforward way; the only major difference lies in the fact that in NB-SAT, the concept of a variable flip needs to be redefined. In *NB-WalkSAT*, the non-Boolean variant of WalkSAT by Frisch and Peugniez [2001], a variable flip corresponds to assigning a different value to a non-Boolean variable such that the literal selected in the corresponding search step, and hence the clause in which it appears, becomes satisfied. (It may be noted that this constitutes an important difference to WSAT($\mathcal{PB}$), in which search steps do not always guarantee the satisfaction of any previously unsatisfied constraints.) Otherwise, NB-WalkSAT is identical to WalkSAT/SKC, although other WalkSAT variants can easily be extended to NB-SAT in an analogous way.

Béjar and Manyà have introduced a similar extension of WalkSAT, called *MV-WalkSAT*, which solves a variant of many-valued SAT that is slightly richer than the non-Boolean CNF formulae underlying NB-SAT [Béjar and Manyà, 1999]. Both, NB-WalkSAT and MV-WalkSAT were applied to many-valued SAT encodings of various combinatorial decision problems, such as graph colouring, where they showed excellent performance. However, to date, the question whether these and other SLS algorithms for many-valued SAT can substantially outperform state-of-the-art SLS algorithms for SAT applied to suitably encoded instances has not been answered conclusively.

The Min Conflicts Heuristic and Variants

There are a number of SLS algorithms for the general finite discrete CSP, although in many cases, the implementations are restricted to binary constraints. Among the most widely known of these are the *Min Conflicts Heuristic (MCH)* [Minton et al., 1990; 1992] and its variants. MCH iteratively modifies the assignment of a single variable in order to minimise the number of violated constraints, which is achieved in the following way: Given a CSP instance P, the search process is initialised by assigning each variable in P a value that is chosen uniformly at random from its domain. Then, in each local search step, first a CSP variable x_i is selected uniformly at random from the *conflict set* $K(a)$, which is the set of all variables that appear in a constraint that is unsatisfied under the current assignment a. A new value v is then chosen from the domain of x_i, such that by assigning v to x_i, the number of unsatisfied constraints (conflicts) is minimised. If there is more than one value of v with that property, one of the minimising values is chosen uniformly at random.

In many ways, MCH is analogous to the SLS algorithms for SAT described earlier in this chapter. Like all SAT algorithms covered here, MCH is based on a

1-exchange neighbourhood. Considering that CNF clauses in SAT play the same role as constraint relations in CSP, the evaluation function underlying MCH, defined as the number of constraints violated under a given assignment, is essentially the same as the one used by GSAT or Novelty. The way in which MCH selects the variable and the value for this variable in each local search step is similar to the two-stage variable selection process underlying the WalkSAT architecture.

Like most iterative improvement methods, MCH is essentially incomplete, since it can get stuck in local minima of the evaluation function. The simplest way to overcome this problem is to use a static restart mechanism analogous to the one found in GSAT. Not surprisingly, however, there are other, substantially more effective solutions. These are mainly derived from mechanisms used in the better performing GSAT and WalkSAT algorithms, which is somewhat surprising, considering that MCH itself predates all of the SLS algorithms for SAT discussed above, including basic GSAT and WalkSAT/SKC.

WMCH is a variant of MCH that uses a random walk mechanism analogous to GWSAT [Wallace and Freuder, 1995]. In each WMCH step, first a variable x_i is chosen uniformly at random from the conflict set (as in MCH). Then, with probability $wp > 0$, x_i is assigned a value from its domain D_i that has been chosen uniformly at random; this kind of search step is called a random walk step. In the remaining cases, that is, with probability $1 - wp$, a conflict-minimising value is chosen and assigned, as in a conventional MCH step. As in the case of GWSAT, this random walk mechanism renders WMCH probabilistically approximately complete for $wp > 0$. Furthermore, WMCH has been empirically observed to perform substantially better than MCH with random restart [Stützle, 1998c].

Note that different from the random walk steps used in SLS algorithms for SAT, such as GWSAT, random walk steps in WMCH do not necessarily have the effect of satisfying a previously unsatisfied constraint. WMCH can be varied slightly such that in each random walk step, after choosing a variable x_i involved in a currently violated constraint C, x_i is assigned a value v such that C becomes satisfied; if no such v exists, a value is chosen at random. This variant, however, performs only marginally better than the random walk mechanism used in WMCH [Stützle, 1998c].

Analogous to GSAT and WalkSAT, MCH can be extended with a tabu search mechanism [Stützle, 1998c; Steinmann et al., 1997]. In *TMCH*, after each search step, that is, after the value of variable x_i is changed from v to v', the variable/value pair (x_i, v) is declared tabu for the next tt steps. While (x_i, v) is tabu, value v is excluded from the selection of values for x_i, except if assigning v to x_i leads to an improvement over the incumbent assignment (aspiration criterion). TMCH appears to generally perform better than WMCH. Interestingly, a tabu tenure setting of $tt := 2$ was found to consistently result in good performance for CSP instances of different types and sizes [Stützle, 1998c].

A Tabu Search Algorithm for CSPs

The tabu search algorithm by Galinier and Hao [1997], *TS-GH*, is currently one of the best performing SLS algorithms for the CSP. TS-GH is based on the same neighbourhood and evaluation function as MCH, but uses a different mechanism for selecting the variable/value pair involved in each search step: Amongst all pairs (x, v') such that variable x appears in a currently violated constraint and v' is any value from the domain of x, TS-GH chooses the one that leads to a maximal decrease in the number of violated constraints. If multiple such pairs exist, one of them is selected uniformly at random. As in MCH, the actual search step is then performed by assigning v' to x. This best-improvement strategy is augmented with the same tabu mechanism used in TMCH: After changing the assignment of x from v to v', the variable value pair (x, v) is declared tabu for tt search steps. As in TMCH, an aspiration criterion is used to override the tabu status of variable/value pairs corresponding to search steps that lead to improvements over the incumbent assignment.

In order to achieve competitive performance of TS-GH, it is crucial to avoid computing evaluation function values for every variable/value pair that may potentially be involved in a search step. Instead, in order to implement TS-GH efficiently, a caching and incremental updating technique analogous to the one underlying efficient implementations of GSAT (see in-depth section on page 271) is used [Galinier and Hao, 1997]: After initialising the search, the effects on the evaluation function of changing the assignment of any variable x to any value d' from its domain are computed and stored in a two-dimensional table of size $n \times k$, where n is the number of variables, and k is the size of the largest domain in the given CSP instance. Based on the entries in this table, the (non-tabu) variable/value pair that results in the maximal improvement in the evaluation function value can be selected in time $O(n \cdot k)$ in the worst case. After each search step, only the entries in the table that are affected by the corresponding change in the current assignment need to be updated. For CSP instances with binary constraint relations, initialising the table takes time $O(n^2 \cdot k)$ in the worst case; the update after a search step can be performed in time $O(n \cdot k)$ in the worst case, but is substantially faster for CSP instances with sparse constraint graphs. Using this technique and an efficient implementation of the tabu mechanism, as described for GSAT/Tabu, the search steps of TS-GH are as efficient as those of MCH.

It may be noted that TS-GH was originally introduced as an algorithm for MAX-CSP, the optimisation variant of CSP, in which the objective is to find a variable assignment that satisfies a maximal number of constraints. (SLS algorithms for MAX-CSP will be further discussed in Chapter 7, Section 7.3.) Empirical studies suggest that when applied to the conventional CSP, TS-GH generally achieves better performance than any of the MCH variants, including TMCH, rendering it one of the best known SLS algorithms for the CSP [Stützle, 1998c]. Unlike in the case of TMCH, for TS-GH, the optimal setting of the tabu tenure parameter, tt, increases with instance size, which makes it harder to solve previously unknown CSP instances with peak efficiency [Stützle, 1998c].

6.7 Further Readings and Related Work

SAT and CSP have been extensively studied for several decades, and there is an extensive body of literature on these problems and on algorithms for solving them. The SLS algorithms presented in this chapter have been selected primarily based on their performance and historical significance; however, there are many other SLS algorithms for SAT and CSP that are interesting and fairly widely known.

One of the earliest applications of SLS techniques to SAT is found in Gu's SAT1 algorithm family [Gu, 1992]. Developed independently and published around the same time as the basic GSAT algorithm, the first SAT1 algorithms are based on a simple iterative improvement method augmented with various techniques for overcoming search stagnation. Subsequently, these early SAT1 algorithms have given rise to numerous variants and extensions, including parallel SLS algorithms for SAT, complete SAT algorithms obtained from combining SLS techniques and backtracking algorithms, and special cases of Iterated Local Search. Many of these algorithms have been applied successfully to SAT-encodings of real-world VLSI circuit testing and synthesis and scheduling problems. A good overview of this line of work can be found in Gu et al. [1997]. Both, basic GSAT and the earliest SAT1 algorithms are predated by the Steepest Ascent Mildest Descent (SAMD) algorithm for MAX-SAT [Hansen and Jaumard, 1990], which will be covered in some more detail in Chapter 7 (page 329).

Since the early 1990s, a large number of SLS algorithms for SAT have been introduced and studied in the literature. These include methods based on Simulated Annealing [Spears, 1993; Beringer et al., 1994; Selman et al., 1994], Evolutionary Algorithms [Gottlieb et al., 2002] and Greedy Randomised Adaptive Search Procedures (GRASP) [Resende and Feo, 1996]. While several of these algorithms have been directly compared to some of the SAT algorithms presented in this chapter, there is no evidence that any of them might generally perform better than the best WalkSAT or dynamic local search algorithms.

SLS algorithms also play an important role in the theoretical complexity analysis of SAT. Using a variant of Papadimitriou's Random Walk algorithm [Papadimitriou, 1991] that restarts the search from a randomly chosen assignment after $3n$ variable flips, Schöning [1999; 2002] proved that any k-CNF formula with n variables can be solved in time $poly(n) \cdot (2(k-1)/k)^n$, where $poly(n)$ is a polynomial function over n. By using the same algorithm with a modified search initialisation, which exploits sets of mututally independent clauses, the currently best known lower bound on the time complexity of SAT for 3-CNF formulae of $poly(n) \cdot 1.3303^n$ was obtained [Schuler et al., 2001]. For k-CNF with $k > 3$, the currently best lower bounds on the time complexity of SAT were obtained by Paturi et al. [1997; 1998], based on an algorithm that first calculates the closure of the given formula F under bounded-length resolution, and then performs a simple stochastic iterated construction search in order to find models of the resulting CNF formula. This algorithm forms the basis of another recent SAT solver, Unit-Walk [Hirsch and Kojevnikov, 2001], which has been empirically shown to reach the performance of state-of-the-art SLS algorithms for SAT for various classes of benchmark instances and is provable probabilistically approximately complete.

The survey paper by Gu et al. [1997] provides an excellent overview of the SAT problem, including an interesting classification of SAT algorithms, complexity results, various types of benchmark instances, and a large number of practical applications. It also presents a number of SLS algorithms for SAT, which, however, is somewhat incomplete and now rather outdated, as well as a comprehensive list of references. A more recent study by Hoos and Stützle [2000a] presents a fairly complete and up-to-date overview of GSAT and WalkSAT algorithms, including detailed results on the run-time behaviour and performance of these algorithms.

The GSAT architecture can be generalised to the CSP in a rather straightforward way; a GSAT variant that includes various additional SLS mechanisms, such as random walk, clause weighting, and a dynamic restart strategy, was described by Kask and Dechter [1995], who used it in an empirical study on the efficacy of preprocessing techniques for SAT and CSP. An interesting extension that combines GSAT with a tree search mechanism based on cycle-cutsets, called GSAT+CC, has been applied to Uniform Random Binary CSP [Kask and Dechter, 1996]. Preliminary empirical results suggest that for a limited class of CSP instances (those with small cutsets), using the additional tree search mechanism results in substantially improved performance, while on other subclasses of the CSP, GSAT+CC does not reach the performance of the previously mentioned GSAT variant.

The Breakout Method [Morris, 1993] is an early and relatively widely known dynamic local search method for the CSP. The original Breakout Algorithm used a deterministic first improvement algorithm as its underlying local search procedure. It served as the inspiration for several other DLS algorithms for the CSP, including the previously mentioned GSAT+CC as well as GENET [Davenport et al., 1994], one of the first extensions of MCH and a precursor of the Guided Local Search algorithm by Voudouris and Tsang [Voudouris, 1997;

Voudouris and Tsang, 1999] (see also Chapter 2, Section 2.2). A recent study has produced empirical evidence suggesting that a version of the Breakout Method based on the same type of neighbourhood relation as Galinier and Hao's Tabu Search algorithm performs significantly better than random walk extensions of MCH [Williams Jr. and Dozier, 2001]. This indicates that dynamic local search is a promising approach for future CSP algorithms.

Binary CSP instances are commonly used for the evaluation of Evolutionary Algorithms, where they serve as a benchmark for investigating algorithm behaviour for constrained problems [Eiben, 2001; Marchiori and Steenbeek, 2000b; Craenen et al., 2000; Dozier et al., 1998]. From the published results, however, it is unclear how these algorithms compare to state-of-the-art SLS algorithms for the CSP in terms of performance; given the experience for SAT, it is doubtful that the proposed EAs can reach state-of-the-art performance.

Recently, Solnon developed ant colony optimisation algorithm for the CSP, using a local search procedure based on MCH [Solnon, 2002a; 2002b]. This algorithm was successfully applied to Uniform Random Binary CSP and graph colouring instances; for hard Uniform Random Binary CSP instances from the solubility phase transition region, the ACO algorithm was found to perform better than WMCH. Furthermore, ACO algorithms have been successfully applied to subclasses of the CSP, such as the car sequencing problem [Solnon, 2000].

Encodings of CSP instances into SAT and vice versa have been the subject of a number of studies. Recent work has focused on the impact of different encodings on the performance of SLS algorithms [Hoos, 1999b; Frisch and Peugniez, 2001; Gent, 2002; Prestwich, 2003], as well as on the effects of polynomial preprocessing techniques on the resulting SAT and CSP instances [Walsh, 2000; Gent, 2002].

As general references for the CSP, the interested reader is referred to the book by Tsang [1993] (in parts now somewhat outdated) as well as to the more recent book by Dechter [2003].

6.8 Summary

In this chapter, we presented and discussed SLS algorithms for two important and prominent combinatorial decision problems, the *Propositional Satisfiability Problem (SAT)* and the *Constraint Satisfaction Problem (CSP)*. Both problems are of substantial theoretical interest and have a range of real-world applications.

SAT is one of the most prominent and widely studied $\mathcal{NP}$-complete decision problems. Most SAT algorithms operate on propositional formulae in conjunctive normal form (CNF); because any formula can be transformed into CNF, this is not a serious limitation. Moreover, instances of other combinatorial problems can often be easily encoded into SAT, using reasonably compact and natural CNF representations. While SAT can be formulated as a special case of CSP as well as of 0-1 Integer Linear Programming, the conventional logical formulation appears to provide a much better basis for solving SAT instances efficiently. Polynomial time simplification techniques, such as *unit propagation*, play a crucial role for preprocessing SAT instances before applying a general SAT solver, as well as within systematic search algorithms for SAT; on their own, they can be used for solving several interesting subclasses of SAT efficiently.

We discussed various types of SAT instances, including *Uniform Random k-SAT*, one of the most prominent classes of randomly generated SAT instances, and the *solubility phase transition phenomenon* observed for this subclass of SAT; SAT-encodings of other combinatorial problems; and instances from several practical applications of SAT, such as circuit verification and design. We briefly mentioned a number of generalisations of SAT, including *Multi-Valued SAT*, *MAX-SAT*, and the *Satisfiability Problem for Quantified Boolean Formulae (QSAT)*, as well as problems related to SAT, such as the *Propositional Validity Problem (VAL)*.

We presented three classes of SLS algorithms for SAT: the *GSAT architecture*, the *WalkSAT architecture* and *dynamic local search algorithms* for SAT. While GSAT and related algorithms played a pivotal role in the early development of SLS algorithms for SAT, recent WalkSAT and dynamic local search algorithms, such as Novelty[+] and SAPS, are amongst the best SAT solvers currently known.

The *Constraint Satisfaction Problem (CSP)* can be seen as a generalisation of SAT in which the variable domains can be different from the set $\{\top, \bot\}$ and the constraining conditions that have to be simultaneously satisfied by any solution can be arbitrary relations between a subset of CSP variables. Our discussion was focused on the *finite discrete CSP*, a $\mathcal{NP}$-complete subproblem in which all variable domains are finite and discrete sets. We gave a brief overview of various widely used classes of benchmark instances for the CSP, including *Uniform Random Binary CSP*, as well as instances of the *Graph Colouring* and *Quasigroup Completion* problem.

We discussed three SLS approaches for solving the CSP: (1) Encoding CSP instances into SAT and solving them using SLS algorithms for SAT (or any other type of SAT solver); (2) using direct generalisations of SAT algorithms for solving CSP instances; and (3) applying native SLS algorithms for the CSP. It is presently not clear whether any of these approaches consistently achieves substantially better performance than the others.

For the first approach, different SAT encodings of the CSP can be used. Between the two encodings discussed here, the *sparse encoding* and the *compact encoding*, the former produces SAT instances that appear to be consistently easier to solve for standard SLS algorithms for SAT. In the context of the second approach, we discussed direct generalisations of WalkSAT for two interesting subclasses of the general CSP, *Pseudo-Boolean CSP* (also known as Pseudo-Boolean Programming) and *Many-Valued SAT*. Our discussion of the third approach was focused on the *Min-Conflicts Heuristic (MCH)* and the *tabu search algorithm by Galinier and Hao (TS-GH)*; while the former played a pivotal role in the development of SLS algorithms for SAT and CSP, the latter achieves substantially better performance than MCH and its more recent variants.

Overall, SAT is (and continues to be) an ideal problem for developing and evaluating algorithmic ideas, including SLS techniques, because of its conceptual simplicity as well as its theoretical and practical significance. While many problems are more naturally encoded into CSP than into SAT, it is presently not clear that native CSP algorithms can substantially outperform highly optimised SAT algorithms on suitably chosen encodings.

Generally, the development and understanding of SLS algorithms is significantly further advanced for SAT than for CSP. One of the major reasons for this lies in the fact that SAT (for CNF formulae) — because of its conceptual simplicity compared to the more general finite discrete CSP — facilitates to a larger extent the development, analysis and efficient implementation of SLS algorithms. While there appears to be substantial room for further improvements in native SLS algorithms for the CSP, one particularly promising approach for solving practically relevant types of CSP instances is to use generalisations of high-performance SLS algorithms for SAT augmented with specific methods for handling certain types of complex constraints.

Finally, it may be noted that for both, SAT and CSP, the potential of many advanced SLS methods, such as Iterated Local Search, Variable Depth Search or Ant Colony Optimisation, has not been fully explored, and it is quite likely that by using such advanced techniques, further significant improvements in our ability to solve these problems can be achieved.

Exercises

6.1 [*Easy*] Consider the problem G of colouring the vertices of the graph shown below with four colours such that no two vertices connected by an edge have the same colour.

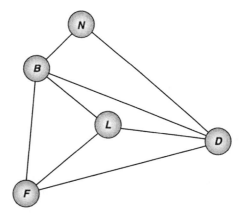

(a) Formulate this problem as a CSP instance.

(b) Formulate this problem as a SAT instance, that is, give a CNF formula F such that any model of F corresponds to a solution of G.

6.2 [*Medium*] When allowing an arbitrary number of tries, GSAT is probabilistically approximately complete. Explain why nevertheless other mechanisms for achieving the PAC property, such as conflict-directed random walk, are preferable over the simple static restart mechanism.

6.3 [*Easy*] Describe how you can use a WalkSAT algorithm to solve the Propositional Validity Problem (VAL) for a given formula in disjunctive normal form (DNF).

6.4 [*Medium*] Discuss the statistical significance of the performance differences shown for various GSAT and WalkSAT algorithms in the two plots from Figure 6.7 (page 280) based on the sample sizes used for these experiments and your knowledge of appropriate statistical tests.

6.5 [*Medium; Hands-on*]

(a) Compare the performance of SAPS and Novelty$^+$ on instances g125.18, ais12, and logistics.c.

(b) Characterise the behaviour of SAPS on logistics.c for varying ρ and α settings.

(These SAT instances are part of the DIMACS/GCP, AIS, and Planning/logistics benchmark sets available from SATLIB [Hoos and Stützle, 2003]; efficient implementations of SAPS and Novelty$^+$ are available from `www.sls-book.net`.)

6.6   [*Easy*]   Formally specify the 6-Queens Problem in the form of a finite discrete CSP instance.

6.7   [*Easy*]   Specify the CSP variable domains as well as the arity of the constraints that arise in the context of encoding a given 3-SAT instance into a CSP instance.

6.8   [*Easy*]   How many clauses and variables are required in the worst case for encoding an NB-SAT instances with n variables and m clauses into semantically equivalent SAT instances using a sparse encoding?

6.9   [*Medium*]   Develop the details of a caching and incremental updating scheme for the evaluation function values in the TS-GH algorithm.

6.10   [*Hard*]   Prove that, when applied to Uniform Random 3-SAT instances, the efficient caching and updating mechanism for variable scores in GSAT, as described in the in-depth section on page 271, has time-complexity $O(1)$ in each search step.

It is the mark of an educated mind
to rest satisfied with the degree of precision
which the nature of the subject admits
and not to seek exactness
where only an approximation is possible.

—Aristotle, Philosopher

7 MAX-SAT AND MAX-CSP

MAX-SAT and MAX-CSP are the optimisation variants of SAT and CSP. These problems are theoretically and practically interesting, because they are among the conceptually simplest combinatorial optimisation problems, yet instances of optimisation problems from many application domains can be represented as MAX-SAT or MAX-CSP instances in an easy and natural way. SLS algorithms are among the most powerful and successful methods for solving large and hard MAX-SAT and MAX-CSP instances.

In this chapter, we first introduce MAX-SAT. Next, we present some of the best-performing SLS algorithms for various types of MAX-SAT instances and give an overview of results on their behaviour and relative performance. In the second part of this chapter, we introduce MAX-CSP and discuss SLS methods for solving the general problem as well as the closely related overconstrained pseudo-Boolean and integer optimisation problems.

7.1 The MAX-SAT Problem

MAX-SAT can be seen as a generalisation of SAT for propositional formulae in conjunctive normal form in which, instead of satisfying all clauses of a given CNF formula F with n variables and m clauses (and hence F as a whole), the objective is to satisfy as many clauses of F as possible. A solution to an instance of this problem is a variable assignment (i.e., a mapping of variables in F to truth values), that satisfies a maximal number of clauses in F.

DEFINITION 7.1 **Unweighted MAX-SAT**

Given a CNF formula $F := \bigwedge_{i=1}^{m} \bigvee_{j=1}^{k_i} l_{ij}$, let $f(F, a)$ be the number of clauses in F that are unsatisfied under variable assignment a. The (Unweighted) Maximum Satisfiability Problem (MAX-SAT) is to find a variable assignment $a^* \in \text{argmin}_{a \in Assign(F)} f(F, a)$ or, equivalently, $a^* \in \text{argmax}_{a \in Assign(F)} (m - f(F, a))$, that is, an assignment a^* that maximises the number of the satisfied clauses in F.

Remark: *Maximising* the number of *satisfied* clauses in F is equivalent to *minimising* the number of *unsatisfied* clauses. Although MAX-SAT is intuitively defined as a maximisation problem, it is often formally more convenient to consider the equivalent minimisation problem; this is the reason for using the objective function $f(F, a)$, whose value is to be minimised, in our definition of MAX-SAT. In the following, we will consider MAX-SAT as a minimisation problem.

This definition captures the search variant of MAX-SAT; the evaluation variant and associated decision problems can be defined in a similar way: Given a CNF formula F, in the evaluation variant, the objective is to determine the minimum number of clauses unsatisfied under any assignment; the associated decision problem for a given solution quality bound b is to determine whether there is an assignment that leaves at most b clauses in F unsatisfied. Note that SAT is equivalent to the decision variant of unweighted MAX-SAT with solution quality bound $b := 0$, that is, to deciding whether for a given CNF formula F there exists an assignment a, such that the number of clauses in F unsatisfied under a is equal to zero.

EXAMPLE 7.1 **A Simple MAX-SAT Instance**

Let us consider the following propositional formula in CNF:

$$F := (\neg x_1)$$
$$\wedge (\neg x_2 \vee x_1)$$
$$\wedge (\neg x_1 \vee \neg x_2 \vee \neg x_3)$$
$$\wedge (x_1 \vee x_2)$$
$$\wedge (\neg x_4 \vee x_3)$$
$$\wedge (\neg x_5 \vee x_3)$$

The minimum number of clauses in F that are unsatisfied under any assignment, $f(F, a^*)$, is one; two of the (many) assignments that achieve this optimal solution quality are $x_1 := x_2 := x_3 := x_4 := x_5 := \perp$ and $x_1 := \perp, x_2 := x_3 := x_4 := x_5 := \top$.

It may be noted that while the SAT problem is defined for arbitrary propositional formulae, the definition of MAX-SAT is restricted to formulae in CNF. Furthermore, different from SAT, MAX-SAT is not invariant under certain logical equivalence transformations, that is, there exist MAX-SAT instances whose underlying CNF formulae are logically equivalent but whose optimal solutions are different. In particular, the solutions of a MAX-SAT instance can change when introducing multiple copies of clauses in the given CNF formula; in unweighted MAX-SAT, the number of copies of a clause can be used to express its importance relative to other clauses. As a consequence, standard simplification techniques for SAT, including unit propagation and pure literal reduction, are not directly applicable to MAX-SAT.

Weighted MAX-SAT

In many applications of MAX-SAT, the constraints represented by the CNF clauses are not all equally important. These differences can be represented explicitly and compactly by using weights associated with each clause of a CNF formula.

DEFINITION 7.2 **Weighted CNF Formula**

A weighted CNF formula is a pair (F, w), where F is a CNF formula $F := \bigwedge_{i=1}^{m} c_i$ with $c_i = \bigvee_{j=1}^{k_i} l_{ij}$, and $w : \{c_1, \ldots, c_m\} \mapsto \mathbb{R}^+$ is a function that assigns a positive real value to each clause of F; $w(c_i)$ is called the weight *of clause c_i. (Without loss of generality, we assume that all clauses in F are pairwise different.)*

Intuitively, the clause weights in a weighted CNF formula reflect the relative importance of satisfying the respective clauses; in particular, appropriately chosen clause weights can indicate the fact that satisfying a certain clause is considered more important than satisfying several other clauses. Weighted MAX-SAT is a straightforward generalisation of unweighted MAX-SAT, in which the objective is to minimise the total weight of the unsatisfied clauses, rather than just their total number.

DEFINITION 7.3 **Weighted MAX-SAT**

Given a weighted CNF formula $F' = (F, w)$, let $f(F', a)$ be the total weight of the clauses of F that are unsatisfied under assignment a, that is, $f(F', a)$ $:= \sum_{c \in CU(F,a)} w(c)$, where $CU(F, a)$ is the set of all clauses of F unsatisfied under a. The Weighted Maximum Satisfiability Problem (Weighted MAX-SAT) *is to find a variable assignment a^* that maximises the total weight of the satisfied clauses in F, that is, $a^* \in \text{argmin}_{a \in Assign(F)} f(F', a)$, or, equivalently, $a^* \in \text{argmax}_{a \in Assign(F)} (\tilde{f} - f(F', a))$, where $\tilde{f} := \sum_{i=1}^{m} w(c_i)$ is the total weight of all clauses in F.*

Although the definition allows for real-valued clause weights, it is easy to show that integer clause weights are sufficient for expressing arbitrary relative importance relations between clauses. Primarily for historically motivated efficiency reasons, many implementations of MAX-SAT algorithms support only integer clause weights. (Many older types of microprocessors performed integer operations substantially faster than floating point operations; this is not the case for modern CPUs.) However, because in most programming languages the ranges of integer data types are very limited compared to floating point data types, such implementations can sometimes not handle certain types of MAX-SAT instances.

Many combinatorial optimisation problems contain logical conditions that have to be satisfied for any feasible candidate solution; these conditions are often called *hard constraints*, while constraints whose violation does not preclude feasibility are referred to as *soft constraints*. When representing such problems as weighted MAX-SAT instances, the hard constraints can be captured by choosing the weights of the corresponding CNF clauses high enough that no combination of soft constraint clauses can outweigh a single hard constraint clause. The decision problem with solution quality bound b associated with such a weighted MAX-SAT instance, where b is lower than the weight of a single hard constraint clause, but at least as high as the combined weight of any set of soft constraint clauses, then accurately represents the given problem; in particular, any solution to such a weighted MAX-SAT instance corresponds to a feasible candidate solution of the underlying combinatorial optimisation problem.

EXAMPLE 7.2 **A Simple Weighted MAX-SAT Instance**

Consider the formula F from Example 7.1 with the following clause weights:

$$
\begin{aligned}
w(c_1) &:= w(\neg x_1) &&:= 2 \\
w(c_2) &:= w(\neg x_2 \vee x_1) &&:= 1
\end{aligned}
$$

$$
\begin{aligned}
w(c_3) &:= w(\neg x_1 \vee \neg x_2 \vee \neg x_3) &&:= 7 \\
w(c_4) &:= w(x_1 \vee x_2) &&:= 3 \\
w(c_5) &:= w(\neg x_4 \vee x_3) &&:= 7 \\
w(c_6) &:= w(\neg x_5 \vee x_3) &&:= 7
\end{aligned}
$$

The total weight of the clauses unsatisfied under assignment $x_1 := \bot, x_2 := x_3 := x_4 := x_5 := \top$ is 1, which is the optimal solution quality for the weighted MAX-SAT instance (F, w).

Furthermore, when considering this weighted MAX-SAT instance with solution quality bound 6, clauses c_3, c_5 and c_6 can be seen as hard constraints, while all other clauses represent soft constraints. The assignment $x_1 := x_2 := x_3 := x_4 := x_5 := \bot$, which was optimal for the unweighted MAX-SAT instance F, has objective function value 7 and is hence not a feasible candidate solution in this context.

Complexity and Approximability Results

MAX-SAT (unweighted as well as weighted) is an $\mathcal{NP}$-hard optimisation problem, since SAT can be reduced to MAX-SAT in a straightforward way. Interestingly, while 2-SAT, the restriction of SAT to CNF formulae with clauses of length 2, can be solved in polynomial time, MAX-2-SAT, the corresponding restriction of MAX-SAT, is known to be $\mathcal{NP}$-hard, as is MAX-3-SAT (i.e., MAX-SAT for CNF formulae with clause length 3).

However, there are polynomial-time algorithms for MAX-SAT that are guaranteed to find solutions within a certain range of the optimum for arbitrary MAX-SAT instances. The first such approximation algorithm was a relatively simple greedy construction method that has been shown to solve any weighted MAX-SAT instance within a factor (approximation ratio) of at most 2 from the respective maximum total weight of the clauses satisfied under any variable assignment [Johnson, 1974]. (More recently, it has been shown that Johnson's algorithm guarantees an approximation ratio of 1.5 [Chen et al., 1997].)

Since 1994, a series of polynomial-time algorithms with substantially improved approximation ratios has been introduced [Yannakakis, 1994; Goemans and Williamson, 1994; 1995; Feige and Goemans, 1995; Mahajan and Ramesh, 1995; Asano, 1997; Asano and Williamson, 2000]; the most recent of these guarantees an approximation ratio of 1.275 [Asano and Williamson, 2000]. (Assuming the correctness of a conjecture by Zwick [1999], which is supported by numerical evidence, this latter result can be improved to 1.201 [Asano and Williamson, 2000].) For the special cases MAX-3-SAT and MAX-2-SAT, the best approximation algorithms guarantee solutions within $8/7 \approx 1.1429$ [Karloff and Zwick, 1997] and 1.075 [Feige and Goemans, 1995; Mahajan and Ramesh, 1995], respectively. It is interesting to note that a simple iterative improvement algorithm with a non-oblivious evaluation function (see Section 7.2) has been proven to achieve a worst-case approximation ratio of $2^k/(2^k - 1)$ for MAX-k-SAT [Khanna et al., 1994].

There are limitations on the theoretical performance guarantees that can be obtained from polynomial-time algorithms for MAX-SAT: If $\mathcal{P} \neq \mathcal{NP}$, there exists no polynomial-time approximation algorithm for MAX-3-SAT, and hence for MAX-SAT, with a (worst-case) approximation ratio lower than $8/7 \approx 1.1429$; for MAX-2-SAT, an analogous result rules out approximation ratios lower than 1.0472 [Håstad, 1997; 2001]. Arbitrarily improved approximation ratios α can be obtained at the cost of run-times that are exponential in instance size and depend on the desired value of α [Dantsin et al., 1998]. It is worth noting that approximation algorithms for MAX-SAT, such as the ones mentioned here, can be empirically shown to achieve much better solution qualities for many types of MAX-SAT instances; however, their performance is usually substantially inferior to that of high-performance SLS algorithms for MAX-SAT (see, e.g., Hansen and Jaumard [1990]).

Randomly Generated MAX-SAT Instances

As in the case of SAT, empirical studies play a prominent role in the analysis of the performance and behaviour of MAX-SAT algorithms. In this context, various classes of randomly generated MAX-SAT instances are commonly used, in particular Uniform Random 3-SAT instances, which are typically sampled from the overconstrained region of the respective solubility phase transition, that is, the clauses/variable ratio is larger than the critical value of approximately 4.3, and the instances are unsatisfiable with very high probability (see also Chapter 6, page 262*f.*).

A number of empirical studies have used test-sets obtained from the random clause length model, in which each of the possible $2 \cdot n$ literals over n variables is included with a fixed probability in any clause (see Chapter 6, page 261). A well-known set of such instances is part of the DIMACS collection of SAT benchmark instances; these jnh instances have 100 variables and between 800 and 900 clauses each, including satisfiable and unsatisfiable instances. Sets of weighted MAX-SAT instances have been derived from test-sets of random clause length formulae, including the jnh instances, by determining for each clause an integer weight between 1 and 1 000 uniformly at random [Resende et al., 1997; Yagiura and

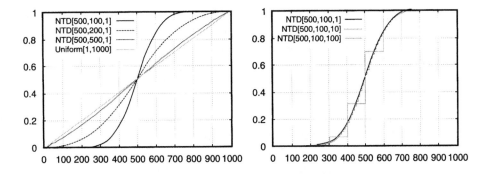

Figure 7.1 Truncated discretised Gaussian distributions used for generating clause weights for weighted Uniform Random 3-SAT test-sets; cumulative distribution functions *NDT*[μ, σ', δ] for $\mu = 500$ and various values of σ' (with uniform distribution over $[1, \ldots, 1\,000]$ for comparison) *(left)* and δ *(right)*.

Ibaraki, 1998; 2001]. Particularly the weighted jnh (wjnh) instances have been widely used for evaluating the performance of MAX-SAT algorithms.

A range of test-sets of weighted MAX-SAT has been introduced by Hoos; these consist of Uniform Random 3-SAT instances with truncated discretised Gaussian clause weight distributions. The weight distributions are characterised by three parameters μ, σ' and δ, where μ is the mean, σ' the standard deviation (before truncation) and δ the granularity of the underlying Gaussian probability distribution, which is symmetrically truncated to the interval $[1, 2\mu - 1]$ (see Figure 7.1); the granularity specifies the minimum difference between non-identical clause weights and hence determines, together with the range $[1, 2\mu - 1]$, the number of distinct values that clause weights can take. It may be noted that for $\sigma' = 0$ or $\delta > 2\mu$, all clause weights are identical, which renders the respective instances equivalent to unweighted MAX-SAT instances. Furthermore, for large values of σ'/μ, the clause weight distributions approach a uniform distribution over the interval $[1, 2\mu - 1]$. These test-sets have been designed and used for investigating the impact of clause weight variance and granularity on the performance of MAX-SAT algorithms [Smyth et al., 2003].

MAX-SAT Encodings of Other Combinatorial Problems

Many $\mathcal{NP}$-hard combinatorial optimisation problems can be quite easily and naturally encoded into MAX-SAT. A good example for this is the following *Minimum-Cost Graph Colouring Problem (Min-Cost GCP)*: Given an (undirected) edge-weighted graph $G := (V, E, w)$ and an integer k, determine a minimum cost k-colouring of G, where a *k-colouring of G* is a mapping a that assigns an integer from the set $\{1, \ldots, k\}$ to each vertex in V, and the cost of a colouring a is the sum of all edge-weights $w(e)$ for which e is an edge whose two incident vertices are assigned the same colour under a.

Any instance G of this problem can be transformed into a weighted MAX-SAT instance $F(G)$ as follows: For each edge $e = (v, v')$ and colour l, we create a clause $c_e := \neg x_{v,l} \vee \neg x_{v',l}$ with weight $w(e)$ (the weight of edge e in G). Furthermore, for each vertex v in G, we create a clause $c_v := \bigvee_{l=1}^{k} x_{v,l}$ with weight $\widehat{w} := \max\{\widehat{w_v} \mid v \in V\} + 1$, where $\widehat{w_v} := \sum_{e \in E(v)} w(e)$, and $E(v)$ is the set of all edges in E that are incident with v; intuitively, $\widehat{w}$ is defined in such a way that it just exceeds the maximum total weight of all edges incident to any particular vertex in G. Finally, for each vertex v in G and for each pair of different colours l, l', we create a clause $c_{v,l,l'} := \neg x_{v,l} \vee \neg x_{v,l'}$ with weight $\widehat{w}$. It is easy to see that the optimal solutions of the weighted MAX-SAT instance $F(G)$ thus obtained correspond exactly to the optimal solutions of the given Min-Cost GCP instance G. Furthermore, under the 1-flip neighbourhood, the locally optimal candidate solutions of $F(G)$ correspond exactly to the k-colourings of G. MAX-SAT-encoded Min-Cost GCP instances with integer weights have been used in several studies on SLS algorithms for MAX-SAT [Yagiura and Ibaraki, 1998; 2001; Hoos et al., 2002].

Another problem that can be easily encoded into weighted MAX-SAT is the *Weighted Set Covering Problem (Weighted SCP)*, in which, given a set A, a collection $F := \{A_1, \ldots, A_m\}$ of subsets $A_j \subseteq A$ and a weight function $w : F \mapsto \mathbb{R}^+$, the objective is to find a minimal weight set cover of A, where a set cover of A is a subset of F such that the sets in C cover all elements of A, that is, $\bigcup C = A$, and the weight of C is the total weight of its elements, that is, $\sum_{A' \in C} w(A')$. This problem is $\mathcal{NP}$-hard and has applications, for example, in Boolean circuit optimisation. MAX-SAT encodings of Weighted SCP instances from the ORLIB benchmark library [Beasley, 2003] have been used for evaluating the performance of MAX-SAT algorithms [Yagiura and Ibaraki, 1998; 2001; Smyth et al., 2003]. Set covering problems are discussed in more detail in Chapter 10, Section 10.3.

Other hard combinatorial optimisation problems that have been encoded into MAX-SAT and used in the context of various studies on MAX-SAT algorithms include time-tabling problems [Yagiura and Ibaraki, 1998; 2001; Hoos et al., 2002], the problem of finding most probable explanations in Bayesian networks (MPE) [Park, 2002], and the problem of minimising the number of crossings that arise when embedding level-graphs into a plane (LGCMP) [Randerath et al., 2001; Smyth et al., 2003]. In almost all cases, these problems contain hard and soft constraints, which are captured by appropriately chosen weights of the respective CNF clauses. Furthermore, all of these problems have real-world applications in diverse areas, such as system diagnosis and database design.

7.2   SLS Algorithms for MAX-SAT

Many SLS methods have been applied to MAX-SAT, resulting in a large number of algorithms for unweighted and weighted MAX-SAT. In this section, we present some of the most prominent and best-performing algorithms, including straightforward applications of SLS algorithms for SAT to unweighted MAX-SAT, variants of WalkSAT, Dynamic Local Search and Tabu Search, as well as Iterated Local Search algorithms. Additionally, we discuss some SLS algorithms that are based on larger neighbourhoods and non-oblivious evaluation functions; these approaches are rather specific to MAX-SAT. MAX-SAT algorithms based on other SLS methods, such as Simulated Annealing, GRASP or Ant Colony Optimisation, will be briefly mentioned in Section 7.4.

Solving MAX-SAT Using SLS Algorithms for SAT

Any SLS algorithm for SAT can be applied to unweighted MAX-SAT in a straightforward way. The only modification required in this context is the addition of a simple mechanism that keeps track of the incumbent candidate solution and returns it at the end of the search process, provided its solution quality meets a given bound, if such a bound has been specified as an input to the algorithm. Hence, in principle any of the SLS algorithms for SAT described in Chapter 6 can be used for solving unweighted MAX-SAT.

It is not clear that SLS algorithms that are known to perform well on SAT can be expected to show equally strong performance on unweighted MAX-SAT. There is some empirical evidence that for long run-times, GWSAT obtains consistently higher solution qualities than a number of earlier SLS algorithms for MAX-SAT, including algorithms based on Simulated Annealing and Tabu Search, when applied to Uniform Random 3-SAT instances of varying constrainedness [Hansen and Jaumard, 1990; Selman et al., 1994]; however, the different termination criteria used in these comparative studies render these results somewhat inconclusive (see also Battiti and Protasi [1997b]). Similar results have been obtained for GSAT/Tabu [Battiti and Protasi, 1997a]; these will be discussed in more detail later.

More recent results show that Novelty[+], one of the best-performing SLS algorithms for SAT known to date (cf. Chapter 6, page 276*ff.*), typically does not reach the performance of state-of-the-art SLS algorithms for MAX-SAT on Uniform Random 3-SAT instances; this is particularly the case for highly constrained instances [Hoos et al., 2003]. Intuitively, WalkSAT algorithms such as Novelty[+] have difficulties in selecting effective search steps in situations where a relatively large number of clauses is unsatisfied. In each search step, they select the variable to be flipped from an unsatisfied clause that is uniformly chosen at random. However, with many unsatisfied clauses, only a few of which contain variables whose flip leads to improved candidate solutions, selecting an unsatisfied clause from which such a variable can be chosen is rather unlikely. This is particularly the case for highly constrained instances, in which all candidate solutions — including optimal quality solutions — have a high number of unsatisfied clauses. GSAT algorithms, on the contrary, do not suffer from this problem, since they are able to choose the variable whose flip achieves the maximal improvement in solution quality with a probability that is independent of the number of unsatisfied clauses and instance constrainedness.

There are very few results on the performance obtained by applying dynamic local search algorithms for SAT to unweighted MAX-SAT; recent empirical results suggest that SAPS, a state-of-the-art SAT algorithm (cf. Chapter 6, page 291*f.*), outperforms GLS [Mills and Tsang, 2000] in terms of the CPU time required for finding quasi-optimal (i.e., best known) solutions for overconstrained Uniform Random 3-SAT instances, but does not reach the performance of IRoTS (a state-of-the-art Iterated Local Search algorithm for MAX-SAT described later in this section) on these instances [Tompkins and Hoos, 2003]. Interestingly, based on RTD analyses it seems that GLS frequently suffers from search stagnation, whereas this does not appear to be the case for SAPS, which typically shows regular exponential RTDs.

WalkSAT Algorithms for Weighted MAX-SAT

GSAT and WalkSAT algorithms can be generalised to weighted MAX-SAT by using the objective function for weighted MAX-SAT — that is, the total weight of the clauses unsatisfied under a given assignment — as the evaluation function based on which the variable to be flipped in each search step is selected.

A WalkSAT variant for weighted MAX-SAT with explicit hard and soft constraints, *WalkSAT-JKS*, has been proposed by Jiang, Kautz and Selman [1995]. Applied to standard weighted MAX-SAT, this algorithm closely resembles WalkSAT/SKC, but differs in that it allows random walk steps even in situations where 'zero damage' flips are available (cf. Chapter 6, page 274). When hard constraints are explicitly identified (via a lower bound on the weights of CNF clauses that are to be treated as hard constraints), this WalkSAT algorithm restricts the clause selection in the first stage of the variable selection mechanism to unsatisfied hard constraint clauses, unless all hard constraints are satisfied by the current candidate assignment. The WalkSAT-JKS algorithm for weighted MAX-SAT has been shown to achieve impressive results on various sets of MAX-SAT-encoded Steiner tree problems [Jiang et al., 1995]; it should be noted, however, that these results crucially rely on a particularly effective encoding of the original Steiner tree problems into MAX-SAT.

In principle, the 2-stage variable selection mechanism underlying all Walk-SAT algorithms can be extended to MAX-SAT in two different ways: (i) by using the objective function for weighted MAX-SAT in the second stage as in WalkSAT-JKS ('*we*' mechanism), and (ii) by considering clause weights in the selection of an unsatisfied clause in the first stage ('*wcs*' mechanism) [Hoos et al., 2002; 2003]. The motivation behind the latter mechanism is based on the following observations: In situations where many clauses are unsatisfied, the probability for selecting the best clause, that is, the unsatisfied clause that contains one of the variables whose flip leads to a maximal improvement in the objective function value, can be very small when basing this selection on a uniform distribution, as used in standard WalkSAT. By selecting an unsatisfied clause c with a probability proportional to the weight of c, the WalkSAT search process becomes more focused on satisfying clauses with high weights. (This probabilistic clause selection method is analogous to the well-known roulette wheel selection used in many Evolutionary Algorithms.)

The *we* and *wcs* mechanisms can be used individually or combined, which leads to three weighted MAX-SAT variants of any WalkSAT algorithm for SAT. A recent empirical study indicates that these variants of WalkSAT/SKC are typically outperformed by the respective Novelty$^+$ variants (note that an analogous situation holds for the SAT versions of these WalkSAT algorithms). Furthermore, Novelty$^+$/*wcs+we* typically performs better than the two other variants and standard Novelty$^+$, except for satisfiable weighted MAX-SAT instances (i.e., instances (F, w) where F is a satisfiable CNF formula), for which standard Novelty$^+$ tends to outperform the *wcs* and *we* variants. Novelty$^+$/*wcs+we* tends to find optimal solutions to the wjnh instances faster (both in terms of CPU time and search steps) than other state-of-the-art algorithms for weighted MAX-SAT, including the GLS and IRoTS algorithms described later (see also Example 7.3). On other types of weighted MAX-SAT instances, in particular on heavily over-constrained weighted Uniform Random 3-SAT instances, none of the Novelty$^+$ variants appears to reach state-of-the-art performance.

However, for various types of MAX-SAT-encoded instances of other problems, including well-known classes of minimum-cost graph colouring and set covering instances, Novelty$^+$/*wcs+we* appears to find quasi-optimal (i.e., best known) solutions in significantly less CPU time than other high-performance algorithms for MAX-SAT, such as IRoTS or GLS, and appears to be the best-performing MAX-SAT algorithm known to date [Hoos et al., 2003].

Dynamic Local Search Algorithms for Weighted MAX-SAT

Generalising DLS algorithms for SAT to weighted MAX-SAT raises an interesting issue: How should the dynamically changing clause penalties used within DLS interact with the fixed clause weights that are part of any weighted MAX-SAT instance?

The first algorithm for weighted MAX-SAT based on the 'discrete Langrangian method', *DLM-SW*, proposed by Shang and Wah [1997], uses an evaluation function of the form $g'(a) := \sum_{c \in CU(a)} (clp(c) + w(c))$, where $CU(a)$ is the set of clauses unsatisfied under assignment a, $clp(c)$ is the penalty associated with clause c, which is dynamically adjusted during the search process, as in the basic DLM algorithm for SAT (cf. Chapter 6, page 288*f.*), and $w(c)$ is the weight of clause c as specified in the given weighted MAX-SAT instances (F, w). Different from basic DLM for SAT, the local search procedure underlying this DLM algorithm for weighted MAX-SAT is an iterative first improvement algorithm (based on the standard 1-flip neighbourhood relation). There is some evidence that DLM-SW performs better than WalkSAT-JKS, but does not reach the performance of either, Novelty$^+$/*wcs* or Novelty$^+$/*wcs+we*, on the `wjnh` instances w.r.t. to the solution quality reached after a fixed number of search steps [Mills and Tsang, 1999b].

Another approach for integrating clause penalties and clause weights has been followed in a straightforward generalisation of DLM-99-SAT (cf. Chapter 6, page 288*f.*) to weighted MAX-SAT [Wu and Wah, 1999]; this variant of DLM differs from the SAT version only in the initialisation of the clause penalties and in the parameter settings δ^+, δ^- and δ_s. The weighted MAX-SAT variant initialises the clause penalties to the weights $w(c)$ of the respective clauses, and chooses the parameters δ^+, δ^- and δ_s, which control the modification of the clause penalties during the search, individually for each clause c proportional to its weight $w(c)$. This approach for handling clause weights in the context of a dynamic local search algorithm differs notably from the one followed in the earlier DLM-SW algorithm. When applied to the `wjnh` instances, DLM-99-SAT for weighted MAX-SAT appears to perform better than DLM-SW [Wu and Wah, 1999], but there is empirical evidence that it typically does not reach the performance of Novelty$^+$/*wcs* and Novelty$^+$/*wcs+we* [Hoos et al., 2003].

Like DLM, GLSSAT, another high-performance dynamic local search algorithm for SAT, has been extended to weighted MAX-SAT [Mills and Tsang, 1999b; 2000]. The resulting GLSSAT variant considers the clause weights of the given weighted MAX-SAT instance only in the utility value of a clause c, defined as $util(a, c) := w(c)/(1 + clp(c))$ if clause c is unsatisfied under assignment a and zero otherwise. Otherwise, the algorithm is identical to GLSSAT (see also Chapter 6, page 287*f.*). It is worth noting that this approach for handling clause weights is conceptually similar to the one underlying Novelty$^+$/wcs. In both cases, the clause weights are not reflected directly in the evaluation function underlying the search process, but influence the search trajectory in a different way. In GLS for MAX-SAT, only the penalty values of clauses with maximal utility are increased after each local search phase; hence, clauses with high weights will typically receive high penalties, which biases the subsidiary local search algorithm towards preferentially satisfying them.

On the wjnh instances, this GLS variant performs substantially better than the previously discussed DLM and WalkSAT algorithms in terms of solution quality reached after a fixed number of iterations [Mills and Tsang, 1999b; 2000]. However, when comparing the CPU time required for finding optimal solutions, both Novelty$^+$/wcs and Novelty$^+$/wcs+we typically show better performance [Hoos et al., 2003]. For weighted Uniform Random 3-SAT instances, GLS for MAX-SAT generally outperforms Novelty$^+$/wcs+we in terms of search steps required for finding quasi-optimal solutions; but in many cases, this performance advantage is insufficient to amortise the substantially higher time complexity of search steps in GLS. For certain types of weighted MAX-SAT instances, such as Uniform Random 3-SAT instances with low variance clause weight distributions, GLS appears to be the best-performing MAX-SAT algorithm known to date [Hoos et al., 2003].

However, GLS for MAX-SAT does not reach the state-of-the-art performance of Novelty$^+$/wcs+we on various types of MAX-SAT-encoded instances of other problems, such as minimum-cost graph colouring or weighted set covering [Hoos et al., 2002]. Furthermore, limited RTD analyses indicate that, different from other state-of-the-art MAX-SAT algorithms, such as Novelty$^+$/wcs+we and IRoTS (described below), GLS for MAX-SAT tends to suffer from stagnation behaviour, which often compromises the robustness of its performance; this appears to be even the case (although to a lesser extent) when all penalty values are regularly decayed, as in GLSSAT2 [Hoos et al., 2003].

EXAMPLE 7.3 **Performance Results for SLS Algorithms for MAX-SAT**

This example illustrates the performance differences between Novelty$^+$/ws+we and GLSSAT2, two of the best-performing MAX-SAT algorithms known to date. All CPU times reported in this example have been measured on PCs with dual 2.4GHz Intel Xeon processors, 512KB cache, and 1GB RAM running Red Hat Linux, Version 2.4smp.

As can be seen from the left side of Figure 7.2, there is typically a clear probabilistic domination relationship between the two algorithms: The qualfied RTDs for reaching optimal quality solutions, as determined using a complete MAX-SAT algorithm, are very similar in shape and do not intersect.

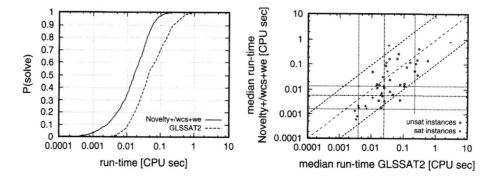

Figure 7.2 *Left:* Qualified RTDs for GLSSAT2 and Novelty$^+$/*wcs+we* for reaching optimal solution quality on a typical unsatisfiable weighted MAX-SAT instance (`wjnh304`) from the `wjnh` benchmark set. *Right:* Correlation of search cost required by GLSSAT2 *vs* Novelty$^+$/*wcs+we* for reaching the optimal solution qualities on the `wjnh` benchmark set; the search cost for each instance is measured as median run-time over 100 runs. The horizontal and vertical lines indicate the median as well as the $q_{0.1}$ and $q_{0.9}$ quantiles of the search cost for the respective algorithm across the test-set; the diagonal lines indicate 10 times, equal, and 1/10th CPU time of Novelty$^+$/*wcs+we* compared to GLSSAT2.

When evaluated across the entire test-set `wjnh`, a well-known and widely used set of randomly generated weighted MAX-SAT instances (cf. Section 7.1), Novelty$^+$/*wcs+we* does not always perform better than GLSSAT2. However, especially for unsatisfiable instances, Novelty$^+$/*wcs+we* tends to find optimal quality solutions up to more than 20 times faster than GLSSAT2.

Table 7.1 summarises the results of a performance comparison between GLSSAT2 and Novelty$^+$/*wcs+we* across a number of well-known test-sets, including the previous studied `wjnh` set and several sets of weighted Uniform Random 3-SAT instances (`rnd`n-m/`w`μ-σ', where n and m denote the number of clauses and variables, and μ and σ' are the parameters of the truncated discretised Gaussian weight distributions $NDT[\mu, \sigma', 1]$). The `wjnh` test-set comprises 14 satisfiable and 30 unsatisfiable instance, and each of the `rnd`n-$*$ test-sets contains 100 unsatisfiable instances.

The instances from these test-sets are much harder for state-of-the-art complete MAX-SAT solvers, such as `wmaxsat-lb2-moms` [Alsinet et al., 2003], than for the SLS algorithms studied here. For example, the run-time required by `wmaxsat-lb2-moms` for solving instance `wjnh304` is more than ten times higher than the median run-time of Novelty$^+$/*wcs+we* (0.29 CPU seconds *vs* 0.023 CPU seconds; interestingly, more than 85% of the run-time of `wmaxsat-lb2-moms` is needed to find an optimal solution rather than for

Test-set	GLSSAT2	Novelty$^+$/wcs + we	fd
wjnh	0.0242 (2 977)	0.0131 (9 261)	0.02 / 0.68
rnd200-1000/w1000-200	0.4589 (69 523)	0.9995 (747 376)	0.49 / 0.07
rnd200-1000/w1000-1000	1.3637 (217 961)	1.2041 (1 033 582)	0.38 / 0.35
rnd200-1400/w1400-1400	3.1840 (408 888)	–	1 / 0

Table 7.1 Performance of GLSSAT2 *vs* Novelty$^+$/wcs+we on selected sets of randomly generated weighted MAX-SAT instances; the table entries are median search cost values over the respective test-sets, where the search cost for a given problem instance is defined as the median run-time required for finding a (quasi-)optimal solution and is reported in CPU seconds (search steps); '–' indicates that no such solutions could be found within more than 1 000 CPU seconds. The search cost values for each algorithm were determined from 100 runs per instance, and algorithms were always run (without restart) until a (quasi-) optimal solution was found. The two values in the *fd* column indicate the fraction of instances from the respective test-set on which GLSSAT2 probabilistically dominated Novelty$^+$/wcs+we (first value) and vice versa (second value).

proving its optimality). Most of the other instances considered here cannot be solved by complete MAX-SAT algorithms within reasonable amounts of CPU time. Consequently, we used multiple, very long runs of state-of-the-art SLS algorithms for MAX-SAT, including GLS and Novelty$^+$/wcs+we to determine the best possible solution qualities. More precisely, we made sure that in each run, after the final solution quality was reached at some run-time t^*, the respective algorithm continued searching for at least time $10t^*$ without finding another improvement. For all instances where provably optimal solution qualities are known, these were shown to be correctly determined by this protocol. (See also Smyth et al. [2003].)

The results shown in Table 7.1 indicate that often, but not always, GLSSAT2 tends to perform better than Novelty$^+$/wcs+we; this is particularly pronounced for the highly overconstrained rnd200-1400/w1400-1400 instances, which Novelty$^+$/wcs+we fails to solve within 1 000 CPU seconds. Note that GLSSAT2 typically requires a substantially lower number of search steps; however, although optimised implementations of both algorithms were used, the time complexity of individual search steps is substantially higher for GLSSAT2 than for Novelty$^+$/wcs+we, which is due to the inherent differences between the two underlying SLS methods (see also Chapter 6, Sections 6.3 and 6.4).

Figure 7.3 (left side) shows the solution quality distributions (SQDs) obtained for GLSSAT2 and Novelty$^+$/wcs+we and run-times of 1 and 10 CPU seconds on a MAX-SAT-encoded instance of the Level Graph Crossing

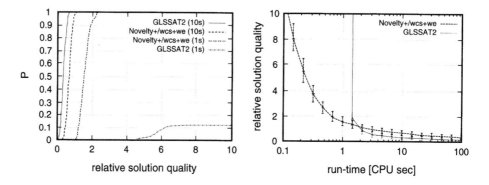

Figure 7.3 *Left:* Solution quality distributions (SQDs) for GLSSAT2 *vs* Novelty$^+$/*wcs+we* on MAX-SAT-encoded LGCMP instance `lgcmp75-1` for run-times of 1 and 10 CPU sec. *Right:* Development of solution quality over time for the same instance; the curves for each algorithm correspond to the median values of the underlying SQDs, while the error bars indicate the respective $q_{0.1}$ and $q_{0.9}$ quantiles. The data underlying these graphs were obtained from 100 runs of each algorithm.

Minimisation Problem (LGCMP). This weighted MAX-SAT instance has 7 500 variables and 128 306 clauses, and all clause weights are either 1 (for clauses derived from the optimisation objective) or 13 307 (for clauses that correspond to hard constraints). The SQD graphs show relative solution quality values defined as $sq/sq^* - 1$, where sq^* is the best known solution quality for the given weighted MAX-SAT instance. Each SQD is based on 100 runs of the respective algorithm. As can be clearly seen, for short run-times, the performance of Novelty$^+$/*wcs+we* drastically dominates that of GLSSAT2 for the given instance, while for longer run-times, GLSSAT2 tends to find higher-quality solutions.

Not surprisingly, allowing more run-time leads to substantial improvements in the solution qualities reached by both algorithms; furthermore, the variation in solution quality obtained from different runs decreases. (This behaviour is rather typical for SLS algorithms for optimisation problems.) These observations are further confirmed by the SQT curves shown on the right side of Figure 7.3, which confirm that for run-times up to 1 CPU sec, the solution qualities reached by Novelty$^+$/*wcs+we* are significantly higher than those obtained by GLSSAT2, while for longer tun-times, GLSSAT2 finds higher-quality solutions.

As can be seen in Table 7.2, slightly different performance results are obtained for other sets of MAX-SAT-encoded set covering and graph colouring instances (`scp4` and `gcp-yi`), as well as for smaller LGCMP instances

Test-set	GLSSAT2			Novelty$^+$/wcs+we		
	$t = 0.1s$	$t = 1s$	$t = 10s$	$t = 0.1s$	$t = 1s$	$t = 10s$
scp4	1.33	0.02	0.01	0.02	0.01	0
gcp-yi	11.24	0.22	0.15	0.12	0.05	0.01
lgcmp75	$8.1 \cdot 10^6$	49.3	0.20	3.27	0.09	0
lgcmp100	$1.1 \cdot 10^6$	248.82	0.25	10.02	1.39	0.51

Table 7.2 Performance of GLSSAT2 *vs* Novelty$^+$/wcs+we on selected sets of MAX-SAT-encoded instances of other problems; each set contains 10 instances. The table entries are median relative solution quality values for various run-times; relative solution quality is defined as $sq/sq^* - 1$, where sq is an absolute solution quality value for an instance with optimal (or best known) solution quality sq^*. For scp4, the sq^* values are the known optimal solution qualities, while for the other test-sets, they are the best solution qualities ever observed by any of the algorithms studied here within 100 runs of 100 CPU seconds each. The medians are taken from the distributions of median relative solution quality over the respective test-set; the underlying solution quality distributions for each problem instance are based on 100 runs of the respective algorithm.

(lgcmp75); for these, Novelty$^+$/wcs+we tends to find significantly higher quality solutions than GLSSAT2 for a wide range of run-times.

Tabu Search Algorithms for MAX-SAT

Hansen and Jaumard's *Steepest Ascent Mildest Descent (SAMD)* algorithm for unweighted MAX-SAT can be seen as one of the earliest applications of Tabu Search to MAX-SAT or SAT [Hansen and Jaumard, 1990]. (The name of the algorithm is derived from a formulation of MAX-SAT as a maximisation problem.) SAMD can be seen as a variant of GSAT/Tabu that imposes a tabu tenure of tt steps only on variables flipped in non-improving steps; variables flipped in improving steps are not declared tabu. Furthermore, SAMD terminates if after a fixed number of search steps no improvement in the objective function value has been achieved. SAMD has been shown to outperform a standard Simulated Annealing algorithm for MAX-SAT as well as various approximation algorithms with theoretical performance guarantees (see also Section 7.1) on a number of Uniform Random k-SAT instances with $k \in \{2, 3, 4\}$ and varying constrainedness [Hansen and Jaumard, 1990]. Although GWSAT has been reported to achieve better solution qualities than SAMD [Selman et al., 1994], the differences in the underlying termination criteria and run-times make a meaningful comparison very difficult [Hansen and Jaumard, 1990; Battiti and Protasi, 1997b].

A tabu search algorithm that is equivalent to GSAT/Tabu without random restart has been applied to unweighted MAX-SAT; experimental results on Uniform Random 3-SAT instances suggest that this variant performs slightly better than SAMD and possibly exceeds the performance of GWSAT [Battiti and Protasi, 1997a]. There is also some indication that a variant of this tabu search algorithm achieves further slight performance improvements; this variant uses an aspiration criterion (which allows a search step to be performed regardless of the tabu status of the corresponding variable if it achieves an improvement in the incumbent candidate solution) and a slightly modified tie-breaking rule for choosing one of several search steps that lead to an identical improvement in objective function value.

A further variant of tabu search for MAX-SAT, *TS-YI*, is based on a first improvement search strategy [Yagiura and Ibaraki, 1998; 2001]. Like all SLS algorithms for MAX-SAT discussed so far, it is based on the 1-flip neighbourhood relation and uses the objective function for evaluating search steps. The search is started from a randomly chosen assignment, and none of the variables are tabu. Then, in each step, the neighbourhood of the current variable assignment is scanned in random order, and the first variable flip that leads to an improving neighbouring variable assignment is executed. If no improving search step is possible, a minimally worsening step (w.r.t. to the standard evaluation function) is performed. Any variable that is flipped is declared tabu for a fixed number tt of subsequent search steps. The search process is terminated after a fixed amount of CPU time or a fixed number of search steps.

TS-YI has been applied to various types of unweighted and weighted MAX-SAT instances. There is some empirical evidence that for unweighted instances generated according to the random clause length model, this tabu search algorithm appears to perform better than WalkSAT/SKC and substantially better than basic GSAT. However, for various test-sets of weighted MAX-SAT instances, particularly for MAX-SAT-encoded minimum-cost graph colouring, weighted set covering, and time tabling problems, its performance appears to be worse than that of WalkSAT/SKC, but substantially better than that of basic GSAT [Yagiura and Ibaraki, 2001]. While it is not clear how the performance of TS-YI compares to that of the previously discussed tabu search algorithms for MAX-SAT, there is no evidence that it generally reaches or exceeds the performance of GLS or of the *wcs* variants of Novelty[+].

Finally, Robust Tabu Search (RoTS; see also Chapter 2, page 80) has recently been applied to MAX-SAT [Smyth et al., 2003]. The RoTS algorithm for MAX-SAT is closely related to GSAT/Tabu for weighted MAX-SAT. In each search step, one of the non-tabu variables that achieves a maximal improvement in the total weight of the unsatisfied clauses is flipped and declared tabu for the next tt steps. Different from GSAT/Tabu, RoTS uses an aspiration criterion that allows a variable to be flipped regardless of its tabu status if this leads to an improvement in the incumbent candidate solution. Additionally, RoTS forces any variable whose value has not been changed over the last $10 \cdot n$ search steps to be flipped (where n is the number of variables appearing in the given MAX-SAT instance). This diversification mechanism helps to avoid stagnation of the search process. Finally, instead of using a fixed tabu tenure, every n search steps, RoTS randomly chooses the tabu tenure tt from $[tt_{min}, \ldots, tt_{max}]$ according to a uniform distribution. The tabu status of variables is determined by comparing the number of search steps that have been performed since the most recent flip of a given variable with the current tabu tenure; hence, changes in tt immediately affect the tabu status and tenure of all variables. An outline of RoTS for MAX-SAT is given in Figure 7.4. Note that if several variables give the same best improvement for the evaluation function, one of these variables is randomly chosen.

Limited empirical results indicate that on the wjnh instances, RoTS requires generally more search steps but in many cases less CPU time for finding optimal solutions than the weighted MAX-SAT version of GLS; however, it does not reach the performance of the *wcs* variants of Novelty$^+$ on these instances. On Weighted Uniform Random 3-SAT instances, RoTS typically shows significantly better performance than Novelty$^+$/*wcs+we*, both in terms of search steps and CPU time required for finding (quasi-)optimal solutions. In terms of CPU time, it typically also exceeds the performance of GLS for MAX-SAT for both weighted and unweighted Uniform Random 3-SAT instances; this performance advantage appears to be particularly pronounced for highly constrained instances [Hoos et al., 2002; 2003].

Iterated Local Search for MAX-SAT

Yagiura and Ibaraki [1998; 2001] proposed and studied a simple ILS algorithm for MAX-SAT, *ILS-YI*, which initialises the search at a randomly chosen assignment and uses a subsidiary iterative first improvement search procedure based on the 1-flip neighbourhood, as well as a perturbation phase that consists of a fixed number of (uninformed) random walk steps; the acceptance criterion always selects the better of the two given candidate solutions. While ILS-YI generally appears to perform better than GSAT in terms of solution quality reached after a fixed amount of CPU time, for various sets of benchmark instances, including MAX-SAT-encoded minimum cost graph colouring problems, its performance is weaker than that of WalkSAT/SKC or TS-YI. However, there are some cases, in particular a large MAX-SAT encoded real-world time-tabling instance, for which

procedure $RoTS(F', tt_{min}, tt_{max}, maxNoImpr)$

    input: *weighted CNF formula F',*

               positive integers $tt_{min}, tt_{max}, maxNoImpr$

    output: *variable assignment $\hat{a}$*

    $n :=$ number of variables in F';

    $a :=$ randomly chosen assignment of the variables in F';

    $\hat{a} := a$;

    $k := 0$;

    repeat

        if (k **mod** $n = 0$) **then**

            $tt := random([tt_{min}, \ldots, tt_{max}])$;

        end

        $v :=$ randomly selected variable whose flip results

               in a maximal decrease in $g(a)$;

        if $g(a$ with v flipped$) < g(\hat{a})$ **then**

            $a := a$ with v flipped;

        else if $\exists$ variable v that has not been flipped for $\geq 10 \cdot n$ steps **then**

            $a := a$ with v flipped;

        else

            $v :=$ randomly selected non-tabu variable whose flip results

                 in the maximal decrease in $g(a)$

            $a := a$ with v flipped;

        end

        if $g(a) < g(\hat{a})$ **then** $\hat{a} := a$; **end**

        $k := k + 1$;

    until no improvement in $\hat{a}$ for *maxNoImpr* steps

    return $\hat{a}$

end *RoTS*

Figure 7.4 Algorithm outline of Robust Tabu Search for MAX-SAT; $g(a)$ denotes the total weight of the clauses in the given formulae that are unsatisfied under a; a variable is tabu if, and only if, it has been flipped during the last tt search steps.

ILS-YI appears to perform better than TS-YI and WalkSAT/SKC [Yagiura and Ibaraki, 2001].

    Another ILS algorithm for MAX-SAT has been recently proposed by Smyth, Hoos and Stützle [2003]. This algorithm, IRoTS, uses the same random initialisation as ILS-YI. Its subsidiary local search and perturbation phases are both based on the previously described RoTS algorithm. Each local search phase executes RoTS steps until no improvement in the incumbent solution has been achieved for a given number of steps. The perturbation phase consists of a fixed number of RoTS search steps with tabu tenure values that are substantially higher than the ones used in the local search phase. At the beginning of each local search and perturbation phase, all variables are declared non-tabu, irrespectively of their previous tabu status. If applying perturbation and subsequent local search to a candidate solution s results in a candidate solution s' that is better than the incumbent candidate solution, the search is continued from s'. If s and s' have the same solution quality, one of them is chosen uniformly at random. In all other cases, the worse of the two candidate solutions s and s' is chosen with probability 0.9, and the better one otherwise.

Empirical results show that when comparing the CPU time required for finding optimal or quasi-optimal solutions, IRoTS typically performs significantly better than GLS and Novelty$^+$/wcs+we on weighted and unweighted Uniform Random 3-SAT instances; the performance advantage of IRoTS is particularly pronounced for highly constrained instances with low-variance clause weight distributions. Overall, IRoTS appears to be the best-performing MAX-SAT algorithm for these types of instances. On the wjnh instances, IRoTS does not reach the performance of the *wcs* variants of Novelty$^+$. Furthermore, while IRoTS finds optimal solutions for a significant fraction of the unsatisfiable instances faster (in terms of CPU time) than GLS, it does not reach the performance of GLS on many of the satisfiable wjnh instances. Similarly, for several classes of MAX-SAT-encoded instances of other combinatorial optimisation problems, such as minimal cost graph colouring and weighted set covering, IRoTS performs significantly worse than GLS for weighted MAX-SAT in terms of finding solutions of optimal or best known quality [Smyth et al., 2003].

Limited experimentation suggests that using a perturbation phase consisting of a sequence of random walk steps instead of the previously described robust tabu search procedure typically results in a decrease in performance.

EXAMPLE 7.4 **Performance Results for SLS Algorithms for MAX-SAT (2)**

The following experiments illustrate the performance differences between IRoTS and GLSSAT2 on weighted and unweighted Uniform Random 3-SAT instances. All CPU times reported in this example have been measured on PCs with dual 2.4GHz Intel Xeon processors, 512KB cache, and 1GB RAM running Red Hat Linux, Version 2.4smp.

First, in order to assess the differences in the solution qualities achieved by both algorithms, solution-quality distributions (SQDs) were measured for IRoTS and GLSSAT2 applied to an unweighted Uniform Random 3-SAT instance with 500 variables and 5 000 clauses over a range of run-times. The left side of Figure 7.5 shows the development of solution quality over time (SQT curves) obtained from the SQD data for both algorithms. Clearly,

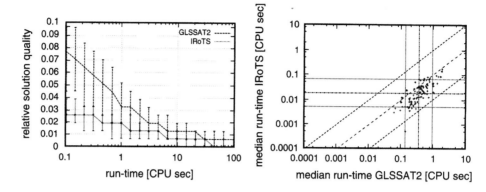

***Figure* 7.5** *Left:* Development of the relative solution quality over time for IRoTS *vs* GLSSAT2 on a hard unweighted Random 3-SAT instance with 500 variables and 5 000 clauses; the curves for each algorithm correspond to the median values of the underlying SQDs, while the error bars indicate the respective $q_{0.1}$ and $q_{0.9}$ quantiles. The data underlying these graphs were obtained from 100 runs of each algorithm. *Right:* Correlation of the search cost required by IRoTS *vs* GLSSAT2 for reaching (quasi-)optimal solution qualities on the rnd200-2000/u benchmark set; the search cost for each instance is measured as median run-time over 100 runs. The horizontal and vertical lines indicate the median, $q_{0.1}$, and $q_{0.9}$ of the search cost for the respective algorithm across the test-set; the diagonal lines indicate equal, 1/10th and 1/100th CPU time of IRoTS compared to GLSSAT2.

IRoTS tends to reach solutions of better quality than GLSSAT2 for any given run-time; this is particularly the case for short runs.

In the next step, the performance of the two algorithms was compared across the test-set rnd200-2000/u, comprising 100 unweighted Uniform Random 3-SAT instances with 200 variables and 2 000 clauses each, all of which are unsatisfiable. Since state-of-the-art systematic search algorithms for MAX-SAT were found to be unable to find provably optimal solutions for the MAX-SAT instances used in this example within a reasonable amount of CPU time, quasi-optimal solutions were determined using the same method as described in Example 7.3. For each of the 100 instances from the test-set, we measured qualified RTDs for finding a quasi-optimal solution over 100 runs of IRoTS and GLSSAT2, respectively. (Both algorithms were always run — without restart — until the desired solution quality was reached.) The right side of Figure 7.5 shows the correlation between the median run-times of IRoTS and GLSSAT2 over the test-set. Clearly, IRoTS performs substantially better than GLSSAT2 across the entire test-set and generally tends to find quasi-optimal solutions up to 80 times faster.

As can be seen from Table 7.3, similar results are obtained for other unweighted and weighted Uniform Random 3-SAT test-sets. As in Example 7.3,

Test-set	IRoTS	GLSSAT2	fd
rnd200–1000/u	0.0132 (6 655)	0.0384 (5 665)	0.99 / 0
rnd200–1400/u	0.0141 (6 584)	0.2525 (32 473)	0.98 / 0
rnd200–2000/u	0.0171 (6 927)	0.3487 (31 510)	0.99 / 0
rnd200–1000/w1000–200	0.0729 (47 523)	0.4589 (69 523)	0.57 / 0
rnd200–1000/w1000–1000	0.5242 (318 897)	1.3637 (217 961)	0.35 / 0.11
rnd200–1400/w1400–1400	0.1021 (57 905)	3.1840 (408 888)	0.97 / 0

Table 7.3 Performance of IRoTS *vs* GLSSAT2 on selected benchmark instances for unweighted and weighted MAX-SAT; the table entries are median search cost values over the respective test-sets, where the search cost for a given problem instance is defined as the median run-time required for finding a quasi-optimal solution and reported as CPU seconds (search steps). The search cost values for each algorithm were determined from 100 runs per instance, and algorithms were always run (without restart) until a quasi-optimal solution was found. The values in the *fd* column indicate the fraction of instances from the respective test-set on which IRoTS probabilistically dominated GLSSAT2 (first value) and vice versa (second value).

$\text{rnd}n\text{-}m/\text{w}\mu\text{-}\sigma'$ denotes a set of 100 Uniform Random 3-SAT instances with n variables and m clauses, with clause weights drawn from a truncated discretised Gaussian weight distribution $NDT[\mu, \sigma', 1]$. (It may be noted that the only difference between the instances in test-sets $\text{rnd}n\text{-}m/\text{u}$ and $\text{rnd}n\text{-}m/\text{w*}$ are the clause weights.) Interestingly, as the constrainedness of the instances (i.e., the number of clauses per variable) is increased, the performance of GLSSAT2 deteriorates, while the performance of IRoTS remains relatively unaffected. This effect is more pronounced for the weighted instances, which also tend to be harder for both algorithms.

MAX-SAT Algorithms Based on Larger Neighbourhoods

While all prominent and high-performance SLS algorithms for SAT are based on the 1-flip neighbourhood, there are very successful SLS algorithms for MAX-SAT that are based on larger neighbourhoods. Yagiura and Ibaraki [1998; 1999; 2001] studied various such algorithms, ranging from simple iterative first improvement to iterated local search methods. The key to the success of these algorithms is a combination of a clever reduction of the 2- and 3-flip neighbourhoods with an efficient caching scheme for evaluating moves in these larger neighbourhoods. This reduction is done in such a way that no possible improving neighbour is lost, that is, local optimality remains invariant under the neighbourhood reduction. Furthermore, under realistic assumptions, each local search step requires time $O(n+m)$ for the 2-flip neighbourhood and time $O(m+t^2 n)$ for the 3-flip neighbourhood in the average case, given an input formula with n variables, m clauses, and no more than t occurrences of each variable (see also in-depth section below); this result has been empirically confirmed for a range of weighted Uniform Random 3-SAT test-sets [Yagiura and Ibaraki, 1998; 1999].

Empirical results for variants of TS-YI and ILS-YI that use the reduced 2- and 3-flip neighbourhoods indicate that on various test-sets of weighted MAX-SAT instances, these larger neighbourhoods lead to significant performance improvements in terms of the solution quality reached after a fixed amount of CPU time. Particularly for MAX-SAT-encoded minimum-cost graph colouring and weighted set covering instances, as well as for a big, MAX-SAT-encoded real-world time-tabling instance, the 2-flip variant of ILS-YI performs better than the other versions of ILS-YI and any of the TS-YI variants. It is presently not clear whether other, state-of-the-art MAX-SAT algorithms can reach or exceed the performance of ILS-YI (or TS-YI) on these types of instances; however, there is some preliminary evidence that GLS and Novelty$^+$/*wcs+we* may perform better in various cases. It is also unclear whether the use of larger neighbourhoods might lead to performance improvements in state-of-the-art SLS algorithms for MAX-SAT, such as Novelty$^+$/*wcs+we*, GLS or IRoTS.

IN DEPTH EFFICIENT EVALUATION OF k-FLIP NEIGHBOURHOODS FOR MAX-SAT

The key to efficiently implementing algorithms for SAT or MAX-SAT that are based on performing iterative improvement steps in a multiflip neighbourhood (i.e., a k-flip neighbourhood with $k > 1$) lies in a combination of two techniques. Both of these make use of the fact that the effect of a multiflip on the evaluation function can be decomposed into the effects of single-variable flips and correction terms. In the following, we explain these techniques primarily for the special case of the 2-flip neighbourhood, and mention generalisations for $k > 2$ only briefly (for details, see Yagiura and Ibaraki [1999]).

Given a weighted CNF formula with clauses $c_1, \ldots, c_m$, consider a search step in the 2-flip neighbourhood, in which the truth values of variables x and y are flipped. Let $\Delta g(a, \{x, y\})$ be the change in evaluation function value caused by the 2-flip of x and y in a. (Recall that we defined MAX-SAT as a minimisation problem; hence, an improving search step corresponds to $\Delta g(a, \{x, y\}) < 0$.) This value can be computed as

$$\Delta g(a, \{x, y\}) := \sum_{i=1}^{m} \Delta g_i(a, \{x, y\}) = \sum_{i=1}^{m} \left(\Delta g_i(a, \{x\}) + \Delta g_i(a, \{y\}) - h_i(a, \{x, y\}) \right),$$

where $\Delta g_i(a, \{z\})$ captures the effect of flipping a single variable z on the satisfaction of clause c_i, and $h_i(a, \{x, y\})$ is an adjustment term that captures the possible interference of flipping x and y simultaneously.

Clearly, if a clause c_i does not contain occurrences of both x and y, we have $h_i(a, \{x, y\}) := 0$. Otherwise, for a clause c_i that contains x as well as y, the following cases can be distinguished:

- Clause c_i is unsatisfied under a. In this case, we use the adjustment term $h_i(a, \{x, y\}) := w(c_i)$; since the flip of either variable in $\{x, y\}$ renders the clause satisfied, this adjustment is needed to prevent double-counting the satisfaction of the clause as a result of the 2-flip.

- Clause c_i is critically satisfied under a, that is, it contains exactly one satisfied literal l. In this case, if l is not equal to x or y or either of their negations, we use $h_i(a, \{x, y\}) := 0$. Otherwise, we need to use $h_i(a, \{x, y\}) := -w(c_i)$ to account for the fact that the 2-flip leaves the satisfaction status of the clause unchanged.

- Clause c_i contains exactly two satisfied literals under a. If those two literals contain both variables x and y, we use $h_i(a, \{x, y\}) := -w(c_i)$ since flipping both variables would render c_i unsatisfied otherwise, $h_i(a, \{x, y\}) := 0$.

- Clause c_i contains more than two satisfied literals under a. In this case, neither any single flip nor the 2-flip can render c_i unsatisfied, and we can use $h_i(a, \{x, y\}) := 0$.

The values $\Delta g(a, \{x, y\})$ that provide the basis for comparing neighbouring candidate solutions and for incrementally updating the evaluation function value after each search step can be easily determined from the values $h(a, \{x, y\}) := \sum_{i=1}^{m} h_i(a, \{x, y\})$ and $\Delta g(a, \{z\}) := \sum_{i=1}^{m} \Delta g_i(a, \{z\})$. These values can be efficiently cached and updated after each search step using a simple extension of the mechanism described in the in-depth section on the efficient implementation of GSAT (page 271). It is important to note that only those values $h(a, \{x, y\})$ that are not equal to zero need to be memorised. Based on this observation, it can be shown that for a formula with n variables, m clauses, and maximal clause length l, in which no variable has more than t occurrences, this mechanism requires $O(n + \min\{m \cdot l^2, n^2\})$ memory in the worst case; the worst-case time complexity for a single search step is $O(n^2 + t \cdot l^2)$, under the (practically realistic) assumption that each value $h(a, \{x, y\})$ can be stored and retrieved in constant time.

This mechanism can be generalised to k-flip neighbourhoods with $k > 2$; in that case, the worst-case memory requirement is $O(n + \min\{m \cdot l^k, n^k\})$ and the worst-case time complexity for a single search step is $O(2^k \cdot n^k + k \cdot t \cdot l^k)$. Under some additional assumptions, it can be shown that the expected memory requirement is only $O(n + m)$ and the expected time complexity of a search step is $O(2^k \cdot n^k + t)$ for any constant k.

A second technique, which achieves further improvements in efficiency when using the 2-flip neighbourhood, is based on the following observation. If both, $\Delta g(a, \{x\}) := \sum_{i=1}^{m} g_i(a, \{x\})$ and $\Delta g(a, \{y\}) := \sum_{i=1}^{m} g_i(a, \{y\})$, are larger or equal to zero, then flipping x and y can only result in an improvement in evaluation function value if $h(a, \{x, y\}) := \sum_{i=1}^{m} h_i(a, \{x, y\}) < 0$. Hence, any search for improving neighbours in the 2-flip neighbourhood can be restricted to all those pairs of variables x, y for which $h(a, \{x, y\}) < 0$. It can be shown that this restriction reduces the size of the neighbourhood to $O(n + m \cdot l)$ in the worst case; an average-case analysis indicates that, under certain conditions, the expected size of this restricted neighbourhood is at most $n + 3/4 \cdot m$. This leads to a further reduction of the worst-case time complexity to $O(n + m \cdot l + t \cdot l^2)$ for a single search step.

A similar approach leads to a reduced 3-flip neighbourhood of size $O(m \cdot l^3 + n \cdot t^2 \cdot l^2\}$ in the worst case and of $O(m + n \cdot t^2\}$ in the average case, with a worst-case time complexity of $O(m \cdot l^3 + n \cdot l^2 \cdot t^2)$ for a single search step.

Non-Oblivious SLS Algorithms for MAX-SAT

All SLS algorithms for SAT and MAX-SAT discussed so far use evaluation functions that are *oblivious* in the sense that they are not affected by the degree of satisfaction of any given clause c, that is, by the number of literals that are satisfied in c under a given assignment. *Non-oblivious evaluation functions*, in contrast, reflect the degree of satisfaction of the clauses satisfied by a given variable assignment.

Theoretical analyses have shown that iterative improvement local search achieves better worst-case approximation ratios for unweighted MAX-SAT when using non-oblivious evaluation functions instead of the standard, oblivious evaluation function that counts the number of clauses unsatisfied under a given assignment [Alimonti, 1994; Khanna et al., 1994]. In particular, using the non-oblivious evaluation functions $g_2(a) := 3/2 \cdot w(S_1(a)) + 2 \cdot w(S_2(a))$ and $g_3(a) := w(S_1(a)) + 9/7 \cdot w(S_2(a)) + 10/7 \cdot w(S_3(a))$, where $w(S_i(a))$ is the total weight of the set of all clauses satisfied by exactly i literals under assignment a, in conjunction with iterative improvement algorithms leads to worst-case approximation ratios of $4/3$ and $8/7$ for MAX-2-SAT and MAX-3-SAT, respectively. (Similar non-oblivious evaluation functions and respective approximation results exist for MAX-k-SAT with $k > 3$.)

Battiti and Protasi [1997a; 1997b] proposed and studied a number of SLS algorithms for MAX-SAT that make use of these non-oblivious evaluation functions. The simplest of these is an iterative best improvement algorithm; it can be seen as a variant of basic GSAT that terminates as soon as a local minimum state is reached. For this algorithm (applied to MAX-3-SAT), using the non-oblivious evaluation function g_3 instead of the standard GSAT evaluation function leads to improved solution qualities; however, both of these algorithms perform significantly worse than GWSAT and SAMD, except when applied to weakly constrained Uniform Random 3-SAT instances. Furthermore, GSAT, GWSAT and GSAT/Tabu perform significantly worse when using a non-oblivious evaluation function [Battiti and Protasi, 1997a]. These observations suggest that the theoretical advantage of using non-oblivious evaluation functions manifests itself only in the worst case, or (more likely) that it does not apply to SLS methods more powerful than iterative improvement.

Note that the non-oblivious and oblivious evaluation functions discussed above have different local minima. Based on this observation, Battiti and Protasi designed a hybrid SLS algorithm that first performs non-oblivious iterative best improvement until a local minimum w.r.t. to the non-oblivious evaluation function is reached, followed by an oblivious iterative best improvement phase that is continued beyond its first local minimum. This hybrid SLS algorithm reaches better solution qualities than SAMD for various Uniform Random 3-SAT testsets, but its performance is inferior to GWSAT for long run-times [Battiti and Protasi, 1997b].

Better performance is achieved by *H-RTS*, a complex hybrid SLS algorithm that combines non-oblivious and oblivious iterative best improvement with an oblivious reactive tabu search procedure [Battiti and Protasi, 1997a]. H-RTS starts the search from a randomly chosen variable assignment; next, non-oblivious iterative best improvement (BIN) steps are performed until a local minimum (w.r.t. the non-oblivious evaluation function) is reached. Then, phases of oblivious iterative best improvement (BI) search and tabu search (TS) are alternated until the total number of variable flips performed since initialising the search reaches $10 \cdot n$, where n is the number of variables in the given MAX-SAT instance, at which point the search is re-initialised (see Figure 7.6). Each BI search phase ends when a local minimum w.r.t. the standard oblivious evaluation function is reached. The subsequent TS phase performs $2 \cdot (tt+1)$ steps of oblivious iterative best improvement tabu search with fixed tabu tenure tt.

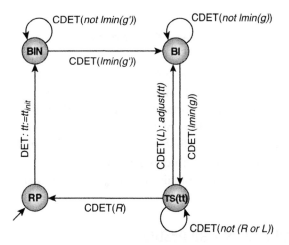

Figure 7.6 GLSM representation of the H-RTS algorithm; $R := \mathsf{mcount}(10 \cdot n + 1)$, $L := \mathsf{scount}(2 \cdot (tt + 1))$, and the transition action $adjust(tt)$ adjusts the tabu tenure setting. (For details, see text.)

When the search is initialised (or restarted), tt is set to a fixed value tt_{init}. After each TS phase, tt is adjusted based on the Hamming distance covered within that search phase (i.e., the number of variables that are assigned different truth values immediately before and after the $2 \cdot (tt + 1)$ TS steps): if that distance is large, the tabu tenure is increased in order to diversify the search; if the distance is big, the tabu tenure is decreased to keep the search process focused on promising regions of the search space. Additionally, an upper and lower bound on the tabu tenure are imposed (for details, see Battiti and Protasi [1997a]).

H-RTS has been applied to various sets of unweighted Uniform Random 3-SAT and Uniform Random 4-SAT instances. In terms of solution quality achieved after a fixed number of search steps (variable flips), H-RTS performs significantly better than basic GSAT, GWSAT and GSAT/Tabu, especially for large, highly constrained problem instances [Battiti and Protasi, 1997a]. Furthermore, H-RTS shows substantially more robust performance w.r.t. to the initial tabu tenure setting tt_{init} than GSAT/Tabu w.r.t. to its tabu tenure parameter, tt. When it was first proposed, H-RTS was one of the best-performing algorithms for unweighted MAX-SAT; however, it typically seems to be unable to reach the performance of some more recent SLS algorithms for MAX-SAT algorithms, such as IRoTS. Interestingly, there is some evidence that the performance of H-RTS does not significantly depend on the initial non-oblivious local search phase, but is rather due to the (oblivious) reactive tabu search procedure.

7.3 SLS Algorithms for MAX-CSP

MAX-CSP generalises CSP analogous to the way in which MAX-SAT generalises SAT: Given a CSP instance, the objective is to satisfy as many constraints as possible. The importance of MAX-CSP resides in the fact that it is one of the simplest extensions of CSP to constraint optimisation problems; as such, it is typically used as a first step for extending algorithmic techniques for CSP solving to optimisation problems. As in Chapter 6, we will focus on finite discrete MAX-CSP, where the domains of all CSP variables are finite and discrete.

The MAX-CSP Problem

The simplest case of MAX-CSP gives all constraints the same importance, and the goal is to maximise the number of satisfied constraints.

DEFINITION 7.4 **Unweighted MAX-CSP**

Given a CSP instance $P := (V, \mathcal{D}, \mathcal{C})$ as in Definition 6.2 (page 293), let $f(a)$ be the number of constraints violated under variable assignment a, and let m be the number of constraints in P. The (Unweighted) Maximum Constraint Satisfaction Problem (MAX-CSP) is to find $a^* \in \operatorname{argmin}_{a \in Assign(P)} f(a)$ or, equivalently, $a^* \in \operatorname{argmax}_{a \in Assign(P)} (m - f(a))$, that is, a variable assignment a^* that maximises the number of the satisfied constraints in P.

As in MAX-SAT, *maximising* the number of *satisfied* constraints is equivalent to *minimising* the number of *unsatisfied* constraints; in the following we consider MAX-CSP as a minimisation problem. Note that CSP is the decision variant of MAX-CSP in which the objective is to determine whether there is a CSP variable assignment that simultaneously satisfies all constraints. The evaluation variant and the associated decision problems are defined as in the case of MAX-SAT.

The MAX-CSP problem arises in the context of overconstrained CSP instances, in which it is impossible to satisfy all given constraints simultaneously. In such cases, the assumption underlying unweighted MAX-CSP that all constraints are equally important is often not appropriate. To address this issue, the MAX-CSP formalism can be extended, similar to MAX-SAT, to include constraint weights that explicitly represent the importance of satisfying specific constraints of a given CSP instance.

DEFINITION 7.5 **Weighted CSP Instance**

A weighted CSP instance is a pair (P, w), where P is a CSP instance and $w : \{C_1, \ldots, C_m\} \mapsto \mathbb{R}^+$ is a function that assigns a positive real value to each constraint C_i of P; $w(C_i)$ is called the weight of constraint C_i. (Without loss of generality, we assume that all constraints in P are pairwise different.)

The objective in weighted MAX-CSP is to find a CSP variable assignment for a given weighted CSP instance that minimises the total weight of the unsatisfied constraints.

DEFINITION 7.6 **Weighted MAX-CSP**

Given a weighted CSP instance $P' := (P, w)$, let $f(a)$ be the total weight of the constraints of P violated under CSP variable assignment a, that is,

$f(a) := \sum_{C \in CU(a)} w(C_i)$, *where $CU(a)$ is the set of all constraints of P violated under assignment a. The* Weighted Maximum Constraint Satisfaction Problem (Weighted MAX-CSP) *is to find a variable assignment a^* that maximises the total weight of the satisfied constraints in P, that is, $a^* \in$* $\mathrm{argmin}_{a \in Assign(P)} f(a)$ *or, equivalently, $a^* \in \mathrm{argmax}_{a \in Assign(P)}(\tilde{f} - f(a))$, where $\tilde{f} := \sum_{i=1}^{m} w(C_i)$ is the total weight of all constraints in P.*

The constraint weights reflect the different priorities in satisfying the respective constraints. They can be used to encode problems that involve *hard constraints*, which must be satisfied in any feasible solution, as well as *soft constraints*, which represent an optimisation goal; problems of this type occcur in many real-world applications.

MAX-CSP is an $\mathcal{NP}$-hard problem because it is a generalisation of CSP, which itself is $\mathcal{NP}$-complete (cf. Chapter 6). As might be expected, even finding high-quality suboptimal solutions for MAX-CSP is difficult in the worst case: for k-ary MAX-CSP with domains of size d, achieving approximation ratios of $d^{k-2\sqrt{k+1}+1} - \epsilon$ is $\mathcal{NP}$-hard for any constant $\epsilon > 0$ [Engebretsen, 2000]. The best theoretical worst-case performance guarantees known to date have been shown for an algorithm based on linear programming and randomized rounding, which achieves an approximation ratio of d^{k-1} [Serna et al., 1998].

Randomly Generated and Structured MAX-CSP Instances

Algorithms for MAX-CSP have mostly been evaluated on instances that are randomly generated according to the Uniform Random Binary CSP model described in Chapter 6, Section 6.5. This generative model has four parameters: the number of CSP variables, n; the domain size for each CSP variable, k; the constraint graph density, α; and the constraint tightness, β. In the context of MAX-CSP, these parameters are typically chosen in such a way that the resulting instances are unsatisfiable [Wallace, 1996a; Galinier and Hao, 1997]. Random weighted MAX-CSP instances are obtained by assigning randomly chosen weights to the constraints; these are typically sampled from a uniform distribution over a given range of integers [Lau, 2002].

Other combinatorial optimisation problems from a wide range of application areas can be encoded into MAX-CSP in a straightforward way. One example for such a problem is *university examination timetabling*: Given a set of examinations, a set of time-slots, a set of students, and for each student, a set of examinations to be taken, the objective is to assign a set of examinations to a set of time slots such that certain hard constraints are satisfied and additional criteria are optimised. (For simplicity's sake, this version of the problem does not capture room assignments.) A typical hard constraint is to forbid any temporal overlaps between the examinations taken by the same student; a typical example of a soft constraint is to maintain a minimum temporal distance between any pair of examinations for the same student (for an extensive list of possible constraints found in real life examination timetabling problems, see Burke [1996]).

A set of benchmark instances that has commonly been used to evaluate algorithms for examination timetabling with exactly these two types of constraints has been defined by Carter et al. [1996]. In particular, the soft constraints penalise timetables in which the temporal distance Δt between two exams taken by the same student is less than six time slots; the penalty is $6 - \Delta t$ if $\Delta t < 6$ and zero otherwise. In the weighted MAX-CSP formulation, this penalty is represented by five constraints for every student; each of these is violated if the temporal distance between two examinations is equal to Δt time slots, where $0 < \Delta t < 6$, and has a weight of $6 - \Delta t$. The hard constraints, which forbid overlapping time slots for exams taken by the same student, are assigned a weight larger than the sum of the weights of all soft constraints.

Another example of a problem that can be represented as a MAX-CSP arises in the context of frequency assignment in wireless communication networks. In the *Frequency Assignment Problem (FAP)*, a set of wireless communication connections is given along with a limited number of available frequencies; the objective is to assign a frequency to each communication connection subject to additional constraints. These constraints typically concern the reduction or the avoidance of electromagnetic interference; for example, a minimum separation may be required between any of the frequencies assigned to two physically close connections. For more details on the FAP and its variants, we refer to Aardal et al. [2003]. One particular variant that can be cast directly into the MAX-CSP framework is described in the following example.

EXAMPLE 7.5 **The Radio Link Frequency Assignment Problem**

A widely used set of FAP benchmark instances stems from the *Radio Link Frequency Assignment Problem (RLFAP)*, a particular FAP variant that has been extensively studied in the EUCLID CALMA project (see Eisenblätter and Koster [2003]). In this project, 11 RLFAP instances stemming from a military communications application were provided by CELAR (Centre d'Electronique de'l Armement, France); these instances are all based on simplified data from a real network of field phones.

Each of the CELAR instances specifies a set of radio links that need to be established between pairs of sites, where each site has an associated list of possible frequencies it may use. The goal is to assign frequencies to the sites such that interference is avoided or, if that is impossible, minimised. For this purpose, for each pair of links (i, j), a separation constraint $|freq(i) - freq(j)| \geq d_{ij}$ is given, where d_{ij} is the minimum distance required between the frequencies $freq(i)$ and $freq(j)$ assigned to links i and j in order to avoid interference; in reality, these distance values depend on the position of the sites as well as on the physical environment. (Note that the separation constraints can be seen as an extension of the binary inequality constraints found in graph colouring; see also Chapter 10, page 477 for a discussion of such extensions.) In the CELAR instances, there are additional binary constraints representing the fact that each communication connection between two sites requires two radio links, one for each direction of communication between the given sites, which are separated by a fixed difference in frequency.

Depending on the nature of specific CELAR instances, three different optimisation criteria are considered. These amount to finding a frequency assignment that

1. causes no interference and uses a minimal number of different frequencies (and hence, maximises the number of unused frequencies that may later be utilised for additional links); or

2. causes no interference and minimises the maximum frequency used (and hence, maximises the unused portion of the frequency spectrum); or

3. minimises the weighted sum of the violated constraints.

For the third objective, which is used for overconstrained, unsatisfiable instances, each constraint is assigned one of four weights, according to the priority of the respective links. The quality of a frequency assignment is measured based on respective violations of the interference constraints by using the objective function

$$f(a) := w_1 \cdot nc_1(a) + w_2 \cdot nc_2(a) + w_3 \cdot nc_3(a) + w_4 \cdot nc_4(a),$$

where $w_1, \ldots, w_4$ are the four different constraint weights, and $nc_i(a)$ is the number of constraints with weight w_i that are violated under assignment a. For some instances, additional mobility constraints are considered, which model the situation that some connections have pre-assigned frequencies whose modification is either costly or impossible.

The size of the CELAR instances ranges from 200 connections and 1 235 constraints to 916 connections and 5 744 constraints. The structure of each instance can be described by an *interference graph* $G := (V, E)$, whose vertices corresponds to the given links and whose edges represent the non-trivial

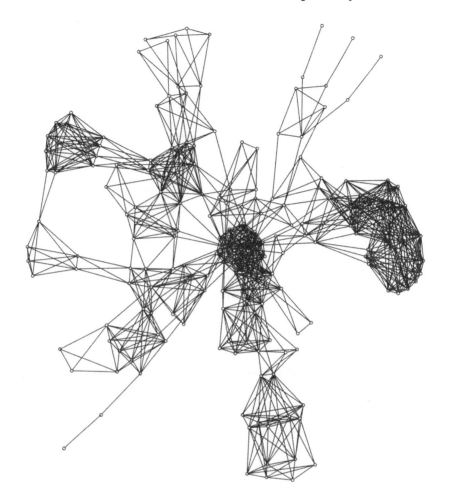

Figure 7.7 Interference graph for the RLFAP instance CELAR06 (for details, see text).

separation constraints between the respective links. (A separation constraint is trivial if, and only if, it is satisfied by all possible combinations of frequencies available for the respective links.) Figure 7.7 shows the interference graph for instance CELAR06, an overconstrained instance with 200 links and 1 332 separation constraints, in which the objective is to minimise the weighted sum of violated constraints. Through a series of research efforts over several years, most of the CELAR instances have been optimally solved; in this context, a number of SLS methods have been used (for details, see Eisenblätter and Koster [2003]).

SLS Algorithms for Unweighted MAX-CSP

Because of the way SLS algorithms for CSP evaluate and minimise constraint violations in order to find solutions to a given CSP instance, these algorithms can generally be applied directly to unweighted MAX-CSP instances.

Variants of the Min-Conflicts Heuristic (MCH; see Chapter 6, Section 6.6) were amongst the first SLS algorithms applied to unweighted MAX-CSP. Empirical results on a set of randomly generated MAX-CSP instances show that WMCH performs better than basic MCH and basic MCH with random restart [Wallace and Freuder, 1995]. Interestingly, a parametric study of WMCH's performance indicates that the performance-optimising setting of wp (the probability for executing random walk steps rather than basic MCH steps) depends on the number of constraints violated in the optimal solutions to the MAX-CSP: higher optimal solution quality values require smaller wp settings.

In a further experimental study, the performance of the same three MCH variants was compared with that of three other CSP algorithms:

- the Breakout Method, an early and relatively widely known dynamic local search method for the CSP [Morris, 1993];

- EFLOP, a hybrid SLS algorithm that combines iterative improvement with value propagation techniques [Yugami et al., 1994];

- weak commitment search, a complete method that uses the min-conflicts heuristic as a value ordering heuristic when extending a partial candidate solution, and that uses a restart mechanism if a partial candidate solution cannot be extended further without violating constraints [Yokoo, 1994]. (In the optimisation case, a partial candidate solution is typically abandoned if it has a higher weight than the incumbent candidate solution.)

While on a set of small randomly generated MAX-CSP instances (with 30 variables and domain size 5) WMCH did not perform significantly better than these three methods, it did achieve better performance on larger instances [Wallace, 1996a].

The best results for randomly generated MAX-CSP instances obtained so far were reported for the tabu search algorithm by Galinier and Hao (TS-GH) [Galinier and Hao, 1997]. Different from MCH variants, which in each step choose a variable involved in a conflict and then consider changing the value of this variable, TS-GH determines each search step by considering the set of all variable-value pairs (v, y) for which v occurs in a currently violated constraint (cf. Chapter 6, page 305; it may be noted that TS-GH had been applied to MAX-CSP before it was evaluated on soluble CSP instances).

On randomly generated MAX-CSP instances with up to 500 CSP variables and domain size 30, TS-GH has been shown to outperform WMCH: TS-GH reached the same solution quality as WMCH in about three to four times fewer search steps and found better quality solutions when allowed the same run-time (in terms of search steps). Due to the speed-up techniques used in efficient implementations, the search steps of TS-GH are only slightly more expensive than those of WMCH; the difference in CPU time per search step has been measured at about 15% [Galinier and Hao, 1997].

One may conjecture that the better performance of TS-GH compared to WMCH is a result of the larger neighbourhood searched by TS-GH in each single step. However, limited empirical results indicate that a variant of TS-GH that uses random walk instead of tabu search for escaping from local optima performs significantly worse than WMCH [Galinier and Hao, 1997]. On the other hand, it is known that the restriction of the neighbourhood to variables involved in conflicts is important for obtaining TS-GH's high performance. There is some empirical evidence suggesting that the performance of TS-GH drops significantly if all variable–value pairs are considered (including those containing variables not involved in currently violated constraints) [Hao and Pannier, 1998].

SLS Approaches to Weighted MAX-CSP

The previously described algorithms for unweighted MAX-CSP can be easily extended to weighted MAX-CSP. Somewhat surprisingly, so far this approach has remained largely unexplored. An exception is the work of Lau and Watanabe [1996] and Lau [2002], who developed an approximation algorithm for weighted MAX-CSP based on semidefinite programming and randomised rounding; for domain sizes two and three, this algorithm achieves an approximation ratio of 2.451.

A variant of this algorithm that applies iterative improvement to the solution obtained from the approximation algorithm (APII) has been empirically compared to an SLS algorithm that consists of a greedy construction heuristic followed by an iterative improvement procedure (GII), as well as to an extension of MCH to weighted MAX-SAT. Applied to 'forced' instances, which are randomly generated in a way that guarantees their solubility, APII achieved substantially better solution qualities than GII and MCH. In these experiments, MCH and APII were allotted approximately the same run-time, while GII terminated in a local minimum within roughly 5% of this time. Furthermore, on the forced instances the approximation algorithm without the subsequent local search phase performed better than GII. However, on randomly generated instances that were not soluble by construction, the performance advantages observed for APII were less pronounced [Lau, 2002], which suggests that the excellent performance of APII on forced instances may be an artifact induced by the instance generation process.

Overconstrained Pseudo-Boolean CSP

Pseudo-Boolean CSP can be seen as a restriction of CSP in which all variables have domains $\{0, 1\}$, but more expressive constraints are supported than the CNF clauses used in SAT or MAX-SAT (see also Chapter 6, page 300*ff.*). The Pseudo-Boolean CSP formalism can be extended to consider optimisation objectives in addition to the conventional, hard constraints. In the resulting *Overconstrained Pseudo-Boolean CSP (OPB-CSP)*, optimisation goals are encoded as competing soft constraints. As in Pseudo-Boolean CSP, the hard constraints in OPB-CSP are of the form

$$\sum_{i=1}^{n} a_{ij} \cdot a(x_i) \geq b_j,$$

where the a_{ij} as well as b_j are rational numbers and $a(x_i)$ is the value of constraint variable x_i under assignment a. (Note that analogous constraints that use any of the relations '$\leq$', '$<$', '$>$', '$=$', '$\neq$' instead of '$\geq$' can be represented using '$\geq$' constraints only.) However, now we additionally consider a set of soft constraints of the same form, but with constants c_{ij} and d_j instead of a_{ij} and b_j. Given such an OPB-CSP instance P with m hard constraints and m' soft constraints, the goal is to determine a variable assignment a that satisfies all hard constraints and minimises the total degree of soft-constraint violation, that is, the objective function $f(a) := \sum_{j=1}^{m'} \sum_{i=1}^{n} \max\{0, d_j - c_{ij} \cdot a(x_i)\}$ instead of the number of violated soft constraints [Walser, 1999; Walser et al., 1998]. Note that the definition of the objective function f relies on the algebraic structure of the constraints in conjunction with the use of numeric variables, and hence cannot be applied to weighted MAX-CSP, which allows arbitrary constraint relations.

It may be noted that OPB-CSP with m hard constraints and m' soft constraints is equivalent to the following formulation as an integer programming problem:

$$\text{Minimise} \qquad f(x_1, \ldots, x_n) := \sum_{j=1}^{m'} \sum_{i=1}^{n} \max\{0, d_j - c_{ij} \cdot x_i\} \qquad (7.1)$$

$$\text{subject to} \qquad a_{ij} \cdot x_i \geq b_j \qquad (j = 1, \ldots, m) \qquad (7.2)$$

$$x_i \in \{0, 1\} \qquad (i = 1, \ldots, n) \qquad (7.3)$$

Using this formulation, it can be relatively easily shown that every OPB-CSP instance can be converted into an equivalent integer linear program, and hence that OPB-CSP can be seen as a special case of integer linear programming [Walser, 1999].

The most straightforward way of applying SLS methods to this problem is to use an evaluation function that captures the violation of hard as well as soft constraints. The basic version of WSAT($\mathcal{PB}$), a well-known SLS algorithm for Pseudo-Boolean CSP (cf. Chapter 6, page 301), can be easily extended to OPB-CSP by using the evaluation function

$$f(x_1, \ldots, x_n) := \sum_{j=1}^{m} \sum_{i=1}^{n} |a_{ij} \cdot x_i - b_j| + \sum_{j=1}^{m'} \sum_{i=1}^{n} d(c_{ij} \cdot x_i, d_j) \cdot w_j, \qquad (7.4)$$

where $d(x, y) := \max\{0, y - x\}$ and the constraint weights w_j are positive real numbers that can be used to bias the search process towards satisfying certain hard constraints [Walser, 1999; Walser et al., 1998]. (Using such weights has been shown to be important for achieving good performance on certain types of OPB-CSP instances.)

To handle hard constraints efficiently, the WSAT($\mathcal{PB}$) variable selection strategy is extended by first randomly selecting an unsatisfied hard constraint with probability wp_h, while a violated soft constraint is chosen with probability $1 - wp_h$, and then selecting from this constraint the variable to be flipped, according to the strategy described in Chapter 6 (page 301f.).

OPB-CSP can be extended by allowing ranges of integers instead of variable domains $\{0, 1\}$. The resulting *overconstrained integer programming (OIP)* instances can be solved using WSAT(OIP), a generalisation of the WSAT($\mathcal{PB}$) algorithm that can handle integer variables. (Like OPB-CSP, OIP can be seen as a special case of integer programming.) Different from WSAT($\mathcal{PB}$), WSAT(OIP) allows modifications of the current value v of a given integer variable to values v' with $|v' - v| \leq 2$. An implementation of WSAT(OIP) is available from Walser's WSAT(OIP) webpage [Walser, 2003]; this supersedes the earlier implementation of WSAT($\mathcal{PB}$), which can be seen as a restricted variant of WSAT(OIP).

A large number of practically relevant problems can be formulated easily and naturally within the OPB-CSP and OIP frameworks. WSAT(OIP) has been tested on a variety of problems that can be encoded using Boolean variables. These problems include radar surveillance problems (which include soft constraints) and the Progressive Party Problem [Smith et al., 1996]. For both problems, WSAT(OIP) showed significantly improved performance over a state-of-the-art commercial integer programming package (CPLEX) and other methods for solving these problems. WSAT(OIP) also achieved excellent performance on capacitated production planning and AI planning problems, which were represented using non-Boolean integer variables [Walser et al., 1998; Kautz and Walser, 1999].

7.4 **Further Readings and Related Work**

MAX-SAT is one of the most widely studied simple combinatorial optimisation problems, and a wide range of SLS algorithms for MAX-SAT has been proposed and evaluated in the literature, including algorithms based on Simulated Annealing, GRASP, Variable Neighbourhood Search, Ant Colony Optimisation and Evolutionary Algorithms.

Hansen and Jaumard studied a Simulated Annealing algorithm that uses a Metropolis acceptance criterion and a standard geometric annealing schedule [Hansen and Jaumard, 1990]. This algorithm was found to perform worse than SAMD on various sets of unweighted Uniform Random k-SAT instances; however, in some cases, it reaches better quality solutions than SAMD with substantially higher run-times (both algorithms are terminated when no improvement in the incumbent solution has been observed for a specified number of search steps).

GRASP was one of the first SLS algorithms for weighted MAX-SAT [Resende et al., 1997]. It was originally evaluated on the wjnh instances described in Section 7.1, but was later found to be substantially outperformed by the first DLM algorithm for weighted MAX-SAT [Shang and Wah, 1997] and other state-of-the-art algorithms. Recently, Variable Neighbourhood Search (VNS) has been applied to weighted MAX-SAT [Hansen and Mladenović, 1999; Hansen et al., 2000]. A variant called *Skewed VNS*, which accepts worse candidate solutions depending on the amount of deterioration in solution quality and the Hamming distance from the incumbent solution, was shown to perform much better than a basic version of VNS and a basic Tabu Search algorithm. However, it is not clear how Skewed VNS performs compared to state-of-the-art algorithms for weighted MAX-SAT, such as GLS or IRoTS.

Roli et al. have studied various Ant Colony Optimisation algorithms for CSP and MAX-CSP [Roli et al., 2001]. They mainly investigated different ways of using pheromones and presented limited computational results for their algorithms on a small set of MAX-SAT instances. These results indicate that their ACO algorithms (without using local search) perform substantially worse than state-of-the-art algorithms for MAX-SAT. Evolutionary Algorithms can be easily applied to MAX-SAT, because the candidate solutions can naturally be represented as binary strings, and standard crossover and mutation operators can be applied in a straightforward way. Some insights into the behaviour of genetic algorithms for MAX-SAT have been obtained, but pure evolutionary algorithms (without local search) were found to perform relatively poorly [Rana, 1999; Rana and Whitley, 1998; Bertoni et al., 2000].

There are relatively few complete algorithms for MAX-SAT. Most of these are based either on branch & bound extensions of backtracking algorithms derived from the Davis-Logeman-Loveland (DLL) procedure [Davis et al., 1962] or on branch & cut approaches. A comparison of a DLL-based MAX-SAT solver by Borchers and Furman [1999b], which uses short runs of GWSAT for obtaining upper bounds on the solution quality, and a branch & cut algorithm by Joy, Mitchell and Borchers [1997] showed that the DLL-based approach performed significantly better than the branch & cut algorithm on MAX-3-SAT, while the branch & cut algorithm was found to be superior on MAX-2-SAT instances and MAX-SAT-encoded Steiner tree problems. The DLL algorithm of Borchers and Furman has recently been significantly improved by including more powerful lower-bounding techniques and variable selection heuristics [Alsinet et al., 2003]. It may be noted that the wjnh instances as well as some weighted MAX-SAT instances with up to 500 variables that were used to evaluate VNS [Hansen et al., 2000] were solved to optimality with CPLEX, a well known general-purpose integer programming software. However, all of these methods appear to be substantially less efficient in finding high-quality solutions for large and hard MAX-SAT instances than state-of-the-art SLS algorithms [Resende et al., 1997; Hoos et al., 2003].

MAX-CSP has received considerable attention from the constraint programming community as a straightforward extension of CSP to optimisation problems. MAX-CSP is a special case of the *Partial Constraint Satisfaction Problem*, which involves finding values for a subset of variables satisfying only a subset of the constraints [Freuder, 1989; Freuder and Wallace, 1992]. More recently, two general frameworks for constraint satisfaction and optimisation have been introduced, *Semi-Ring Based CSP* [Bistarelli et al., 1997] and *Valued CSP* [Schiex et al., 1995]. So far, most research concentrated on establishing formal comparisons of these frameworks or adapting propagation techniques and complete algorithms to solve problems formulated within these frameworks; we are not aware of SLS algorithms for the latter two frameworks.

For MAX-CSP, significant research efforts have been directed towards the development of efficient complete algorithms. Since the first branch & bound algorithm for unweighted MAX-CSP [Freuder and Wallace, 1992], especially the lower bounding techniques have been significantly refined, leading to much better performing branch & bound algorithms [Wallace, 1994; 1996b; Larrosa et al., 1999; Kask and Dechter, 2001; Larrosa and Dechter, 2002; Larrosa and Meseguer, 2002].

Few results are available for SLS algorithms for MAX-CSP other than the ones described in this chapter. Kask compared the performance of an algorithm based on the Breakout Method with that of a state-of-the-art branch & bound algorithm and found that the former outperforms the latter on Random MAX-CSP with dense constraint graphs, while for sparse constraint graphs, the branch & bound algorithm is slightly faster than the breakout algorithm [Kask, 2000]. Hao and Pannier compared TS-GH to a Simulated Annealing algorithm for MAX-CSP; their computational results suggest that Simulated Annealing is clearly inferior to TS-GH [Hao and Pannier, 1998]. Battiti and Protasi extended H-RTS to the Maximum k-Conjunctive Constraint Satisfaction problem (MAX-k-CCSP), a special case of MAX-CSP with Boolean variables, in which each constraint corresponds to a conjunction of k literals [Battiti and Protasi, 1999].

As previously stated, the Overconstrained Integer Programming Problem (OIP) introduced by Walser is a special case of integer linear programming (ILP). There exist several SLS algorithms for 0–1 ILP (i.e., pseudo-Boolean optimisation) and general ILP. Computational results by Walser suggest that the 'general-purpose' Simulated Annealing strategy (GPSIMAN) by Conolly [1992] is outperformed by WSAT(OIP) on a variety of problems [Walser, 1999]. Extensions of GPSIMAN were later applied by Abramson to set partitioning problems [Abramson et al., 1996]. Abramson and Randall applied Simulated Annealing to encodings of optimisation problems into general ILP problems and later introduced a modelling environment based on dynamic list structures [Abramson and Randall, 1999]. Furthermore, there are adaptations of evolutionary algorithms and GRASP for integer linear programming [Pedroso, 1999; Neto and Pedroso, 2001].

Several SLS algorithms have been developed for solving the more general *Mixed Integer Linear Programming Problem (MILP)*, which allows $\{0, 1\}$ variable domains as well as continuous domains in the form of intervals over real numbers. Obviously, these algorithms can also be applied to pure ILP problems, which can be seen as a special case of MILP in which no continuous variables occur. For an overview of SLS algorithms for MILP we refer to a recent article by Løkketangen [2002].

7.5 **Summary**

The Maximum Satisfiability Problem (MAX-SAT) is the optimisation variant of SAT in which the goal is to find a variable assignment that maximises the number or total weight of satisfied clauses. As one of the conceptually simplest hard combinatorial optimisation problems, MAX-SAT is of considerable theoretical interest. Furthermore, a diverse range of hard combinatorial optimisation problems, many of which have direct real-world applications, can be efficiently and naturally encoded into MAX-SAT. By using appropriately chosen clause weights and solution quality bounds, combinatorial optimisation problems with hard and soft constraints can be represented by weighted MAX-SAT instances.

Considerable effort has been spent in designing efficient (i.e., polynomial-time) approximation algorithms for MAX-SAT that have certain worst-case performance guarantees. For widely used types of benchmark instances, including test-sets of randomly generated MAX-SAT instances as well as MAX-SAT encodings of other combinatorial optimisation problems (such as set covering and time-tabling), these approximation algorithms do not reach the performance of even relatively simple SLS algorithms. Furthermore, different from the situation for SAT, systematic search algorithms for MAX-SAT are substantially less efficient than SLS algorithms in finding high-quality solutions to typical MAX-SAT instances.

The most successful SLS algorithms for MAX-SAT fall into four categories: (i) tabu search algorithms, in particular *Robust Tabu Search (RoTS)* and *Reactive Tabu Search (H-RTS)*; (ii) dynamic local search (DLS) algorithms, particularly *Guided Local Search (GLS)*; (iii) iterated local search (ILS) algorithms, particularly *Iterated Robust Tabu Search (IRoTS)*; and (iv) generalisations of high-performance SAT algorithms, in particular *Novelty$^+$ with weighted clause selection*. Some of these algorithms achieve state-of-the-art performance on mildly overconstrained instances whose optimal solutions leave relatively few clauses unsatisfied (GLS as well as Novelty$^+$ and its variants for weighted MAX-SAT seem to fall into this category), while others, such as IRoTS, appear to be state-of-the-art for highly overconstrained instances.

All of these algorithms make use of information on the search history, mainly in the form of a tabu list or dynamically adjusted clause penalties. There is some evidence that by using large neighbourhoods, such as *reduced versions of the 2-flip and 3-flip neighbourhoods*, high-performance ILS and Tabu Search algorithms for MAX-SAT can be further improved; these improvements, however, critically rely on efficient mechanisms for searching these larger neighbourhoods. On the other hand, although the use of *non-oblivious evaluation functions*, that is, evaluation functions that are not indifferent w.r.t. to the number of literals that are simultaneously satisfied in a given clause, leads to theoretical and practical improvements in the performance of simple iterative improvement methods for unweighted MAX-SAT, there is little evidence that non-oblivious evaluation functions are instrumental in reaching state-of-the-art SLS performance on MAX-SAT instances of any type.

Although a wide range of other SLS methods has been applied to MAX-SAT, including Simulated Annealing, GRASP, Ant Colony Optimisation and

Evolutionary Algorithms, there is currently no evidence that any of these can achieve state-of-the-art performance.

The Maximum Constraint Satisfaction Problem (MAX-CSP) can be seen as a generalisation of CSP where the objective is to find a CSP variable assignment that maximises the number or total weight of satisfied constraints. Current empirical results suggest that the best performing SLS algorithms for CSP are also best for MAX-CSP; in particular, the tabu search algorithm by Galinier and Hao, *TS-GH*, appears to be the most efficient algorithm for unweighted MAX-CSP known to date. However, most existing experimental studies on SLS algorithms for MAX-CSP are limited to particular classes of randomly generated MAX-CSP instances. Furthermore, the potential of many advanced SLS methods, such as Dynamic Local Search or Iterated Local Search, in the context of MAX-CSP is largely unexplored. Overall, considerably more research is necessary to obtain a more complete picture on the relative performance and behaviour of SLS algorithms for MAX-CSP.

On the other hand, generalisations of WalkSAT to *overconstrained pseudo-Boolean CSP* and *integer programming problems*, which can be seen as special cases of MAX-CSP, have been used successfully to solve various application problems, and in many cases, they have been shown to achieve substantially better performance than specialised algorithms and state-of-the-art commercial optimisation tools. However, compared to state-of-the-art complete integer or constraint programming algorithms, SLS methods for these problems have been much less explored, very likely leaving considerable room for further improvement.

Exercises

7.1 [**Easy**] How can an implementation of a standard SLS algorithm for SAT, such as GSAT, be used (without modifications) for solving weighted MAX-SAT instances with integer clause weights? Discuss potential drawbacks of this approach to solving weighted MAX-SAT instances.

7.2 [**Easy**] Give an example that illustrates how duplication of clauses in a CNF formula can affect the optimal solutions of the corresponding MAX-SAT instance.

7.3 [**Medium; Hands-On**] Implement the SAMD algorithm and the TS-YI algorithm described in Section 7.2 (page 329*f.*) for weighted MAX-SAT (you can use the UBCSAT code available from `www.sls-book.net` as a convenient implementation framework). Analyse the performance of the two algorithms on the test-sets `rnd100-500/w500-100`, `rnd100-500/w500-200` and `rnd100-500/w500-500` from `www.sls-book.net`, using appropriate computational experiments and empirical evaluation methods. Based on your observations, formulate a hypothesis on the dependency of the algorithms' performance on the variability of the clause weight distributions; describe further experiments that could be conducted in order to test your hypothesis.

7.4 **[*Easy*]** Consider the instance of the Weighted Set Covering Problem specified by the following diagram

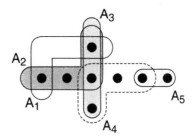

and let $w(A_i) := w_i$ $(i \in \{1, \ldots, 5\})$ be arbitrary weights. Represent this problem

    (a)   as a weighted (discrete finite) MAX-CSP instance;

    (b)   as a weighted MAX-SAT instance.

7.5 **[*Medium*]** Given a Min-Cost GCP instance G, prove that (i) the optimal solutions of the MAX-SAT instance $F(G)$, obtained from the encoding described in Section 7.1 (page 319*f.*), correspond exactly to the optimal solutions of G and that (ii) under the 1-flip neighbourhood, the locally optimal candidate solutions of $F(G)$ correspond exactly to the k-colourings of G.

7.6 **[*Medium*]** Is it the case that any weighted MAX-CSP instance can be encoded into an equivalent, reasonably compact weighted MAX-SAT instance? If so, how big is the difference between the size of the original MAX-CSP instance and the MAX-SAT encoding?

7.7 **[*Medium*]** Give an extended definition of weighted MAX-SAT in which weights can be attached to arbitrary subformulae of a propositional formula. Discuss potential advantages and disadvantages of such a generalised version of weighted MAX-SAT, particularly w.r.t. to solving such problems with SLS algorithms.

7.8 [*Easy*] Explain how an examination timetabling problem can be represented as an instance of weighted MAX-CSP. Exemplify your encoding by applying it to a small examination timetabling instance with four exams, six students and four time slots.

7.9 [*Medium*] Extend the definition of weighted MAX-CSP such that penalties can be specified for assignments of particular values to CSP variables. Does this increase the representational power of weighted MAX-CSP?

7.10 [*Medium*] Formulate the Frequency Assignment Problem (introduced in Section 7.3, page 343*ff.*)

    (a)   as a pseudo-Boolean optimisation problem;

    (b)   as a weighted (discrete finite) MAX-CSP instance.

Traveller, there is no path,
paths are made by walking.
—Antonio Machado, Poet

8 TRAVELLING SALESMAN PROBLEMS

The Travelling Salesman Problem (TSP) is probably the most widely studied combinatorial optimisation problem and has attracted a large number of researchers over the last five decades. Work on the TSP has been a driving force for the emergence and advancement of many important research areas, such as stochastic local search or integer programming, as well as for the development of complexity theory. Apart from its practical importance, the TSP has also become a standard testbed for new algorithmic ideas.

In this chapter we first give a general overview of TSP applications and benchmark instances, followed by an introduction to the most basic local search algorithms for the TSP. Based on these algorithms, several SLS algorithms have been developed that have greatly improved the ability of finding high quality solutions for large instances. We give a detailed overview of iterated local search algorithms, which are currently among the most successful SLS algorithms for large TSP instances, and present several prominent, high-performance TSP algorithms that are based on population-based SLS methods. While most of this chapter focuses on symmetric TSPs, we also discuss aspects that arise in the context of solving asymmetric TSPs.

8.1 TSP Applications and Benchmark Instances

Given an edge-weighted, completely connected, directed graph $G := (V, E, w)$, where V is the set of $n := \#V$ vertices, E the set of (directed) edges, and $w : E \mapsto \mathbb{R}^+$ a function assigning each edge $e \in E$ a weight $w(e)$, the Travelling

Salesman Problem (TSP) is to find a minimum weight Hamiltonian cycle in G, that is, a cyclic path that contains each vertex exactly once and has minimal total weight (a formal definition was given in Chapter 1, page 20*ff.*). Following one of the most intuitive applications of the TSP, namely, finding optimal round trips through a number of geographical locations, the vertices of a TSP instance are often called 'cities', the paths in G are called 'tours' and the edge weights are referred to as 'distances'. In this chapter we focus mainly on the symmetric TSP, that is, the class of TSP instances in which for each pair of edges (v_i, v_j) and (v_j, v_i) we have $w((v_i, v_j)) = w((v_j, v_i))$. We will also highlight some of the issues that arise when dealing with the asymmetric TSP (ATSP), where for at least one pair of vertices the directed edges (v_i, v_j) and (v_j, v_i) have different weights.

TSP as a Central Problem in Combinatorial Optimisation

The TSP plays a prominent role in research as well as in a number of application areas. The design of increasingly efficient TSP algorithms has provided a constant intellectual challenge, and many of the most important techniques for solving combinatorial optimisation problems were developed using the TSP as an example application. This includes cutting planes in integer programming [Dantzig et al., 1954], which led to the modern, high performing *branch & cut* methods [Grötschel and Holland, 1991; Padberg and Rinaldi, 1991; Applegate et al., 1998; 2003a], polyhedral approaches [Grötschel and Padberg, 1985; Padberg and Grötschel, 1985], *branch & bound* algorithms [Little et al., 1963; Held and Karp, 1971], as well as early local search algorithms [Croes, 1958; Flood, 1956; Lin, 1965; Lin and Kernighan, 1973]. Additionally, many of the general SLS methods presented in Chapter 2, such as Simulated Annealing or Ant Colony Optimisation, were first tested on the TSP. The TSP also played an important role in the development of computational complexity theory [Garey and Johnson, 1979]. In fact, several books are entirely devoted to the TSP [Gutin and Punnen, 2002; Lawler et al., 1985; Reinelt, 1994], and an enormous number of research articles cover the various aspects of TSP solving. For details on the history of TSP solving, we refer to Schrijver's overview paper on the history of combinatorial optimisation [Schrijver, 2003], the book chapter by Hoffmann and Wolfe [1985] and the web page by Applegate et al. [2003b].

There are various reasons for this central role of the TSP in combinatorial optimisation. Firstly, it is a conceptually simple problem, which is easily explained and understood, but as an $\mathcal{NP}$-hard problem, it is difficult to solve [Garey and Johnson, 1979]. Secondly, the design and analysis of algorithms for the TSP are not obscured by technicalities that arise from dealing with side constraints, which are often difficult to handle in practice. Thirdly, the TSP is now established as a standard testbed for new algorithmic ideas, which are often assessed based on their performance on the TSP. Fourthly, given the significant amount of interest in the research community, new contributions to TSP solving or insights into the problem structure are likely to have a large impact. Finally, the TSP arises in a variety of applications and is therefore of significant practical relevance.

Benchmark Instances

Extensive computational experiments have always played an important role in the history of the TSP. These experiments involve several types of TSP instances. In many cases, these are predominantly *metric TSP* instances, that is, instances in which the vertices correspond to points in a metric space and the edge weights correspond to metric distances between pairs of points. Metric TSP instances for which the distances are based on the standard Euclidean metric are also called *Euclidean*. Regardless of whether they are metric or not, almost all available TSP benchmark instances use integer distances. One main reason is that on older computers, integer computations were much faster than computations using floating point numbers; high precision approximations of the true distances in metric spaces can be achieved by multiplying all floating point distances in a given TSP instance with a constant factor and subsequent rounding.

A well known and widely used collection of TSP instances is available through TSPLIB, a benchmark library for the TSP [Reinelt, 2003]. TSPLIB comprises more than 100 instances with up to 85 900 cities. For all except four TSPLIB instances, optimal solutions have been determined; as of September 2003, the largest instance solved provably to optimality has 15 112 cities. Most of the TSPLIB instances stem from influential studies on the TSP; many of them originate from practical applications, such as minimising drill paths in printed circuit board manufacturing, positioning detectors in X-ray crystallography or finding the shortest round trip through all the Biergärten in Augsburg, Germany.[*] Many of the remaining TSPLIB instances are of geographical nature, where the inter-vertex distances are derived from the distances between cities and towns with given coordinates. Two examples of TSPLIB instances are shown in the upper part of Figure 8.1.

A set of TSP instances derived from problems in VLSI design, ranging from 131 to 744 710 vertices, is available from the web page by Applegate et al. [2003b].

---

[*]Augsburg is close to one of the authors' (T. S.) home town; however, T. S. never managed to visit all Biergärten in one night. The reason for this failure may well be that at the time he lived there, he was not yet aware of the shortest tour.

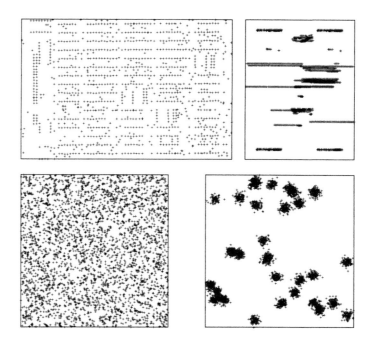

Figure **8.1** Four (Euclidean) TSP benchmark instances. The two instances in the top row stem from an application in which drill paths in manufacturing printed circuit boards are to be minimised in length (left side: TSPLIB instance pcb1173 with 1 173 vertices and right side: fl1577 with 1 577 vertices, the latter instance shows a pathological clustering of vertices). The bottom row shows a Random Uniform Euclidean instance (left side) and a Randomy Clustered Euclidean instance (right side); both instances have 3 162 vertices each.

From the same web page, a TSP instance of potential interest for globetrotters is available; this *World TSP* instance comprises all 1 904 711 populated cities or towns registered in the National Imagery and Mapping Agency database and the Geographic Names Information System; several additional TSP instances comprising the towns and cities of individual countries are available from the same site. An overview of further practical applications of the TSP can be found on the web page by Applegate et al. [2003b] or in the book by Reinelt [1994].

A large part of the experimental (and also theoretical) research on the TSP has used randomly generated instances with the most widely used classes being *Random Euclidean (RE) instances* and *Random Distance Matrix (RDM) instances*. In RE instances, the vertices correspond to randomly placed points in an l-dimensional hypercube, and the edge weights are given by the pairwise Euclidean distances between these points. (The Euclidean distance between two points

$x := (x_1, \ldots, x_l)$ and $y := (y_1, \ldots, y_l)$ is defined as $d(x, y) := \sqrt{\sum_{i=1}^{l} (x_i - y_i)^2}$. Commonly, real valued distances are scaled by a constant factor α and subsequently rounded or truncated to obtain integer values.)

Most experimental studies involving RE instances have focused on two-dimensional instances (i.e., $l = 2$), in which the points are uniformly distributed in a square; we refer to these as *Random Uniform Euclidean (RUE) instances*. An example for an RUE instance is shown in Figure 8.1 (bottom left plot). The class of RUE instances has the interesting property that, as the instance size approaches infinity, the ratio of the optimal tour length to $\sqrt{n}$ (where n is the number of vertices) converges towards a constant γ [Beardwood et al., 1959]; for squares with sides of length one, the value of γ is approximately 0.721 [Johnson et al., 1996; Percus and Martin, 1996].

Another type of two-dimensional RE instances, which have been used in the 8th DIMACS Implementation Challenge on the TSP [Johnson et al., 2003a], places the points in clusters within a square area. More precisely, these *Random Clustered Euclidean (RCE) instances* are obtained by first distributing the cluster centres uniformly at random; then, each actual point is placed by choosing a cluster centre uniformly at random, and then adding to each coordinate a displacement sampled from a normal distribution. An example of an RCE instance is shown in Figure 8.1 (bottom right plot). RCE instances are interesting because it is known that various local search algorithms are negatively affected by the clustering of the vertices.

RDM instances are symmetric, non-Euclidean instances in which the edge weights are randomly chosen integers from a given interval. This distribution of TSP instances is known to pose a considerable challenge for many SLS algorithms [Johnson and McGeoch, 1997]. However, since they are conceptually and structurally far-removed from any application, these instances are mainly of theoretical interest.

Lower Bounds on the Optimal Solution Quality

A large amount of research efforts on the TSP have been dedicated to finding good lower bounds on the optimal solution quality for given instances. Lower bounds are used in complete search methods, such as branch & bound algorithms, to estimate the minimum cost incurred for completing partial solutions and to prune search trees if the cost estimation is larger than or equal to the best solution encountered earlier in the search process. In this context, it is important to have estimates that are close to the real costs, because better estimates facilitate more extensive pruning of the search tree. In general, lower bounds are also useful for assessing the quality of solutions obtained from incomplete algorithms. This is particularly the case for lower bounds obtained by deterministic methods, which are often used for the evaluation of SLS algorithms in cases where optimal solution qualities are unknown.

A general approach for obtaining lower bounds is to solve a relaxation of the original problem, which is typically obtained by removing some problem constraints. Feasible solutions to the original problem then correspond to a subset of the solutions to the relaxed problem, and an optimal solution to the relaxed problem is therefore always a true lower bound for the solution quality of the original problem. Efficient lower-bounding techniques are based on relaxations resulting in problems that can be solved quickly but at the same time have an optimal solution quality close to that of the original problem.

One of the simplest lower bounds for a TSP instance G is based on the following observation: By removing a single edge from an optimal tour s^* with weight $w(s^*)$, a spanning tree t of the graph G is obtained. (Recall that a spanning tree in a weighted graph is a subgraph that contains all vertices and has no cycles; the weight of a spanning tree is defined to be the sum of the weights of the edges it contains.) Clearly, a minimum weight spanning tree t^* has a total edge weight $w(t^*) \leq w(t)$, and hence $w(t^*)$ is a lower bound for $w(s^*)$. This lower bound is also quick to compute. In fact, the well known algorithms of Kruskal and Prim run in time $O(m \log n)$, where m is the number of edges; by using Fibonacci heaps, the time-complexity of Prim's algorithm can be reduced to $O(m + n \log n)$. (For the best lower bounds on the complexity of computing minimum spanning trees see Chazelle [2000].) However, this spanning tree bound can still be relatively far from the optimal solution value.

Tighter bounds than the minimum spanning tree lower bound can be obtained as follows: Let $G \setminus \{v_1\}$ be the graph obtained from G by deleting vertex v_1 and all the edges incident to v_1. A one-tree is a spanning tree on the vertices $v_2, v_3, \ldots, v_n$ plus two edges incident to vertex v_1 (see Figure 8.2 for an example of a one-tree). We get a minimum weight one-tree for G by computing a minimum spanning tree of $G \setminus \{v_1\}$ and adding the two minimum weight edges incident to v_1. The weight of the resulting one-tree is a lower bound for $w(s^*)$, because every minimum weight tour s^* of G is a one-tree. This lower bound could be improved by choosing several or all vertices to play the role of v_1 and then taking the maximum weight of the corresponding one-trees as a lower bound; however, this does not result in significant gains and is quite time consuming [Reinelt, 1994; Cook et al., 1997].

Luckily, there exist other techniques to improve upon the one-tree bound. These are based on the following observation: We can assign a value p_i to each vertex v_i and add these values to the weights of all edges incident to a vertex $v_i \in V$, resulting in a graph $G' := (V, E, w')$ with edge weights

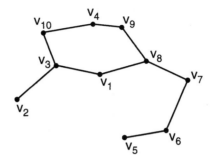

Figure 8.2 Example of a one-tree in a graph of ten vertices. The subtree on vertices v_2 to v_{10} forms a minimum spanning tree.

$w'((v_i, v_j)) := w((v_i, v_j)) + p_i + p_j$ (recall that the edges are not oriented). This has the effect of increasing the weight of each tour in G by a constant amount of $2 \cdot \sum_{i=1}^{n} p_i$ in G'. Clearly, this transformation preserves the optimality of tours. It may, however, result in different optimal one-trees [Held and Karp, 1971; Cook et al., 1997]; then, by subtracting $2 \cdot \sum_{i=1}^{n} p_i$ from the weight of an optimal one-tree of G', a lower bound on the minimum weight tour in G is obtained. The quality of this lower bound depends on the values of the vertex penalties $p_1, \ldots, p_n$. The *Held-Karp (HK) lower bound* is obtained from the set of vertex penalties $\hat{p}_1, \ldots, \hat{p}_n$ that maximises the value of the resulting lower bound. The exact Held-Karp bound can be computed by algorithms based on linear programming. As an alternative, Held and Karp proposed an algorithm that iteratively modifies the penalty assignments $p_1, \ldots, p_n$ [Held and Karp, 1971]. Roughly speaking, this algorithm iteratively increases the penalties for vertices with degree one in the current optimal one-tree, and decreases the penalties for vertices with degree greater than two; the search is terminated when a minimum weight one-tree is obtained in which all vertices have degree two (this corresponds to a feasible solution of G), or when a maximum number of iterations has been performed. The lower-bounding technique of Held and Karp is an example of the more general method called *Lagrangian Relaxation*, and the iterative penalty adjustment algorithm is an instance of a *subgradient optimisation* method [Fisher, 1981; 1985; Held et al., 1974].

Experimental results suggest that for many types of TSP instances, the Held-Karp bounds are very tight. For RUE instances, the HK lower bound is typically within less than one percent of the actual optimal solution value [Johnson et al., 1996]. For TSPLIB instances, the gap between the HK bound and the respective optimum solution quality is often slightly larger, but for almost all instances the HK bound is still within two percent of the optimal solution quality.

State-of-the-Art Methods for TSP Solving

The algorithmic techniques for TSP solving have reached a very high level of sophistication. This is true for complete algorithms as well as for SLS algorithms.

Currently, the best performing complete algorithms for the TSP are *branch & cut methods*, which are based on solving a series of linear programming relaxations of an integer programming problem [Mitchell, 2002]. These relaxations are typically obtained by allowing the binary variables typically used in integer programming formulations of the TSP [Nemhauser and Wolsey, 1988; Reinelt, 1994] to take arbitrary values from the interval [0, 1], instead of constraining them to integer values from the set {0, 1}. Cutting plane methods are used to make the relaxation more closely approximate the optimum solution of the original integer programming problem. This is done by finding linear inequalities that are satisfied by all integer feasible solutions but not by the non-integer optimal solution to the current relaxation. These so-called cuts are then added to obtain the next linear optimisation problem, which again is solved to optimality. This process is iterated until finding 'good' cuts becomes hard. Then, it may become preferable to branch by splitting the current problem into two subproblems; this is done by forcing one edge to be part of any solution in one subproblem and not to appear in any solution of the other subproblem [Grötschel and Holland, 1991; Padberg and Rinaldi, 1991]. For a detailed description of state-of-the-art branch & cut algorithms for the TSP, we refer to Applegate et al. [2003a].

Within modest computation times, efficient implementations of state-of-the-art branch & cut algorithms for the TSP can routinely solve to optimality small to medium size symmetric TSP instances, ranging from a few hundred to around 1 000 to 3 000 vertices. Given substantially longer run-times, these algorithms can also solve much larger instances; as previously mentioned, the largest TSPLIB instance that has been solved (provably) optimally (as of September 2003), instance d15112, has 15 112 vertices. (Partly motivated by the availability of such powerful complete algorithms, many studies on symmetric TSP algorithms now focus on solving large instances with thousands of vertices.)

Despite these impressive successes, complete algorithms suffer from some limitations. Firstly, the computation times quickly become prohibitively large with increasing instance size. For example, finding an optimal solution and proving its optimality for instance d15112 required a total estimated computation time of 22.6 CPU years on a Compaq EV6 Alpha processor running at 500 MHz (The actual computation was performed on a network of up to 110 workstations [Applegate et al., 2003b].) Secondly, the computation times of complete methods vary strongly among instances: while TSPLIB instance pr2392 (with 2 392 vertices) was solved within 116 CPU seconds on a 500 MHz Compaq XP1000 workstation, solving TSPLIB instance d2103 with 2 103 vertices required a total run-time of about 129 CPU days. (In the latter case, the actual computations were performed on a network of 55 Alpha 21164 processors running at 400 and 500 MHz [Applegate et al., 2003b].)

Given the limitations of complete algorithms, there is considerable interest in SLS algorithms for the TSP. If reasonably high quality solutions are required very quickly, heuristic construction search algorithms are very useful. Construction methods that give a reasonable trade-off between computation time and solution quality, such as the Savings Heuristic or Farthest Insertion (both are described in Section 8.2), can find tours for RUE instances with a few thousand vertices within about 11 to 16 percent of the Held-Karp lower bounds (which, as explained above, are known to be close to the optimal solution quality) in fractions of a CPU second on a 500MHz Alpha processor [Johnson and McGeoch, 2002]. When allowing run-times of a few CPU seconds, the same relative solution quality can be obtained for instances with several hundred thousand vertices [Johnson and McGeoch, 2002]. Constructive search methods are also important in the context of generating good initial solutions for iterative improvement algorithms. The performance of iterative improvement algorithms for the TSP also depends crucially on the underlying neighbourhood relation and on various details of the search process. In Section 8.2, we give an overview of commonly used iterative improvement algorithms and report some illustrative performance results.

Optimal or close to optimal solutions can typically be obtained by using hybrid SLS algorithms at the cost of higher computation times. State-of-the-art SLS algorithms for the TSP can find optimal solutions for symmetric instances with thousands of cities within seconds or minutes of CPU times on modern workstations (as of 2003) [Cook and Seymour, 2003; Helsgaun, 2000]; significantly larger instances can typically be solved optimally or almost optimally within CPU hours. For example, the best performing SLS algorithm identified in a recent extensive experimental study, ILK-H (which is presented in Section 8.3), found a candidate solution of TSPLIB instance d15112 whose quality is only 0.0186 percent away from the known optimum in about seven hours of CPU time on a 500MHz Alpha processor [Johnson and McGeoch, 2002]. For the same instance, other SLS algorithms obtained solution qualities within one percent of the optimum in less than seven CPU seconds. The impressive performance of SLS algorithms when applied to very large TSP instances is exemplified by the results obtained for a RUE instance comprising 25 million cities, for which after 8 CPU days on a IBM RS6000 machine, Model 43-P 260, a solution quality within 0.3 percent of the estimated optimal solution quality was reached by a high-performance iterated local search algorithm [Applegate et al., 2003c]. (The performance of various SLS algorithms for the TSP is further illustrated by the results of the 8th DIMACS Implementation Challenge on the TSP [Johnson et al., 2003a]).

Asymmetric TSPs

Empirical results indicate that asymmetric TSP (ATSP) instances, in which a given graph has at least one pair of vertices for which $w((v, v')) \neq w((v', v))$, are typically harder to solve than symmetric TSP instances of comparable size [Johnson et al., 2002]. TSPLIB includes 27 ATSP instances ranging from 17 to 443 vertices. A large number of additional instances has been recently generated in the context of empirical studies of ATSP algorithms [Cirasella et al., 2001; Johnson et al., 2002]; these include several classes of randomly generated ATSP instances that model real-world problems, such as moving drills along a tilted surface, scheduling read operations on computer disks, collecting coins from pay phones or finding shortest common super-strings for a set of genomic DNA sequences (a problem that arises in genome reconstruction). There are also some individual instances directly taken from practical applications of the ATSP, such as stacker crane problems, vehicle routing [Fischetti et al., 1994], robot motion planning, scheduling read operations on a tape drive or code optimisation [Young et al., 1997]. These instances and random instance generators are available online at the web site of the 8th DIMACS Implementation Challenge [Johnson et al., 2003a].

ATSP instances can be solved by means of a native ATSP algorithm or, alternatively, by a transformation into symmetric TSP instances, which are then solved using a high-performance algorithm for the symmetric TSP. One such transformation works as follows [Jonker and Volgenant, 1983]. Given a directed graph $G :=(V, E, w)$ with vertex set $V := \{v_1, \ldots, v_n\}$, edge set E, and weight function w, we define an undirected graph $G' := (V', E', w')$ with $V' := V \cup \{v_{n+1}, v_{n+2}, \ldots, v_{n+n}\}$, $E' := V' \times V'$, and w' as

$$w'((v_i, v_{n+j})) := w'((v_{n+j}, v_i)) := w((v_i, v_j)) \qquad \text{for } i, j \in \{1, \ldots, n\} \text{ and}$$
$$(v_i, v_j) \in E$$
$$w'((v_{n+i}, v_i)) := w'((v_i, v_{n+i})) := -M \qquad \text{for } i \in \{1, \ldots, n\}$$
$$w'((v_i, v_j)) := M \qquad \text{otherwise,}$$

where M is a sufficiently large number, for example, $M := \sum_{v,v' \in V} w((v, v'))$. An example of this transformation is shown in Figure 8.3. It is easy to see that for each Hamiltonian cycle with weight w_a of an asymmetric TSP instance there is a Hamiltonian cycle in the symmetric instance with weight $w_a - n \cdot M$. Once a solution for the symmetric TSP instance G' is obtained, it can easily be transformed into a solution for the ATSP instance G. There exist other transformations that replace each vertex in G by three vertices, but avoid that the resulting symmetric TSP instance has negative edge weights (which may cause problems with some existing implementations of TSP algorithms).

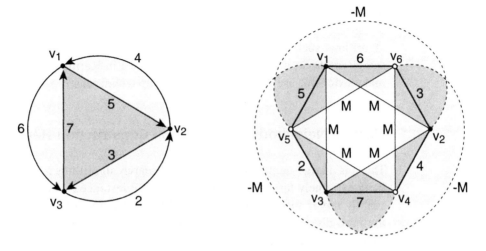

Figure **8.3** Transformation of an ATSP instance (left side) into a symmetric TSP instance (right side). (For details, see text.)

Although symmetric TSP instances obtained from such transformations have twice or thrice as many vertices as the respective original ATSP instances, solving these using algorithms for the symmetric TSP is often more effective than solving the original ATSP instances using native ATSP algorithms [Johnson et al., 2002]. Furthermore, empirical results indicate that the Held-Karp (HK) lower bounds on the optimal solution quality for the symmetric TSP instances obtained from such transformations are often tighter than the widely used Assignment Problem (AP) lower bounds for the respective ATSP instances [Johnson et al., 2002]. Recent empirical results indicate that ATSP instances with a relatively large gap between the AP and the HK bound are most efficiently solved by transforming them into symmetric TSP instances and solving these using state-of-the-art symmetric TSP algorithms, such as Helsgaun's Lin-Kernighan variant [Helsgaun, 2000]. However, ATSP algorithms that are guided by information from the AP bound, such as Zhang's heuristic [Zhang, 1993] (a truncated branch & bound algorithm which uses the AP lower bound), tend to show better performance for ATSP instances for which both bounds are relatively close to each other [Johnson et al., 2002].

8.2 'Simple' SLS Algorithms for the TSP

Much of the early research on incomplete algorithms for the TSP has been focused on construction heuristics and iterative improvement algorithms.

These techniques are important, because they are at the core of many more advanced SLS algorithms. They range from extremely fast constructive search algorithms, such as the Nearest Neighbour Heuristic, to complex variable depth search methods, in particular, variants of the Lin-Kernighan Algorithm, which make extensive use of a number of speedup techniques.

Nearest Neighbour and Insertion Construction Heuristics

There is a large number of constructive search algorithms for the TSP, ranging from extremely fast methods for metric TSP instances, whose run-time is only slightly larger than the time required for just reading the instance data from the hard-disk (see, for example, Platzman and Bartholdi III [1989]), to more sophisticated algorithms with non-trivial bounds on the solution quality achieved in the worst case. In the context of SLS algorithms, construction heuristics are often used for initialising the search; iterative improvement algorithms for the TSP typically require fewer steps for reaching a local optimum when started from higher-quality tours obtained from a good construction heuristic.

One particularly intuitive and well-known constructive search algorithm has been already discussed in Chapter 1, Section 1.4: The *Nearest Neighbour Heuristic (NNH)* starts tour construction from some randomly chosen vertex u_1 in the given graph and then iteratively extends the current partial tour $p = (u_1, \ldots, u_k)$ with an unvisited vertex u_{k+1} that is connected to u_k by a minimum weight edge (u_{k+1} is called a *nearest neighbour* of u_k); when all vertices have been visited, a complete tour is obtained by extending p with the initial vertex, u_1. The tours constructed by the NNH are called *nearest neighbour tours*.

For TSP instances that satisfy the triangle inequality, nearest neighbour tours are guaranteed to be at most by a factor of $1/2 \cdot (\lceil \log_2(n) \rceil + 1)$ worse than optimal tours in terms of solution quality [Rosenkrantz et al., 1977]. In the general case, however, there are TSP instances for which the Nearest Neighbour Heuristic returns tours that are by a factor of $1/3 \cdot (\log_2(n+1) + 4/3)$ worse than optimal, and hence, the approximation ratio of the NNH for the general TSP cannot be bounded by any constant [Rosenkrantz et al., 1977]. In practice, however, the NNH typically yields much better tours than these worst-case results may suggest; for metric and TSPLIB instances, nearest neighbour tours are typically only 20–35% worse than optimal. In most cases, nearest neighbour tours are locally similar to optimal solutions, but they include some very long edges that are added towards the end of the construction process in order to complete the tour (two examples are shown in Figure 8.4). This effect is avoided to some extent by a variant of the NNH that penalises insertions of such long edges [Reinelt, 1994]. Compared to the standard NNH, this variant requires only slightly more

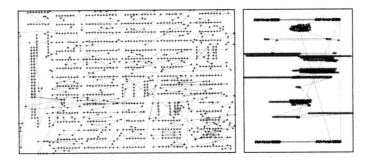

Figure 8.4 Two examples of nearest neighbour tours for TSPLIB instances. *Left:* pcb1173 with 1 173 vertices, *right:* fl1577 with 1 577 vertices. Note the long edges contained in both tours.

computation time, but when applied to TSPLIB instances, it finds tours that are about 5% closer to optimal.

Insertion heuristics construct tours in a way that is different from that underlying the NNH; in each step, they extend the current partial tour p by inserting a heuristically chosen vertex at a position that typically leads to a minimal length increase. Several variants of these heuristics exist, including

   (i)   *nearest insertion* construction heuristics, where the next vertex to be inserted is a vertex u_i with minimum distance to any vertex u_j in p;

   (ii) *cheapest insertion*, which inserts a vertex that leads to the minimum increase of the weight of p over all vertices not yet in p;

   (iii) *farthest insertion*, where the next vertex to be inserted is a vertex u_i for which the minimum distance to a vertex in p is maximal;

   (iv) *random insertion*, where the next vertex to be inserted is chosen randomly.

For TSP instances that satisfy the triangle inequality, the tours constructed by nearest and cheapest insertion are provably at most twice as long as an optimal tour [Rosenkrantz et al., 1977], while for random and farthest insertion, the solution quality is only guaranteed to be within a factor $O(\log n)$ of the optimum [Johnson and McGeoch, 2002]. In practice, however, the farthest and random insertion heuristics perform much better than nearest and cheapest insertion, yielding tours that, in the case of TSPLIB and RUE instances, are on average between 13% and 15% worse than optimal [Johnson and McGeoch, 2002; Reinelt, 1994].

The Greedy, Quick-Borůvka and Savings Heuristics

The construction heuristics discussed so far build a complete tour by iteratively extending a connected partial tour. An alternative approach is to iteratively build several tour fragments that are ultimately patched together into a complete tour. One example for a construction heuristic of this type is the *Greedy Heuristic*, which works as follows. First, all edges in the given graph G are sorted according to increasing weight. Then, this list is scanned, starting from the minimum weight edge, in linear order. An edge e is added to the current partial candidate solution p if inserting it into G', the graph that contains all vertices of G and the edges in p, does not result in any vertices of degree greater than two or any cycles of length less than n edges.

There exist several variants of the Greedy Heuristic that use different criteria for choosing the edge to be added in each construction step. One of these is the *Quick-Borůvka Heuristic* [Applegate et al., 1999], which is inspired by the minimum spanning tree algorithm by Borůvka [1926]. First, the vertices in G are sorted arbitrarily (e.g., for metric TSP instances, the vertices can be sorted according to their first coordinate values). Then, the vertices are processed in the given order. For each vertex u_i of degree less than two in G', all edges incident to u_i that appear in G but not in G' are considered. Of these, the minimum weight edge that results neither in a cycle of length less than n nor in a vertex of degree larger than two, is added to G'. Note that at most two scans of the vertices have to be performed to generate a tour.

Another construction heuristic that is based on building multiple partial tours, is the *Savings Heuristic*, which was initially proposed for a vehicle routing problem [Clarke and Wright, 1964]. It works by first choosing a base vertex u_b and $n - 1$ cyclic paths (u_b, u_i, u_b) that consist of two vertices each. As long as more than one cyclic path is left, at each construction step two cyclic paths p_1 and p_2 are combined by removing one edge incident to u_b in both, p_1 and p_2, and by connecting the two resulting paths into a new cyclic path p_{12}. The edges to be removed in this operation are selected such that a maximal reduction in the cost of p_{12} compared to the total combined cost of p_1 and p_2 is achieved.

Regarding worst-case performance, it can be shown that greedy tours are at most $(1 + \log n)/2$ times longer than an optimal tour, while the length of a savings tour is at most a factor of $(1 + \log n)$ above the optimum [Ong and Moore, 1984]; no worst-case bounds on solution quality are known for Quick-Borůvka tours. Empirically, the Savings Heuristic produces better tours than both Greedy and Quick-Borůvka; for example, for large RUE instances, the length of savings tours is on average around 12% above the Held-Karp lower bounds, while Greedy and Quick-Borůvka find solutions around 14% and 16% above these lower bounds, respectively [Johnson and McGeoch, 2002]. Computation times are modest, ranging for 1 million vertex RUE instances from 22 (for Quick-Borûvka) to around 100 seconds (for Greedy and Savings) on a 500 MHz Alpha CPU.

Construction Heuristics Based on Minimum Spanning Trees

Yet another class of construction heuristics builds tours based on minimum-weight spanning trees (MSTs). In the simplest case, such an algorithm consists of the following four steps: First, an MST t for the given graph G is computed; then, by doubling each edge in t, a new graph G' is obtained. In the third step, a Eulerian tour p of G', that is, a cyclic path that uses each edge in G' exactly once, is generated; a Eulerian tour can be found in $O(e)$, where e is the number of edges in the graph [Cormen et al., 2001]. Finally, p is converted into a Hamiltonian cycle in G by iteratively short-cutting subpaths of p (see Chapter 6 in Reinelt [1994] for an algorithm for this step). This last step, however, does not increase the weight of a tour if the given TSP instance satisfies the triangle inequality; hence, in this case, the final tour is at most twice as long as an optimal tour. However, empirically this construction heuristic performs rather poorly, with solution qualities that are on average around 40% above the optimal tour lengths for TSPLIB and RUE instances [Reinelt, 1994; Johnson and McGeoch, 2002].

Much better performance is obtained by the *Christofides Heuristic* [Christofides, 1976]. The central idea behind this heuristic is to compute a minimum weight perfect matching of the odd-degree vertices of the MST (there must be an even number of such vertices), which can be done in time $O(k^3)$, where k is the number of odd-degree vertices. (A minimum perfect matching of a vertex set is a set of edges such that each vertex is incident to exactly one of these edges; the weight of the matching is the sum of the weights of its edges.) This is sufficient for converting the MST into an Eulerian graph, that is, a graph containing an Eulerian tour. As described previously, in a final step, this Eulerian tour is converted into a Hamiltonian cycle. For TSP instances that satisfy the triangle inequality, the resulting tours are guaranteed to be at most a factor 1.5 above the optimum solution quality.

While the standard version of the Christofides Heuristic appears to perform worse than both, the Savings and Greedy Heuristics [Reinelt, 1994; Johnson and McGeoch, 2002], its performance can be substantially improved by additionally using greedy heuristics in the conversion of the Eulerian tour into a Hamiltonian cycle. The resulting variant of the Christofides Heuristic appears to be the best-performing construction heuristic for the TSP in terms of the solution quality achieved; however, its run-time is higher than that of the Savings Heuristic by a factor that increases with instance size from about 3.2 for RUE instances with

1 000 vertices to about 8 for RUE instances with 3.16 million vertices [Johnson et al., 2003a].

k-Exchange Iterative Improvement Methods

Most iterative improvement algorithms for the TSP are based on the k-exchange neighbourhood relation, in which candidate solutions s and s' are direct neighbours if, and only if, s' can be obtained from s by deleting a set of k edges and rewiring the resulting fragments into a complete tour by inserting a different set of k edges. For iterative improvement algorithms for the TSP that use a fixed k-exchange neighbourhood relation, $k = 2$ and $k = 3$ are the most common choices. Current knowledge suggests that the slight improvement in solution quality obtained by increasing k to four and beyond is not amortised by the substantial increase in computation time [Lin, 1965].

The most straightforward implementation of a k-exchange iterative improvement algorithm considers in each step all possible combinations for the k edges to be deleted and replaced. After deleting k edges from a given candidate solution s, the number of ways in which the resulting fragments can be reconnected into a candidate solution different from s depends on k; for $k = 2$, after deleting two edges (u_i, u_j) and (u_k, u_l), the only way to rewire the two partial tours into a different complete tour is by introducing the edges (u_i, u_k) and (u_l, u_j). Note that after a 2-exchange move, one of the two partial tours is reversed. (For an illustration, see Figure 1.6, page 44.)

For $k = 3$, there are several ways of reconnecting the three tour fragments obtained after deleting three edges, and in an iterative improvement algorithm based on this neighbourhood, all of these need to be checked for possible improvements. Figure 8.5 shows two of the four ways of completing a 3-exchange move after removing a given set of three edges. Furthermore, 2-exchange moves can be seen as special cases of 3-exchange moves in which the sets of edges deleted from and subsequently added to the given candidate tour have one element in common. Allowing an overlap between these two sets has the advantage that any tour that is locally optimal w.r.t. a k-exchange neighbourhood is also locally optimal w.r.t. to all k'-exchange neighbourhoods with $k' < k$.

Based on the 2-exchange and 3-exchange neighbourhood relations, various iterative improvement algorithms for the TSP can be defined in a straightforward way; these are generally known as 2-opt and 3-opt algorithms, because they produce tours that are locally optimal w.r.t. the 2-exchange and 3-exchange neighbourhoods, respectively. In particular, different pivoting rules can be used (these determine the mechanism used for selecting an improving neighbouring candidate solution; see also Chapter 2, Section 2.1). In general, first-improvement

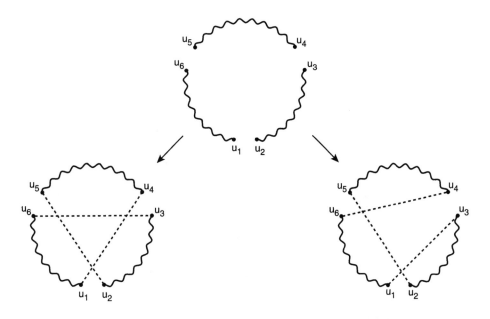

Figure 8.5 Two possible ways of reconnecting partial tours in a 3-exchange move after edges (u_1, u_2), (u_3, u_4) and (u_5, u_6) have been removed from a complete tour. Note that in the left result, the relative direction of all three tour fragments is preserved.

algorithms for the TSP can be implemented in such a way that the time complexity of each search step is substantially lower than for best-improvement algorithms. But even first-improvement 2-opt and 3-opt algorithms need to examine up to $O(n^2)$ and $O(n^3)$ neighbouring candidate solutions in each step, which leads to a significant amount of CPU time per search step when applied to TSP instances with several hundreds or thousands of vertices. Fortunately, there exists a number of speedup techniques that result in significant improvements in the time complexity of local search steps [Bentley, 1992; Johnson and McGeoch, 1997; Martin et al., 1991; Reinelt, 1994].

Fixed Radius Search

For any improving 2-exchange move from a tour s to a neighbouring tour s', there is at least one vertex that is incident to an edge e in s that is replaced by a different edge e' with lower weight than e. This observation can be exploited for speeding up the search for an improving 2-exchange move from a given tour s. For a vertex u_i, two searches are performed that consider each of the two tour neighbours of u_i as a vertex u_j, respectively. For a given u_j, a search around u_i is performed for vertices u_k that are closer to u_i than $w((u_i, u_j))$, the radius of the search. For each vertex u_k found in this *fixed radius near neighbour* search, removing one of its two incident edges in s leads to a feasible 2-exchange move. The first such 2-exchange move that results in an improvement in solution quality is applied to s, and the iterative improvement search is continued from the resulting tour s' by performing a fixed radius near neighbour search for another vertex. If fixed radius near neighbour searches for all vertices do not result in any improving 2-exchange move, the current tour is 2-optimal.

The idea of fixed radius search can be extended to 3-opt [Bentley, 1992]. In this case, for each search step, two fixed radius near neighbour searches are required, one for a vertex u_i as in the case of 2-opt (see above), resulting in a vertex u_k, and the other for the tour neighbour u_l of u_k with radius $w((u_i, u_j)) + w((u_k, u_l)) - w((u_i, u_k))$.

Candidate Lists

In the context of identifying candidates for k-exchange moves, it is useful to be able to efficiently access the vertices in the given graph G that are connected to a given vertex u_i by edges with low weight, for example, in the form of a list of neighbouring vertices u_k that is sorted according to edge weight $w((u_i, u_k))$ in ascending order. By using such *candidate lists* for all vertices in G, fixed radius near neighbour searches can be performed very efficiently; this is illustrated by the empirical results reported in Example 8.1 on page 376*f.*. Interestingly, the use of candidate lists within iterative first-improvement algorithms, such as 2-opt, often leads to improvements in the quality of the local optima found by these algorithms. This suggests that the highly localised search steps that are evaluated first when using candidate lists are more effective than other k-exchange steps.

Full candidate lists comprising all $n-1$ other vertices require $O(n^2)$ memory and take $O(n^2 \log n)$ time to construct. Therefore, especially to reduce memory requirements, it is often preferable to use bounded-length candidate lists; in this case, a fixed radius near neighbour search for a given vertex u_i is aborted when the candidate list for u_i has been completely examined, if the radius criterion did not stop the search earlier. As a consequence, the tours obtained from an iterative improvement algorithm based on this mechanism are no longer guaranteed to be locally optimal, because some improving moves may be missed.

Typically, candidate lists of length 10 to 40 are used, although shorter lengths are sometimes chosen. Simply using short candidate lists that consist of the vertices connected by the k lowest weight edges incident to a given vertex can be problematic, especially for clustered instances like those shown on the right side of Figure 8.1 (page 360). For metric TSP instances, alternative approaches to constructing bounded-length candidate lists include so-called quadrant-nearest neighbour lists [Pekny and Miller, 1994; Johnson and McGeoch, 1997] and candidate lists based on Delaunay triangulations [Reinelt, 1994].

Helsgaun proposed a more complex mechanism for constructing candidate lists that is based on an approximation to the Held-Karp lower bounds (see Section 8.1) [Helsgaun, 2000]. This mechanism works as follows: Based on the modified edge weights $w'((u_i, u_j))$ obtained from an approximation to the Held-Karp lower bounds, so-called α-values are computed for each edge (u_i, u_j) as $\alpha((u_i, u_j)) := w'(t^+(u_i, u_j)) - w'(t)$, where $w'(t)$ is the weight of a minimum weight one-tree t and $w'(t^+(u_i, u_j))$ is the weight of a minimum weight one-tree $t^+(u_i, u_j)$ that is forced to contain edge (u_i, u_j). For each edge, $\alpha(u_i, u_j) \geq 0$, and $\alpha(u_i, u_j) = 0$ if the edge (u_i, u_j) is contained in some minimum weight one-tree. A candidate list for a vertex u_i can now be obtained by sorting the edges incident to u_i according to their α-values in ascending order and bounding the length of the list to a fixed value k or by accepting only edges with α-values that are below some given threshold. The vertices contained in these candidate lists are called α-*nearest neighbours*.

Empirically it was shown that compared to the candidate lists obtained by the other methods mentioned above, candidate lists based on α-values can be much smaller and still cover all edges contained in an optimal solution. For example, for TSPLIB instance att532, candidate lists consisting of 5 α-nearest neighbours cover an optimal solution, while list length 22 is required when using standard candidate lists based on the given edge weights [Helsgaun, 2000].

Don't Look Bits

Another widely used mechanism for speeding up iterative improvement search for the TSP is based on the following observation. If in a given search step, no improving k-exchange move can be found for a given vertex u_i (e.g., in a fixed radius near neighbour search), it is unlikely that an improving move involving u_i will be found in future search steps, unless at least one of the edges incident to u_i in the current tour has changed.

This can be exploited for speeding up the search process by associating a single *don't look bit (DLB)* with each vertex; at the start of the iterative improvement search, all DLBs are turned off (i.e., set to zero). If in a search step no improving move can be found for a given vertex, the respective DLB is turned on (i.e., set to one). After each local search step, the DLBs of all vertices incident to edges that were modified (i.e., deleted from or added to the current tour) in this step are turned off again. The search for improving moves is started only at vertices whose DLB is turned off. In practice, the DLB mechanism significantly reduces the time complexity of first-improvement search, since after a few neighbourhood scans, most of the DLBs will be turned on. The speedup that can be achieved by using DLBs is illustrated by the empirical results for various variants of 2-opt shown in Example 8.1.

The DLB mechanism can be easily integrated into more complex SLS methods, such as Iterated Local Search or Memetic Algorithms. One possibility is to set only the DLBs of those vertices to zero that are incident to edges that were deleted by the application of a tour perturbation or a recombination operator; this approach is followed in various algorithms described in Sections 8.3 and 8.4 and typically leads to a further substantial reduction of computation time when compared to resetting all DLBs to zero. Furthermore, DLBs can be used to speed up first-improvement local search algorithms for combinatorial problems other than TSP.

EXAMPLE 8.1 **Effects of Speedup Techniques for 2-opt**

To illustrate the effectiveness of the previously discussed speedup techniques, we empirically evaluated three variants of 2-opt: a straight-forward implementation that in each search step evaluates every possible 2-exchange move (2-opt-std); a fixed radius near neighbour search that uses candidate lists of unbounded length (2-opt-fr+cl); and a fixed radius near neighbour search that uses candidate lists of unbounded length as well as DLBs (2-opt-fr+cl+dlb). For all variants, the search process was initialised at a random permutation of the vertices, and it was terminated as soon as a local minimum was encountered. These algorithms were run 1 000 times on several benchmark instances from TSPLIB using an Athlon 1.2 GHz MP CPU with 1 GB RAM running Suse Linux 7.3. (The 2-opt implementation used for these experiments is available from `www.sls-book.net`.)

The results reported in Table 8.1 show that the speedup techniques achieve substantial decreases in run-time over a standard implementation, an effect that increases strongly with instance size. The most significant speedup seems to be due to the combination of fixed-radius nearest neighbour search with candidate lists, while the additional use of DLBs can sometimes reduce the computation times by another factor of two. In addition, the bias in the local search towards first examining the most promising moves that is introduced by the use of candidate lists results in a significant improvement in the solution quality obtained by 2-opt; the use of DLBs diminishes this effect only slightly. When bounded length candidate lists are used, very

Instance	2-opt-std		2-opt-fr + cl		2-opt-fr + cl + dlb		3-opt-fr + cl	
	Δ_{avg}	t_{avg}	Δ_{avg}	t_{avg}	Δ_{avg}	t_{avg}	Δ_{avg}	t_{avg}
rat783	13.0	93.2	3.9	3.9	8.0	3.3	3.7	34.6
pcb1173	14.5	250.2	8.5	10.8	9.3	7.1	4.6	66.5
d1291	16.8	315.6	10.1	13.0	11.1	7.4	4.9	76.4
fl1577	13.6	528.2	7.9	21.1	9.0	11.1	22.4	93.4
pr2392	15.0	1 421.2	8.8	47.9	10.1	24.9	4.5	188.7
pcb3038	14.7	3 862.4	8.2	73.0	9.4	40.2	4.4	277.7
fnl4461	12.9	19 175.0	6.9	162.2	8.0	87.4	3.7	811.6
pla7397	13.6	80 682.0	7.1	406.7	8.6	194.8	6.0	2 260.6
rl11849	16.2	360 386.0	8.0	1 544.1	9.9	606.6	4.6	8 628.6
usa13509	—		7.4	1 560.1	9.0	787.6	4.4	7 807.5

Table 8.1 Computational results for different variants of 2-opt and 3-opt. Δ_{avg} denotes the average percentage deviation from the optimal solution quality over 1 000 runs per instance, and t_{avg} is the average run-time for 1 000 runs of the respective algorithm, measured in CPU milliseconds on an Athlon 1.2 GHz CPU with 1GB of RAM. (For further details, see text.)

similar results were obtained for most instances (not shown here); only on the pathologically clustered instance fl1577, the solution quality decreases to an average of almost 60% above the optimum, while the computation time is reduced by about 10% (these observations were made for a length bound of 40).

3-opt achieves better-quality solutions than the previously mentioned 2-opt variants at the cost of substantially higher computation times; this is illustrated by the results for 3-opt with a fixed-radius search using candidate lists of a length limited to 40 shown in the last column of Table 8.1. Interestingly, using unbounded candidate lists for 3-opt leads to computation times that can be substantially higher than the ones reported in Table 8.1. This illustrates that bounding the length of candidate lists becomes increasingly important in the context of local search algorithms based on larger neighbourhoods.

The Lin-Kernighan (LK) Algorithm

Empirical evidence suggests that iterative improvement algorithms based on k-exchange neighbourhoods with $k > 3$ return better tours, but the computation times required for searching these large neighbourhoods render this approach ineffective. Variable depth search algorithms overcome this problem by partially exploring larger neighbourhoods (see also Chapter 2, page 67*ff.*).

The best-known variable depth search method for the TSP is the Lin-Kernighan (LK) Algorithm that was described from a high level perspective in Chapter 2 (page 68*ff.*); it is an iterative improvement method that uses complex search steps obtained by iteratively concatenating a variable number of elementary 1-exchange moves. In each complex step, which we also call an LK step, a set of edges $X := \{x_1, \ldots, x_r\}$ is deleted from the current tour, and another set of edges $Y := \{y_1, \ldots, y_r\}$ is added to it. The number of edges that are exchanged, r, is determined dynamically and can vary for each complex search step. (This is explained in more detail below.) For an overview of the general idea underlying the construction of the LK steps, we refer to Figure 2.4 (page 69) and the text description given there.

The two sets X and Y are constructed iteratively, element by element, such that edges x_i and y_i as well as y_i and x_{i+1} must share an endpoint, respectively; a complex step that satisfies this criterion is called *sequential*. Based on this criterion, the edges in X and Y can be represented as $x_i = (u_{2i-1}, u_{2i})$ and $y_i = (u_{2i}, u_{2i+1})$, respectively. Furthermore, at any point during the iterative construction of a complex step, that is, for any $X = \{x_1, \ldots, x_i\}$ and $Y = \{y_1, \ldots, y_i\}$, there needs to be an alternative edge y_i' such that the complex step defined by $X := \{x_1, \ldots, x_i\}$ and $Y' := \{y_1, \ldots, y_i'\}$ applied to the current tour yields a valid tour, (i.e., a Hamiltonian cycle in the given graph G); there is only one exception to this rule for the case $i = 2$, which is treated in a special way [Lin and Kernighan, 1973].

The Lin-Kernighan Algorithm initialises the search process at a randomly chosen Hamiltonian cycle (i.e., a vertex permutation) of the given graph G. The search for each improving (complex) LK step starts with selecting a vertex u_1; next, an edge $x_1 := (u_1, u_2)$ is selected for removal, then an edge $y_1 := (u_2, u_3)$ is chosen to be added, etc. At each stage of this construction process, the length $w(p_i)$ of the tour p_i obtained by applying the constructive search step determined by $X := \{x_1, \ldots, x_i\}$ and $Y' := \{y_1, \ldots, y_i'\}$ (as defined above) is computed as well as $g_i := \sum_{j=1}^{i} w(y_j) - w(x_j)$, the total gain for $X = \{x_1, \ldots, x_i\}$ and $Y = \{y_1, \ldots, y_i\}$. The construction process is terminated whenever the total gain g_i is smaller than $w(p) - w(p_{i^*})$, where p is the current tour and p_{i^*} is the best tour encountered during the construction, that is, $i^* := \operatorname{argmin}_i w(p_i)$. At this point, if the complex step corresponding to $X := \{x_1, \ldots, x_{i^*}\}$ and $Y' := \{y_1, \ldots, y_{i^*}'\}$ leads to an improvement in solution quality, this step is executed and p_{i^*} becomes the current tour.

To bound the length of the search for an improving complex step, in the original Lin-Kernighan Algorithm the sets X and Y are required to be disjoint; this means that an edge that has been removed cannot be added back later in the same complex search step and vice versa. A limited amount of backtracking is allowed if a sequence of elementary moves does not yield an improved tour. In LK, backtracking is triggered when during the construction of an LK step no improving one has been found; backtracking is applied only at the first two levels, that is, for the choices of x_1, y_1, x_2 and y_2. During backtracking, alternatives for edge y_2 are considered in order of increasing (or equal) weight $w(y_2)$. If this is unsuccessful, the alternative choice for x_2 is considered; since this leads to a temporary violation of the sequentiality criterion (see above), it needs to be handled in a special way. If none of these alternatives for x_2 and y_2 can be extended into an improving complex step, backtracking is applied to the choice of y_1. When all alternatives for y_1 are exhausted without finding an improving complex step, the other edge incident to the starting vertex u_1 is considered as a choice for x_1. Only after all these attempts at finding an improving step by a search centred at vertex u_1 have failed, is an alternate choice for u_1 considered. This backtracking mechanism ensures that all 2- and 3-exchange moves are checked when searching for improving search steps; consequently, the tours obtained by LK (when run to completion) are locally optimal w.r.t. to the 2- and 3-exchange neighbourhoods.

In addition to the complex LK steps, Lin and Kernighan also proposed to consider some specially structured, non-sequential 4-exchange moves as candidates for improving search steps. (An example of a non-sequential 4-exchange move is the double-bridge move illustrated in Figure 2.11 on page 88.) However, Lin and Kernighan noted that the improvement obtained by the additional check of these moves depends strongly on the given TSP instance.

The Lin-Kernighan Algorithm uses several techniques for pruning the search. Firstly, the search for edges (v, v') to be added to Y is limited to the five shortest edges incident to vertex v. Secondly, for $k \geq 4$, no edge in the current tour can be removed if it was contained in a collection of previously found high-quality tours. Furthermore, several mechanisms are provided for guiding the search. These include a rule that the edges to be added to Y are chosen such that $w(x_{i+1}) - w(y_i)$ is maximised (a limited form of look-ahead) and a preference for the longer of two alternative edges in the context of choosing edge x_4, one of the edges that is removed from the current tour.

Lin and Kernighan applied LK to various TSP instances ranging from 20 to 110 vertices. For all of these instances, LK found optimal solutions; however, the success probability (i.e., the probability that one run of LK finds an optimal solution) dropped from 1 for small instances with about 20 cities to approximately 0.25 for instances with about 100 cities.

Variants of the LK Algorithm

The details of the original LK Algorithm can be varied in many ways, and the design choices made by Lin and Kernighan do not necessarily lead to optimal performance. These design choices include the depth and the width of backtracking, the rules used for guiding the search (e.g., look-ahead), the use of a bound on the length of complex LK steps and the choice of 2-exchange moves as elementary search steps. Additional room for variation exists w.r.t. algorithmic details that are not specific to LK, such as the type and length of neighbourhood list or the search initialisation procedure.

Some alternatives for these design choices are realised in the four well-known LK variants by Johnson and McGeoch [1997; 2002], Applegate et al. [1999], Neto [1999] and Helsgaun [2000]. For a detailed discussion of these LK algorithms and their performance we refer to the original papers.

A particularly noteworthy LK variant is *Helsgaun's LK (LK-H)*, which differs from the original Lin-Kernighan Algorithm in several key features and typically performs substantially better. In LK-H, the complex moves correspond to sequences of sequential 5-exchange moves; these are iteratively built using candidate lists based on α-values (see page 375). If at any point during the construction of a complex step a tour improvement can be achieved, the corresponding search step is executed immediately. In some sense this corresponds to a first-improvement search mechanism within the construction sequence for a single complex search step, while the original LK algorithm uses a best-improvement strategy in this context. Finally, LK-H uses only backtracking on the choice of x_1, the first edge to be removed from the current tour.

EXAMPLE 8.2 **Performance of LK Algorithms**

In this example, we illustrate the performance obtained by current LK algorithms on a number of TSPLIB instances. On the left side of Figure 8.6, we show the asymptotic solution quality distributions (asymptotic SQDs, for details see Section 4.2, page 162*ff.*) for two prominent, publically available LK implementations, the LK-H algorithm (version 1.2) and the LK algorithm of Applegate, Bixby, Chvatál and Cook (LK-ABCC, 99.12.15 release). In addition, we present the asymptotic SQD for the 3-opt algorithm used for generating the results from Table 8.1 (page 377). Each asymptotic SQD is based on 1 000 runs starting from random initial tours.

On the right side of Figure 8.6 we show the distribution of the computation times required for reaching a local optimum. Additional summary results for the two LK variants applied to the TSPLIB instances from Example 8.1 are shown in Table 8.2 together with the results for the 3-opt algorithm

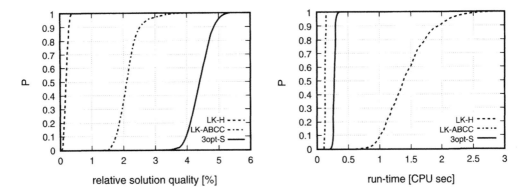

Figure 8.6 *Left side:* Asymptotic solution quality distributions of LK-H, LK-ABCC and the 3-opt algorithm from Example 8.1 on TSPLIB instance pcb3038. *Right side:* Distribution of the computation times to run a local search from a random initial solution for the three algorithms on instance pcb3038.

Instance	LK-ABCC		LK-H		3-opt-fr + cl	
	Δ_{avg}	t_{avg}	Δ_{avg}	t_{avg}	Δ_{avg}	t_{avg}
rat783	1.85	21.0	0.04	61.8	3.7	34.6
pcb1173	2.25	45.3	0.24	238.3	4.6	66.5
d1291	5.11	63.0	0.62	444.4	4.9	76.4
fl1577	9.95	114.1	5.30	1 513.6	22.4	93.4
pr2392	2.39	84.9	0.19	1 080.7	4.5	188.7
pcb3038	2.14	134.3	0.19	1 437.9	4.4	277.7
fnl4461	1.74	239.3	0.09	1 442.2	3.7	811.6
pla7397	4.05	625.8	0.40	8 468.6	6.0	2 260.6
rl11849	6.00	1 072.3	0.38	9 681.9	4.6	8 628.6
usa13509	3.23	1 299.5	0.19	13 041.9	4.4	7 807.5

Table 8.2 Computational results for LK-ABCC and LK-H and the 3-opt algorithm from Example 8.1 on various TSPLIB instances. Δ_{avg} denotes the average percentage deviation from the optimum solution quality and t_{avg} is the average run-time over 1 000 runs of the respective algorithm, measured in CPU milliseconds on an Athlon 1.2 GHz CPU with 1GB of RAM. (For further details, see text.)

from Example 8.1. (All experiments were performed on an Athlon 1.2 GHz MP CPU with 1 GB RAM running Suse Linux 7.3.)

As can be observed from these results, LK-H finds significantly better tours than LK-ABCC and 3-opt. However, it does so at the cost of significantly higher computation times. In fact, the run-times shown in Figure 8.6 and Table 8.2 do not include the preprocessing times required by LK-H for generating the candidate sets (LK-H uses the α-nearest neighbours described on page 375, which are time-intensive to compute). In the case of instance pcb3038, this preprocessing requires 4.41 CPU seconds and it increases to about 131 CPU seconds for instance usa13509. The preprocessing times for the two other algorithms are much smaller; for example, LK-ABCC requires a preprocessing time of 0.53 CPU seconds for instance usa13509. This example shows that different LK algorithms can vary substantially w.r.t. their performance, as measured by solution quality or computation time.

IN-DEPTH Efficiently Implementing SLS Algorithms for the TSP

In order to obtain the performance results reported in most empirical studies on SLS algorithms for the TSP, efficient implementations are required that make use of fairly sophisticated data structures. This is particularly true for state-of-the-art LK variants. In general, data structures used within SLS algorithms for the TSP need to support the following operations:

   (i) determine where a given vertex is located within a tour;

   (ii) determine the successor and the predecessor of a vertex within a given tour;

   (iii) check whether a vertex u_k is visited between vertices u_i and u_j for a given tour and orientation; and

   (iv) execute a k-exchange move, which includes swaps and inversions of tour segments.

For TSP instances up to around 1 000 vertices, the standard array representation for tours appears to be most efficient [Fredman et al., 1995]. In this representation, a cyclic path $(u_{\phi(1)}, \ldots, u_{\phi(n)}, u_{\phi(1)})$ in the given graph G is stored in two arrays, which hold the permutation of vertex indices $\phi := (\phi(1), \ldots, \phi(n))$ and its inverse, $\psi := (\psi(1), \ldots, \psi(n))$, where $\psi(i)$ is the position of vertex index i in ϕ. Clearly, the predecessor, successor and 'between' queries can be answered in constant time. The time-complexity of the move operation, however, has been empirically determined as $O(n^{0.7})$ [Bentley, 1992]; this operation is therefore a bottleneck for large instances, and more advanced data structures are required to reduce its time complexity.

    One widely used alternative to the array representation is based on two-level trees [Chrobak et al., 1990; Fredman et al., 1995]. In this representation, a tour is divided into roughly $\sqrt{n}$ segments of length between $1/2 \cdot \sqrt{n}$ and $2 \cdot \sqrt{n}$ each; these segments are represented by vertices at the first level of a tree whose root corresponds to the entire tour, while the leaves are the vertices of G. (For details on the implementation of this data structure we refer to Fredman et al. [1995] and Applegate et al. [1999].) When using two-level trees, the successor and predecessor of a vertex can be determined in constant time and the same holds for answering 'between' queries; the respective constants, however, are slightly larger than for the array representation. The worst-case complexity of the move operation, on the other hand, is only $O(\sqrt{n})$. Based on extensive computational experiments, Fredman et al. [1995] recommend the use of the two-level tree representation when solving TSP instances with up to around one million vertices. For larger instances, they recommend to use a tour representation based on splay-trees [Sleator and Tarjan, 1985], which allows each operation to be performed in $O(\log n)$ in the worst case.

It should be noted that LK algorithms are not easy to implement efficiently. Neto estimated that the development of a high-performance LK implementation that uses most of the techniques described here requires around eight man-months [Neto, 1999]; this estimate has been confirmed by other researchers [Merz, 2002]. Fortunately, at least three very efficient implementations of LK variants are publicly available. These are the LK implementation by Applegate, Bixby, Chvátal, and Cook, which is a part of the Concorde library [Applegate et al., 2003b], Helsgaun's LK variant [Helsgaun, 2003] and Neto's LK implementation [Neto, 2003].

Local Search for the Asymmetric TSP

ATSP algorithms are generally much less studied than algorithms for symmetric TSP instances; in particular, this is true for construction heuristics as well as for 'simple' local search methods. Although most construction heuristics are directly applicable to the ATSP, few computational results are available [Johnson et al., 2002]. Empirical results show that for the ATSP, in contrast to what is observed on symmetric TSP instances, the Nearest Neighbour Heuristic typically performs much better than the Greedy Heuristic. However, for many classes of ATSP instances, even better results (in terms of solution quality) are obtained by construction heuristics that are based on the assignment problem lower bound for the ATSP. These constructive search methods generate a tour by iteratively merging a set of vertex-disjoint simple directed cycles that forms a minimum cost vertex cover of the given graph G; such sets are obtained as a side product of the computation of the assignment problem lower bound. An early heuristic for the merging step has been developed by Karp [1979]; a variant of this approach was recently proposed by Glover et al. [2001].

When applying iterative improvement algorithms to the ATSP, a slight complication arises from the fact that sub-tour reversals lead to changes in solution quality. While 2-exchange moves always involve sub-tour reversals, there is a specific 3-exchange move that preserves the direction of all partial tours. The iterative improvement methods based on this type of move are called reduced 3-opt algorithms; these are amongst the simplest iterative improvement algorithms for the ATSP. The speedup techniques described above for symmetric TSP algorithms can be directly applied to reduced 3-opt.

A variable depth search algorithm for the ATSP has been developed by Kanellakis and Papadimitriou [1980]. This KP algorithm can be seen as an adaptation of the Lin-Kernighan Algorithm to the ATSP; it makes use of double-bridge moves, a special type of non-sequential 4-exchange moves. An implementation of this method by Cirasella et al. [2001] has been shown to yield significantly better solution qualities than reduced 3-opt, albeit at the cost of substantially increased run-times [Johnson et al., 2002].

8.3 Iterated Local Search Algorithms for the TSP

Iterated Local Search (ILS), as introduced in Chapter 2, Section 2.3, offers a straight-forward, yet flexible way of extending simple local search algorithms (see also the algorithm outline on page 86 and the GLSM model on page 136). ILS applications for the TSP have a long history, and some of the hybrid SLS algorithms thus obtained are amongst the best-performing TSP algorithms currently known.

Iterated Descent

Historically, *Iterated Descent* by Baum [1986a; 1986b] was the first ILS method for the TSP. Variants of Iterated Descent use different first-improvement methods as their subsidiary local search procedure, including 2-opt, a limited form of 3-opt that examines only a part of the 3-exchange neighbourhood, and a first-improvement algorithm that is based on a 2-exchange neighbourhood on vertices under which two candidate solutions are direct neighbours if, and only if, the corresponding vertex permutations differ in exactly two positions. The perturbation phase of Iterated Descent consists of a random 2-exchange step, and its acceptance criterion always selects the candidate solution with the better solution quality.

Although Iterated Descent performs better than pure 2-opt and 3-opt local search, from today's perspective, its performance is not impressive. Most likely, the most substantial weakness of Iterated Descent is its perturbation mechanism. It is now also known that the 2-opt and 3-opt local search procedures used in Iterated Descent perform poorly on the RDM instances used in Baum's empirical evaluation.

Large-Step Markov Chains (LSMC)

The *Large-Step Markov Chains (LSMC)* algorithm by Martin, Otto and Felten [1991; 1992] is the first high-performance ILS algorithm for the TSP. The name of this approach reflects the fact that the behaviour of LSMC (like that of many other ILS algorithms) can be modelled as a Markov chain (see, e.g., Papoulis [1991]) on the locally minimal candidate solutions obtained at the end of each local search phase, where the segment of the search trajectory between any two such subsequent local minima corresponds to a 'large step'.

One important contribution of the LSMC approach is the exploitation of a particular 4-exchange step, the so-called *double-bridge move*. A double-bridge move first removes four edges from the tour, resulting in a decomposition into four segments A, B, C, D. Then, these segments are reconnected in the order A, D, C, B by adding four new edges (for a graphical illustration, see Figure 2.11 on page 88). The double-bridge move was originally introduced by Lin and Kernighan [1973] in their LK algorithm, where this type of search step is applied within the iterative improvement process. However, the double-bridge move is typically not used in current variants of LK, with the notable exception of LK-H. In LSMC, a random double-bridge move is used to perturb a locally optimal tour. LSMC considers only double-bridge moves for which the combined weight of the four new edges is lower than a constant k times the average edge weight in the current locally optimal candidate solution. Originally, a value of $k := 10$ was used; but experimental results suggest that the performance of the algorithm is not very sensitive to the value of k, as long as it is not too small [Martin et al., 1992].

As its subsidiary local search procedure, the first LSMC algorithm initially used a 3-opt first-improvement search, which was later replaced by a more power-ful LK algorithm. The local search procedure exploits three speed-up techniques, namely: (i) a type of fixed radius search that uses the minimum and the maximum weight edge in the current candidate solution for pruning the search for improv-ing 3-exchange steps; (ii) a so-called *change list*, a concept that is equivalent to the use of don't look bits; and (iii) a hash table for storing 3-opt candidate solutions, which is consulted for checking whether a tour has been previously identified to be locally optimal.

The acceptance criterion in LSMC is taken from Simulated Annealing: Given two candidate tours s and s'', where s'' has been obtained from s by perturbation and subsequent local search, s'' is always accepted if its solution quality is better than that of s; otherwise, s'' is accepted with probability $\exp(f(s) - f(s''))/T)$, where T is a parameter called temperature. However, in a later variant an accep-tance criterion was used that always accepts the better of the two candidate solu-tions. The resulting zero-temperature LSMC algorithm is also known as *Chained Local Optimisation* (CLO) [Martin and Otto, 1996].

LSMC with a subsidiary 3-opt local search procedure has been shown to solve small random Euclidean TSP instances with up to 200 cities in less than one CPU hour on a SUN SPARC 1 workstation (a very slow computer compared to cur-rent PCs). Relatively good performance was also observed on several TSPLIB instances. For example, LSMC found an optimal solution of instance lin318 in about four CPU hours on the SUN SPARC 1; by using a LK algorithm as the subsidiary local search procedure, the time required for solving this instance optimally was reduced by a factor of about four. In this particular case, however, it was shown to be essential to use non-zero temperatures in the LSMC acceptance criterion. LSMC with the LK subsidiary local search procedure also found optimal solutions to several larger TSPLIB instances, including `att532` and `rat783` [Martin et al., 1991].

Iterated Lin-Kernighan

An early variation on LSMC is the *Iterated Lin-Kernighan (ILK)* algorithm developed by Johnson and McGeoch [Johnson, 1990; Johnson and McGeoch, 1997]. There are several key differences between the LSMC algorithm and Johnson's ILK. Firstly, the acceptance criterion used in ILK always selects the better of the two locally optimal candidate solutions. Secondly, the perturbation phase does not make use of the limiting condition on the edge weight of a double-bridge move imposed in the LSMC approach, but it applies random double-bridge moves instead (i.e., the four cut-points are chosen uniformly at random). Thirdly, the local search is initialised with a randomised version of the Greedy Heuristic; in the randomised version, instead of deterministically selecting the minimum weight feasible edge, between the two shortest edges the one with less weight is chosen with a probability of $2/3$, and the other one is selected in the remaining cases. Finally, ILK uses a substantially more efficient implementation of LK, and consequently ILK performs better than LSMC, especially when considering computation time. Early results for ILK (as of 1990) were quite promising: Applied to TSPLIB instances with 318 to 2392 vertices, optimal solutions were obtained (for the 2392 vertex instance, this required about 55 CPU hours on a Sequent computer [Johnson, 1990], an extremely slow machine compared to current PCs).

The ILK algorithm was further fine-tuned and extensively tested for an overview article on the state-of-the-art in incomplete TSP algorithms [Johnson and McGeoch, 1997]. The main differences between the 1997 variant and the earlier ILK algorithm appear to be the exploitation of don't look bits after the double-bridge move and the use of a bound on the depth of the LK search. This 'production-mode ILK' was shown to achieve optimal or close to optimal solutions on a variety of random Euclidean instances and TSPLIB instances. Difficulties in finding solutions within less than one percent of the optimum solution quality were only reported on TSPLIB instance `fl3795`, which shows a pathological clustering similar to instance `fl1577` depicted in Figure 8.1 on page 360. Running times were rather modest; for example, on 10 000 vertex RUE instances, n iterations of ILK took approximately 1 570 seconds on a SGI

Challenge 196 MHz machine [Johnson and McGeoch, 1997]. The ILK algorithm was for a considerable time the state-of-the-art SLS algorithm for the TSP.

Chained Lin-Kernighan

Like ILK, the *Chained Lin-Kernighan algorithm (CLK-ABCC)* by Applegate et al. [1999] uses the LK algorithm as its subsidiary local search procedure. CLK-ABCC differs from ILK in various details of the LK local search, including its use of smaller candidate sets (by default it uses quadrant nearest neighbour sets of size 12); it also uses a different perturbation mechanism that affects only a locally restricted part of a candidate solution, and initialises the search using the Quick-Borůvka construction heuristic (see Section 8.2). For further details on the LK local search procedure used in CLK-ABCC, we refer to Applegate et al. [1999] and to the original CLK-ABCC code, which is available from the web page of Applegate et al. [2003b]. In the following, we focus on some of the other algorithmic features of CLK-ABCC, particularly the perturbation mechanism.

The standard CLK-ABCC algorithm uses so-called *geometric double-bridge moves* as perturbation steps; these are based on the following method for selecting the four edges to be removed from the current candidate tour s. For convenience, a direction is imposed on s and we denote the arc between vertex u and its direct tour successor $succ(u)$ as $(u, succ(u))$. In a first step, a set U of $\min\{0.001 \cdot n, 10\}$ vertices are randomly sampled from the given graph $G := (V, E, w)$. Then, among the edges $(u, succ(u))$ contained in s with $u \in U$, the one with the maximal difference $w((u, succ(u))) - w((u, u^*))$, where u^* is the nearest neighbour of u, — that is, the vertex in V that minimises $w((u, u^*))$ — is removed from s. In a second step, three vertices are chosen uniformly at random from the k nearest neighbours of vertex u. Let u_1, u_2 and u_3 denote these vertices; then, the edges $(u_i, succ(u_i))$, $i = 1, 2, 3$ are removed. The four edges chosen in this process determine the double-bridge move used for perturbation. The value of k controls the locality of the perturbation. For small k, a geometric double-bridge move results in a localised perturbation that only affects edges close to one specific vertex, while for large k, less localised perturbations are obtained.

An interesting detail of the CLK-ABCC algorithm is the resetting strategy for the don't look bits (DLBs) that is applied after a perturbation. Many ILS algorithms for the TSP, including LSMC and ILK, reset only the DLBs of vertices that are incident to edges changed by the perturbation; then, only these vertices are considered as starting points for the search for an improving move. CLK-ABCC additionally resets the DLBs of all vertices that are at most ten edges away from the end-points of the modified edges in the current tour, as well as the DLBs of the vertices in the neighbour sets of these endpoints. Experimental results suggest that among several alternative DLB resetting strategies, this mechanism leads to the best overall performance of CLK-ABCC [Applegate et al., 1999]. (Similar observations have been made independently by Stützle [1998c] in the context of Iterated 2-opt and 3-opt for the TSP.)

The Chained Lin-Kernighan algorithm by Applegate, Cook and Rohe (CLK-ACR) is a variant of CLK-ABCC that differs from this earlier algorithm in two main aspects [Applegate et al., 2003c]. Firstly, CLK-ACR uses a different mechanism for selecting the double-bridge move in the perturbation step. The first edge to be deleted, $(u, succ(u))$, is chosen as described for CLK-ABCC. The three other edges to be removed are of the form $(u_i, succ(u_i))$, $i = 1, 2, 3$, where each vertex u_i is obtained as the endpoint of a random walk of length l in the neighbourhood graph, starting from vertex u. The neighbourhood graph is defined as (V, E'), where V is the set of vertices of the given instance, and E' is the set of all edges of the form (u, v) such that v is in the candidate list of u in the LK heuristic. (As in CLK-ABCC, by default, the candidate lists consist of the 12 quadrant nearest neighbours.) The locality of the perturbation can be controlled by varying the length l of the random walk. It was found that for long runs, better results are obtained when using higher values of l [Applegate et al., 2003c]. Secondly, CLK-ACR is based on a subsidiary LK search procedure that differs from the one used in CLK-ABCC w.r.t. the depth and the breadth of the backtracking mechanism.

A performance comparison of CLK-ABCC and CLK-ACR in the context of the 8th DIMACS Implementation Challenge on the TSP revealed that CLK-ACR is slightly superior on the largest instances tested (see Johnson and McGeoch [2002] as well as the DIMACS challenge web pages by Johnson et al. [2003a]). It is somewhat unclear how the performance of those algorithms compares to that of ILK. When running both algorithms for n iterations, ILK finds better-quality tours than CLK-ACR or CLK-ABCC on most instances, however the run-times required by ILK are several times larger than those of CLK-ACR and CLK-ABCC. (The difference corresponds to a factor of two to five for RUE instances, and is substantially larger for TSPLIB and RCE instances.)

The CLK-ACR code is capable of handling extremely large TSP instances and has been applied to instances up to 25 million vertices, where it reached a solution quality within 1% of the estimated optimum in 24 CPU hours on an IBM RS6000 Model 43-P 260 workstation with 4 GB RAM; given a total run-time of 8 CPU days, a solution quality within 0.3% from the estimated optimum was obtained.

Iterated Helsgaun (ILK-H)

Given the excellent performance of LK-H, Helsgaun's variant of the LK algorithm (see also Section 8.2), using this local search procedure as the core of an ILS

procedure *constructionILK-H*$(G, \hat{s})$
    input: *weighted graph* $G := (V, E, w)$, *incumbent candidate solution* $\hat{s}$
    output: *candidate solution* $s \in S(\pi')$
    $p :=$ *empty tour*;
    $u_i :=$ *selectVertexRandomly*(V);
        append vertex u_i *to partial tour* p;
    while p *is not a complete tour* **do**
        $C := \{u_j \mid (u_i, u_j)$ *is a candidate edge* $\wedge \, \alpha((u_i, u_j)) = 0 \wedge (u_i, u_j) \in \hat{s}\}$;
        if $C = \emptyset$ **then**
            $C := \{u_j \mid (u_i, u_j)$ *is a candidate edge*$\}$;
        end
        if $C = \emptyset$ **then**
            $C := \{u_j \mid u_j$ *not chosen yet*$\}$;
        end
        $u_j :=$ *choosePseudoRandomVertex*(C);
            append vertex u_j *to partial tour* p;
        $u_i := u_j$;
    end
    return p
end *constructionILK-H*

Figure 8.7 The construction procedure used in the perturbation phase of the ILK-H algorithm. (For details, see text.)

algorithm is a fairly obvious idea. This leads to the *Iterated Helsgaun Algorithm (ILK-H)* [Helsgaun, 2000], which is one of the best-performing SLS algorithms for the TSP currently known in terms of the solution quality reached [Helsgaun, 2000; Johnson and McGeoch, 2002].

Since the subsidiary local search procedure of LK-H may use double-bridge moves, a perturbation mechanism based on this type of move can be expected to be insufficient. Instead, the perturbation mechanism used in ILK-H is based on a construction heuristic that is strongly biased by the incumbent candidate solution. This constructive search procedure, shown in Figure 8.7, iteratively builds a candidate solution for a given TSP instance in a manner similar to the Nearest Neighbour Heuristic. Starting from a randomly selected vertex, in each step the partial tour is extended with a vertex u_j that is not contained in the current incomplete path p. Let u_i be the current end point of path p. Then, vertex u_j is chosen for extending the partial tour p if the edge (u_i, u_j) is contained in the incumbent candidate solution, (u_i, u_j) is contained in the candidate list for vertex u_i, and $\alpha((u_i, u_j)) = 0$ (see Section 8.2, page 375 for the definition of α-values). If at any stage of the search process no such vertex exists, a vertex u_j contained in

u_i's candidate list is chosen, if possible. In particular, the vertex with the smallest index in u_i's candidate list that is not contained in the current partial tour p is chosen. If no such vertex can be found, a list of all vertices is traversed until a vertex u_j is found that is not contained in p.

The acceptance criterion used in ILK-H only accepts tours that lead to an improvement in the incumbent candidate solution. In addition to standard speedup techniques, such as don't look bits, ILK-H uses hashing techniques (originally described by Lin and Kernighan [1973]) to efficiently check whether a solution has been earlier found to be a local optimum. Further details of ILK-H can also be directly checked in the source code [Helsgaun, 2003].

ILK-H finds optimal solutions for many TSPLIB instances with up to several thousand vertices within relatively short run-times (CPU minutes on a high-performance PC). Longer runs of ILK-H resulted in improvements over the best known solutions to the largest unsolved TSPLIB instances, for many of the instances from VLSI design, and for the World TSP instance mentioned in Section 8.1; in the latter case, lower bound computations have shown that the solution found by ILK-H deviates at most 0.098% from the optimal solution quality (see Applegate et al. [2003b]).

EXAMPLE 8.3   **Performance of ILS Algorithms for the TSP**

Applied to large TSP instances, state-of-the-art ILS algorithms can find solutions whose quality is within fractions of a percent from the optimum within reasonable computation times. Figure 8.8 shows qualified run-time distributions (QRTDs) for CLK-ABCC and ILK-H on TSPLIB instance pcb3038 with 3 038 vertices, based on 100 independent runs of each of the two algorithms on an Athlon 1.2 GHz MP CPU with 1GB RAM running Suse Linux 7.3; each run was terminated as soon as either the given target solution quality or a cutoff time of 10 000 CPU seconds had been reached. The parameters of both algorithms were set to the respective default values in the publicly available codes. (For the experiments reported here, the 99.12.15 release of CLK-ABCC and version 1.2 of ILK-H were used.)

These QRTDs show that both algorithms very quickly reach solution qualities that are at most 0.5% above the optimum (the median and maximum computation times to find solutions of this quality were measured as 1.4 and 15.7 CPU seconds for CLK-ABCC and 1.5 and 2.5 CPU seconds for ILK-H, respectively). Furthermore, CLK-ABCC can reach higher solution qualities (e.g., within 0.1% of the optimum), with a reasonably high probability, but it starts showing severe stagnation behaviour, as can be seen from the exponential distribution on the left side of Figure 8.8, which indicates the expected run-time behaviour when using an optimal static restart strategy.

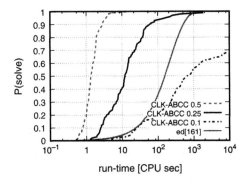 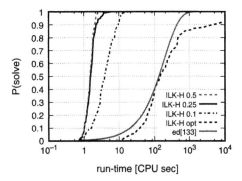

Figure 8.8 *Left:* Qualified run-time distributions for CLK-ABCC on TSPLIB instance pcb3038, using various target solution quality values (shown as percentage deviations from the optimum). *Right:* Analogous results for ILK-H. The exponential distributions $ed[m]$ are shown to illustrate stagnation behaviour of the respective algorithm. (For further details, see text.)

(We refer to Chapter 4, Section 4.4 for details on how to use exponential distributions for detecting stagnation behaviour.) CLK-ABCC has difficulties in reaching extremely high quality solutions, as witnessed by the fact that in our experiment, it found an optimal solution of pcb3038 in only one of 100 runs of 10 000 CPU seconds each. In contrast, within the same time, ILK-H solves instance pcb3038 optimally with a probability of more than 0.9. However, ILK-H is also affected by stagnation behaviour, as can be seen from the comparison with the exponential distribution shown on the right side of Figure 8.8.

ILS algorithms can reach close to optimal solution qualities for much larger instances. In Table 8.3, we give some indicative performance results of an iterated-3-opt algorithm (an implementation of this algorithm is available from www.sls-book.net), CLK-ABCC and ILK-H for several TSPLIB instances ranging from 4 461 to 13 509 vertices. Here, each algorithm was run 10 times with a time limit of 600 CPU seconds; the table shows basic statistics of the solution quality distributions obtained for this run-time. On the smallest of these instances, ILK-H found the known optimal solution in 6 of 10 runs; neither of the three algorithms found optimal solutions of the three other instances within the given computation time limit.

High-performance ILS algorithms for the TSP have been shown to be able to find high-quality solutions for even larger problem instances. For example, when running CLK-ABCC for 600 seconds on the largest TSPLIB instance, pla85900, the average solution quality measured across 10 runs was 0.31% above the best known lower bound (as of September 2003).

Instance	*Iterated*-3-opt			*CLK-ABCC*			*ILK-H*		
	Δ_{min}	Δ_{avg}	Δ_{max}	Δ_{min}	Δ_{avg}	Δ_{max}	Δ_{min}	Δ_{avg}	Δ_{max}
fnl4461	0.20	0.28	0.35	0.03	0.08	0.11	0.0	0.0016	0.0027
pla7397	0.22	0.35	0.48	0.040	0.19	0.33	0.011	0.029	0.054
rl11849	0.44	0.66	0.76	0.16	0.25	0.39	0.044	0.062	0.095
usa13509	0.39	0.51	0.58	0.097	0.13	0.17	0.029	0.038	0.055

Table 8.3 Solution quality statistics for an iterated-3-opt algorithm, CLK-ABCC and ILK-H on several large TSPLIB instances. Δ_{min}, Δ_{avg} and Δ_{max} denote the minimum, average and maximum percentage deviation from the known optimal solution qualities over 10 runs of 600 CPU seconds each on an Athlon 1.2 GHz MP CPU with 1GB of RAM.

Other Perturbation Mechanisms

Generally, the perturbation mechanism and its relation to the subsidiary local search procedure can have a significant impact on the performance of an ILS algorithm; consequently, a wide range of perturbation mechanisms has been proposed and studied in the context of ILS algorithms for the TSP.

Hong et al. [1997] have studied ILS algorithms based on 2-opt, 3-opt and LK local search that use single random k-exchange steps with fixed values of k between 2 and 50 for perturbation. The resulting algorithms have been empirically evaluated on TSPLIB instances lin318 and att532, as well as on a RDM instance with 800 vertices. Although there is some indication that on TSPLIB instances perturbations with $k > 4$ result in better solution qualities after a fixed number of local search steps, it is unclear whether these results also hold when using a termination criterion based on a bound on CPU time, or whether these observations generalise to other TSP instances.

Perturbations can be more complex than simple (random) k-exchange steps. One example for a complex perturbation is the mechanism proposed by Codenotti et al. [1996] which involves modifications of the instance data. Their perturbation procedure works as follows: First, the given metric TSP instance G is slightly modified by introducing small perturbations in the edge weights. (For Euclidean TSP instances, this is achieved by changing the coordinates of some vertices.) Note that as a result of this modification, a previously locally optimal tour s may no longer be a local minimum. Then, the subsidiary local search procedure is run on G' until a local minimum s' is found (see Figure 8.9 for an illustration of this procedure). At this point, the modified instance G' is discarded, and s' is returned as the overall result of the perturbation, which provides the

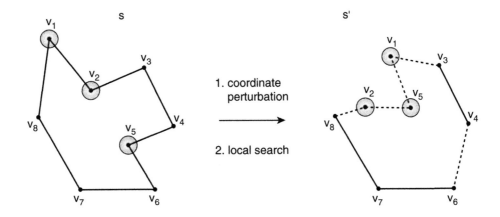

Figure 8.9 Example of a perturbation based on instance data modifications for a small Euclidean TSP instance. Here, the data perturbation corresponds to a modification of the coordinates of vertices v_1, v_2 and v_5. *Left:* Locally optimal tour s for the original coordinates; *right:* locally optimal tour s' for the perturbed coordinates.

starting point of the subsequent local search phase for the original instance G. There is some indication that in a standard ILS algorithm, using this perturbation mechanism — despite its relatively high time-complexity — can result in slightly better performance than using simple double-bridge perturbation [Codenotti et al., 1996]. However, state-of-the-art ILS algorithms for the TSP, such as CLK-ABCC, typically achieve much better performance [Applegate et al., 2003c]. (It should be noted that this general perturbation approach has been proposed and successfully applied in the context of a very early ILS algorithm for a location problem [Baxter, 1981].)

Another interesting perturbation mechanism is the *genetic transformation (GT)* by Katayama and Narisha [1999], which introduces ideas from Evolutionary Algorithms into ILS. The GT procedure is based on the intuition that sub-tours that are common between the current incumbent tour, $\hat{s}$, and the current locally optimal tour, t, should be preserved; it works as follows: First, all common sub-tours between $\hat{s}$ and t are determined; this can be achieved in time $O(n)$, where n is the number of vertices in the given TSP instance. Then, the perturbation result is obtained by connecting these sub-tours using a procedure that works analogously to the Nearest Neighbour Heuristic. The starting vertex u_i is chosen at random from the set of vertices that have zero or one incident edges; in the former case, the sub-tour consists of a single vertex, while in the latter case, u_i is the end of a subtour with at least two vertices. Then, in each construction step, u_i is connected to an eligible vertex for which $w((u_i, u_j))$ is minimal. The construction is continued from u_j if that vertex has degree one, or otherwise from the free

```
procedure GILS(G)
    input: weighted graph G
    output: candidate tour ŝ
    s := init(G);
    s := localSearch(G, s);
    t := init(G);
    t := localSearch(G, t);
    if (f(s) < f(t)) then
        ŝ := s;
    else
        ŝ := t;
    end
    while not terminate(G, ŝ) do
        t' := GT(G, ŝ, t);
        t'' := localSearch(G, t');
        if (f(t'') < f(ŝ)) then
            ŝ := t'';
        end
        t := t'';
    end
    return ŝ
end GILS
```

Figure 8.10 Algorithm outline of Genetic Iterated Local Search (GILS) for the TSP; the function *GT* implements the GT perturbation mechanism. (For further details, see text.)

endpoint of the subtour starting at u_j. (This construction process is similar to the mechanism underlying the DPX recombination operator, which has been used in an earlier memetic algorithm [Freisleben and Merz, 1996; Merz and Freisleben, 1997].)

The GT perturbation mechanism is embedded into the *Genetic Iterated Local Search (GILS)* algorithm for the TSP, outlined in Figure 8.10. Empirical comparisons between this algorithm and a variant that uses the standard double-bridge move for perturbation have shown that using the GT perturbation can result in significant improvements in solution quality [Katayama and Narihisa, 1999].

Other Acceptance Criteria

While various choices for the subsidiary local search and perturbation procedures have been studied in the literature, much less attention has been paid to the acceptance criterion, although it can have a strong impact on the balance between diversification and intensification of the search process. In fact, most ILS implementations for the TSP accept only improving tours. However, other acceptance criteria are used occasionally. For example, the previously mentioned LSMC algorithms by Martin, Otto and Felten use a Metropolis acceptance criterion (see page 384*ff.* for details on LSMC), where non-zero temperatures have been shown to improve the performance of their algorithms on some instances. Simulated annealing-type acceptance criteria were also examined by Rohe [1997], who found that for one TSPLIB instance, d18512, when using a carefully tuned annealing schedule and long computation times, slight improvements in solution quality could be obtained compared to a standard ILK algorithm.

Hong, Kahng and Moon have studied a variant of ILS they call *hierarchical LSMC*, which is based on the same Metropolis acceptance criterion as LSMC and uses the following mechanism for controlling the temperature parameter. By default, the temperature is set to zero, that is, only improving candidate solutions are accepted. However, when search stagnation is detected, the temperature is set to $f(s_i)/200$ for the following 100 iterations, that is, a deterioration of the tour length by 0.5 percent is accepted with a probability of $1/e$; after 100 iterations, the temperature is reset to zero. The criterion used for detecting search stagnation is satisfied if, and only if, no improvement in the incumbent tour has been obtained for i_r iterations. The value of i_r is chosen depending on the subsidiary local search procedure; for 2-opt or 3-opt local search, $i_r := 2 \cdot n$ is used, while in hierarchical LSMC with LK local search, i_r is set to 100. Limited empirical results suggest that on some TSPLIB instances, hierarchical LSMC may perform slightly better than other ILS algorithms for the TSP, but it is unclear whether this performance advantage is statistically significant.

A more detailed study of different acceptance criteria has been undertaken by Stützle and Hoos [Stützle, 1998c; Stützle and Hoos, 2001], who performed an empirical analysis of the run-time distributions of ILS algorithms that accept only improving solutions. (This type of analysis is illustrated in Example 8.3, page 390*f.*) Based on the severe stagnation behaviour that was observed in this study for very high target solution qualities, two acceptance criteria were introduced that had been designed to increase search diversification. The first of these is a simple *dynamic restart criterion* (see also Section 4.4), which restarts ILS from a new initial solution, if no improved solution has been found for i_r iterations. This dynamic restart acceptance criterion can be defined as

$$
acceptSR(G, s, s'') := \begin{cases} s'' & \text{if } w(s'') < w(s) \\ s & \text{if } w(s'') \geq w(s) \text{ and } i - \hat{\imath} < i_r \\ init(G) & \text{otherwise} \end{cases}
$$

where i denotes the current iteration number, $\hat{i}$ is the iteration number of the most recent improvement in the incumbent candidate solution, and $init(G)$ is a newly generated initial tour for the given TSP instance G.

Note that every time the search is restarted from a new initial candidate solution, some initial, instance-dependent time t_{init} has to be spent before there is a reasonable chance to encounter high-quality candidate solutions. To avoid this disadvantage, a less radical and more directed *fitness-distance diversification mechanism* can be used. This mechanism is based on a variant of the dynamic restart criterion $acceptSR(G, s, s'')$ in which the restart function $init(G)$ is replaced by $fdd(\hat{s}, G)$, a function that attempts to find a high-quality candidate solution beyond a certain minimum distance from the incumbent tour $\hat{s}$. In the following, we use $d(s, s')$ to denote the bond distance between two tours s and s', that is, the number of edges contained in s but not in s' or vice versa. The function $fdd(\hat{s}, G)$ works as follows:

1. Generate a set P comprising p copies of $\hat{s}$.

2. Apply one perturbation step followed by a subsidiary local search phase to each candidate solution in P.

3. Let Q be the set of the q highest quality candidate solutions from P (where $1 \leq q \leq p$).

4. Let $\tilde{s}$ be a candidate solution from Q with maximal distance to $\hat{s}$. If $d(\tilde{s}, \hat{s}) \leq d_{min}$ or if a given maximal number of iterations has not been exceeded, then go to step 2; otherwise return $\tilde{s}$.

Note that step 3 ensures that high-quality candidate solutions are obtained, while the goal of step 4 is increased search diversification. In Stützle and Hoos [2001], the parameter d_{min} is estimated by first computing the average distance d_{avg} between a number of local optima; each time fdd is called, d_{min} is alternatingly set to $0.25 \cdot d_{avg}$ and $0.5 \cdot d_{avg}$. Several variations of this fitness-distance diversification mechanism have also been studied, but none of those was found to obtain significantly better performance.

EXAMPLE 8.4   **Effectiveness of Acceptance Criteria**

The performance results in Table 8.4 illustrate the influence of various acceptance criteria on the performance of an iterated 3-opt algorithm. (Similar results for ILS algorithms based on 2-opt and LK local search procedures can be found in Stützle and Hoos [2001].) The perturbation procedure of this ILS algorithm is a slight variation of the geometric double-bridge perturbation used in CLK-ABCC. ILS-Descent, ILS-Restart and ILS-FDD denote three variants of the algorithm that differ solely in their acceptance criterion; while ILS-Descent accepts only improving candidate solutions, ILS-Restart and

Instance	ILS-Descent			ILS-Restart			ILS-FDD		
	f_{opt}	Δ_{avg}	t_{avg}	f_{opt}	Δ_{avg}	t_{avg}	f_{opt}	Δ_{avg}	t_{avg}
rat783	0.71	0.029	238.8	0.51	0.018	384.9	1.0	0	159.5
pcb1173	0	0.26	461.4	0	0.040	680.2	0.56	0.011	652.9
d1291	0.08	0.29	191.2	0.68	0.012	410.8	1.0	0	245.4
fl577	0.12	0.52	494.6	0.92	<0.01	477.6	1.0	0	294.1
pr2392	0	0.23	1 538.5	0	0.22	2 008.5	0.3	0.027	2 909.1
pcb3038	0	0.22	3 687.6	0	0.20	4 824.3	0	0.099	5 535.9
fl3795	0.2	0.36	3 601.9	0.2	<0.01	3 080.9	0.9	<0.01	3 506.7

Table 8.4 Performance results for ILS-Descent, ILS-Restart and ILS-FDD on symmetric instances from TSPLIB, based on 10 to 100 runs of each algorithm on every instance; f_{opt} denotes the frequency with which a provably optimal solution was reached, Δ_{avg} is the average percentage deviation from the optimum solution quality, t_{avg} is the average CPU time for finding the best tour encountered in a run. The maximal CPU times for each run were limited to 900 sec for rat783, 1 200 sec for pcb1173 and d1291, 2 400 sec for fl1577, 3 600 sec for pr2392 and 7 200 for pcb3038 and fl3765. All run-times were measured in CPU seconds on a 266MHz Pentium II processor with 512 MB RAM running Red Hat Lunix.

ILS-FDD use the previously described dynamic restart and fitness-distance diversification acceptance criteria, respectively. As can be seen from the empirical results in Table 8.4, ILS-Restart and ILS-FDD perform significantly better than ILS-Descent in terms of the solution qualities reached within a given amount of run-time and the frequency of finding optimal solutions. Furthermore, the performance advantage of the most complex variant, ILS-FDD, over both, ILS-Descent and ILS-Restart, becomes increasingly pronounced for larger instances.

Tour Merging

The performance differences between ILS-Restart and ILS-Descent illustrated in Example 8.4 suggest that multiple runs of an ILS algorithm for the TSP can lead to significant improvements over one single long run of the same algorithm. Yet, when randomly restarting the search process, all information contained in high-quality tours from previous runs are lost. *Tour merging* is a hybrid algorithm that attempts to utilise information collected from multiple runs in order to obtain better tours [Applegate et al., 1999; Cook and Seymour, 2003].

Tour merging is a two-phase approach. The first phase consists of generating a set T of very high quality tours for a given TSP instance defined by a graph

$G := (V, E, w)$. This can be done by applying a high-performance ILS algorithm, such as ILK, CLK-ABCC or ILK-H. From T, a subgraph $G' := (E', w')$ of G is formed, where E' is the set of edges that are contained in at least one of the tours in T, and w' is the original weight function w restricted to edges in E'. In the second phase, the objective is to find an optimal tour for the TSP instance defined by G'. Note that an optimal tour in G' corresponds to the best combination of any tour fragments contained in the elements of G'. The optimal solution of G' can be determined by either using a general purpose complete TSP algorithm or by complete algorithms that exploit characteristics of T. The former approach was chosen by Applegate et al. [1999], while the latter has been explored by Cook and Seymour [2003]; as may be expected, the second approach, which we discuss in more detail in the following, was found to achieve better performance.

The tour-merging algorithm by Cook and Seymour [2003], *TM-CS*, uses for its second phase a specialised dynamic programming procedure that exploits the fact that G' is a sparse graph with low branch-width [Robertson and Seymour, 1991]. The performance of several variants of TM-CS has been evaluated on a large number of TSPLIB instances ranging from 1 000 to 33 810 vertices as well as on several geographic TSP instances with up to 10 639 vertices, which correspond to cities located in various countries. These variants use either CLK-ABCC or ILK-H for generating the tours in the set T; specifically, T is obtained from k runs of the respective ILS algorithm of n iterations each, where n is the number of vertices of an instance and k is a parameter of the algorithm.

From these experiments, a number of observations were made. Firstly, the final solution quality (but also the overall computation time) is substantially higher when using ILK-H rather than CLK-ABCC, mainly because of the significantly better quality of the tours generated by ILK-H. Secondly, the final solution quality improves when increasing the number k of tours collected in T. Thirdly, compared to the time required to generate the tours in T, the second phase requires only relatively small computation times. Finally, on a few instances, the second phase failed because the branch-width of G' exceeded a given limit of 20 imposed by the dynamic programming procedure, or because that procedure exceeded the 2GB memory limit of the computer used for the experiments. Note that the density of the graph G' grows with the number k of tours in T, which increases the danger that the dynamic programming algorithm fails. Very good performance was obtained when building G' based on the 10 best out of 40 ILK-H tours in T. This variant of TM-CS found the optimal solution for TSPLIB instance d15112 in about 22 CPU days on a 500 MHz Alpha processor; interestingly, the second phase of the algorithm required only 107 seconds, while the remaining time was used for generating the 40 ILK-H tours in the first phase. The same algorithm also achieved improvements in the best known solutions for TSPLIB instances brd14051 and d18512 [Cook and Seymour, 2003].

Iterated Local Search for the Asymmetric TSP

Most research efforts on designing efficient ILS algorithms have focused on the symmetric TSP, and much less work has been done for the asymmetric TSP (ATSP). ILS algorithms for the ATSP are typically very similar to those for symmetric TSP instances; the main difference lies in the use of an ATSP-specific subsidiary local search procedure, such as reduced 3-opt or the Kanellakis-Papadimitriou variable depth search algorithm (KP), in order to avoid the high overhead involved in computing the effects of sub-tour reversals. Since double-bridge moves do not involve any subtour reversals, they can be used for perturbation in exactly the same way as they are used in ILS algorithms for symmetric TSP.

Only few ILS algorithms have been developed for the ATSP. These include iterated reduced 3-opt algorithms by Johnson et al. [2002] and Stützle & Hoos [1998c; 2001] as well as an iterated KP (IKP) algorithm by Johnson et al. [2002]. Experimental results suggest that IKP finds better quality solutions than the Johnson et al.'s version of iterated reduced 3-opt, however at the cost of substantially higher run-times on certain classes of instances [Johnson et al., 2002]. In fact, for some types of large ATSP instances, n iterations of IKP take up to 100 times more CPU time than $10 \cdot n$ iterations of iterated reduced 3-opt. Published results for an ILS algorithms using HyperOpt local search by Burke et al. [2001] suggest that in most cases the solution qualities reached by this algorithm are worse than for iterated reduced 3-opt while requiring larger computation times [Burke et al., 2001; Johnson et al., 2002].

Somewhat surprisingly, in a recent comparative study of several ATSP algorithms, the best quality tours were obtained by transforming ATSP instances into symmetric TSP instances (see Section 8.1, page 366f.) which are then solved using a state-of-the-art SLS algorithm for the symmetric TSP, such as ILK-H [Johnson et al., 2002]. In fact, for various classes of ATSP instances, this approach produced the best solutions known to date; however, the computation times required for ILK-H are in the range of several hours for the largest ATSP instance with 3 162 vertices, which corresponds to a symmetric TSP instance with 6 324 vertices.

8.4 Population-Based SLS Algorithms for the TSP

Many variants of population-based SLS algorithms have been applied to the TSP. In fact, with the tour-merging approach, we have already presented an extreme case of a population-based algorithm: The collection of high-quality tours can be interpreted as a population of solutions from which a 'perfect offspring' is generated. However, unlike the algorithms we present in this section, tour merging does not generate a new population that undergoes further modifications. In the following, we discuss TSP algorithms that iteratively manipulate populations of solutions, ranging from extensions of Iterated Local Search to inherently population-based approaches, such as Memetic Algorithms and Ant Colony Optimisation (see also Chapter 2, Section 2.4). Given the large number of TSP algorithms based on the latter two approaches, we focus on some prominent algorithms that illustrate the main considerations arising in the context of applying these methods to the TSP.

Population-Based ILS

Iterated Local Search can be easily extended into a population-based SLS method by applying the main search steps of a standard ILS algorithm to each element of the population — that is, to each candidate solution — in parallel; additionally, limited interaction among the population elements may be allowed. Such extensions have strong similarities to other population-based search metaphors, such as Evolutionary Algorithms and, in particular, Memetic Algorithms (see Chapter 2, page 101*ff.*). Both approaches make use of selection mechanisms to focus the search on promising regions of the search space, and the perturbation mechanism of ILS can be seen as a (possibly complex) mutation operator. However, compared to general Evolutionary Algorithms, population-based extensions of ILS are conceptually simpler, because they do not make use of recombination mechanisms to generate new candidate solutions from two or more elements of the population.

Generally, the interaction between population elements is fairly limited in population-based ILS algorithms, which facilitates rather straight-forward and efficient parallel implementations. Martin and Otto described a parallel implementation of a population-based extension of their Chained Local Optimisation (CLO) algorithm for the TSP, in which a single ILS process (which runs the CLO algorithm on an individual candidate solution) runs on each processor, and a simple selection mechanism is used to guide the search [Martin and Otto, 1996]. In their approach, every 10 to 100 CLO steps per processor, the best tour within the current population is broadcast to all processors and replaces all population elements on the other processors. Martin and Otto claimed that this population-based CLO algorithm performs well, but did not present detailed empirical results.

Hong, Kahng and Moon introduced a similar population-based ILS algorithm that uses a slightly more complex selection mechanism [Hong et al., 1997]. To each candidate solution of the population, l ILS iterations are applied, where l is a parameter of the algorithm. Then, one candidate solution s_j is probabilistically selected based on its solution quality; the probability of selecting a solution is inversely proportional to its rank in the current population scaled in such a way that the best tour is four times as likely to be selected as the currently worst tour. Perturbation and local search are then applied to s_j, resulting in a new tour s_j''. If s_j'' is better than the best tour found so far during the search, it replaces s_j, otherwise, s_j'' replaces the worst tour s_k in the current population, unless it has lower quality than s_k, in which case s_j'' is discarded. Note that for $l = \infty$, this algorithm is equivalent to performing m independent runs of standard ILS, where m is the population size (i.e., the number of candidate solutions in the population).

Similar population-based extensions have been studied by Stützle [1998c], who examined the following three population-based ILS algorithms for the TSP with varying degrees of interaction between the members of the population:

(i) A variant that does not use any interaction within the population and effectively performs a fixed number of independent ILS runs.

(ii) A variant called *Replace-Worst*, in which after every l iterations a copy of the best tour in the population replaces the worst tour. This approach gradually focuses the search around the best found tours, where the parameter l controls the 'convergence-rate' of this process.

(iii) A variant *ss-ILS*, in which standard ILS is only applied to one tour s_0 selected from the current population sp; if after j iterations of ILS an improvement over s_0 has been achieved, the improved tour replaces one of the tours in sp (for a detailed outline, see Figure 8.11).

Computational experiments with the population-based ILS algorithms for the TSP by Hong et al. [1997] and Stützle [1998c] suggest that these can achieve substantial performance improvements over conventional ILS algorithms that accept only better-quality candidate solutions. For fixed run-time, the population-based algorithms were shown to find candidate solutions whose average deviation from the optimum is roughly half of that obtained from ILS-Descent on a number of TSPLIB instances [Stützle, 1998c]. Surprisingly, the variants without interaction within the population were often found to be similarly effective as the variants with interaction; further experimentation is required to determine whether interaction within the population is more advantageous for larger TSP instances. Furthermore, empirical evidence suggests that for the TSP current population-based ILS extensions are inferior in performance to ILS algorithms such as ILS-FDD which, while not population-based, use more complex acceptance criteria (see Section 8.3, page 396*f.*). However, for other problems, such as the Quadratic Assignment Problem (see Chapter 10, Section 10.2), population-based ILS algorithms reach state-of-the-art performance.

```
procedure ss-ILS(G)
    input: weighted graph G
    output: tour ŝ
    sp := initPopulation(G);
    ŝ := best(G, sp);
    while (not terminate(G, sp)) do
        s₀ := select(G, sp);
        s := s₀;
        for i := 1 to j do
            s' := perturb(G, s);
            s'' := localSearch(G, s');
            if (f(s'') < f(ŝ)) then
                ŝ := s'';
            end
            s := accept(G, s, s'');
        end
        if (f(s) < f(s₀)) then
            sp := replace(G, sp, s);
        end
    end
    return ŝ
end ss-ILS
```

Figure 8.11 Algorithm outline for ss-ILS; $best(G, sp)$ denotes the individual from population sp with the best objective function value, $select(G, sp)$ selects a candidate solution from sp based on its objective function value, and $replace(G, sp, s)$ returns the result of replacing one individual in sp with candidate solution s. (For details, see text.)

Evolutionary Algorithms for the TSP

The TSP has been the target of a large amount of research on Evolutionary Algorithms, and many EAs for the TSP have been proposed and studied in the literature. An important general issue in the design of EAs for the TSP is the representation of candidate solutions. Most commonly, tours are represented as permutations of vertex indices. Several other representations have been studied [Homaifar et al., 1993; Whitley et al., 1989; Walters, 1998], but it is not clear whether compared to the permutation representation, any of these offers particular advantages. Most research efforts on EAs have been focused on the design of recombination operators [Potvin, 1996; Merz and Freisleben, 2001] and the development of highly effective hybrid EAs that include efficient subsidiary local search algorithms to improve candidate solutions — so-called Memetic Algorithms (see Chapter 2, page 101*ff.*).

If one general conclusion can be drawn from all these research efforts, then it is that Memetic Algorithms (MAs), that is, combinations of EAs with efficient subsidiary local search algorithms, are generally superior to EAs that do not use subsidiary local search. Furthermore, it can be generally observed that different types of recombination operators can result in significant performance differences. One important property of recombination operators for the TSP is *respectfulness* [Radcliffe and Surry, 1994; Merz and Freisleben, 2001]: respectful recombination operators ensure that solution components (here: edges) that are common to all parents are also present in all offspring. Intuitively, the importance of respectful recombination is tightly connected to the typical search space structure for TSP instances, which is characterised by high fitness-distance correlations and small distances between local optima (see Example 5.4 on page 218*ff.* as well as Example 5.5 on page 223*f.*).

The Memetic Algorithm by Merz and Freisleben

The memetic algorithm by Merz and Freisleben (MA-MF) is probably one of the best studied and most effective MAs for the TSP. Initially proposed in 1996 [Freisleben and Merz, 1996], MA-MF has been continuously improved through the incorporation of more efficient subsidiary local search procedures and better recombination operators, as well as through the addition of diversification mechanisms based on partial restarts [Merz and Freisleben, 1997; Merz, 2000; Merz and Freisleben, 2001]. Here, we describe the main features of the latest and best performing MA-MF variant [Merz and Freisleben, 2001].

The initial population in MA-MF is generated by a randomised variant of the Greedy Heuristic (see Section 8.2). The constructive search procedure first iteratively inserts $n/4$ edges (where n is the number of vertices in the given graph), which are selected as follows: after choosing uniformly at random a vertex v of the given graph G that is not contained in the current partial tour, the shortest feasible edge incident to v is selected with probability $2/3$ and the second-shortest feasible edge otherwise; edges are feasible if, and only if, one of their endpoints is not contained in the current partial tour. The partial tour obtained after this initial random edge selection is then completed using the standard construction mechanism of the Greedy Heuristic. As usual in MAs, the search initialisation is completed by applying the subsidiary local search procedure (here, an LK variant) to all tours in the population.

Various recombination operators are used in different variants of MA-MF; of these, a greedy recombination operator GX, which is based on ideas from the

Greedy Heuristic, achieves the best performance. The GX operator generates one offspring from two parent tours and consists of four phases:

1. Copy edges that are common to the two parents to the offspring; the fraction of common edges to be copied is determined by a parameter p_e.

2. Add new short edges that are not contained in any of the parents. For a vertex u_i one of the five nearest neighbours is chosen such that edge (u_i, u_j) is not contained in any of the parents and the addition of edge (u_i, u_j) is feasible; the fraction of the edges to be chosen in this way is determined by another parameter, p_n.

3. Copy edges from the parents, where edges are ordered according to increasing length. Only edges that do not lead to a violation of the TSP constraints are considered and edges not common to the parents may be included; the fraction of the edges added in this way is determined by a third parameter, p_c.

4. If necessary, the candidate tour is completed using a randomised version of the Greedy Heuristic.

Experimental results show that the best performance is obtained when GX is maximally respectful (i.e., for $p_e = 1$) [Merz and Freisleben, 2001]. Good settings for the parameters p_n and p_c vary between TSP instances. Recombination is applied to $n/2$ pairs of tours that are chosen uniformly at random from the current population.

The mutation operator used in MA-MF is the standard double-bridge move that is also used in most ILS algorithms. The candidate solutions to which mutation is applied are chosen uniformly at random. MA-MF uses a $(\mu + \lambda)$ selection strategy for determining the new population after each generation. If μ is the population size and λ the number of new tours obtained through mutation or recombination and subsequent application of local search, the new population comprises the μ best tours from the $\mu + \lambda$ tours obtained by adding the newly generated tours to the current population; in assembling the new population, it is ensured that μ distinct tours are chosen. Finally, an additional restart operator is applied conditionally, to maintain a certain diversity of the population. Whenever the average distance between the tours in the population (measured in terms of bond distance) falls below 10, or when the average solution quality of the population has remained unchanged for 30 iterations, a random k-exchange move with $k := 0.1 \cdot n$ and subsequent local search is applied to all tours in the population with the exception of the incumbent candidate solution.

Computational results confirm that MA-MF is among the best-performing memetic algorithms for the TSP [Merz and Freisleben, 2001]. For all tested

TSPLIB instances with up to 1 002 vertices, the known optimal solutions could be reached in every single run within an average time of about two CPU minutes on a Pentium III 500 MHz processor. When applied to instances with up to 3 795 vertices, the probability for finding optimal solutions dropped significantly, although high average solution qualities with a deviation of less than 0.08% from the optimum were still achieved. Additional experiments have shown that even for the largest TSPLIB instances, MA-MF finds high-quality solutions within reasonable computation times; solution qualities within 1% of the optimum were reached within an average run-time of one CPU hour on a Pentium III 500 MHz processor, except for the largest instance with 85 900 vertices, where the computation required to find such high-quality solutions took roughly four hours.

A performance comparison between MA-MF and high-performance ILS variants, such as ILK-H, ILK-JM and CLK-ABCC, suggests that, especially for instances with more than 5 000 cities, MA-MF does not reach state-of-the-art performance [Johnson et al., 2003a]. However, it is conceivable that by replacing the subsidiary local search procedure with a more effective LK algorithm, MA-MF could become competitive with state-of-the-art SLS algorithms for the TSP.

The Repair-Based Memetic Algorithm by Walters

The memetic algorithm by Walters, *MA-W* [Walters, 1998], differs in several key aspects from many other memetic algorithms for the TSP. In particular, MA-W does not make use of the permutation representation of tours, and it uses a standard recombination operator instead of specialised operators for the TSP.

The candidate solution representation used in MA-W is based on nearest neighbour indexing. For a given tour orientation, the successor of u_i is encoded by its index in the (sorted) nearest neighbour list of u_i. A tour p is represented by a vector $s := (s_1, \ldots, s_n)$ such that $s_i = k$ if, and only if, the successor of u_i in p is the kth nearest neighbour of u_i. As a side-effect, for symmetric TSPs this representation leads to some redundancy because of the directionality imposed by the encoding of successors: depending on which of the two possible directions is imposed on the tour, the same tour can be represented by two different candidate solutions.

While most recent evolutionary algorithms for the TSP make use of specialised recombination and mutation operators that are tailored to the permutation representation of tours and the known solution characteristics of the TSP, MA-W uses a (slightly modified) two-point crossover operator, which is derived from a generic, problem-independent recombination operator [Walters, 1998]. The mutation operator modifies the nearest neighbour indices of randomly selected vertices. The new index is selected according to a probability distribution, which is also used to initialise the population. The indices corresponding to the three nearest neighbours are selected with a probability of 0.45, 0.25 and 0.15, respectively; in the remaining cases, an index between four and ten is chosen uniformly at random. (For further details on MA-W, including parameter settings, we refer to Walters' original paper [Walters, 1998].)

The use of standard mutation and recombination operators, however, can lead to infeasible candidate solutions that do not correspond to any valid tour. MA-W uses a repair mechanism to restore valid tours from such infeasible paths. This repair mechanism preserves as many edges of the infeasible path as possible; if for some vertex, an outgoing edge e needs to be replaced in order to make the path 'more feasible', it is replaced by an edge e' such that $|w(e) - w(e')|$ is as small as possible.

In more detail, the repair process works as follows: First, a working list is created that comprises all edges contained in the infeasible path p. Next, this list is sorted according to edge weight plus a small amount of random noise (about 20% of the edge weight); the use of the noise randomises the order and thus helps to prevent domination of the end result — that is, the feasible tour returned by the repair process — by specific short edges. Then, the working list is traversed in ascending order, and edges are included into the new path p' if they do not lead to any cycle with less than n edges. If an edge e cannot be included into p', it is replaced in the working list by another edge e' that is incident to the same vertex and whose weight is as close as possible to that of e, while e is moved to a list of failed edges. If e' can be included in p' without violating feasibility, this is done. If after a full traversal of the working list, p' is not a valid tour, the repair process is repeated, starting with path p' and the current working list.

As its subsidiary local search procedure, MA-W uses an efficient 3-opt algorithm that uses the standard speed-up techniques described in Section 8.2. Compared to state-of-the-art TSP algorithms that are based on variants of the LK algorithm, the empirical performance results for MA-W are promising [Walters, 1998; Stützle et al., 2000]. For example, MA-W requires an average run-time of 572.7 CPU seconds on a Pentium II 450 MHz processor for finding an optimal solution to TSPLIB instance pr2392, a result that is only surpassed by the best-performing SLS algorithms for the TSP known to date. However, on many other instances, MA-W does not reach the performance of current state-of-the-art ILS algorithms, and it is generally not clear how MA-W's performance scales to instances with more than around 4 000 cities.

ACO Algorithms for the TSP

The TSP has played a central role in the development of Ant Colony Optimisation (ACO) algorithms, because the first ACO algorithm, Ant System

[Dorigo et al., 1991; Dorigo, 1992; Dorigo et al., 1996] and most of its successors, including Ant Colony System [Dorigo and Gambardella, 1997], $\mathcal{MAX}$–$\mathcal{MIN}$ Ant System [Stützle, 1998c; Stützle and Hoos, 2000], Rank-based Ant System [Bullnheimer et al., 1999] and Best-Worst Ant System [Cordón et al., 2000] were developed using the TSP as the first example application. All of these algorithms follow the same basic outline shown in Figure 2.14 on page 97, but they differ in several important algorithmic details concerning the update of the pheromone trails or additional means of diversifying the search. A detailed overview of the available ACO algorithms is given in Chapter 3 of the book by Dorigo and Stützle [2004].

The currently most successful ACO algorithms for the TSP share two important features: Firstly, they include effective mechanisms for achieving a good balance between intensification and diversification of the search and, secondly, before updating the pheromone trails, a subsidiary local search procedure is applied to the tours constructed by the ants [Dorigo and Stützle, 2004]. Hence, these ACO algorithms are hybrid SLS techniques that combine probabilistic solution construction with standard local search techniques.

$\mathcal{MAX}$–$\mathcal{MIN}$ Ant System

One of the most effective ACO algorithms for the TSP is $\mathcal{MAX}$–$\mathcal{MIN}$ Ant System ($\mathcal{MMAS}$) [Stützle, 1998c; Stützle and Hoos, 2000]. $\mathcal{MMAS}$ is closely related to Ant System (AS) [Dorigo et al., 1991; Dorigo, 1992; Dorigo et al., 1996] (see Chapter 2, Section 2.4), from which it mainly takes the mechanism for constructing candidate solutions: Starting from a randomly chosen vertex u_0, in each construction step, the current partial tour p is extended with a vertex u_j that is adjacent to its endpoint u_i; u_j is chosen probabilistically from a candidate list comprising the cl nearest neighbours of u_i excluding all vertices already contained in p; the probability of choosing vertex u_j is computed from the pheromone trail strength τ_{ij} for edge (u_i, u_j) and a heuristic value $\eta_{ij} := 1/w((u_i, u_j))$ according to Equation 2.3 on page 95. Only if all the vertices of u_i's candidate list are already visited in p, the next vertex is chosen to be the one connected to u_i by an edge with least possible weight.

$\mathcal{MMAS}$ introduces three major modifications to AS. Firstly, it strongly exploits the best candidate solutions found during the search — a feature it has in common with a variety of other ACO algorithms [Dorigo and Gambardella, 1997; Bullnheimer et al., 1999; Cordón et al., 2000]. $\mathcal{MMAS}$ uses the modified pheromone update rule

$$\tau_{ij} := \tau_{ij} + \Delta\tau_{ij}^{best},$$

where $\Delta\tau_{ij}^{best} := 1/w(p^{best})$, and only one ant is allowed to update the pheromone trail levels according to its tour, p^{best}; this may either be the 'globally best' ant, that is, the ant corresponding to the best candidate solution found since the start of the algorithm, or the 'iteration-best' ant, that is, the ant that represents the best tour obtained in the current iteration. The pheromone update mechanism ensures that $\tau_{ij} = \tau_{ji}$ for all i and j; this is motivated by the fact that the entire algorithm is designed for symmetric TSP instances. The choice between the iteration-best and globally best pheromone update influences the greediness of the search. When the globally best ant always deposits pheromone, the search focuses quickly around the respective tour, whereas when allowing the iteration-best ant to deposit pheromone, over time, a potentially larger number of edges are reinforced, resulting in a less directed search. Empirical results indicate that while for rather small TSP instances it may be best to use iteration-best pheromone update only, for large TSPs with several hundreds of vertices the best performance is obtained by giving a stronger emphasis to the globally best ant. This can be achieved, for example, by choosing the globally best ant for trail update with a frequency that gradually increases over time [Stützle, 1998c; Stützle and Hoos, 2000].

One disadvantage of the greedier trail update procedure is an increased danger of search stagnation. Therefore, a second important feature of $\mathcal{MMAS}$ is its use of lower and upper limits, τ_{min} and τ_{max}, on the pheromone trail level for each individual edge, which effectively avoids search stagnation. In particular, these limits have the effect of bounding the probability of selecting a vertex u_j when an ant is in vertex u_i according to Equation 2.3 (page 94) to an interval $[p_{min}, p_{max}]$, with $0 < p_{min} \leq p_{ij} \leq p_{max} \leq 1$. Only if an ant has visited all but one of the vertices in the given graph will it deterministically choose that vertex for its next and penultimate step, such that $p_{min} = p_{max} = 1$. (Note that as a result of the pheromone trail limits, $\mathcal{MMAS}$ is probabilistically approximately complete [Stützle and Dorigo, 2002].)

It is easy to show that in the limit, the pheromone trail level on any edge is bounded from above by $w(p^*)/\rho$, where ρ is the pheromone trail evaporation (see Equation 2.4, page 95) and $w(p^*)$ is the weight of an optimal tour for the given TSP instance. Based on this result, $\mathcal{MMAS}$ uses an estimate of this value, $w(\hat{p})/\rho$, where $\hat{p}$ is the incumbent tour, to define τ_{max}. In fact, each time a new incumbent tour is found, the value of τ_{max} is adapted. The lower pheromone trail limit is set to $\tau_{min} := \tau_{max}/a$, where a is a parameter of the algorithm [Stützle, 1998c; Stützle and Hoos, 2000]. Experimental results suggest that for effectively avoiding search stagnation, the lower pheromone trail limits are more important than the upper limits [Stützle, 1998c].

Thirdly, at the start of the algorithm, $\mathcal{MMAS}$ sets the initial pheromone trails to an estimate of $\tau_{max} := w(p_i)/\rho$, where the tour p_i is obtained by applying local search to a nearest neighbour tour. Together with small values for the pheromone evaporation parameter ρ, this ensures that during its initial search phase, $\mathcal{MMAS}$ is very explorative. As a further means for increasing the exploration of paths that have only a small probability of being chosen, $\mathcal{MMAS}$ occasionally re-initialises the pheromone trails; trail re-initialisation is triggered when stagnation behaviour is detected (as measured by some statistics on the pheromone trails or the number of iterations during which no improvement of the incumbent tour has occurred [Stützle and Hoos, 2000]).

While early versions of $\mathcal{MMAS}$ used 2-opt as a subsidiary local search procedure [Stützle and Hoos, 1996; 1997], more recent variants use an efficient 3-opt procedure [Stützle, 1998c; Stützle and Hoos, 2000]. Results for these latter variants suggest that $\mathcal{MMAS}$ can find optimal solutions for TSP instances with a few hundred vertices up to slightly more than 1 000 vertices rather efficiently in a few CPU minutes to around one CPU hour on an UltraSparc I 167 MHz processor. There is some experimental evidence that by using an LK algorithm as a subsidiary local search procedure, the performance of $\mathcal{MMAS}$ can be significantly improved [Stützle and Hoos, 2000]. Yet, even this variant of $\mathcal{MMAS}$ was found not to be competitive with the currently best ILS algorithms for the TSP, which may partly be due to the fact that the underlying LK procedure was not as highly optimised as the LK algorithms used in state-of-the-art ILS algorithms. However, the fact that the perturbation in one of the currently best performing ILS algorithms for the TSP, ILK-H, is actually based on a constructive mechanism that is not too different from the one used in ACO algorithms (see also Section 8.3, page 388*ff.*), suggests that combinations of $\mathcal{MMAS}$ and the best available LK algorithms, such as LK-H, could reach or improve over state-of-the-art performance.

Population-Based Algorithms for the ATSP

All previously described, population-based algorithms for the symmetric TSP can be extended to the ATSP in a rather straightforward way. In most cases, one major difference concerns the use of a local search procedure that is suitable for the ATSP, such as reduced 3-opt. Additionally, some operators used in the context of Evolutionary Algorithms require adaptations, and for ACO algorithms, asymmetric pheromone trails need to be supported, that is, one may have $\tau_{ij} \neq \tau_{ji}$.

Empirical results suggest that the ATSP algorithms thus obtained can typically find optimal solutions of asymmetric TSPLIB instances with up to 170 vertices within reasonable run-times [Merz and Freisleben, 1997; Stützle and Hoos, 2000; Stützle, 1998c; Walters, 1998]. When taking into account the differences between the machines that were used for these experiments, the best performance is probably achieved by the ATSP version of Walters' repair-based MA, closely followed by the population-based ILS variants and $\mathcal{MAX}$–$\mathcal{MIN}$ Ant System. Very good results have also been reported for the memetic algorithm of Buriol et al. [2004], which uses a specialised local search algorithm for the ATSP as its subsidiary local search procedure. However, there is some indication that none of these population-based algorithms reaches the performance of the best known SLS algorithms for the ATSP [Johnson et al., 2002]; however, further empirical analysis is required for a conclusive performance comparison.

8.5   Further Readings and Related Work

The literature on the TSP is vast, and it is practically impossible to adequately cover in a single book chapter all relevant work on SLS algorithms for the TSP, not to mention other types of TSP algorithms. From this perspective, it is not surprising that there are a number of books that are entirely devoted to the TSP. A classic in the TSP literature is the book edited by Lawler et al. [1985], which covers many aspects of TSP research up to the year 1985. The book by Reinelt [1994] provides an extensive computational study of construction heuristics and local search algorithms. The most recent entry in this list is the book edited by Gutin and Punnen [2002], which covers many aspects of TSP solving, including exact algorithms, SLS methods, empirical analysis of heuristic TSP algorithms and problems related to the TSP.

Here, we provide a few selected references to important work related to the SLS methods covered in this chapter, starting with iterative improvement algorithms. The 2.5-opt algorithm extends 2-opt by additionally checking whether a tour can be improved by inserting a vertex between the two vertices incident to the first edge that is removed in a 2-exchange move. 2.5-opt was shown to find substantially better tours than 2-opt within only slightly higher computation times [Bentley, 1992]. The Or-opt algorithms generalises 2.5-opt by allowing the tour segment that is inserted between two tour neighbours to be of maximal length three (the length of the segment is one in the case of 2.5-opt) [Or, 1976]; however, no results for Or-opt using current speed-up techniques are available, and therefore its performance potential is unclear.

The Generalised Insertion Procedure (GENI) combines a construction heuristic with a local search algorithm that is applied each time after adding a vertex to the current partial tour [Gendreau et al., 1992]. This procedure is extended by a post-optimisation phase, during which iteratively the following steps are applied: remove a vertex, apply local search and then re-insert the vertex into the tour [Gendreau et al., 1992]. The Hyperopt algorithm combines enumerative algorithms with local search. It is based on the deletion of a set of edges and subsequent enumeration of all possible ways of reconnecting the resulting tour segments and individual vertices [Burke et al., 2001]. However, the computational results obtained so far are not promising, neither for the symmetric TSP [Johnson et al., 2003a] nor for the ATSP [Johnson et al., 2002].

There are several other SLS algorithms for the TSP that are based on (exponentially) large neighbourhoods. Computational results for dynasearch algorithms (see Chapter 2, Section 2.1) based on the 2-exchange, 2.5-exchange and 3-exchange neighbourhoods and iterated versions of Dynasearch show promising performance, but do not appear to be competitive with LK or iterated LK [Congram, 2000]. Ejection chains are another type of variable depth search algorithm that is closely related to LK [Glover, 1996; Rego and Glover, 2002]. Compared to LK, they allow more flexibility in building complex search steps (for more details, we refer to [Rego and Glover, 2002]).

Apart from tour merging (see Section 8.3), a few other approaches have been proposed that make use of ILS algorithms as a sub-routine. Tamaki proposed a different way of using the essential ideas of tour merging to enhance local search algorithms for the TSP and achieved promising results [Tamaki, 2003]. Schilham used a collection of high-quality tours obtained from effective ILS algorithms (in particular, from CLK-ABCC) to probabilistically bias the choice of initial tours for subsequent ILS runs [Schilham, 2001]. Finally, let us mention the dynamic programming algorithm of Balas and Simonetti [2001], which tries to find a best way of locally reordering a tour, and the multi-level approach by Walshaw [2002].

Basically all of the 'simple' SLS techniques covered in Chapter 2, Section 2.2, have been applied to the TSP. Although in some cases, reasonably good performance has been reported (mainly for TSPLIB instances), none of these SLS algorithms appears to be competitive with the best-performing ILS or population-based algorithms [Johnson and McGeoch, 1997; Johnson et al., 2003a]. However, the interested reader may learn more about these TSP algorithms from the following references.

Simulated Annealing is mainly covered in the book chapter by Johnson and McGeoch [1997], although several earlier approaches exist. Simple tabu search algorithms have been described by Knox [1994] and Malek et al. [1989], while more advanced tabu search strategies are covered by Fiechter [1994] as well as by Dam and Zachariason [1996]. Among the dynamic local search methods, two approaches have been applied to the TSP: Guided Local Search by Voudouris and Tsang [1999] and the Noising Method by Charon and Hudry [2000].

There is a large body of literature on population-based SLS algorithms for the TSP and in particular, on Evolutionary Algorithms. In addition to the two evolutionary algorithms presented in Section 8.4, a few other recent algorithms are worth mentioning. (For an overview of earlier approaches we refer to

Johnson and McGeoch [1997], Merz and Freisleben [2001] and Potvin [1996].) Nagata and Kobayashi introduced an EA with edge assembly crossover, which constructs offspring based on the union of the edge sets of the two parents and then applies a greedy construction algorithm to merge sub-tours [Nagata and Kobayashi, 1997]. This EA does not use explicit local search; however, its recombination operator incorporates some local search features. Very good performance has been reported for the memetic algorithm by Seo and Moon [2002], which uses a particular recombination operator called Voronoi Quantised Crossover, and the MA by Nguyen et al. [2002], which uses a greedy recombination operator that tries to inherit as many edges from the parents as possible when generating an offspring. A modification of this latter MA actually found, as of September 2003, new best tours for several large VLSI instances [Applegate et al., 2003b].

Several other population-based algorithms share some connection to Evolutionary Algorithms, but introduce additional ideas. One of these is the Iterative Partial Transcription (IPT) approach by Möbius et al. [1999], which can be seen as a local search method that is based on information exchanges between pairs of solutions. Houdayer and Martin proposed a population-based algorithm that iteratively generates offspring by choosing k parents, freezing the common edges among the k parents, and solving a TSP in which only edges that are not frozen may be changed [Houdayer and Martin, 1999]. For an overview of various ant colony optimisation algorithms to the TSP, we refer to the overview article by Stützle and Dorigo [1999b] or Chapter 3 of the book by Dorigo and Stützle [2004].

A number of extensive computational studies on heuristic TSP algorithms are now available. First and foremost, the 8th DIMACS Implementation Challenge on the TSP [Johnson et al., 2003a] provides an on-line collection of empirical results for many implementations of construction heuristics, local search algorithms and more complex SLS algorithms. A detailed summary of the results as of mid-2001 is available in a book chapter by Johnson and McGeoch [2002]; the same book also includes a chapter on computational results for SLS algorithms for the ATSP [Johnson et al., 2002]. For the most recent results, we refer the interested reader to the DIMACS Challenge web pages [Johnson et al., 2003a], which provide pairwise performance comparisons of algorithms on sets of TSP instances as well as results for all algorithms on each single instance available for the implementation challenge. However, some care has to be taken because of the way results are presented. For many stochastic algorithms, only a single run was performed on each problem instance and the computation times vary strongly between the algorithms. Hence, it may be difficult to draw final conclusions regarding the relative advantages of many advanced SLS algorithms, whose performance is often not dominated by any other algorithm on every single instance. Nevertheless, a first indication of the performance trade-offs of state-of-the-art SLS algorithms for the TSP can be obtained, and the use of empirical methods like those presented in Chapter 4 may enhance the comparability of some of the SLS algorithms' performance.

Although the 8th DIMACS Implementation Challenge on the TSP provides extensive empirical results, there are a number of earlier, extensive studies that are still quite relevant. These include the study of construction heuristics and local search algorithms by Bentley [1992], the previously mentioned book by Reinelt [1994] and a book chapter on local search for the TSP by Johnson and McGeoch [1997].

8.6 **Summary**

The TSP is a central problem in combinatorial optimisation with many theoretical and practical applications; it also has been, and still is, at the core of numerous efforts to push the limits on the size of practically tractable optimisation problems. State-of-the-art complete TSP algorithms can solve instances up to several thousand vertices in reasonable computation times (CPU hours to several CPU days), while the best SLS algorithms can find solutions whose quality is within fractions of a percent of the optimum for much larger instances with up to millions of vertices.

Construction heuristics can find reasonably good solutions for TSP instances extremely fast. They also play an important role as initialisation procedures for various SLS algorithms. 2-opt and 3-opt are the most prominent *iterative improvement algorithms* based on *k-exchange neighbourhoods*. Various *speed-up techniques*, including *fixed-radius near neighbour search*, *candidate lists* and *don't look bits*, play a crucial role in the design of efficient SLS algorithms for the TSP, particularly in the case of iterative improvement algorithms. *Variable depth search methods*, such as the *Lin-Kernighan Algorithm*, can find higher quality solutions, but typically require longer run-times. They also require considerable fine-tuning and are substantially harder to implement than simpler iterative improvement algorithms, such as 2-opt and 3-opt.

Somewhat surprisingly, current empirical evidence suggests that one of the conceptually simplest hybrid SLS methods, *Iterated Local Search*, gives rise to some of the best-performing TSP algorithms currently known. If very high quality solutions need to be found, methods different from standard ILS may be preferable, such as the *tour merging* approach, which uses an ILS algorithm as a subroutine in a more complex, hybrid search algorithm. *Population-based SLS methods*, such as *Memetic Algorithms* or *Ant Colony Optimisation*, appear to be highly promising as the basis for high-performance TSP algorithms.

However, recent results indicate that these population-based algorithms need to be very carefully designed for the specific task of TSP solving in order to reach (and eventually surpass) the performance of current state-of-the-art TSP algorithms.

Generally, there is substantial evidence that the SLS methods underlying the best-performing TSP algorithms often achieve excellent or even state-of-the-art performance for many other combinatorial optimisation problems. This is the case for many ACO algorithms as well as evolutionary algorithms (and in particular, for memetic algorithms). Furthermore, because the TSP is easy to understand and does not involve side constraints that complicate algorithm design, it is ideally suited for gaining practical experience in the design and implementation of high-performance SLS algorithms for combinatorial optimisation problems.

Exercises

8.1 **[Medium]** In Section 8.1, we stated that adding a penalty p_i to each vertex v_i of a given TSP instance and defining modified edge weights $w'((v_i, v_j)) := w((v_i, v_j)) + p_i + p_j$ preserves the optimality of tours, but it may result in different one-trees. Show that the optimality of tours is preserved. Give an example of a weighted graph where the optimum one-tree changes after the addition of appropriate vertex penalties.

8.2 **[Medium]** In Section 8.1, we described a transformation of ATSP instances into symmetric TSP instances. Show first how a solution to a symmetric TSP instance obtained from this transformation can be decoded into a solution of the corresponding ATSP instance. Then prove that any optimal solution of the symmetric instance is also an optimal solution of the original asymmetric TSP instance.

8.3 **[Medium]** A *Hamiltonian path* in a graph G is a path that visits every vertex of G exactly once. The *Hamiltonian Path Problem* is to determine whether a Hamiltonian path exists in a given graph. In an extension of this problem the objective is, given a weighted graph G' and two vertices u and v in G', to find a Hamiltonian path p between u and v with minimum weight (i.e., u and v are the first and the last node of the path p, respectively); like the Hamiltonian Path Problem, this problem is $\mathcal{NP}$-hard. Show how the problem of finding a Hamiltonian path with minimum weight between two vertices in a given graph can be solved using a TSP algorithm.

8.4 [*Easy*] Specify all the possible ways of rewiring the tour segments in a 3-exchange step after three edges have been deleted. How many ways are there for rewiring tour segments in a 4-exchange step?

8.5 [*Medium*] Prove the following statement. If fixed-radius near neighbour searches for all vertices do not result in any improving 2-exchange step, the current tour represents a local optimum with respect to the 2-exchange neighbourhood.

8.6 [*Medium*] Explain how the complex steps in the Lin-Kernighan algorithm, which were introduced as sequences of 1-exchange moves, can also be interpreted as sequences of 2-exchange moves.

8.7 [*Easy*] The performance of ILS algorithms that differ only in perturbation strength, which can be measured as the number of edges that are deleted from a given tour during a perturbation phase, can be compared based on solution quality distributions measured from a number of runs with a fixed number of iterations. However, this has the disadvantage that the resulting CPU times may differ strongly between the algorithms. Explain the reason(s) underlying these differences in run-time.

8.8 [*Medium*] In the original Lin-Kernighan algorithm, the set of deleted edges X and the set of added edges Y are required to be disjoint. Why may the variable depth search be unbounded if this criterion is dropped?

What happens if we require instead that either no deleted edge can be added subsequently or that no added edge can be deleted subsequently? Can either of these criteria lead to unbounded searches?

8.9 [*Easy; Hands-On*] Study the influence of the don't look bit resetting strategy after a perturbation in ILS. Consider the following three resetting strategies: (i) reset all don't look bits to zero (ii) reset only don't look bits of end-points of broken edges and (iii) reset don't look bits of the end-points of broken edges plus the 25 tour neighbours of those endpoints.

An implementation of ILS for the TSP is available from `www.sls-book. net`.

8.10 [*Medium; Hands-On*] One possibility to obtain a stochastic local search procedure from an, at least in principle, deterministic k-opt algorithm is to generate a random permutation of the vertex indices and to choose starting vertices for the searches of improving moves according to this random order. (This is a possibility also considered in the implementation available from `www.sls-book.net`.)

Study the solution quality distribution of the local optima returned by such a randomised local search procedure by following these steps:

1. Generate one random tour, one nearest neighbour tour, one greedy tour and one tour by random insertion.

2. Starting from each of these tours, perform 10 000 runs of a randomised 2-opt algorithm (with all the available speed-up techniques) on instances from TSPLIB that are larger than 1 000 vertices.

3. Generate the resulting empirical asymptotic solution quality and termination-time distributions, and try to approximate these with known distributions.

4. Compare the location and shape of these solution quality distributions with those obtained for a random restart heuristic that starts a *deterministic* local search from 10 000 random initial solutions.

An implementation of a 2-opt algorithm for the TSP is available from `www. sls-book.net`.

There is a time for some things,
and a time for all things;
a time for great things,
and a time for small things.
—Miguel de Cervantes Saavedra, Writer

9 SCHEDULING PROBLEMS

Scheduling is a ubiquitous task in a wide range of real-world settings and forms one of the most important classes of combinatorial problems. SLS algorithms are commonly and very successfully used for solving scheduling problems in practice. We begin this chapter with an introduction to scheduling problems and an overview of the different types of problems that fall into the scheduling domain. We then present and discuss stochastic local search algorithms for various important and prominent types of scheduling problems: single-machine, flow shop and group shop problems. As we will show, some approaches, issues and results are similar for the various types of scheduling problems, while others differ considerably. Given the variety of scheduling problems and SLS approaches for solving them, this chapter can merely provide an introduction and highlight some important issues. The interested reader will find pointers to the literature on scheduling problems in the 'Further Readings and Related Work' section at the end of this chapter.

9.1 Models and General Considerations

Scheduling problems arise in virtually all situations where performing a given set of actions or operations requires the allocation of resources and time slots subject to certain feasibility and optimisation criteria. Scheduling problems are often difficult to solve, because resources are usually scarce and complex dependencies may exist between the actions. As an example, consider the scheduling of landings and takeoffs at an airport. Here, the (typically scarce) resources are the runways.

Each arriving or departing flight has a time window indicating the allowable take-off or landing times (which is determined by a pre-planning process or a given flight schedule). Furthermore, there are constraints, for example, on the minimum time interval between successive take-offs and landings. A typical optimisation objective is to minimise the total (weighted) delay over all flights. Other examples of real-world scheduling problems include the scheduling of batch processes in chemical plants, the allocation of processing time to processes in multi-processor computing systems and tournament scheduling in popular sports.

Basic Concepts and Problem Types

Scheduling problems can be abstractly defined as follows: Given a set of *jobs*, $\mathcal{J} := \{J_1, \ldots, J_n\}$, that have to be processed by a set of *machines*, $\mathcal{M} := \{M_1, \ldots, M_m\}$, find a *schedule*, that is, a mapping of jobs to machines and processing times subject to *feasibility constraints* and *optimisation objectives*. In the airport example given above, the jobs are the incoming and outgoing flights and the machines represent the runways; a schedule would assign each flight a runway and a starting (or landing) time; feasibility constraints may include the maintenance of minimal times and distances between subsequently starting and landing aircrafts; and the optimisation criterion is the minimisation of the total (weighted) delays.

In many problems it is assumed that schedules are only feasible if they respect the following constraints: (i) each machine can only process one job at a time, and (ii) each job can only be processed by one machine at any time. Furthermore, in many cases it is additionally assumed that (iii) once a machine has started processing a job, it will continue running on that job without interruption until the job is finished; schedules that satisfy this property are called *non–pre-emptive*.

Concrete scheduling problems can be classified according to the given job characteristics, machine environment and optimisation objective. Generally, we distinguish between *single-machine problems* and *multi-machine problems*. Furthermore, a fundamental distinction is made between *single-stage problems*, in which each job consists of one atomic operation that is executed by exactly one machine in a single processing stage, and *multi-stage problems*, in which jobs can consist of multiple operations, which may have to be performed on different machines. In this latter case, the previously mentioned restriction to non–pre-emptive schedules is applied at the level of operations instead of jobs.

In single-stage, single-machine problems, for each job J_i a *processing time* p_i is given. In multi-stage problems, processing times are given for each operation of each job, that is, for a complex job $J_i := \{o_{i1}, \ldots, o_{i\,m(i)}\}$ a processing time

p_{ij} is specified for each operation o_{ij}. In multi-machine problems, the processing time of a job or operation may additionally depend on the machine to which it is assigned. (Generally, we assume that processing times are independent of the time at which a machine processes a job.)

In many scheduling problems, there are constraints on the time during which a job is available for processing. These include *release dates* r_i, which indicate the earliest time at which job J_i is available for processing. Conversely, *due dates* d_i indicate deadlines for the completion of job J_i. In the airport example above, a flight cannot be scheduled to land earlier than it reaches the airport, and it should not be assigned a slot that has it land later than the fuel reserves of the aircraft last.

Additional job characteristics include: (i) the assignment of *weights* w_i, which indicate the relative importance of a job J_i compared to other jobs; (ii) *setup times* t_{ij} that are incurred when job J_j is processed immediately after job J_i on the same machine (these reflect the amount of time required for resetting or reconfiguring the machine and may also depend on the respective machine); and (iii) *precedence constraints* between jobs, which determine the order in which certain pairs of jobs can be processed — if a job J_i has precedence over another job J_j, processing of the latter cannot commence before processing of the former has finished. Precedence constraints can also be defined on the operations of a single job; these restrict the order in which the operations comprising a job can be executed.

In our airport example, the processing time of a departing flight can be defined as the time required by the aircraft from entering a critical zone of the runway until it reaches a minimum security distance after take-off. During this time interval, no other aircraft is allowed to enter the runway. Release dates indicate the times when flights are ready for takeoff or landing, while due dates correspond to the scheduled departure and arrival times (adjusted by the time required for taxiing). Finally, flights can be assigned weights that indicate their relative importance (for example, short-haul regional flights may have a lower weight than intercontinental flights).

Multi-machine scheduling problems can be further classified according to other properties of the given machine environment. Single-stage scheduling problems with multiple machines are also known as *parallel machine problems*; these include the special cases of *identical parallel machine problems*, where the processing time of a job is independent of the machine on which it is processed; *uniform parallel machine problems*, where the machines may have different speeds that uniformly affect the processing times of all jobs; and *unrelated parallel machine problems*, where the processing time of the jobs may depend on the machines to which they are assigned in a non-uniform way, that is, the speed of a machine may depend on the job being processed.

Furthermore, multi-stage, multi-machine problems can be classified based on the order in which the operations of a complex job are to be performed. In *flow shop problems*, the operations of all jobs are performed on the machines $M_1, M_2, \ldots, M_m$ in that order. In *open shop problems*, there are no restrictions on the order in which the operations of a job are performed by the machines. In *job shop problems*, the routing of each job through the machines is fixed, but may differ between jobs. This last condition is relaxed in the more general *group shop environment*, where only a partial order is given for the operations of each job. In all of these cases, it is assumed that each operation can be performed on only one machine, and that this assignment of operations to machines is given as part of the specification of a problem instance.

Depending on the number of runways, the airport example can fall into the class of either single-machine or parallel machine scheduling problems. When modelling the same problem in a more detailed way, each flight (i.e., job) may require a sequence of operations, including, for example, various stages of taxiing from the gate to the runway as well as the actual take-off or landing phases. In this case, the gate areas as well as various taxiways are modelled as additional machines, and the respective scheduling problem becomes substantially more complex.

Finally, various optimisation objectives are commonly used in the context of scheduling problems; these are mostly based on the minimisation of a function of the completion times of the given jobs, where the *completion time* C_i of job J_i is the earliest time at which J_i is completely processed. For scheduling problems with due dates, additionally the *lateness* of job J_i can be computed as $L_i := C_i - d_i$, its *tardiness* as $T_i := \max\{C_i - d_i, 0\}$ and its *earliness* as $E_i := \max\{d_i - C_i, 0\}$. Based on these measures, the most commonly used objective functions to be minimised are: (i) the maximum completion time $C_{max} := \max\{C_1, \ldots, C_n\}$, which is also called the *makespan*; (ii) the sum of the (weighted) completion times, computed as $\sum_{i=1}^{n} w_i \cdot C_i$; and (iii) the total weighted tardiness, given by $\sum_{i=1}^{n} w_i \cdot T_i$.

All of these are non-decreasing functions of the job completion times $C_1, \ldots, C_n$; objective functions with this property are also called *regular*. Non-regular objective functions arise, for example, in scheduling problems with due dates for which tardiness as well as earliness is penalised. In practical applications, earliness penalties can, for example, reflect storage costs in a just-in-time production environment.

EXAMPLE 9.1   **A Simple Single-Stage Scheduling Problem**

Consider an (ficticious) airport with two runways (i.e., machines) and the following seven flights (i.e., jobs) to be scheduled such that the total weighted tardiness is minimised. Each job can be processed by any of the two machines.

Job	J_1	J_2	J_3	J_4	J_5	J_6	J_7
Processing time	10	15	15	10	20	15	10
Due date	15	20	20	30	40	50	50
Release date	0	0	5	10	15	30	35
Weight	1	2	3	2	1	1	3

Consider the sequence $\phi := (J_1, J_3, J_5, J_7; J_2, J_4, J_6)$, where the two sub-sequences $\phi_1 := (J_1, J_3, J_5, J_7)$ and $\phi_2 := (J_2, J_4, J_6)$ on each of the two machines are separated by a semicolon. (This sequence was obtained by assigning a job as soon as it became available so that it starts at the earliest possible starting time on any of the two machines.) From this sequence, the starting and the completion times of all jobs can be computed easily. In this example, care has to be taken, because the starting time of a job is the maximum of its release time and the completion time of the job directly preceeding it. Once the completion times are available, the tardiness and the weighted tardiness of each job can be computed. The results of these computations are given in the following table.

Job	J_1	J_2	J_3	J_4	J_5	J_6	J_7
Completion time	10	15	25	25	45	45	55
Tardiness	0	0	5	0	5	0	5
Weighted tardiness	0	0	15	0	5	0	15

The total weighted tardiness of this schedule is $15 + 5 + 15 = 35$. The schedule itself is illustrated in Figure 9.1 as a *Gantt-chart*, which is a widely used graphical representation for schedules. The x-axis of the Gantt-chart represents the time and the bars on the y-axis the various machines. Each job is indicated by a rectangle whose width corresponds to the processing time, and whose left and right borders correspond to the starting and completion time of the respective job. In addition, different colours may be used to indicate specific job attributes or properties, such as a job being completed after its due date. It may be noted that the sequence ϕ is not optimal. For example, the total weighted tardiness can be reduced by processing job J_7 on machine M_2 before job J_6, which reduces the total weighted tardiness by 5.

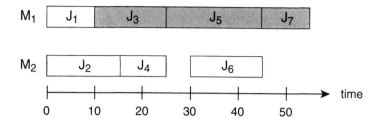

Figure 9.1 Gantt chart illustrating the schedule for the sequence $\phi_1 := (J_1, J_3, J_5, J_7)$ on machine M_1 and $\phi_2 := (J_2, J_4, J_6)$ on machine M_2. Note that, although job J_4 finishes on machine M_2 at time 25, job J_6 can only start at time 30, because of its release date of 30.

Symbolic Notation for Scheduling Problems

Given the large variety of scheduling problems arising in different contexts, a symbolic notation is commonly used for the precise definition of the particular problem under consideration. Widely used is the three field notation $\alpha|\beta|\gamma$, where α, β and γ represent the machine environment, the job environment and the objective function to be optimised, respectively [Graham et al., 1979]. Possible values for the three fields are defined as follows.

The α field is composed of two subfields, that is, $\alpha := \alpha_1\alpha_2$, where α_1 denotes the particular environment and α_2 the number of machines. Let ϵ be the empty symbol; we have that $\alpha_1 \in \{\epsilon, P, Q, R, F, O, J, \}$, where

- ϵ (i.e., empty) corresponds to a single machine,

- P corresponds to identical parallel machines,

- Q corresponds to uniform parallel machines,

- R corresponds to unrelated parallel machines,

- F indicates a flow shop environment,

- O indicates an open shop environment,

- J indicates a job shop environment.

The second subfield α_2 can have value ϵ for problems with an arbitrary number of machines, or $m \in \mathbb{N}$ for problems with a fixed number of machines, m. In particular, single-machine problems are indicated by $m = 1$.

The job environment specifier is composed of four subfields, that is, $\beta :=$ $\beta_1\beta_2\beta_3\beta_4$, which specify whether release dates, deadlines, setup times and precedence constraints are given. In all four cases, ϵ denotes that the respective feature is not present. Furthermore, $\beta_1 = r_j$ indicates that jobs have release dates, $\beta_2 = d_j$ means that jobs have deadlines (due dates), and $\beta_4 = prec$ signifies that there exist precedence constraints. For β_3, a number of possibilities exist: t_{jk} indicates general sequence-dependent job setup times, t_j sequence-independent job setup times s_{fg} sequence-dependent family setup times and s_f sequence-independent family setup times.

Finally, γ, which represents the optimisation objective, can take various values, including the following: C_{max} indicates makespan minimisation, L_{max} maximum lateness minimisation, $\sum_j w_j \cdot C_j$ minimisation of the sum of the weighted completion times, and $\sum_j w_j \cdot T_j$ minimisation of the total weighted tardiness.

EXAMPLE 9.2 **Symbolic Notation for Various Scheduling Problems**

The following examples illustrate the three field notation.

$1|t_{ij}|C_{max}$ is a single-machine scheduling problem with sequence dependent setup times, where the objective is to minimise the makespan. This problem actually corresponds to finding a Hamiltonian path in an edge-weighted, asymmetric graph.

$Fm|prec|\sum_j w_j C_j$ is a problem of scheduling jobs in an m machine flow shop environment subject to precedence constraints among the jobs, where the goal is to minimise the weighted sum of the jobs' completion times.

$Jm|d_j|\sum_j w_j T_j$ is a problem of scheduling jobs in an m machine job shop environment with given due dates for the jobs, where the goal is to minimise the total weighted tardiness.

Candidate Solutions and Neighbourhood Relations

There are several ways of representing candidate solutions for scheduling problems. The most straightforward representation, which explicitly assigns a time-slot (and machine, in the case of multi-machine problems) for processing to each job or operation, has a number of disadvantages compared to other representations. Particular disadvantages concern the definition and the efficient search of neighbourhood relations that maintain the feasibility of candidate solutions.

For many types of scheduling problems, it is convenient and sufficient to represent candidate solutions as permutations or sequences of jobs. Given such a permutation, it is typically straightforward to derive the start and end times of the jobs, if needed. Problems for which such permutations capture all essential features of candidate solutions are known as *permutation scheduling problems* or, more generally, *sequencing problems*; examples for permutation scheduling problems include many single-machine scheduling problems as well as flow shop problems for which the order in which the jobs are processed is required to be the same on all machines.

One advantage of the permutation representation is that it makes it easy to maintain the feasibility of candidate solutions throughout a local search process and gives rise to several natural and commonly used neighbourhood relations, including the following (see Figure 9.2 for graphical illustrations):

Transpose neighbourhood N_t: Two permutations ϕ, ϕ' are transpose neighbours if, and only if, one can be obtained from the other by swapping the positions of two adjacent jobs, that is, $(\phi, \phi') \in N_t$ if, and only if, there is a position i such that $\phi = (\phi(1), \ldots, \phi(i), \phi(i+1), \ldots, \phi(n))$ and $\phi' = (\phi(1), \ldots, \phi(i+1), \phi(i), \ldots, \phi(n))$.

Exchange neighbourhood N_e: Two permutations ϕ, ϕ' are 2-exchange neighbours if, and only if, one can be obtained from the other by exchanging two jobs at arbitrary positions, that is, $(\phi, \phi') \in N_e$ if, and only if, there are positions i, j such that $\phi = (\phi(1), \ldots, \phi(i), \ldots, \phi(j), \ldots, \phi(n))$ and $\phi' = (\phi(1), \ldots, \phi(j), \ldots, \phi(i), \ldots, \phi(n))$. The exchange neighbourhood is also sometimes referred to as the *swap* or *interchange neighbourhood*.

Insertion neighbourhood N_i: Two permutations ϕ, ϕ' are insertion neighbours if, and only if, one can be obtained from the other by removing a job from one position and inserting it at another position, that is, $(\phi, \phi') \in N_i$ if, and only if, there are positions i, j such that $\phi = (\phi(1), \ldots, \phi(i-1),$ $\phi(i), \phi(i+1), \ldots, \phi(j-1), \phi(j), \phi(j+1), \ldots, \phi(n))$, and $\phi' = (\phi(1), \ldots,$ $\phi(i-1), \phi(i+1), \ldots, \phi(j-1), \phi(j), \phi(i), \phi(j+1), \ldots, \phi(n))$ or $\phi' = (\phi(1), \ldots, \phi(i-1), \phi(j), \phi(i), \phi(i+1), \ldots, \phi(j-1), \phi(j+1), \ldots, \phi(n))$. This neighbourhood is occasionally also called the *shift neighbourhood*.

The sizes of these neighbourhoods are $n-1$ for transpose, $n \cdot (n-1)/2$ for exchange and $(n-1)^2$ for insertion, where n is the number of jobs. Note that the transpose neighbourhood is a subset of the larger exchange and insertion neighbourhoods. Because of its small size, it can be evaluated much faster, but local optima w.r.t. the transpose neighbourhood are typically much worse than local optima for the exchange and insertion neighbourhoods. One might assume that the

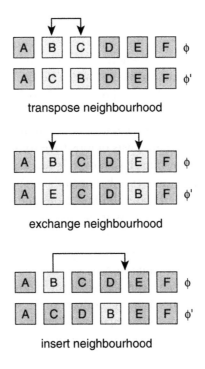

transpose neighbourhood

exchange neighbourhood

insert neighbourhood

Figure **9.2** Illustration of the typical neighbourhood relations used in SLS algorithms for permutation scheduling problems.

largest of these neighbourhoods, insertion, also leads to the highest quality local optima. However, this is not necessarily the case; in general, the question whether the insertion or the exchange neighbourhood leads to better SLS performance needs to be decided on a case-by-case basis, typically by means of empirical analyses. Other neighbourhood relations, such as k-exchange neighbourhoods for $k > 2$, can be defined in a straightforward way (see also Chapter 1, Section 1.5); however, in the context of scheduling problems, such larger neighbourhoods are not commonly used.

In many cases, SLS algorithms for scheduling problems use specific pruning techniques for reducing the effective size of the neighbourhood evaluated in each search step, which can lead to substantial speed-ups of the search process [Congram et al., 2002; Nowicki and Smutnicki, 1996b; Woodruff and Spearman, 1992]. In many cases, these techniques strongly exploit characteristics of the particular scheduling problem to be solved, in particular, characteristics of the given objective function. Examples of such pruning techniques will be given later in this chapter.

Many scheduling problems require candidate solution representations that are different from simple job permutations. For example, consider the airport example in the case where there are multiple runways, or, more generally, any parallel machine problem with release and due dates. In this case, a natural representation for candidate solutions is based on the idea of partitioning the given set of jobs into m subsets, one for each machine, and to specify the sequence of jobs for each machine. Candidate solutions can then be represented by using m ordered lists of jobs (one for each machine).

SLS algorithms based on this representation need to be able to move jobs between machines (i.e., between the respective lists), which requires generalisations of the neighbourhood relations defined above. In this context, often different types of search steps are used for modifying the partial schedules of particular machines and for moving jobs between machines. The latter can be based, for example, on the idea of moving one or more jobs from one machine to another machine, or of symmetrically exchanging jobs between machines.

Yet another representation for candidate solutions will be described in detail in Section 9.4, in the context of SLS algorithms for group shop scheduling problems, where schedules are often represented by specifying a permutation of the operations of the jobs for each machine.

9.2 Single-Machine Scheduling

Single-machine scheduling problems are important for several reasons. Firstly, they occasionally occur in practice, like in the example of scheduling take-offs and landings for an airport with only one runway, or when scheduling jobs for processing on a single CPU. Secondly, in many production environments that give rise to multi-machine scheduling problems, there is one machine that forms a bottleneck for the entire production process; one approach for attacking such a problem is to first solve the scheduling problem that arises for the bottleneck machine before the remaining machines are scheduled. Thirdly, single-machine problems occasionally occur as subproblems in algorithms for solving more complex scheduling problems [Adams et al., 1988].

Some Single-Machine Problems and Complexity Results

Early research in scheduling has shown that a variety of single-machine problems can be solved efficiently (i.e., deterministically in polynomial time). As an example, consider the *Single-Machine Maximum Lateness Problem (SMMLP)*, in which, given n jobs $J_1, \ldots, J_n$ and respective due dates $d_1, \ldots, d_n$, the goal is to minimise the maximum lateness $L_{max} := \max\{L_1, \ldots, L_n\}$. (Using the symbolic notation for scheduling problems, this problem is denoted as $1|d_j|L_{max}$.) An optimal solution for this problem can be found by sequencing the jobs in non-decreasing order of their due dates; this simple scheduling mechanism is known as the *earliest due date (EDD) rule* [Jackson, 1955]. Hence, an optimal solution is determined by sorting the jobs according to their due dates, which can be achieved in time $O(n \cdot \log n)$. Since any schedule that minimises L_{max} also minimises the maximal tardiness, T_{max}, the sequence given by the EDD rule is also optimal for the *Single-Machine Maximum Tardiness Problem (SMMTP)*.

EXAMPLE 9.3 **Minimisation of Maximum Lateness**

Consider the following instance of the SMMLP.

Job	J_1	J_2	J_3	J_4	J_5	J_9
Processing time	3	2	2	3	4	3
Due date	6	13	4	9	7	17

The optimal sequence according to the EDD rule is $\phi^* := (J_3, J_1, J_5, J_4, J_2, J_9)$. From this sequence we can compute the completion time C_i for each job $\phi^*(i)$ by first summing the processing times of job $\phi^*(i)$ and those sequenced before it, that is, $C_i := \sum_{j=1}^{i} p_{\zeta^*(j)}$, where $\zeta^*(j)$ is the index of job $\phi^*(j)$ — i.e., $J_{\zeta^*(j)} = \phi^*(j)$ — and then calculating the corresponding lateness (and tardiness) of each job.

Job	J_3	J_1	J_5	J_4	J_2	J_9
C_i	2	5	9	12	14	17
L_i	−2	−1	2	3	1	0
T_i	0	0	2	3	1	0

Hence, we have that $L_{max} = T_{max} = 3$.

Another way of solving the SMMLP efficiently is by means of a simple construction heuristic that in each step selects the unscheduled job with the earliest due date and appends it to the current partial sequence. Such construction heuristics play an important role in the scheduling literature. In this context, the heuristic mechanism used for selecting the job to be scheduled next is called a *priority rule* or *dispatching rule*; a large variety of these priority rules have been proposed in the literature (for an extensive overview see Haupt [1989]).

Polynomial time algorithms exist for a number of other single-machine scheduling problems [Pinedo, 1995]. However, in most of these cases, minor generalisations of the job characteristics or objective function lead to $\mathcal{NP}$-hard single-machine problems that are difficult to solve in practice. For example, the variant of the SMMTP in which release dates for each job are given (using the formal notation, this problem can be specified as $1|d_j r_j|T_{max}$), has been proven to be (strongly) $\mathcal{NP}$-hard [Lenstra et al., 1977].

Similarly, if we consider the variant of the SMMTP in which the optimisation objective is to minimise the total tardiness instead of the maximum tardiness, we are faced with the $\mathcal{NP}$-hard *Single-Machine Total Tardiness Problem (SMTTP)* [Du and Leung, 1990], formally denoted as $1|d_j|\sum_i T_i$. However, from a complexity theory point of view, the SMTTP is somewhat easier than the SMMTP with release dates, since the SMTTP can be optimally solved by a pseudo-polynomial algorithm with run-time $O(n^4 \cdot \sum_{i=1}^{n} p_i)$, where p_i is the processing time of job J_i [Lawler, 1977].

The Single-Machine Total Weighted Tardiness Problem

In a straightforward extension of the Single-Machine Total Tardiness Problem, for each job an integer weight is specified which represents its relative importance, and the objective is to minimise the sum of the weighted tardiness values over all jobs. This scheduling problem is known as the *Single-Machine Total Weighted Tardiness Problem (SMTWTP)*. More formally, an instance of the SMTWTP consists of n jobs that have to be processed on a single machine. For each job J_i, a processing time p_i, a weight w_i and a due date d_i are given, and all jobs become available for processing at time zero. The goal is to find a schedule — that is, a permutation of the n jobs — that minimises the total weighted tardiness $\sum_{i=1}^{n} w_i \cdot T_i$.

EXAMPLE 9.4 **A Simple SMTWTP Instance**

Consider an extension of Example 9.3, where each job has an additional weight w_j; the problem data are given as:

Job	J_1	J_2	J_3	J_4	J_5	J_6
Processing time	3	2	2	3	4	3
Due date	6	13	4	9	7	17
Weight	2	3	1	5	1	2

When using the optimal sequence for the SMMLP instance from Example 9.3, $\phi := (J_3, J_1, J_5, J_4, J_2, J_6)$, the following completion time, tardiness and weighted tardiness values are obtained:

Job	J_3	J_1	J_5	J_4	J_2	J_6
C_i	2	5	9	12	14	17
T_i	0	0	2	3	1	0
$w_i \cdot T_i$	0	0	4	15	3	0

Hence, the total weighted tardiness for this sequence is $\sum_{i=1}^{6} w_i \cdot T_i = 0 + 0 + 4 + 15 + 3 + 0 = 22$. Note that the given sequence ϕ is not optimal for this SMTWTP instance. The optimal solution is $\phi^* := (J_3, J_1, J_4, J_2, J_5, J_6)$, which has a total weighted tardiness of 7.

Since the SMTWTP contains the SMTTP as a special case, it is $\mathcal{NP}$-hard; however, unlike the unweighted case, the SMTWTP is $\mathcal{NP}$-hard in the strong sense, and hence, there exists no pseudo-polynomial time algorithm for optimally solving this problem [Lenstra et al., 1977].

Most recent empirical studies of SMTWTP algorithms use a set of randomly generated instances that are available from OR-Library [Beasley, 2003]. This benchmark set comprises instances with 40, 50 and 100 jobs, which were generated for an empirical study of SLS algorithms for the SMTWTP by Crauwels et al. [1998]. The instances were obtained by generating the processing times p_i and the weights w_i for each job J_i randomly according to a uniform distribution from the integer intervals $\{1, \ldots, 100\}$ and $\{1, \ldots, 10\}$, respectively. The due dates are randomly drawn integers from the set $\{\max\{0, (1 - \text{TF} - \text{RDD}/2)\} \cdot \sum_{i=1}^{n} p_i, \ldots, (1 - \text{TF} + \text{RDD}/2) \cdot \sum_{i=1}^{n} p_i\}$, where TF, the tardiness factor, and RDD, the relative range of due dates, are two parameters of this family of instance distributions. The benchmark set comprises instances for various values of TF and RDD between 0.2 and 1.

In general, high TF values result in due dates that are relatively small compared to the completion time of the last job, which corresponds to $\sum_{i=1}^{n} p_i$. Hence, in this case many jobs can be expected to be late in an optimal solution. A low TF value results in relatively loose due dates, which means that in an optimal solution many jobs can be expected to be finished before their due date. The RDD value mainly determines the variability of the jobs' due dates; the lower RDD, the more similar are the values of the due dates among the jobs. Different TF and RDD values lead to substantial differences in the hardness of the respective SMTWTP instance. In particular, instances with small TF values appear to be generally easily solved by SLS algorithms.

Empirical evidence indicates that instances from these random SMTWTP instance distributions are hard to solve for complete algorithms. In particular, a specific branch & bound algorithm [Potts and Wassenhove, 1985] has been found to be unable to solve a substantial number of instances with 50 jobs within 120 CPU seconds on an HP 9000-G50 computer; furthermore, it has been reported that the computation times for solving size 100 instances with this branch & bound algorithm are prohibitive [Crauwels et al., 1998]. In the case of the SMTTP, however, there are complete algorithms that can solve non-trivial instances of up to 500 jobs [Szwarc et al., 2001], which — in agreement with the previously mentioned complexity results — suggests that this problem is considerably easier than the SMTWTP. As a consequence, most of the research in SLS algorithms for single-machine tardiness scheduling has been focused on the SMTWTP.

An empirical study of various types of SLS algorithms for the SMTWTP, including Tabu Search, Simulated Annealing and Genetic Algorithms, indicates that by using restart strategies, very high-quality schedules can be obtained [Crauwels et al., 1998]. In the following we describe two recent SLS algorithms that incorporate diversification mechanisms similar to multiple restarts and are among the best-performing algorithms for the SMTWTP known to date. But first, we give some details on the effectiveness of construction heuristics and iterative improvement algorithms for the SMTWTP.

Simple Hybrid Local Search Algorithms for the SMTWTP

One of the simplest approaches to generating good candidate solutions to sequencing problems such as the SMTWTP is to use construction heuristics. For the greedy selection in the construction steps, the following three dispatching rules are commonly used:

Earliest due date (EDD). As previously explained, jobs are sequenced in non-decreasing order of their due dates, d_j. Recall that this is the heuristic rule that guarantees minimisation of T_{max}.

Modified due date (MDD). This dispatching rule sequences jobs in non-decreasing order of their modified due dates $mdd_j := \max\{C + p_j, d_j\}$, where C is the sum of the processing times of jobs that have already been sequenced [Baker and Bertrand, 1982; Bauer et al., 1999].

Apparent urgency (AU). Under this rule, jobs are sequenced in non-decreasing order of their apparent urgency $au_j := (w_j/p_j) \cdot e^{-(\max\{d_j - C_j, 0\})/k\bar{p}}$ [Morton et al., 1984]. Here, $\bar{p}$ denotes the average processing time of the remaining jobs, and k is a parameter, and $C_i := C + p_i$.

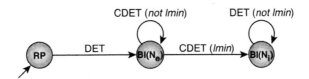

Figure 9.3 GLSM model for the 2-phase local search algorithm that first performs IBI(N_e) and then IBI(N_i). GLSM state RP initialises the search at a randomly selected permutation, BI(N_e) performs a best improvement step in neighbourhood N_e, and BI(N_i) performs a best improvement step in neighbourhood N_i.

Note that unlike EDD, MDD and AU select the next job to be scheduled based on a criterion that depends on the partial sequence constructed so far. The latter dispatching rules have higher time complexity, but they typically achieve better solution qualities (see also Example 9.5).

The candidate solutions obtained from a construction heuristic can often be substantially improved using local search methods, such as iterative improvement. For the SMTWTP, various neighbourhoods can be used in this context, in particular the previously discussed exchange and insertion neighbourhoods, N_e and N_i. In the following, we denote the iterative best improvement methods based on N_e and N_i as IBI(N_e) and IBI(N_i), respectively; both of these terminate when they reach a local minimum w.r.t. to their underlying neighbourhood relation.

Since the local minima of the evaluation function (here: the total weighted tardiness of a given candidate schedule) for the two neighbourhood relations generally do not coincide, further improvements in solution quality can often be achieved by performing an iterative improvement search based on N_e after reaching a local mininum for N_i or vice versa. The simplest way of exploiting this observation is by means of a 2-phase local search algorithm that either performs first IBI(N_e) and then IBI(N_i), or vice versa. A GLSM model for the local search algorithm that results when first applying IBI(N_e) and then IBI(N_i) is shown in Figure 9.3. Note that this approach can be seen as a particular variant of Variable Neighbourhood Descent (VND) [Mladenović and Hansen, 1997; Hansen and Mladenović, 1999]; see Section 2.1 (page 66*f.*).

EXAMPLE 9.5 **Effectiveness of 2-Phase Local Search for the SMTWTP**

In order to illustrate the performance differences between the three construction heuristics, EDD, MDD and AU, when used alone or in combination

Construction Heuristic	No Local Search			$+IBI(N_e)$			$+IBI(N_i)$		
	Δ_{avg}	n_{opt}	t_{avg}	Δ_{avg}	n_{opt}	t_{avg}	Δ_{avg}	n_{opt}	t_{avg}
EDD	135	24	0.001	0.62	33	0.140	1.19	38	0.64
MDD	61	24	0.002	0.65	38	0.078	1.31	36	0.77
AU	21	21	0.008	0.92	28	0.040	0.56	42	0.26

Construction Heuristic	$+IBI(N_e) + IBI(N_i)$			$+IBI(N_i) + IBI(N_e)$		
	Δ_{avg}	n_{opt}	t_{avg}	Δ_{avg}	n_{opt}	t_{avg}
EDD	0.24	46	0.20	0.47	48	0.67
MDD	0.40	46	0.14	0.44	42	0.79
AU	0.59	46	0.10	0.21	49	0.27

Table **9.1** Performance comparison of construction heuristics alone and in combination with various local search methods for the SMTWTP; Δ_{avg} denotes the average percentage deviation from the best known solution, n_{opt} is the number of instances (out of 125) for which the best known solution quality was reached, and t_{avg} is the average run-time in CPU seconds on a Pentium III 450MHz CPU with 256MB RAM running Red Hat Linux 6.1. (For further details, see text.)

with several simple and hybrid iterative best improvement methods on the previously described randomly generated SMTWTP instances, the following experiment was performed. All combinations of the single-phase iterative best improvement on the exchange and insertion neighbourhoods, IBI(N_e) and IBI(N_i), and of the two 2-phase algorithms, IBI(N_e)+IBI(N_i) and IBI(N_i)+IBI(N_e), with the three construction heuristics, EDD, MDD and AU, as well as the construction heuristics without any subsequent local search were run on a test-set comprising the 125 SMTWTP instances with 100 jobs from OR-Library [Beasley, 2003]; the test set contains 5 instances for each of the 25 combinations of parameter values TF and RDD from the set $\{0.2, 0.4, 0.6, 0.8, 1.0\}$.

Each algorithm was run once on each of the 125 problem instances and the solution quality reached as well as the run-time were recorded. This evaluation method is justified by the fact that the algorithms studied here are deterministic except for random tie-breaking in the case of multiple best heuristic or evaluation function values, which occurs relatively rarely. Solution qualities were then compared to the best solution qualities known for each respective instance.

Table 9.1 shows the results of this experiment. As indicated before, some of the instances are very easy, and the best known solutions are found even by the construction heuristics alone. Yet, when averaging the results over the entire test-set, all three construction heuristics achieve only rather poor solution qualities, which can be substantially improved by a subsequent local search phase. As expected, the 2-phase algorithms perform consistently better than the underlying simple iterative best improvement methods at the cost of only a relatively modest increase in run-time.

An ACO Algorithm for the SMTWTP

The ACS-BSD algorithm, developed by den Besten, Stützle and Dorigo [2000], is essentially an application of Ant Colony System (ACS) [Dorigo and Gambardella, 1997], a particular ACO algorithm, to the SMTWTP. The excellent performance obtained by ACS-BSD crucially depends on its use of the construction heuristics and subsidiary 2-phase local search methods described in the previous section.

In ACS-BSD, pheromone values τ_{ij} are associated with each assignment of a job J_j to a sequence position i. During the construction phase, each ant builds candidate solutions (i.e., permutations of jobs), starting with an empty sequence, by iteratively appending unscheduled jobs to the partial sequence constructed so far. The job to be appended next is selected based on the pheromone values τ_{ij} and heuristic values η_{ij}. The latter are obtained from the previously discussed dispatching rules; more precisely, ACS-BSD uses $\eta_{ij} := 1/au_j$ if the empirically measured tardiness factor (TF) of the given instance is larger than 0.3, and $\eta_{ij} := 1/mdd_j$ otherwise. This mechanism is based on empirical results indicating that for low TF, better performance is obtained by using heuristic values derived from the MDD rule.

The construction process of ACS-BSD is based on the following mechanism: Given a current partial candidate sequence of length $i - 1$, with probability q the job J_j to be selected for position i is chosen such that $\tau_{ij} \cdot \eta_{ij}^\beta$ is maximised; in the remaining cases (i.e., with probability $1 - q$), job J_j is chosen randomly with probability

$$
p_{ij} := \begin{cases} \dfrac{\tau_{ij} \cdot \eta_{ij}^\beta}{\sum_{l \in U} \tau_{il} \cdot \eta_{il}^\beta} & \text{if } j \in U \\ 0 & \text{otherwise,} \end{cases} \tag{9.1}
$$

where U is the index set of unscheduled jobs. Hence, with probability q, the ant makes the best choice as indicated by the combination of pheromone trails and the heuristic information (exploitation), while with probability $1 - q$, it performs a probabilistically biased exploration.

Upon closer examination of the construction process, it was noted that if a job J_j is not placed at the position i for which $\tau_{ij} \cdot \eta_{ij}^{\beta}$ is maximal, it often ends up in one of the last positions in the sequence. Because the absolute position of a job in the sequence is important, ACS-BSD uses a candidate list, from which in each construction step the job to be added to the current partial sequence is chosen [Dorigo and Gambardella, 1997; Stützle, 1998a]. This candidate list is built as follows: The incumbent candidate solution $\widehat{\phi}$ is scanned starting from the first position, and each job that is not contained in the current partial sequence is included in the candidate list; this process is stopped when the candidate list has reached a maximum length of *cand* jobs. The intuition behind this mechanism is that jobs appearing early in some high-quality sequences should also occur early in other high-quality sequences. Using *cand* := 20 was found to result in reasonably good empirical performance of the algorithm.

Experiments with variants of ACS-BSD that use different subsidiary local search algorithms have shown that the previously observed performance advantage of the 2-phase hybrid algorithms over simpler iterative improvement algorithms carries over to the respective ACO algorithms. However, which of the two 2-phase local search algorithms, IBI(N_e)+IBI(N_i) and IBI(N_i)+IBI(N_e) results in the best performance was observed to be strongly dependent on the given problem instance [den Besten et al., 2000]. Therefore, the final ACS-BSD algorithm uses a heterogeneous colony of ants, where each of these two local search algorithms is applied by one half of the ants.

In ACS-BSD (as in Ant Colony System in general), two forms of pheromone update are applied. Immediately after each construction step of an individual ant, the pheromone trail values corresponding to the respective job assignment are reduced by applying the *local pheromone update rule*:

$$\tau_{ij} := (1 - \xi) \cdot \tau_{ij} + \xi \cdot \tau_0,$$

where $\xi, 0 < \xi \leq 1$ and τ_0 are two parameters. In ACS-BSD, τ_0 is set to a small constant value $\tau_0 := 1/(n \cdot T_{MDD})$, where T_{MDD} is the weighted tardiness of the (partial) sequence as determined by applying the MDD rule. The effect of this local pheromone update is to make the decision of assigning job J_j to position i less desirable for the other ants, which diversifies the set of sequences generated in the construction phase.

After all ants have completed the sequence construction and the subsequent local search phase, a global pheromone update is performed during which the pheromone values that correspond to the incumbent solution are modified as follows:

$$\tau_{ij} := (1 - \rho) \cdot \tau_{ij} + \rho \cdot \Delta\tau_{ij},$$

where $\rho, 0 < \rho \leq 1$, is the pheromone evaporation rate, and $\Delta\tau_{ij} := 1/\widehat{T}$, where $\widehat{T}$ is the total weighted tardiness of the incumbent solution. (Notice that all other pheromone values remain unchanged; in particular, evaporation is not applied to all τ_{ij}.)

Empirical results indicate that ACS-BSD is one of the best-performing SMTWTP algorithms currently known; when applied to the randomly generated SMTWTP instances from OR-Library described above, it consistently found provably optimal or best known solutions for all instances within at most 1200 CPU seconds on a PC with a 450MHz Pentium III CPU and 256MB RAM running Red Hat Linux 6.1.

Recently, another ACO algorithm for the SMTWTP has been developed by Merkle and Middendorf [2003]. This algorithm, ACO-MM, differs in several important aspects from ACS-BSD, including the use of a pheromone summation rule that uses the sum of the pheromone values τ_{lj} for $l < i$ when choosing a job for position i, as well as different heuristic values. However, the subsidiary local search procedure used in ACO-MM performs much worse than the 2-phase local search approach described above; it is therefore not surprising that ACO-MM does not reach the performance of ACS-BSD. Nevertheless, it is likely that by using some of the features of ACO-MM, the performance of ACO-BSD can be further improved.

Iterated Dynasearch for the SMTWTP

Iterated local search (ILS) algorithms are known to be among the best-performing SLS methods for a variety of combinatorial problems. Hence, it is not too surprising that ILS can be successfully applied to the SMTWTP; based on a powerful dynasearch local search algorithm, the ILS algorithm by Congram, Potts and van de Velde, *ILS-CPV*, is one of the state-of-the-art methods for solving the SMTWTP [Congram et al., 2002].

Recall from Section 2.1 (page 70*f.*) that Dynasearch is based on the idea of performing complex iterative improvement steps that are assembled from a set of mutually independent, simple search steps. In the case of the dynasearch algorithm used in ILS-CPV, the simple steps are based on the standard exchange neighbourhood, N_e. Given a candidate sequence ϕ, the steps that exchange job

$\phi(i)$ with $\phi(j)$ and $\phi(k)$ with $\phi(l)$ are independent w.r.t. their effect on the total weighted tardiness if $\max\{i, j\} \leq \min\{k, l\}$ or $\min\{i, j\} \geq \max\{k, l\}$. Dynasearch effectively performs iterative best improvement in the neighbourhood consisting of the sequences reachable from the current candidate sequence by any set of independent improving exchange steps. Details on the methods used for efficiently searching and evaluating this neighbourhood can be found in the in-depth section on the next page.

ILS-CPV starts from an initial candidate solution generated by the AU construction heuristic [Morton et al., 1984] and uses an efficient implementation of the previously mentioned dynasearch algorithm as its subsidiary local search procedure. During each perturbation phase, first, six random exchange steps are performed, in which the positions of the two jobs to be exchanged are determined according to a uniform distribution. Then, in order to obtain a better starting point for the subsequent local search phase, the resulting sequence is scanned, starting from the first position, and any adjacent non-late jobs that are not in EDD order are transposed. After 100 iterations of the ILS algorithm, in order to achieve further diversification, during this scan any adjacent non-late jobs in EDD order are also transposed with a probability of $1/3$, unless this would result in one of the jobs becoming late.

The acceptance criterion uses a form of backtracking to the incumbent candidate solution. In particular, for l ILS iterations (where l is a parameter of the algorithm), each new local optimum is accepted regardless of its quality; this corresponds to a random walk phase of the iterated local search process. If in these l iterations the incumbent candidate solution, $\hat{s}$, has not been improved, the random walk phase starts again from $\hat{s}$. The search is terminated either after a fixed number of iterations or when a given limit on the maximum CPU time is reached.

ILS-CPV has been evaluated on the same set of benchmark instances from OR-Library as ACS-BSD and reported to achieve similarly excellent performance in terms of its ability of consistently finding optimal solutions [Congram et al., 2002]. Although so far, a direct comparison between the two algorithms has not been performed, based on the published results for both algorithms, ILS-CPV appears to reach optimal solutions faster than ACS-BSD.

Recently, the dynasearch algorithm of Congram et al. has been further enhanced by additionally considering simple search steps based on the insertion neighbourhood, N_i, as components for complex dynasearch steps [Grosso et al., 2004]. An ILS algorithm based on this enhanced dynasearch procedure achieves substantially better performance than ILS-CPV and typically reaches the same solution quality as ACS-BSD in about $1/10$th of the computation time. This ILS algorithm is currently the best known algorithm for the SMTWTP.

A complex dynasearch step consists of the best possible combination of independent exchange steps that can be applied to a given sequence ϕ. Dynasearch finds this step by means of a dynamic programming algorithm (see, e.g., Bertsekas [1995]), hence the name 'Dynasearch'. Before explaining how this works, we need to introduce some notation. Let $\phi := (\phi(1), \ldots \phi(n))$ be the current candidate sequence, and let ϕ_j be the partial sequence that has minimum total weighted tardiness among the sequences that can be obtained from $(\phi(1), \ldots, \phi(j))$ by applying independent exchange steps. Furthermore, let $\Delta(j)$ be the maximum improvement incurred by independent steps that are applied to $(\phi(1), \ldots, \phi(j))$ to obtain ϕ_j, and let $\delta(i, j)$ denote the cost reduction resulting from an exchange of the jobs at positions i and j in ϕ.

The essential part is now to consider the two possibilites of how ϕ_j can be formed.

- ϕ_j is simply obtained by appending job $\phi(j)$ to ϕ_{j-1}.

- ϕ_j is obtained by appending job $\phi(j)$ to ϕ_{j-1} and immediately exchanging job $\phi(j)$ with job $\phi(i+1)$, where $1 \leq i < j - 1$. Hence, the subsequence $(\phi(i+1), \phi(i+2), \ldots, \phi(j-1), \phi(j))$ becomes $(\phi(j), \phi(i+2), \ldots, \phi(j-1), \phi(i+1))$, which is appended to ϕ_i to form ϕ_j.

In the first case, we have $\Delta(j) := \Delta(j-1)$, while in the second case, we have $\Delta(j) := \Delta(i) + \delta(i+1, j)$. (The independence requirement of individual steps in Dynasearch is necessary to allow for the summation of $\Delta(i)$ and $\delta(i+1, j)$ in the second case.) From these possibilites, the one that results in the largest $\Delta(j)$ is chosen.

The values $\Delta(j)$ can be iteratively computed as follows. Initialise $\Delta(0) := \Delta(1) := 0$; then, $\Delta(j)$, $j = 2, \ldots, n$ is determined using the recursive formula

$$\Delta(j) := \max\{\Delta(j-1), \max\{\Delta(i) + \delta(i+1, j) \mid 0 \leq i \leq j - 1\}\}. \qquad (9.2)$$

Note that $\Delta(n)$ is the maximal improvement that can be achieved by a complex search step; the corresponding complex search step can be determined using a trace-back mechanism that utilises information on the components that contributed the maximum value in Equation 9.2 in each step of the recursion.

When implementing Dynasearch for the SMTWTP, it is very important to compute the values $\delta(i, j)$ as efficiently as possible; in this context, several SMTWTP-specific speedup techniques can be used. The first of these is based on the observation that an exchange between two jobs at positions i and j ($i < j$) can only affect the contribution of these two jobs and the jobs sequenced between them towards the total weighted tardiness value of the given sequence. Hence, only these contributions need to be considered when evaluating the respective search step.

Additionally, before evaluating the effectiveness of a step, a pre-testing is performed, which determines whether a search step can potentially reduce the total weighted tardiness by more than some specified value Δf. As an example of this technique, consider the case where jobs $\phi(i)$ and $\phi(j)$ are exchanged and $p_{\zeta(j)} \geq p_{\zeta(i)}$, where $\zeta(k)$ is the index of job $\phi(k)$ — i.e., $J_{\zeta(k)} = \phi(k)$. Consider first the case of a first-improvement algorithm and the case $\Delta f = 0$, i.e., the pre-test determines whether a step has the potential to improve the current solution. If the combined weighted tardiness of jobs $\phi(i)$ and $\phi(j)$, $w_{\zeta(i)}T_{\zeta(i)} + w_{\zeta(j)}T_{\zeta(j)}$, increases, the step cannot yield an improvement, because the jobs between positions i and j will increase their completion times (recall that $p_{\zeta(j)} \geq p_{\zeta(i)}$), and therefore their contribution to the total weighted tardiness will increase or, at best, stay the same. In the case of a best-improvement pivoting rule, Δf is set to the best improvement found so far, and it needs to be checked whether exchanging jobs $\phi(i)$ and $\phi(j)$ can lead to an improvement larger than Δf. The case of exchanging jobs $\phi(i)$ and $\phi(j)$ if $p_{\zeta(j)} \leq p_{\zeta(i)}$ is handled in a more complicated way by exploiting additional data structures [Congram et al., 2002].

Finally, if the pre-testing is passed, the effect of exchanging jobs $\phi(i)$ and $\phi(j)$ has to be computed exactly. The evaluation of the respective search step is sped-up based on the observation that in the case of $p_{\zeta(j)} \leq p_{\zeta(i)}$, subsequences of consecutive jobs between positions i and j that are non-late will remain non-late, while in the case $p_{\zeta(j)} \geq p_{\zeta(i)}$, with the use of additional data structures, the evaluation of the effect of steps involving consecutive late jobs can be performed more efficiently.

9.3 Flow Shop Scheduling

So far, we have considered only single-machine problems. Now, we turn our attention to more complex scheduling problems, where each job consists of several operations to be processed on a number of different machines. In this section, we focus on flow shop scheduling problems, in which the operations of each job are performed in the same order, that is, the order in which a job passes through the machines is the same for all jobs. Flow shop scheduling problems arise, for example, in steel-rolling mills or in chemical plants. Generally, they mostly occur in application contexts where expensive processing equipment and production processes provide a strong need for optimising the given environment [Dudek et al., 1992].

In flow shop scheduling problems, it is usually assumed that all jobs are available for processing at time zero, pre-emption is not allowed, each operation is to be performed on a specific machine, each machine can process at most one job at a time, and each job can be processed by at most one machine at a time. Additional problem characteristics can vary and give rise to many variants of flow shop scheduling [Pinedo, 1995].

A first, important characteristic is how jobs pass through the buffers that may be required between successive machines for holding jobs that have to wait for being processed. If these are queues that operate based on the *first come – first served* principle, the jobs pass through all machines in the same order. (Note that this condition is stronger than the one stating that the order of the machines through which a job passes is the same for all jobs.) If this condition holds, any optimal schedule corresponds to a permutation of the jobs. Such scheduling problems are known as *permutation flow shop problems*. If, on the other hand, changes in the sequence in which jobs are processed by the machines are allowed, which is a realistic assumption in certain real-world applications, the corresponding flow shop scheduling problems become much harder to solve in practice.

Frequently, in a flow shop environment, limitations on the capacity of the buffer between successive machines occur. If such capacity constraints are given, *blocking* may occur: If the buffer between machines M and M' has reached its maximal capacity, a job that has completed processing on M cannot leave that machine, which prevents M from processing any other job.

An extreme case arises if no buffers exist between the machines and no delays are allowed between processing any two subsequent operations of the same job; this constraint is called the *no-wait requirement*. In order to satisfy this requirement, the processing of jobs on the first machine may have to be delayed, in order to ensure that once they enter this machine, all processing stages can be completed without any delays. The no-wait requirement arises, for example, in the production of steel [Pacciarelli and Pranzo, 2000], where delays between processing stages may affect material characteristics; it is also characteristic for so-called just-in-time production environments, which are gaining increasing importance in many practical applications.

These two features, blocking and no-wait requirements, also occur in other multiple machine environments, such as the group shop problems discussed in the next section. For an overview of scheduling models with blocking and no-wait requirements we refer to Hall and Sriskandrajah [1996].

The Permutation Flow Shop Problem

It is clear that further complicating factors, such as sequence-dependent setup times, due dates, release dates or pre-emption, can easily be added, and that many different objective functions may be considered in the context of flow shop scheduling. In the following, we focus on permutation flow shop problems in which the capacity of the buffer between the machines is unlimited, that is, blocking does not occur and jobs may wait for processing. Furthermore, we consider the optimisation objective of minimising the completion time of the last job (i.e., the makespan C_{max}). This widely studied flow shop scheduling problem is known as the *Permutation Flow Shop Problem (PFSP)*.

Formally, a PFSP instance is given by a set of m machines $M_1, \ldots, M_m$ and a set of n jobs $J_1, \ldots, J_n$, where each job J_i consists of m operations $o_{i1}, \ldots, o_{im}$ that have to be performed on machines $M_1, \ldots, M_m$ in that order, with processing time p_{ij} for operation o_{ij}. The objective is to find a job sequence ϕ that minimises the makespan. (Using the formal notation for scheduling problems, the PFSP is denoted as $Fm\|C_{max}$.)

Computing the completion time of operations and/or jobs in the PFSP is slightly more complicated than in the single-machine case. Formally, given a sequence $\phi := J_{\zeta(1)}, \ldots, J_{\zeta(n)}$ — where $\zeta(k)$ is the index of job $\phi(k)$ — the completion time $C_{\zeta(k)\,j}$ of a job $J_{\zeta(k)}$ on a machine M_j, that is, the completion time of operation $o_{\zeta(k)\,j}$, can be computed using the following set of recursive equations.

$$C_{\zeta(1)\,j} := \sum_{h=1}^{j} p_{\zeta(1)\,h} \qquad\qquad j = 1, \ldots, m$$

$$C_{\zeta(k)\,1} := \sum_{h=1}^{k} p_{\zeta(h)\,1} \qquad\qquad k = 1, \ldots, n \qquad (9.3)$$

$$C_{\zeta(k)\,j} := \max\{C_{\zeta(k-1)\,j}, C_{\zeta(k)\,(j-1)}\} + p_{\zeta(k)\,j} \qquad \begin{array}{l} k = 2, \ldots, n; \\ j = 2, \ldots, m \end{array}$$

EXAMPLE 9.6 **A Simple Permutation Flow Shop Problem Instance**

Consider a PFSP instance with five jobs, three machines and the following processing times:

Job	J_1	J_2	J_3	J_4	J_5
p_{j1}	3	3	4	2	3
p_{j2}	2	1	3	3	1
p_{j3}	4	2	1	2	3

Using the recurrence relations from above, the makespan of the sequence $\phi := (J_1, J_2, J_3, J_4, J_5)$ can be computed to be 21. The resulting schedule is graphically represented as a Gantt-chart in Figure 9.4.

Some special cases of the PFSP are polynomially solvable. The best known example is the 2-machine case (i.e., $m = 2$), which can be solved by Johnson's algorithm in time $O(n \cdot \log n)$ [Johnson, 1954]. This algorithm first partitions the jobs into two sets S_1 and S_2, such that S_1 contains all jobs for which $p_{i1} \leq p_{i2}$ and S_2 contains all remaining jobs (these satisfy $p_{i1} > p_{i2}$). Then, first the jobs in S_1 are sequenced in non-decreasing order of p_{i1}, and then the jobs in S_2 are sequenced in non-increasing order of p_{i2}. This algorithm can be extended to some other special

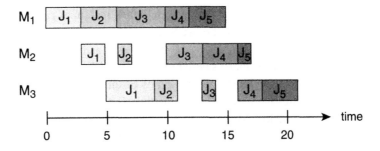

Figure 9.4 Gantt chart illustrating the schedule for the sequence $\phi := (J_1, J_2, J_3, J_4, J_5)$ and the processing times given in Example 9.6. The three machines are numbered M_1, M_2 and M_3; job numbers are indicated in the boxes.

cases [French, 1982]. However, for three or more machines, the PFSP has been proven to be strongly $\mathcal{NP}$-hard [Garey et al., 1976a]. Furthermore, PFSP variants with different optimisation objectives, like minimising the sum or the weighted sum of the completion times, have been shown to be $\mathcal{NP}$-hard even in the two machine case [Garey et al., 1976a]. For an overview of complexity results on the PFSP we refer to Pinedo [1995] (page 348*f*.).

Like in the case of the SMTWTP, algorithms for the PFSP have been tested mainly on randomly generated instances, some of which are available from OR-Library [Beasley, 2003]. Probably the most widely used set of benchmark instances for the PFSP is the one generated by Taillard [1993]. It includes ten instances for each possible combination of $n \in \{20, 50, 100\}$ and $m \in \{5, 10, 20\}$, as well as ten instances each with $n = 200$ and $m \in \{10, 20\}$, and ten instances with $n = 500$ and $m = 20$; the processing times for all operations were randomly generated according to a uniform distribution over $[1, \ldots, 99]$.

A large set of additional benchmark instances has been introduced by Watson et al. [2002; 2003]. While this set also contains instances similar to those in Taillard's benchmark set, it is mainly comprised of randomly generated structured PFSP instances. The structured instances have processing times that were generated taking into account two main features. In *job-correlated* instances, it is assumed that the processing times of a job are similar on the different machines, while in *machine-correlated* instances, the processing times for any operation on a machine are correlated between all jobs. In addition to purely job-correlated and purely machine-correlated instances, instances with a mixture between job- and machine-correlations were generated. This latter class of instances is inspired by the observation that such a mixture is often found in real manufacturing shop floors [Panwalker et al., 1973]. For each class, instances with 20 machines and 20, 50, 100 and 200 jobs were generated.

Not all commonly used benchmark instances can be considered challenging for current state-of-the-art SLS algorithms for the PFSP. Generally, instances with only few machines are relatively easily solved. For example, for all instances from Taillard's benchmark set with five or ten machines, provably optimal solutions have been determined by branch & bound algorithms. In addition, current state-of-the-art SLS algorithms for the PFSP find these optimal solutions rather quickly. The 20 machine instances, on the other hand, are significantly harder, and for many of them, provably optimal solutions are unknown. Interestingly, it has been demonstrated in a recent empirical study that even rather basic SLS algorithms for the PFSP find optimal (or best known) solution qualities on many of the structured instances by Watson et al. [2002]. This indicates that, compared to unstructured instances, structured PFSP instances appear to be relatively easy to solve.

Iterated Local Search for Permutation Flow Shop Scheduling

The fact that iterated local search methods achieve very good performance on single-machine scheduling problems, such as the SMTWTP, suggests that they could also be good candidates for solving permutation scheduling problems that arise in more complex machine environments. Stützle's ILS algorithm for the PFSP, *ILS-S-PFSP*, is a rather straightforward instantiation of the standard ILS framework [Stützle, 1998b].

ILS-S-PFSP initialises the search using the NEH heuristic by Nawaz et al. [1983], which is the best-performing construction heuristic for the PFSP for a wide variety of problem instances [Turner and Booth, 1987; Koulamas, 1998; Ruiz and Maroto, to appear]. NEH is an insertion heuristic similar to those discussed for the TSP (see Section 8.2, page 369). It first computes for each job J_i the sum p_i of the processing times of its operations, and then orders the jobs according to non-increasing p_i. Then, the jobs are considered in this order for inclusion into a partial sequence; at each step, the next job is inserted into the position where it leads to a minimum increase in the makespan. When using basic speed-up techniques for the makespan computations that need to be performed when testing all possible insertion points for the next job, the NEH heuristic runs in time $O(n^2 \cdot m)$ [Taillard, 1990].

As for the SMTWTP, local search methods for the PFSP can be based on any of the commonly used neighbourhoods N_t (transpose), N_e (exchange) and N_i (insertion). The subsidiary local search procedure used in ILS-S-PFSP is based on N_i, which was found to result in significantly better performance than a local search in N_t. Furthermore, in the case of the PFSP, the N_i neighbourhood can be evaluated much faster than N_e [Taillard, 1990]. More precisely, the possible insertion points for any given job $\phi(i)$ can be evaluated in time $O(n \cdot m)$, which leads to an overall run-time of $O(n^2 \cdot m)$ for completely evaluating the entire N_i neighbourhood of any candidate sequence [Taillard, 1990].

For large PFSP instances, the use of iterative best improvement local search, as proposed by Taillard [1990], results in a very high time-complexity for each individual local search step. Therefore, ILS-S-PFSP uses a modified iterative first improvement approach that results in substantially faster search steps. First, a random permutation of the job indices is generated. Then, in each step the next index i from this permutation is selected and all possibilities for inserting job $\phi(i)$ are examined; these two stages are alternated until at least one improving sequence is found during such a partial neighbourhood scan, at which point job $\phi(i)$ is inserted at the sequence position that leads to the maximal reduction in makespan. (Note that this constitutes a type of first improvement mechanism, since, although a best improvement w.r.t. the neighbouring sequence reachable by re-scheduling $\phi(i)$ is achieved, the overall best improving neighbour of the given sequence may require moving a different job.) These steps are iterated until a local minimum is found.

Generally, a good perturbation mechanism often performs simple modifications of a candidate solution that cannot be directly reversed by the local search procedure. In ILS-S-PFSP, random steps in N_t and N_e are used in this context. Small perturbations were empirically found to be sufficient for achieving very good performance of the overall ILS algorithm. However, the optimal number of elementary perturbation steps was observed to be instance dependent. (Interestingly, in the context of this ILS algorithm, the optimal perturbation strength appears to decrease with increasing instance size.) Based on these observations, the perturbation procedure used in ILS-S-PFSP consists of two random steps in N_t, followed by one random step in N_e with the additional restriction that $|i - j| \leq \max\{n/5, 30\}$, where i and j are the positions of the two jobs exchanged by the step in N_e, and n is the number of jobs. This latter restriction limits the strength of the perturbation and avoids too strong a disruption of the current solution.

The acceptance criterion used in ILS-S-PFSP accepts a new candidate solution based on the Metropolis condition as given by Equation 2.1 (page 75) with a constant temperature value $T := \bar{p}/15$, where $\bar{p}$ is the average processing time over all operations. This acceptance mechanism was chosen based on an extensive study of various alternatives.

Empirical results have shown that ILS-S-PFSP finds substantially higher quality solutions than the underlying iterative improvement algorithm [Stützle, 1998b]. Despite the fact that it is a conceptually rather straightforward ILS algorithm, ILS-S-PFSP has been empirically shown to reach very high-quality solutions for all benchmark instances available from OR-Library, similar to fine-tuned implementations of the best-performing SLS algorithms for the PFSP; however, it takes considerably longer than other state-of-the-art PFSP algorithms (such as the tabu search algorithm presented in the following) for reaching these high solution qualities [Stützle, 1998b]. It is likely that the performance of ILS-S-PFSP can be further improved by using a more powerful and efficiently implemented subsidiary local search procedure.

Critical Paths and Block Structure

The subsidiary local search procedure of ILS-S-PFSP is based on a standard neighbourhood relation and does not take into account specific properties of the PFSP. By exploiting such properties, better performing local search algorithms for this specific scheduling problem can be obtained.

It may be noted that in a schedule for the PFSP, some operations cannot be delayed without increasing the makespan of a sequence. Operations for which this is true are called *critical*, while all other operations are called *slack*. Critical operations can be determined by first scheduling every operation as early as possible and computing for every operation o_{ij} its earliest starting time est_{ij}; this can be done analogously to the recursive computation of the completion times C_{ij} based on Equations 9.3 (page 440). The makespan C_{max} can be determined as part of the same computation. Once the makespan is known, in a next step, analogous computations can be performed in the reverse order, starting from the last operation on the last machine, in order to obtain the latest completion time lct_{ij} for each operation that permits the schedule to still finish at the time determined by the makespan. At this point, for each operation o_{ij} the latest start time can be determined as $lst_{ij} := lct_{ij} - p_{ij}$. Finally, the critical operations are precisely those that must be scheduled to start at the earliest possible time to meet the given makespan, that is, an operation o_{ij} is critical if, and only if, $est_{ij} = lst_{ij}$.

A maximal sequence of critical operations is called a *critical path*. Note that for any given candidate schedule, there can be more than just one single critical path. Additionally, we call the maximal subsequence of critical operations on a single machine a *critical block*.

EXAMPLE 9.7 **Critical Path**

Consider the PFSP instance from Example 9.6 (page 440) and the candidate schedule represented by the Gantt chart in Figure 9.4 (page 441). In this schedule, the operations $o_{11}, o_{21}, o_{31}, o_{32}, o_{42}, o_{43}$ and o_{53} are critical operations, while all other operations have some slack time. For example, operation

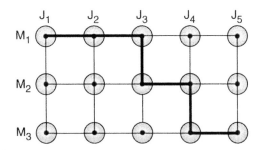

Figure 9.5 Illustration of the critical path for the schedule from Figure 9.4 (page 441). The critical path is marked by the bold lines. Each of the operations is represented by a circle; job and machine numbers are indicated at the top and on the left.

o_{12} could start one time unit later without delaying the start of operation o_{22} or even four time units later without delaying o_{32}, which is one of the critical operations.

In this example, the critical path consists of all seven critical operations in that order. There are three critical blocks $B_1 := (o_{11}, o_{21}, o_{31})$, $B_2 := (o_{32}, o_{42})$ and $B_3 := (o_{43}, o_{53})$. The critical path for this instance is illustrated in Figure 9.5.

Note that critical blocks can also be of length one. However, in the following we will only be interested, for convenience, in critical blocks of size greater than or equal to two. In the case of blocks with at least three jobs, any job that is neither the first nor the last job in the block is called *internal* to that block. The notion of critical blocks can be used in the context of pruning the effective local search neighbourhood based on the following fact [Werner, 1992]:

PROPOSITION 9.1

Consider a critical path in a candidate sequence ϕ with K critical blocks. Let $m(B_i)$ be the machine associated with the critical block B_i, and let $\mathcal{I}_i$ be the set of jobs that are internal to critical block B_i $(1 \le i \le K)$. If $m(B_1) = M_1$, include job $\phi(1)$ in $\mathcal{I}_1$, and if $m(B_K) = M_m$, include job $\phi(n)$ in $\mathcal{I}_K$. Then no reordering of the jobs in any of the sets $\mathcal{I}_i$ can improve the makespan of ϕ.

As a consequence, some of the possible moves can be excluded from consideration when searching the local neighbourhood for an improving candidate solution, which effectively reduces the number of neighbouring candidate solutions that need to be evaluated. Note that in Example 9.7, only operation o_{21} would correspond to an internal job and, hence, no neighbourhood pruning based on Proposition 9.1 can be achieved.

Tabu Search for Permutation Flow Shop Scheduling

The tabu search algorithm of Nowicki and Smutnicki [1996a], *TS-NS-PFSP*, is currently one of the best-performing algorithms for the PFSP. TS-NS-PFSP is based on the insertion neighbourhood N_i, but it makes strong use of block properties and Proposition 9.1 in order to restrict the size of the effectively searched local neighbourhoods. We refer to the resulting neighbourhood relation as N_i^{NS}; details on the definition of this neighbourhood are given in the in-depth section on page 448*f.*

Similar to ILS-S-PFSP, TS-NS-PFSP uses the NEH heuristic [Nawaz et al., 1983] for generating an initial sequence. Then, in each search step, a best-improving non-tabu neighbour of the current sequence w.r.t. to the reduced neighbourhood N_i^{NS} is selected; additionally, TS-NS-PFSP uses a complex aspiration criterion based on which the tabu status of search steps that lead to sufficiently substantial improvements can be overridden [Nowicki and Smutnicki, 1996a].

TS-NS-PFSP associates a tabu status with possible steps in the (reduced) insertion neighbourhood represented by pairs of jobs. The tabu mechanism is based on the idea that after moving a job $\phi(i)$ within a given sequence ϕ to a position $j > i$, any move that would restore the relative order of $\phi(i)$ and its former neighbour, $\phi(i+1)$, is declared tabu for a fixed number of tt search steps. Likewise, after moving job $\phi(i)$ to a position $j < i$, any move that would restore the relative order of $\phi(i)$ and its former neighbour, $\phi(i-1)$, is declared tabu for tt search steps. This mechanism can be realised as follows, using a tabu list consisting of tt pairs of jobs. If a neighbouring candidate solution is obtained from ϕ by removing a job at position i and inserting it at position $j > i$, the pair $(\phi(i), \phi(i+1))$ is added to the tabu list; if $j < i$, then the pair $(\phi(i-1), \phi(i))$ is added. Given a candidate solution ϕ', a search step that moves a job from position k to position $l > k$ is declared tabu if, and only if, at least one pair of jobs $(\phi'(j), \phi'(k))$, $j = k+1, \ldots, l$ is in the tabu list, or, in the case $k > l$, if at least one pair of jobs $(\phi'(k), \phi'(j))$, $j = l, \ldots, k-1$ is in the tabu list.

When searching for an improving search step, TS-NS-PFSP generates a set RM of representative neighbours of the current sequence. The elements of RM are the best neighbouring sequences in N_i^{NS} for each job $\phi(i), i = 1, \ldots, n$ (these representative neighbours have an additional use, as indicated later). Next, a set RM' is built which consists of all those neighbours that are not tabu or whose tabu status is overridden by the aspiration criterion. Then, TS-NS-PFSP selects the best sequence in RM' to replace the current candidate solution. If RM' is empty (i.e., all neighbouring solutions are tabu and not aspired), the oldest element of the tabu list is removed; this is repeated until at least one neighbouring solution becomes eligible (that is, not tabu).

The aspiration criterion of TS-NS-PFSP is rather complex, when compared to the standard aspiration mechanism that accepts a neighbouring candidate solution that is tabu if it represents an improvement over the current incumbent candidate solution. TS-NS-PFSP memorises the evaluation function values over the entire search trajectory. Its aspiration criterion accepts a tabu neighbouring candidate solution ϕ' if it has a makespan smaller than $h(C_{max}(\phi'))$, where $C_{max}(\phi')$ is the makespan of ϕ' and the aspiration function value $h(C_{max}(\phi'))$ is determined as follows. First, the sequence of memorised evaluation function values is searched for values equal to $C_{max}(\phi)$, the makespan of the current candidate solution. (Note that only the value of the makespan is relevant, while the schedule that achieved that makespan value is not used.) Let $C_{max}(\phi^{(l)})$ denote the makespan of the schedule obtained after search step number l, and let l' be number of memorised makespan values, that is, the current length of the search trajectory. Furthermore, let L be the set of all l with $1 < l < l'$ such that $C_{max}(\phi^{(l)}) = C_{max}(\phi)$. Finally, define $h_l := \min\{C_{max}(\phi^{(l-1)}), C_{max}(\phi^{(l+1)})\}$. Then $h(\phi)$ is set to $\min\{h_l \mid l \in L\}$. Hence, a schedule ϕ' is aspired if, and only if, it improves over the best makespan obtained in any iteration directly preceeding or directly succeeding an iteration in which the same makespan as that of the current schedule was achieved.

Finally, TS-NS-PFSP occasionally backtracks to the incumbent candidate solution, $\widehat{\phi}$. To implement this mechanism, each time the incumbent candidate solution is improved, TS-NS-PFSP stores the tabu list TL associated with $\widehat{\phi}$ as well as its set of admissible neighbours $\widehat{RM}$ (i.e., neighbours that are not tabu or that satisfy the aspiration criterion); $\widehat{RM}$ is the original set RM that was used when the neighbourhood of $\widehat{\phi}$ was previously searched minus the neighbour that had been chosen to continue the search after the previous backtrack step. This reduced set is used to ensure that after backtracking, a different search trajectory is followed. The backtracking mechanism is triggered when for a given number of iterations after search initialisation or after the most recent backtracking step no improvement in the incumbent solution has been found. TS-NS-PFSP terminates when either a limit on the maximum number of local search steps has been reached, or when during backtracking an empty set $\widehat{RM}$ is encountered.

Currently, TS-NS-PFSP is considered a state-of-the-art algorithm for the PFSP. In 1996, when this algorithm was first proposed, it was shown to substantially outperform all previously published SLS algorithms for the PFSP. It was also the first algorithm to exploit the reduced insertion neighbourhood N_i^{NS}.

Since then, several other algorithms based on N_i^{NS} have been proposed [Reeves and Yamada, 1998; Yamada, 2001], but it is somewhat unclear whether or by how much any of these improve over the performance of TS-NS-PFSP. Only one more recent tabu search algorithm appears to perform slightly better than TS-NS-PFSP [Grabowski and Wodecki, 2002]. Overall, TS-NS-PFSP provides an excellent example for the crucial role that the exploitation of specific problem characteristics can play in achieving state-of-the-art performance, in particular by means of strong neighbourhood pruning and the use of speed-up techniques.

IN DEPTH NEIGHBOURHOOD RESTRICTIONS IN **TS-NS-PFSP**

To precisely define N_i^{NS}, we first introduce some additional notation. Consider an arbitrary critical path of a sequence ϕ composed of K critical blocks B_j, $1 \leq j \leq K$. By B_j^l and B_j^r we denote the first (left-most) position and the last (right-most) position, respectively, of a block B_j. For convenience we also introduce two dummy blocks B_0 and B_{K+1} with $B_0^l := B_0^r := 1$ and $B_{K+1}^l := B_{K+1}^r := n$. We have that $B_j^l = B_{j-1}^r$ and $B_j^r = B_{j+1}^l$. Recall that in the insertion neighbourhood, two permutations are direct neighbours if one can be obtained from the other by removing a job $J_{\phi(i)}$ from position i and inserting it at another position j.

A first reduction of the neighbourhood size is obtained as follows. Suppose a job $J_{\phi(i)}$ is in block B_j; then, only insertions of this job into positions $[1, \ldots, B_j^l]$ and $[B_j^r, \ldots, n]$ would need to be potentially tested, because all the other positions are internal to block B_j (this reduction exploits Proposition 9.1). However, this restriction leads only to a rather small reduction of the neighbourhood size, depending on the topology of the critical path.

TS-NS-PFSP uses a considerably stronger reduction of the neighbourhood. Let $J_{\phi(i)}$ be a job in block B_j and let B_{j-1} and B_{j+1} be the two blocks immediately preceeding and succeeding block B_j. Then, only positions in blocks B_{j-1} and B_{j+1} are considered as potential insertion positions for job $J_{\phi(i)}$. In order to precisely define the potential insertion positions, we need to distinguish between two possibilities:

1. Job $J_{\phi(i)}$ is internal to some block B_j.

2. Job $J_{\phi(i)}$ is an endpoint of a block B_j.

In the first case, we define $\Delta_r := \lfloor \epsilon(B_{j+1}^r - B_{j+1}^l) \rfloor$ and $\Delta_l := \lfloor \epsilon(B_{j-1}^r - B_{j-1}^l) \rfloor$. If we have $m(B_1) \neq M_1$ and $m(B_K) \neq M_m$ (we consider only blocks of size at least two, so the first block need not be on the first machine; the same holds for the last block w.r.t. the last machine), then the set of possible insertion positions of job $J_{\phi(i)}$ is determined to be $PR(i) := \{k \mid B_j^r \leq k \leq B_j^r + \Delta_r\}$ for moves of $J_{\phi(i)}$ towards the right of position i and $PL(i) := \{k \mid B_j^l - \Delta_l \leq k \leq B_j^l\}$ for moves towards the left of position i. A special case arises if B_1 is associated with machine M_1, in which case it makes no sense to move job $J_{\phi(i)}$ from block B_1 to the left and we have $PL(i) = \emptyset$; similarly, if B_K is on machine M_m, no right moves need to be considered, and we have $PR(i) = \emptyset$.

In the second case, a job belongs to two critical blocks. The computations of $PR(i)$ and $PL(i)$ are performed analogously to the first case, with the only difference that for the computation of the set $PL(i)$, the smaller block index is set to j, while in the right move case the higher block index is set to j. To avoid redundancies, TS-NS-PFSP never considers moving a job $J_{\phi(B_j^r)}$ to position $B_j^r + 1$, because the same effect is achieved by the equivalent move of job $J_{\phi(B_j^l+1)}$ to position B_j^l (actually, such a move constitutes a move in the transpose neighbourhood; see Figure 9.2, page 425).

Nowicki and Smutnicki determined the value for the parameter ϵ based on empirical results; they suggested to use $\epsilon := 0$ if $n/m > 3$, $\epsilon := 0.5$ if $2 \leq n/m \leq 3$, and $\epsilon := 1$ otherwise. Note that when setting $\epsilon := 0$, for each job there are at most two possible positions for insertion, leading to an extremely small neighbourhood and also to the potential loss of improving neighbours.

9.4 Group Shop Problems

We now turn our attention to a class of multi-stage, multi-machine scheduling problems that is more general than the previously considered flow shop scheduling problems. By relaxing the characteristic flow shop requirement that the operations of all jobs need to be performed in the same, fixed order, we obtain the *Group Shop Problem (GSP)*, which contains the Permutation Flow Shop Problem, as well as the widely studied Open Shop Problem and Job Shop Problem as special cases.

An instance of the Group Shop Problem (GSP) consists of a set of machines $\mathcal{M} := \{M_1, \ldots, M_m\}$, a set of jobs $\mathcal{J} := \{J_1, \ldots, J_n\}$, where each job J_i consists of m operations $o_{i1}, \ldots, o_{im}$, and a processing time p_{ij} for each operation o_{ij}. For formal convenience, we assume that operation o_{ij} of each job J_i has to be performed on machine M_j. This assumption can be made without loss of generality, since situations in which a particular job J_i does not need to be processed on a machine M_j can be captured by $p_{ij} := 0$. Unlike in the case of the Permutation Flow Shop Problem, the operations of each given job do not have to be performed in the canonical order $o_{i1}, \ldots, o_{im}$.

Precedence constraints for the operations of a job J_i, that is, constraints on the order in which these operations can be processed, are given by means of a partition of the set $\{o_{i1}, \ldots, o_{im}\}$ in the form of a set of groups $G_i = \{g_{i1}, \ldots, g_{il(i)}\}$ and a canonical total order $g_{i1} \prec g_{i2} \prec \ldots \prec g_{il(i)}$ between those groups. This defines a partial order for the execution of the operations of J_i, according to which all operations of the first group have to be processed before the operations of the second group, etc., but which does not restrict the order in which operations within the same group are performed.

As usual, at any given time each job can be processed by at most one machine, and each machine can process at most one job at a time. Furthermore, pre-emption is not allowed, and all jobs are available for processing at time zero.

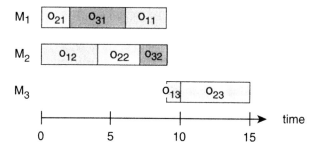

Figure 9.6 Gantt chart for a feasible candidate schedule of the GSP instance from Example 9.8.

Given a GSP instance, the objective is to find a feasible schedule, that is, a schedule that satisfies all the previously stated requirements as well as the precedence constraints, and that has minimal makespan C_{max}.

EXAMPLE 9.8 A Simple Group Shop Problem Instance

Consider the following GSP instance with three machines and three jobs: $\mathcal{M} := \{M_1, M_2, M_3\}$ and $\mathcal{J} := \{J_1, J_2, J_3\}$ with $J_1 := \{o_{11}, o_{12}, o_{13}\}$, $J_2 := \{o_{21}, o_{22}, o_{23}\}$ and $J_3 := \{o_{31}, o_{32}\}$. Furthermore, the following groups and group orderings are specified:

$$G_1 : \{o_{11}, o_{12}\} \prec \{o_{13}\}$$
$$G_2 : \{o_{21}\} \prec \{o_{22}, o_{23}\}$$
$$G_3 : \{o_{31}, o_{32}\}$$

These precedence constraints imply, for example, that in job J_1, operations o_{11} and o_{12} can be processed in any order, but both of them have to be completed before the processing of operation o_{13} can begin. Finally, the processing times p_{ij} are specified as follows:

p_{ij}	$i = 1$	$i = 2$	$i = 3$
$j = 1$	3	2	4
$j = 2$	4	3	2
$j = 3$	1	5	0

A feasible schedule for this GSP instance is shown in Figure 9.6.

The Group Shop Problem has been introduced only recently; it was first stated formally in the context of research conducted within the Metaheuristics Network [Dorigo, 2003]. The original inspiration came from a problem formulated for the Whizzkids contest [Whizzkids, 1997], a mathematics challenge published in the Dutch newspaper 'De Telegraaf' on September 6, 1997. The proposed GSP instance involved scheduling 197 meetings (operations) between 15 teachers (machines) and 20 parents (jobs) during a parents' contact day at a high school subject to constraints on the ordering between some of the meetings.

Special Cases and Complexity Results

The GSP can be seen as a generalisation of many well-known and widely studied scheduling problems, in particular, the *Job Shop Problem (JSP)* and the *Open Shop Problem (OSP)*.

In the JSP (denoted as $J||C_{max}$ in the formal notation), there exists a total order of all the operations belonging to each job. In the GSP formulation, this corresponds to the case that each group comprises only one single operation. Hence, there are precedence constraints among the operations of a job such that a total ordering of the operations is induced. As the best-known variant of the GSP, the JSP has received a considerable amount of attention in the literature (overall, there are several hundreds of papers on the JSP and its variations).

In the OSP (denoted as $O||C_{max}$ in the formal notation), there are no precedence relations among the operations of each job, that is, the operations of each job can be processed in an arbitrary order. Hence, the OSP can be seen as the special case of the GSP in which for each job all its operations form one group. Compared to the JSP, the OSP poses the additional difficulty that in order to find an optimal schedule, the order of the jobs as well as the order of the operations has to be optimised.

The GSP is an $\mathcal{NP}$-hard problem, since it is well known that its special cases, the JSP and the OSP, are both $\mathcal{NP}$-hard [Garey et al., 1976a; Gonzalez and Sahni, 1976]. However, several special cases of the OSP and the JSP can be solved optimally in polynomial time. The OSP with two machines (formally denoted as $O2||C_{max}$) has been shown to be solvable in linear time [Gonzalez and Sahni, 1976]; however, the variants obtained by considering L_{max} or $\sum C_i$ instead of C_{max} as the objective function are strongly $\mathcal{NP}$-hard [Achugbue and Chin, 1982]. The OSP with three or more machines and the standard C_{max} objective function is $\mathcal{NP}$-hard.

For the JSP, even the two machines case with objective function C_{max} is known to be $\mathcal{NP}$-hard [Lenstra and Rinnooy Kan, 1979]; the same holds when the objective functions $\sum C_i$ or L_{max} are used. Only a few special cases of the JSP are known to have polynomial time complexity. One of these arises when the processing times for all operations are either zero or one (the analogous special cases of the OSP can also be solved optimally in polynomial time).

Benchmark Instances

There are a large number of benchmark instances for the JSP and the OSP. In the JSP case, probably the most widely known benchmark instances are three instances formulated by Muth and Thompson [1963]. The best known of these instances, `ft10`, consists of 10 jobs with a total of 100 operations that are to be processed on 10 machines. It has received particular attention, because it has been solved to optimality only about 30 years after it had been introduced by Carlier and Pinson [1989]. However, this instance no longer poses a significant challenge to state-of-the-art complete and incomplete algorithms.

There are several hundreds of other JSP instances available, most of which can be downloaded from OR-Library [Beasley, 2003]. Most of these instances stem from empirical studies of JSP algorithms, and some of them have been specifically designed to be hard. The commonly used benchmark sets include the original set of Muth and Thompson [1963], 40 instances of 8 different sizes from Lawrence [1984], 5 instances from Adams et al. [1988], a set of 10 instances by Applegate and Cook [1991], 20 instances used by Storer et al. [1992], 4 instances by Yamada and Nakano [1992], as well as a large number of randomly generated instances by Taillard [1993].

For many of the smaller instances with up to 10 or 20 jobs and a similar number of machines, provably optimal solutions are known. Generally, it has been observed that quadratic instances, that is, those with an identical number of jobs and machines, appear to be particularly hard. Currently, quadratic instances with 15 jobs and machines pose a significant challenge for both, complete algorithms and SLS methods [Jain and Meeran, 1999].

For the OSP, three sets of benchmark instances are widely used. These are a set of 80 randomly generated instances by Taillard [1993], ranging in size from 16 to 400 operations, a set of 35 quadratic instances by Brucker et al. [1997] with 25 to 64 operations, and a set of 80 instances by Guèret and Prins [1999] with 9 to 100 operations. Recent empirical results suggest that state-of-the-art SLS algorithms have little difficulty in finding provably optimal solutions or matching the best known solutions for Taillard's instances [Blum, to appear; Liaw, 2000]. The two other benchmark sets contain instances that appear to be much harder to solve.

As previously mentioned, the general GSP has been only recently formulated. Besides the previously mentioned benchmark instance from the Whizz-kids competition, additional instances have been generated in the context of the research conducted by members of the Metaheuristics Network; these were obtained by introducing group partitions into the operations of each job for some well-known JSP instances. The respective collection of GSP instances is maintained by Blum [2003a].

procedure *scheduleGeneration* (π')
    input: *GSP instance* π'
    output: *candidate schedule s*
    $O :=$ *operations in* π';
    $O_u := O$;
    $s :=$ *empty schedule*;
    for $i := 1$ **to** $\#O$ **do**
        $Q := \{o \in O_u \mid \not\exists \, o' \in O_u : o' \preceq o\}$;
        $Q' :=$ *generateCandidateOperations*(π', Q, s);
        $o :=$ *selectOperation*(π', Q', s);
        append operation o to schedule s;
        $O_u := O_u \setminus \{o\}$;
    end
    return *s*
end *scheduleGeneration*

Figure **9.7** General algorithm outline for a schedule generation method for the GSP; O denotes the set of all operations for the given GSP instance, and O_u is the set of currently unscheduled operations.

Schedule Generation for Group Shop Scheduling

As for many other types of scheduling problems, construction heuristics play an important role in the context of the GSP and its special cases. Note that, similar to the previously discussed scheduling problems, any candidate schedule for a given GSP instance can be represented as a permutation of the given operations, that is, as a sequence that contains every operation of every given job exactly once. Given a schedule, the corresponding permutation can be obtained by concatenating the sequences of operations executed on each machine. In the case of the GSP instance from Example 9.8, the permutation $(o_{21}, o_{31}, o_{11}; o_{12}, o_{22}, o_{32}; o_{13}, o_{23})$ represents the schedule shown in Figure 9.6 (page 450); for added clarity, we use semicolons to separate the sequences of operations for each given machine.

Schedule generation methods for the GSP are constructive search methods that start with an empty schedule and iteratively append operations. Because of the precedence constraints between the groups within each job, the selection of the operation to be appended in each step is somewhat more complex than in the case of the SMTWTP or the PFSP. In the following, we write $o' \preceq o$ to indicate that the group precedence constraints of a given GSP instance require operation o' to be scheduled no later than operation o. A general outline for a schedule generation method for the GSP can be specified as shown in Figure 9.7.

A concrete schedule generation mechanism for the GSP is specified by the two functions *generateCandidateOperations* and *selectOperation*. There are two main mechanisms for generating the set Q' of candidate operations. The first of these initially computes the earliest completion times of all operations in the set Q of eligible operations [Giffler and Thompson, 1960]. Then, a machine M with minimal completion time, ect_{min}, under the current schedule s is chosen, and Q' is computed as the set of all operations in Q that need to be processed on M and whose earliest starting time is smaller than ect_{min}. This approach for defining the set Q' leads to the construction of *active schedules*, that is, schedules in which no operation can be completed earlier by any change in the processing sequence of any machine without delaying other operations.

A second mechanism for generating candidate operations starts by computing the earliest possible starting time est_{min} of any operation in Q. Then, Q' is computed as the set of all operations in Q that can start at time est_{min}. This second mechanism generates *non-delay schedules*, that is, schedules in which no machine is ever kept idle as long as there is an operation that can be processed. Note that optimal schedules are always active schedules, but not necessarily non-delay schedules.

The second function, *selectOperation*, implements the actual dispatching (or priority) rule that determines which of the operations in Q' is chosen for being appended to the current schedule. Over the years, a substantial amount of research has been devoted to finding good dispatching rules for special cases of the GSP. Many of these rules are relatively simple deterministic schemes, while others combine a number of elementary dispatching rules. An example for the latter is the AU rule [Morton et al., 1984], which is a combination of the weighted shortest processing time rule, which optimally solves the problem $1 || \sum_{i=1}^{n} w_i \cdot C_i$ [Smith, 1956] and a due date oriented rule (see also page 430). Some approaches try to use learning techniques to determine an appropriate combination of dispatching rules [Fisher and Thompson, 1963; Crowston et al., 1963]. Generally, different rules perform best on different types of problem instances, and there is no single dispatching rule that performs best on the full range of GSP instances. (For an overview of commonly used dispatching rules, see Haupt [1989] or Panwalker and Iskander [1977].)

Schedule generation mechanisms play an important role in the context of high-performance SLS algorithms for the GSP and its various special cases. In the following, we will discuss two such algorithms for the two most prominent special cases of the GSP, the OSP and the JSP. For the OSP, we describe a state-of-the-art ACO algorithm that combines the ants' constructive search process with ideas from tree search algorithms, in particular beam search [Ow and Morton, 1988]. For the JSP, we present the high-performance tabu search algorithm by Nowicki and Smutnicki.

A Beam Search-ACO Hybrid Algorithm for Open Shop Scheduling

Ant Colony Optimisation is essentially a constructive search method. As discussed in Chapter 2, Section 2.3, constructive search methods and their iterated variants are closely related to tree search techniques and, therefore, the combination of ACO with concepts from other types of tree search is a relatively natural idea. Blum [to appear] proposed such a hybrid algorithm for the Open Shop Problem; this algorithm combines an ACO method with beam search, a tree search technique that has been originally proposed by Ow and Morton for solving a specific scheduling problem [Ow and Martin, 1988].

Beam search (BS) maintains a set B of at most bw partial candidate solutions. In every construction step, each of the partial candidate solutions from B is extended in up to fw possible ways. These extensions are first proposed based on a preselection mechanism and then evaluated using a heuristic function, based on which the bw best-ranked extended partial candidate solutions are selected to form the new set B of partial candidate solutions. Commonly used heuristic functions are based on lower bounds on the evaluation function value of a complete candidate solution. Complete candidate solutions obtained by extending elements of B are kept in a separate set B_f. The algorithm stops when there are no more partial candidate solutions to be considered for further extension, at which point the best candidate solution in B_f is returned. The parameters bw and fw, which control important aspects of the search process, are called the *beam width* and the *filter width*, respectively. Standard BS algorithms are deterministic and do not use backtracking.

The central idea of Blum's combination of ACO with BS is to let each ant perform a probabilistic beam search. Hence, each ant works on a set of at most bw partial candidate solutions. At each construction step, every ant extends all its partial candidate solutions exactly as in BS in fw possible ways. The main difference to BS is that the partial candidate solutions are extended according to the biased randomised construction mechanism that is characteristic for ACO algorithms. (That is, the possible extensions are selected probabilistically based on pheromone values and heuristic information.) The resulting algorithm is called *BeamACO*. The appeal of BeamACO partly lies in the intuition that (i) as a probabilistic extension of BS, it may perform more robustly, and (ii) ACO algorithms may benefit from properties of BS such as the parallel exploration of search trees and the use of lower bounding techniques.

We now describe Blum's BeamACO algorithm for the Open Shop Problem, BeamACO-OSP [Blum, to appear]. For notational convenience, we will denote the operations of the given OSP instance as $o_1, \ldots, o_l$, where l is the total number of operations for all given jobs. In BeamACO-OSP, pheromone values τ_{pq} are assigned to related operations o_p and o_q; operations are related either if they belong to the same job or if they are processed on the same machine. In this case, τ_{pq} refers to the desirability of scheduling operation o_p before, but not necessarily immediately before, operation o_q. In the following, let R_p be the set of operations related to operation o_p. The heuristic value η_p of choosing operation o_p is based on the earliest start time of o_p, est_p; from est_p, the value $1/(est_p+1)$ is computed and then normalised by the sum of these values over all operations in the current candidate set, which yields the heuristic value of o_p.

Candidate solutions are constructed by making use of a schedule generation method (see Figure 9.7, Page 453). Let $O_u(s)$ be the set of operations that are not scheduled in a partial candidate schedule s. Note that in the OSP, there are no precedence constraints between operations; hence, a partial schedule can be extended by any unscheduled operation. At each construction step, first we consider the set $I_u(s)$ which consists of the operations in $O_u(s)$ minus all operations for which all related operations have already been scheduled in s. Next, from $I_u(s)$ a candidate set of eligible operations is built by computing

$$O'(s) := generateCandidateOperations(\pi', I_u(s), s)$$

where the function *generateCandidateOperations* combines the two previously described mechanisms for generating active and non-delay schedules (see page 454). More precisely, in each construction step, one of the respective functions for generating candidate sets is chosen and applied uniformly at random. Finally, operation o_p is chosen with a probability given by

$$
\phi_p := \begin{cases} \left(\min\{\tau_{pq} \mid o_q \in R_p \cap O(s)\}\right)^\alpha \cdot \eta_p/Z & \text{if } o_p \in O'(s); \\[2ex] 0 & \text{otherwise;} \end{cases}
\tag{9.4}
$$

where Z is a normalising constant obtained by summing the values of $(\min\{\tau_{pq} \mid o_q \in R_p \cap U(r)\})^\alpha \cdot \eta_p$ over all $o_p \in O'(s)$, and α is a parameter that determines the relative impact of the pheromone strengths compared to the heuristic values η_p. Recall that τ_{pq} represents the desirability of scheduling o_p before o_q, and hence the probability of choosing o_p monotonically increases with the minimum pheromone value τ_{pq} over the unscheduled operations related to o_p.

For each partial candidate solution, fw extensions are generated. (A further technicality in the algorithm is that after the first extension of a partial schedule s by an operation o_i has been generated, for the $fw - 1$ other extensions, the candidate set $O'(s)$ is further restricted to include only operations in $O'(s) \cap R_i$. This is done to avoid constructing the same candidate solution more than once; for details we refer to Blum [to appear].) Next, for each new partial candidate schedule, a lower bound is computed (there is a maximum total number of $fw \cdot bw$ such partial schedules). This lower bound is obtained as follows: First, for every job and for every machine, the sum of the processing times of the corresponding unscheduled operations is computed (i.e., for a machine M_i, only unscheduled operations are considered that need to be processed on M_i; an analogous consideration applies to jobs). Then, these sums are added to the earliest completion time of the corresponding job or machine given the partial schedule under consideration; the maximum of these values is the desired lower bound for the completion of the schedule. By using appropriate data structures, this lower bound can be computed incrementally during candidate solution construction. As usual in beam search, only the *bw* partial candidate solutions with the smallest lower bound values are kept.

After all partial schedules have been completed, they are further optimised by an iterative best-improvement algorithm that considers transpositions of operations on a critical path. Details on the neighbourhood relation used in this context can be found in Sampels et al. [2002].

The pheromone management of BeamACO-OSP is realised as follows. At the beginning of the search process, all pheromone values are initialised to 0.5. Subsequently, after each iteration, depending on a stagnation criterion, either the incumbent candidate solution or the best candidate solution obtained since the last initialisation of the pheromone trails is used for updating the pheromone values τ_{pq} according to

$$\tau_{pq} := \tau_{pq} + \rho \cdot (\delta(o_p, o_q, s) - \tau_{pq}), \tag{9.5}$$

where $\delta(o_p, o_q, s) := 1$ if o_p is scheduled before o_q in schedule s, and $\delta(o_p, o_q, s) := 0$ otherwise, and ρ is the pheromone persistence parameter. This update rule effectively limits the pheromone values to the interval $[0, 1]$. Additional mechanisms are used for dynamically tightening the range of the pheromone values, and for occasionally re-initialising all pheromone trails, in order to avoid stagnation of the population-based search. (This type of mechanism for ACO was originally introduced in $\mathcal{MAX}$–$\mathcal{MIN}$ Ant System [Stützle and Hoos, 2000]; see Chapter 8, Section 8.4.)

An empirical performance comparison between BeamACO-OSP and two other SLS algorithms for the OSP [Liaw, 2000; Prins, 2000] as well as an earlier ACO algorithm [Blum, 2003b] based on all commonly used OSP benchmark instances suggests that BeamACO-OSP is the best-performing OSP algorithm known to date.

Tabu Search for Job Shop Scheduling

We conclude our discussion of SLS algorithms for group shop problems with a discussion of Nowicki and Smutnicki's highly efficient tabu search algorithm for the JSP (in the following referred to as *TS-NS-JSP*) [Nowicki and Smutnicki, 1996a]. From a high level perspective, this algorithm is very similar to the TS-NS-PFSP algorithm described in Section 9.3. In the following we therefore focus on some of the key commonalities and differences of the two algorithms.

Similar to TS-NS-PFSP, the main ingredients of TS-NS-JSP are the underlying neighbourhood relation and the pruning mechanisms used for efficiently searching this neighbourhood. The neighbourhood definition makes use of critical paths (as in the PFSP case). A block in the critical path is a maximal sequence of operations on the same machine with length larger or equal to one. An operation of a block is called an *internal operation* if, and only if, it is neither the first nor the last operation of the block.

Based on these notions of critical path and block, several neighbourhood relations have been proposed. Most local search algorithms for the JSP are based on the transpose neighbourhood N_t restricted to operations that are on the same machine; the various neighbourhoods mainly differ in how strongly they restrict the neighbourhood that is searched. (N_t is typically used, because search steps based on the insertion or exchange neighbourhoods defined in Section 9.1 may lead to infeasible candidate solutions.) Nowicki and Smutnicki's neighbourhood relation for the JSP, N_t^{NS}, exploits the following observations.

1. Reversing the order of two operations on a critical path in a candidate schedule s never results in an infeasible candidate solution. (Note that such reversals correspond exactly to search steps in the standard transpose neighbourhood.)

2. Reversing the order of two operations that are not on a critical path cannot lead to a reduction of the makespan.

3. Reversing the order of two internal operations in a block of a critical path cannot lead to a reduction of the makespan.

4. Reversing the order of the first two operations on the first machine block or the last two operations on the last block cannot reduce the makespan.

Based on these observations, $N_t^{NS}(s)$ comprises only schedules that are obtained by transpositions of the first two operations and the last two operations of any block except for the first and the last block. Depending on the block structure, it may even happen that $N_t^{NS}(s)$ is the empty set. However, it can be proven that this can occur only if s is an optimal schedule for the given JSP instance [Nowicki and Smutnicki, 1996a]. It is interesting to note that N_t^{NS} is the most restrictive neighbourhood relation used in any local search algorithms for the JSP [Jain et al., 2000]. Unfortunately, the neighbourhood graph induced by N_t^{NS} may not be connected; this means that certain schedules s' may not be reachable from the current candidate schedule by means of search steps in N_t^{NS}.

As previously stated, TS-NS-JSP is in many aspects similar to the TS-NS-PFSP algorithm from Section 9.3. The tabu status is associated with individual search steps in N_t^{NS} such that after performing a search step, the reverse step is considered tabu for the next tt search steps, where tt is the standard tabu tenure parameter.

The initial candidate solution is generated by an insertion algorithm based on the work of Werner and Winkler [1995]. Then, each search step is selected by first generating a critical path and then choosing the best search step from the local N_t^{NS} neighbourhood that is either not tabu or whose tabu status is overridden by an aspiration criterion. TS-NS-JSP uses an aspiration criterion that overrides the tabu status of a search step if it leads to an improvement over the best makespan encountered so far. As a side effect of the small size of the neighbourhood used in TS-NS-JSP, it may frequently happen that all neighbours are tabu. In this case, the tabu tenure of all steps except for the one that was performed most recently is iteratively decreased by one until the effective local neighbourhood becomes non-empty.

Two additional mechanisms are used to increase the efficiency of the search process in TS-NS-JSP. The first of these is based on the idea of occasionally restoring the search process to a previously stored, highly promising candidate solution, a so-called *elite solution*. This mechanism is triggered whenever a fixed number of iterations has been performed without achieving an improvement in the incumbent candidate solution and works, in principle, in a similar way as previously described for TS-NS-PFSP. The second technique is a cycle detection mechanism, which stops the local search if it is deemed to be stagnating. (For details, see Nowicki and Smutnicki [1996a].)

TS-NS-JSP has been shown to achieve excellent performance, and despite the considerable amount of research on SLS algorithms for the JSP and the number of algorithms developed since its initial publication, TS-NS-JSP is still one of the best known algorithms for the JSP. Its performance appears to be rivalled only by few other algorithms, such as the ones presented in Balas and Vazacopoulos [1998] and Grabowski and Wodecki [2001].

It is interesting to note that two algorithms that are as closely related as TS-NS-PFSP and TS-NS-JSP are amongst the best known methods for solving these two prominent special cases of the GSP. Together with other results from the literature, this suggests that tabu search algorithms may be generally well suited for scheduling problems that are similar in structure to the JSP and the PFSP. However, there are other types of group shop problems, such as the OSP, that are more effectively solved by other SLS methods.

9.5 **Further Readings and Related Work**

Scheduling research has a long history. First major treatments of scheduling problems and computational aspects of scheduling are found in the book edited by Muth and Thompson [1963] and in the text book by Conway, Maxwell and Miller [1967]. Since then, a number of books on the subject have been published, including textbooks for graduate courses, such as the one by French [1982] and the more recent book by Pinedo [1995], which also includes a large part on stochastic scheduling models and applications. Somewhat more research-oriented are the books by Blazewicz et al. [1993] and by Brucker [1998]. A number of applications and case studies are described in the book edited by Zweben and Fox [1994]. Much work on scheduling heuristics is presented in the textbook by Morton and Pentico [1993]. An overview of SLS algorithms for a broad range of scheduling problems with a large number of pointers to literature can be found in Anderson et al. [1997].

An online overview of complexity results for scheduling problems is maintained by Brucker and Knust [2003]. A complexity hierarchy has been developed that describes the relationship between the complexity of a large number of scheduling problems. This complexity hierarchy gives rise to general results, such as the fact that the $\mathcal{NP}$-hardness of scheduling problems for objective $\sum_i C_i$ implies $\mathcal{NP}$-hardness of the respective problems with the objectives $\sum_i w_i \cdot C_i$ or $\sum_i w_i \cdot T_i$. Examples of this hierarchy can be found, for example, in the book by Pinedo [1995] or in Brucker and Knust [2003].

Single-machine scheduling problems have received a significant amount of interest by researchers working with SLS algorithms. In this area, the Single-Machine Total Weighted Tardiness Problem (SMTWTP) is one of the most commonly tackled problems, because it is harder to solve than many other single-machine problems. Early approaches to the SMTWTP include mainly construction heuristics, such as the AU heuristic [Morton et al., 1984] (see also Potts and Van Wassenhove [1991] for an overview). Various SLS methods have been applied to the SMTWTP, including Simulated Annealing [Matsuo et al., 1987; Potts and Wassenhove, 1991], Tabu Search and Genetic Algorithms [Crauwels et al., 1998]. As stated in Section 9.2, the currently best-performing SLS algorithms for the SMTWTP are Iterated Dynasearch [Congram et al., 2002; Grosso et al., 2004], closely followed by other iterated local search algorithms [den Besten et al., 2001] and ant colony optimisation algorithms [den Besten et al., 2000].

The Permutation Flow Shop Problem (PFSP) is a widely studied scheduling problem, although some criticism has been raised against its widespread use in scheduling research [Dudek et al., 1992; McKay et al., 2002]. For the PFSP, a number of good performing constructive algorithms are available [Nawaz et al., 1983].

Early research on SLS algorithms for the PFSP has been mainly focused on Simulated Annealing [Osman and Potts, 1989; Ogbu and Smith, 1990; Ishibuchi et al., 1995], Tabu Search [Widmer and Hertz, 1989; Taillard, 1990; Reeves, 1993a; Nowicki and Smutnicki, 1996b] and Evolutionary Algorithms [Bierwirth and Stöppler, 1992; Reeves, 1995; Reeves and Yamada, 1998]. Ant Colony Optimisation (ACO) and Iterated Local Search (ILS) have also been applied with some success to the PFSP [Stützle, 1998a; 1998b]. One of the best-performing SLS algorithms for the PFSP is the tabu search algorithm by Nowicki and Smutnicki [1996b], and only few more recent approaches appear to be able to reach or surpass its performance [Grabowski and Wodecki, 2002; Reeves and Yamada, 1998; Yamada, 2001]. Recall that ILS performs very well in terms of the solution qualities obtained, but — probably mainly because of the less efficient subsidiary local search procedure — it has been found to require longer computation times than the tabu search algorithm of Nowicki and Smutnicki [Stützle, 1998b]. The excellent performance observed by Stützle was also confirmed in a more recent experimental study of 25 different methods, ranging from simple construction heuristics to memetic algorithms [Ruiz and Maroto, to appear]. An enlightening analysis of PFSP instance characteristics and their impact on SLS performance has been presented by Watson et al. [2002].

The Job Shop Problem (JSP) has a long tradition in scheduling research. A good performing, well known heuristic for the JSP is the *Shifting Bottleneck Procedure* by Adams et al. [1988]. It is currently used in a number of hybrid, state-of-the-art SLS algorithms for the JSP [Balas and Vazacopoulos, 1998; Pezzella and Merelli, 2000]. Virtually all SLS methods described in this book have been applied to the JSP, with varying success. Computational results for SLS algorithms until ca. 1996 are summarised in Vaessens et al. [1996]. Generally, it can be observed that the development of increasingly powerful SLS algorithms has been going hand in hand with the design of more involved neighbourhood relations that increasingly exploit problem characteristics of the JSP. Work on specialised neighbourhood relations evolved from rather simplistic approaches [Laarhoven et al., 1992] over increasingly smaller local neighbourhoods [Matsuo et al., 1988; Dell'Amico and Trubian, 1993] to the one underlying the state-of-the-art tabu search algorithm by Nowicki and Smutnicki [1996a]. The most recent articles in this line of research include the approach by Grabowski and Wodecki [2001], which considers special insert moves in addition to the transpose moves underlying other well-known neighbourhood relations for the JSP. Complex neighbourhoods, as used in Variable Depth Search, are considered in Balas and Vazacopoulos [1998].

An extensive review of the literature on the JSP and the state-of-the-art in JSP solving as of 1999 is given by Jain and Meeran [1999]. Jain et al. [2000] examine in detail several key choices in the tabu search algorithm of Nowicki and Smutnicki. They show that the initial solution has a considerable influence on the performance of the algorithm, and their analysis suggests that effective diversification schemes may be required to obtain better-performing algorithms. An analysis of the difficulty of JSP instances has been presented by Watson et al. [2003], who developed a cost model for the prediction of the search cost for solving JSP instances.

Currently, among the best-performing SLS algorithms for the JSP we find SLS algorithms searching large neighbourhoods [Balas and Vazacopoulos, 1998] and sophisticated tabu search algorithms [Nowicki and Smutnicki, 1996a; Pezzella and Merelli, 2000; Grabowski and Wodecki, 2001]. While it is somewhat unclear which of these algorithms performs best, it is clear that research on the JSP has reached a high level of sophistication in terms of the algorithmic techniques employed.

The Open Shop Problem (OSP) has received considerably less attention in the scheduling literature. Early work that explicitly addresses the OSP includes a genetic algorithm by Fang, Ross and Corne [1994] and an early ACO algorithm by Pfahringer [1996]. More recently, Liaw [1999] presented a tabu search algorithm, which was later shown to be outperformed by a genetic algorithm developed by the same author [Liaw, 2000]. Another genetic algorithm has been developed by Prins [2000]. Computational results by Blum [to appear] suggest that the Bea-mACO approach presented in this chapter improves over the performance of these earlier algorithms, and that BeamACO is a new state-of-the-art algorithm for the OSP.

So far, the only computational studies of the general Group Shop Problem (GSP) have been conducted by members of the Metaheuristic Network [Dorigo, 2003]. In this research, five different GSP algorithms based on Ant Colony Optimisation, Evolutionary Algorithms, Iterated Local Search, Tabu Search and Simulated Annealing were developed and studied [Sampels et al., 2002]. A general observation from these studies indicates that the relative performance of these algorithms changes substantially depending on the type of GSP instances considered. In particular, GSP instances resembling job shop problems appear to be particularly well suited for tabu search algorithms, while instances closer to open shop characteristics seem to be solved most efficiently using ant colony optimisation approaches [Sampels et al., 2002; Blum, 2003b].

As previously stated, there is a vast literature on scheduling problems and algorithms, including research results on complete algorithms, approximation algorithms, SLS algorithms and new, more complex scheduling problems. An excellent reference for research results in scheduling is the Journal of Scheduling. Research articles on applications of SLS algorithms to scheduling problems are also frequently published in many operations research and computer science journals.

9.6 **Summary**

Scheduling problems play an important role in many production environments and occur in a broad range of application areas. Although some types of scheduling problems can be solved easily and efficiently, most of the existing and practically arising scheduling problems are $\mathcal{NP}$-hard and difficult to solve in practice, particularly, when provably optimal solutions are desired. SLS algorithms play a significant role in solving such problems as efficiently as possible.

We have exemplified how SLS algorithms have been applied successfully to a number of different types of scheduling problems with characteristics that occur in many real-world scheduling problems. Clearly, the scheduling problems covered in this chapter are academic versions of real-world scheduling tasks that facilitate the study of general algorithmic techniques and the derivation of conclusions that are not restricted to typically fairly complex and extremely specific applications.

In one of the simplest types of scheduling problems, only one machine is required or forms the bottleneck for processing the jobs in a given environment. Here, we considered the *Single-Machine Total Weighted Tardiness Problem (SMTWTP)*, for which instances of size 100 appear to be beyond the capabilities of branch & bound algorithms. However, SLS algorithms for the SMTWTP are very successful; in particular, this holds for the ACO and ILS algorithms presented here. Despite the differences between these two SLS techniques, they both have in common that they make use of highly effective local search algorithms in conjunction with adaptive restart mechanisms.

More complex environments involve several machines. As a first multi-machine model, we discussed the *Permutation Flow Shop Problem (PFSP)*. Differently from the situation for the SMTWTP, highly efficient tabu search algorithms that exploit a strongly restricted neighbourhood appear to be the best-performing SLS algorithms for the PFSP. However, less specifically tuned SLS algorithms, such as the iterated local search algorithm presented in this chapter, are also able to find very high-quality solutions.

The *Group Shop Problem (GSP)* is a recently introduced, general scheduling problem that contains the PFSP along with the well-known *Open Shop Problem (OSP)* and *Job Shop Problem (JSP)* as special cases. For the latter two problems we have presented two particularily successful SLS algorithms. In the case of the OSP, we presented an algorithm that introduces ideas taken from deterministic tree search algorithms into a SLS algorithm based on Ant Colony Optimisation. This combination led to a new state-of-the-art algorithm for the OSP. For the JSP, we discussed a highly tuned version of a tabu search algorithm that integrates a form of backtracking as a diversification mechanism into the underlying local search process.

Overall, SLS algorithms are presently the method of choice for solving many types of hard scheduling problems. Current state-of-the-art SLS algorithms for difficult scheduling problems, including the ones presented here, are typically hybrid SLS methods that combine various simple search strategies and additional mechanisms, particularly for preventing search stagnation. In many cases, they are strongly based on specific problem characteristics that are used as the basis for realising extremely efficient local search procedures and/or for guiding the search process.

At the same time, as a result of their inherent flexibility and extensibility, SLS methods are particularly promising in light of a recent trend in scheduling research towards considering increasingly complex problems that model the characteristics of real-world problems more accurately.

Exercises

9.1   [*Medium*]   Think about a realistic scheduling problem arising in some application you are familiar with and try to formalise it. Give details on the job characteristics, the machine environment and the objective function.

Once you formulated it, give its three-field notation and try to find literature on or related to your particular problem.

9.2   [*Easy*]   Describe the following scheduling problems in your own words.

(a) $1|r_i|L_{max}$

(b) $1|r_i\, t_{ij}|T_{max}$

(c) $R3|r_i\, prec\,|\sum_i w_i \cdot C_i$

(d) $Om|r_i\, d_i|\sum_i w_i \cdot T_i$

9.3   [*Medium*]   Given an instance of a scheduling problem with n jobs, show that the size of the exchange neighbourhood and the insert neighbourhood, as defined in Section 9.1, are $n \cdot (n-1)/2$ and $(n-1)^2$, respectively.

9.4   [*Medium*]   Several objective functions for scheduling problems are closely related. Show that a schedule that minimises L_{max} also minimises T_{max}. Does the converse also hold, that is, is it the case that a schedule that minimises T_{max} is also guaranteed to minimise L_{max}? (Justify your answer.)

9.5   [*Easy*]   Consider the ILS-CPV algorithm for the SMTWTP, and let Δf_d and Δf_e be the improvements in total weighted tardiness obtained by performing the best dynasearch move and the best exchange move, respectively. Is it possible that $|\Delta f_d| < |\Delta f_e|$?

9.6 [*Medium; Hands-On*] The single-machine scheduling problem with release dates and the objective of minimising the weighted sum of the jobs' completion times ($1|r_i| \sum_i w_i \cdot C_i$ in the formal notation) is an $\mathcal{NP}$-hard problem. Study the effectiveness of 2-phase local search algorithms for this problem analogous to Example 9.5 (page 431*ff.*).

Benchmark instances and a very basic local search code are available from `www.sls-book.net`.

9.7 [*Easy*] Describe a randomised iterative improvement algorithm for the problem of minimising total weighted tardiness in an unrelated parallel machine scheduling problem.

9.8 [*Easy*] Use the recursive formula for the computation of the makespan given in Equation 9.3 (page 440) to verify that the makespan of the sequence given in Example 9.6 (page 440) is correct. Then, compute for all operations their earliest start times and their latest start times; based on these computations, determine the critical path.

9.9 [*Easy*] Show how the Permutation Flow Shop Problem (PFSP) can be formulated as a Group Shop Problem (GSP). Exemplify this formulation by defining the PFSP instance from Example 9.6 (page 440) as a GSP instance.

9.10 [*Hard*] Recall Nowicki and Smutnicki's neighbourhood, N_t^{NS}, for the JSP. Demonstrate how the definition of this neighbourhood can be extended to the GSP, in general, and to the OSP, in particular.

10 OTHER COMBINATORIAL PROBLEMS

The problems covered in the previous chapters are only some of many combinatorial problems to which stochastic local search algorithms have been applied successfully. In this chapter, we present and discuss SLS applications to other combinatorial problems, which have been selected partly because of their fundamental nature, partly because of their relevance for certain application areas. In each of the main sections, we will introduce one combinatorial problem, discuss its applications and commonly used benchmark instances, and present one or more SLS approaches for solving this problem. The problems we cover are: Graph Colouring, Quadratic Assignment, Set Covering, Combinatorial Auctions Winner Determination and DNA Code Design. While the first three problems have been extensively studied in the literature for many years, the latter two have only relatively recently gained their current prominence.

The algorithms presented in this chapter are primarily intended to illustrate the application of SLS methods to the respective problems. Especially for the Graph Colouring Problem, the Quadratic Assignment Problem and the Set Covering Problem, the algorithms we present were chosen from a large number of known SLS algorithms for the respective problem; our selection represents a compromise between our desire to present state-of-the-art algorithms and the need to give reasonably didactic examples. References to other SLS algorithms and more detailed information on the problems covered here are provided in the 'Further Readings and Related Work' section of this chapter.

10.1 Graph Colouring

The *Graph Colouring Problem (GCP)* plays a central role in graph theory [Jensen and Toft, 1994] and lies at the core of many application relevant problems, such as timetabling [Leighton, 1979; de Werra, 1985; Schaerf, 1999] and frequency assignment [Gamst, 1986; Eisenblätter et al., 2002]. The GCP can be defined as follows: a *k-colouring* of an undirected graph $G := (V, E)$, where V is the set of $\#V = n$ vertices and $E \subseteq V \times V$ is the set of edges, is a mapping $a : V \mapsto \{1, 2, \ldots, k\}$ that assigns a positive integer from $\{1, 2, \ldots, k\}$ to each vertex in V, where each integer represents a colour, such that the endpoints of every edge in E are assigned different colours, that is, $\forall (u, v) \in E : a(u) \neq a(v)$. The decision version of the GCP asks whether for a given graph G and integer k, a k-colouring exists. In the optimisation version, the objective is to find the minimum number k such that a k-colouring exists; this minimum number k is also known as the *chromatic number of G*.

EXAMPLE 10.1 **Simple GCP Instance**

Figure 10.1 shows a simple GCP instance with six vertices $v_1, \ldots, v_6$; each vertex is assigned one of the three colours $\{1, 2, 3\}$ (here represented by different shades of grey). In this example we have $a(v_1) := a(v_6) := 1$ (light grey), $a(v_2) := a(v_5) := 2$ (dark grey), $a(v_3) := a(v_4) := 3$ (white). The chromatic number for this graph is three. In fact, the leftmost as well as the rightmost three vertices form a clique (i.e., a fully connected sub-graph), which implies

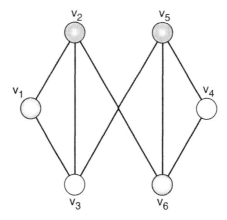

Figure 10.1 Simple GCP instance with six vertices; the given 3-colouring is optimal for the given graph.

a lower bound of three on the chromatic number of the given graph, because any valid colouring of a clique of size k requires at least k colours.

Note that a solution of a given GCP instance does not depend on the particular numbering of the colours, because all permutations of a valid k-colouring are isomorphic solutions. Alternatively to the formulation as an assignment problem, the GCP can be represented as a partitioning problem, in which a k-colouring corresponds to a partition of the set of vertices into k subsets such that for no edge $(u, v) \in E$ the incident vertices u and v belong to the same subset. For example, the partition representation of the 3-colouring of Example 10.1 divides the vertices into three subsets $\{v_1, v_6\}$, $\{v_2, v_5\}$ and $\{v_3, v_4\}$.

The decision version of the GCP is an $\mathcal{NP}$-complete problem that can be seen as a special case of the Constraint Satisfaction Problem (CSP), as discussed in Chapter 6, Section 6.5. In the CSP formulation, all constraints are binary inequality constraints; this special structure can be exploited to derive algorithms that are more efficient than standard CSP algorithms. Considering this as well as the central role of the GCP in graph theory and its numerous applications, graph colouring problems are typically treated independently of more general CSPs.

It is well known that the optimisation version of the GCP is $\mathcal{NP}$-hard [Garey and Johnson, 1979]. Furthermore, the GCP is not efficiently approximable, that is, in general, approximation ratios of $(\#V)^{1-\epsilon}$ cannot be achieved in polynomial time for any $\epsilon > 0$ (unless $\mathcal{NP} = \mathcal{ZPP}$; $\mathcal{ZPP}$ is the class of problems that can be solved in expected polynomial time by a probabilistic algorithm with zero error probability) [Feige and Kilian, 1998]. Therefore, from a theoretical perspective, the GCP is among the hardest combinatorial optimisation problems, since obtaining solution qualities within a fixed constant bound of the optimum is $\mathcal{NP}$-hard. Currently, the best polynomial-time approximation algorithm is only guaranteed to achieve an approximation ratio of $O(\#V \cdot (\log \log \#V)^2 / (\log \#V)^3)$ [Halldórsson, 1993].

Applications and Benchmark Instances

The GCP arises in many practical application domains; one of the most intuitive applications is that of colouring maps, where each region or country to be coloured corresponds to a vertex of a graph whose edges represent the borders between regions. (For an example of a map colouring problem, see Figure 6.9 on page 295). Other applications of the GCP include the determination of lower bounds on the number of time slots in timetabling problems [Leighton, 1979; de Werra, 1985; Carter, 1986]; special cases of frequency assignment problems

[Gamst, 1986]; the register allocation problem, in which variables are to be assigned to a limited number of registers in a CPU [Chaitin et al., 1981; Chaitin, 1982; Chow and Hennessy, 1990; Briggs et al., 1994]; the estimation of sparse Jacobian matrices [Coleman and Moré, 1983; Hossain and Steihaug, 2002]; and the testing for unintended short circuits on printed circuit boards [Garey et al., 1976b].

Recently, a special case of the GCP known as the *Quasigroup Completion Problem (QCP)* or Latin Square Problem has received significant attention. For a discussion of this problem we refer to Chapter 6, page 298, where it is described in more detail.

Various classes of randomly generated GCP instances are often used as benchmark problems for graph colouring algorithms. One of these classes is based on $G_{n,p}$ graphs, which have n vertices and are generated by including each of the $n \cdot (n-1)/2$ possible edges independently at random with probability p. Note that the parameter p determines the density of the resulting graphs, that is, the ratio of the number of edges and the maximum possible number of edges. Sparse graphs contain only a relatively small fraction of the possible edges, while for dense graphs this fraction is relatively large (i.e., close to one).

Another class of widely used GCP benchmark instances is based on the so-called geometric random $U_{n,d}$ graphs; these are generated as follows: first, n vertices are placed at random positions in a two-dimensional unit square in the Euclidean plain, where the x and y coordinates are chosen according to a uniform distribution over the interval $[0, 1]$; then, edges are created between any vertices u and v whose Euclidean distance is smaller than a given value d. (Clearly, the density of the resulting graphs increases with d.) Note that unlike in $G_{n,p}$ graphs, in geometric random $U_{n,d}$ graphs, the edges between different pairs of vertices are not statistically independent. Empirical results suggest that complete GCP algorithms solve geometric random $U_{n,d}$ instances much more efficiently than $G_{n,p}$ graphs [Mehrotra and Trick, 1996].

Other widely used classes of random graphs are generated such that there is a known upper bound on the chromatic number. One such class are the *Leighton graphs*, which are constructed based on cliques of a size between two and the desired chromatic number; the vertices in these cliques are chosen in such a way that the chromatic number will not exceed a given limit. The density of these graphs is generally less than 0.25.

A common way of generating guaranteed k-colourable graphs is to partition the vertices of a graph into k subsets and then to introduce edges only between vertices that are not in a same subset; often, edges are included with a probability p, independently of other edges. Classes of these graphs, which are generally referred to as $G_{n,p,k}$ graphs, can be generated in various ways. For example, the size of the subsets can be chosen as uniform as possible or with a large variance, restrictions on the vertex degrees may be introduced, etc.; for a more detailed discussion, we refer to Culberson et al. [1995]. One popular class of such $G_{n,p,k}$ graphs are the *flat graphs*, where an additional constraint on the degree of vertices is used to generate instances with a small variation in vertex degree, that is, in the number of edges incident to the vertices. This restriction was found to result in difficult GCP instances [Culberson and Luo, 1996]; generators for such graphs are available from Culberson's graph colouring page [Culberson, 2003].

A large collection of GCP instances, which stems in part from applications of the GCP, is available from the COLOR02/03 webpage [Johnson et al., 2003b]. This collection includes instances derived from register allocation problems, course scheduling problems, quasi-group completion problems (with and without pre-coloured squares), and several other combinatorial problems, as well as instances based on randomly generated graphs.

Simulated Annealing for Graph Colouring

One of the first SLS methods that have been applied to GCP is Simulated Annealing (SA) [Chams et al., 1987; Johnson et al., 1991]. The *Penalty Function SA Algorithm* (PFSA) is based on the partition-based formulation of GCP, that is, each candidate solution is a partition of V into k non-empty colour classes $V_1, \ldots, V_k$, which does not necessarily correspond to a valid k-colouring; the number of colour classes, k, may vary throughout the search [Johnson et al., 1991]. The search is initialised by randomly partitioning the vertices into a given number of colour classes. In each search step, a neighbouring candidate solution is generated by moving a vertex v from a non-empty colour class V_i into a new colour class $V_j \in \{V_1, \ldots, V_{k+1}\}$. The proposal mechanism of PFSA generates a neighbouring candidate solution by choosing V_i, $v \in V_i$, and V_j uniformly at random from the respective sets; choosing $j := k + 1$ corresponds to placing vertex v into a new colour class. The acceptance criterion is a probabilistic choice according to the Metropolis condition (see Equation 2.1 on page 75) that uses the evaluation function

$$g(s) := -\sum_{i=1}^{k}(\#V_i)^2 + \sum_{i=1}^{k} 2 \cdot \#V_i \cdot \#E_i,$$

where E_i is the set of all edges with both endpoints in V_i. Although this evaluation function does not explicitly count the number of colours k currently used, the first term gives a bias towards a small number of colour classes, while the second term penalises constraint violations that are incurred when reducing the number of colours too much; the idea is that by using this evaluation function, the algorithm minimises k as a 'side-effect' [Johnson et al., 1991]. It is easy to show that all local minima of this evaluation function correspond to legal colourings.

PFSA uses a geometric cooling schedule under which at each temperature a number of search steps is performed that is proportional to the neighbourhood size (see Example 2.4, page 77 for a similar SA algorithm for the TSP). The search is terminated when the fraction of accepted candidate solutions over a number of successive temperature values falls below a given threshold.

A second SA algorithm for the GCP, the *Kempe Chain SA Algorithm (KCSA)*, searches a restricted space of candidate solutions that correspond to valid colourings, that is, assignments of colours to vertices that always assign different colours to the endpoints of any edge [Johnson et al., 1991]. In principle, KCSA uses the same evaluation function as PFSA, but because of the restriction to valid colourings, this function is reduced to

$$g(s) := -\sum_{i=1}^{k}(\#V_i)^2.$$

The initial solution in KCSA is generated by the *Sequential Colouring Algorithm*, a construction heuristic that iteratively assigns colours to the vertices $v_1, \ldots, v_n$ by placing each vertex in the colour class with the lowest possible index, while maintaining a valid colouring. Different from PFSA, KCSA is based on a *Kempe chain neighbourhood*, under which neighbouring candidate solutions differ in the colouring of a subgraph of G. The proposal mechanism based on this neighbourhood generates a candidate solution as follows:

- Select a non-empty colour class V_i, a vertex $v \in V_i$, and a non-empty colour class V_j uniformly at random;

- let KC_{ij} be the maximally connected subgraph that contains vertex v and only vertices of colour classes V_i and V_j; KC_{ij} is also called a *Kempe chain*;

- swap the colours i and j assigned to the vertices in KC_{ij}.

This type of search step is illustrated in Figure 10.2. All other components of KCSA are identical to those of PFSA.

A third SA algorithm for the GCP, called *Fixed-k SA*, keeps the number of colours constant throughout the search process, while trying to minimise the number of violated edge constraints; this approach is similar to that underlying most SLS algorithms for the general CSP (cf. Chapter 6, Section 6.6). Once a valid k-colouring has been found for a specific value of k, the optimisation version of the GCP can be attacked by iteratively reducing k and applying the same algorithm until a value k is reached for which no k-colouring can be obtained.

Fixed-k SA uses the same search space as PFSA along with a simple 1-exchange neighbourhood relation under which two candidate solutions are direct neighbours if, and only if, one can be obtained from the other by assigning a different colour to exactly one vertex. The search process is initialised by

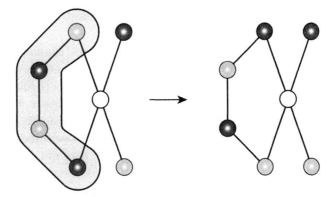

Figure 10.2 Search step in the Kempe chain neighbourhood used in the KCSP algorithm for GCP. The highlighted set of four light and dark grey vertices are connected by a path, forming a Kempe chain, and exchange their colours in one local search step.

randomly partitioning the vertices into k colour classes. Then, in each search step, a candidate solution is proposed by choosing uniformly at random one of the vertices that is involved in violated edge constraints and by assigning that vertex to a different colour class, which is also chosen uniformly at random. (Using a similar proposal mechanism that can select arbitrary vertices which are not necessarily involved in constraint violations leads to substantially inferior performance [Johnson et al., 1991].) The acceptance criterion is the same as used in PFSA and KCSA, except that now an evaluation function is used that simply counts the number of violated edge constraints under a given candidate solution. All other components of Fixed-k SA are identical to those of PFSA and KCSA.

An experimental analysis of these three SA algorithms on a variety of GCP instances based on random $G_{n,p}$, $U_{n,d}$ and $G_{n,p,k}$ graphs has shown that none of them completely dominates the others in terms of performance [Johnson et al., 1991]. However, it has been observed that Fixed-k SA tends to achieve the best performance on sparse graphs, while KCSA performs better for dense graphs; PFSA typically fails to reach the performance of the better of the two other algorithms.

The Memetic Algorithm by Galinier and Hao (MA-GH)

Among the currently most successful SLS algorithms for the GCP is the memetic algorithm by Galinier and Hao [1999], *MA-GH*, which uses a subsidiary local search procedure based on short runs of an effective tabu search algorithm. Like Fixed-k SA, MA-GH searches k-colourings of a graph for fixed k and evaluates candidate solutions by counting the number of violated edge constraints.

MA-GH maintains a small population of candidate solutions that is initialised by a greedy algorithm that constructs (potentially invalid) k-colourings based on a modification of the DSATUR algorithm [Brélaz, 1979]. This construction heuristic is initialised with k empty colour classes. In each construction step, it chooses an uncoloured vertex with a minimal number of valid colour classes and places it in the valid colour class with the lowest index (a colour class V_i is valid for a given vertex v, if including v in V_i does not violate any edge constraint); if there is no valid colour class for a vertex, it remains uncoloured. Typically, this algorithm cannot assign all vertices to colour classes such that a valid k-colouring results. Therefore, in a final step, the uncoloured vertices are added to colour classes that are independently and uniformly chosen at random.

As its subsidiary local search procedure, MA-GH uses a tabu search algorithm, *TS-GH-GCP*, that is based on the same restricted 1-exchange neighbourhood as Fixed-k SA and allows only vertices involved in an edge-constraint violation to be moved into a different colour class. In each search step, a best-improving non-tabu neighbouring candidate solution is selected. If the colour class of vertex v changes from V_i to V_j, it is forbidden to assign colour i to vertex v in the subsequent tt steps, except when such a move would lead to an improvement in the incumbent candidate solution (aspiration criterion). The tabu tenure, tt, for a specific pair (v, i) is chosen randomly in dependence of the current evaluation function value (for details, see Galinier and Hao [1999]). TS-GH-GCP terminates when a valid k-colouring has been found or a given limit on the number of search steps has been reached. TS-GH-GCP can be seen as a variant of TS-GH, a state-of-the-art tabu search algorithm for the more general Constraint Satisfaction Problem (cf. Chapter 6, page 305*f.*). As a stand-alone algorithm, TS-GH-GCP performs better than many other SLS algorithms for the GCP in terms of the probability of finding high-quality colourings within a fixed amount of run-time.

After this tabu search procedure has been applied to all candidate solutions of the initial population, in each iteration of MA-GH, a candidate solution (offspring) is generated by applying a specific recombination operator called *Greedy Partition Crossover (GPX)* to two candidate solutions (parents) that are uniformly chosen at random. As usual in memetic algorithms, the resulting offspring is improved by applying the subsidiary local search procedure. The partition thus obtained replaces the worse of the two parents. Unlike typical evolutionary algorithms, MA-GH does not make use of a mutation operator. MA-GH terminates when a valid k-colouring has been found, a given run-time limit has been reached, or when the population diversity (i.e., the average distance between the candidate solutions in sp in the underlying neighbourhood graph) drops below a given threshold. An outline of the algorithm is shown in Figure 10.3.

```
procedure MA-GH (G,k)
    input: graph G, integer k
    output: valid k-colouring ŝ of G or ∅

    sp := init(G);
    for each s ∈ sp do
        s := tabuSearch(G, s);
    end
    while not terminate(G, sp) do
        (s₁, s₂) := selectParents(G, sp);
        s := GPXrecombination(G, s₁, s₂);
        s' := tabuSearch(G, s);
        sp := updatePopulation(G, sp, s');
    end
    if g(best(G, sp)) = 0 then
        return best(G, sp)
    else
        return ∅
    end
end MA-GH
```

Figure 10.3 Algorithm outline of MA-GH; $best(g, sp)$ denotes the individual from population sp with the best evaluation function value, that is, the fewest violated colouring constraints, for the given graph G. (For details, see text.)

The main innovation in MA-GH is the GPX recombination operator. This crossover operator exploits the partition representation of the GCP (and the candidate solutions) as a partitioning of V into colour classes $V_1, \ldots, V_k$ and tries to greedily transfer colour classes of maximal size from the two parents to the offspring. GPX takes as its input two candidate solutions s_1 and s_2, and builds a new partition by alternatingly selecting colour classes from each parent; in step $i, i = 1, \ldots, k$, a colour class with the maximum number of vertices is chosen from parent s_1 (if i is odd) or from parent s_2 (if i is even) as colour class V_i of the offspring; after each step, the vertices in V_i are removed from both parents. Finally, the remaining unassigned vertices are placed in colour classes that are uniformly and independently chosen at random.

EXAMPLE 10.2   **Greedy Partition Crossover**

Given a GCP instance with ten vertices $\{1, \ldots, 10\}$ and three colours, consider the partitions

$$s_1 := \{\{1, 2, 3, 4\}, \{5, 6, 7\}, \{8, 9, 10\}\}, \ s_2 := \{\{4, 6, 7, 8\}, \{1, 2, 10\}, \{3, 5, 9\}\}$$

as parents in a GPX recombination. In the first step of GPX, the colour class $\{1, 2, 3, 4\} \in s_1$ is copied to the offspring s_o, and the vertices of this set are deleted from s_1 and s_2. This results in (partial) candidate solutions

$$s_1' = \{\{5, 6, 7\}, \{8, 9, 10\}\}, \ s_2' = \{\{6, 7, 8\}, \{10\}, \{5, 9\}\}, \ s_o = \{\{1, 2, 3, 4\}\}.$$

Next, the largest colour class of s_2', $\{6, 7, 8\}$, is chosen and added to s_o, after which again the corresponding vertices are deleted in both parents, resulting in

$$s_1'' = \{\{5\}, \{9, 10\}\}, \ s_2'' = \{\{10\}, \{5, 9\}\}, \ s_o = \{\{1, 2, 3, 4\}, \{6, 7, 8\}\}.$$

The next step leads, after copying colour class $\{9, 10\}$ to s_o, to

$$s_1''' = \{\{5\}\}, \ s_2''' = \{\{5\}\}, \ s_o = \{\{1, 2, 3, 4\}, \{6, 7, 8\}, \{9, 10\}\}.$$

At this point, the only remaining vertex 5 is assigned to a randomly chosen colour class in s_o, resulting, for example, in the final offspring $\{\{1, 2, 3, 4\}, \{6, 7, 8\}, \{5, 9, 10\}\}$.

This GPX recombination mechanism is based on the intuition that the important information to be transmitted to the offspring is which sets of vertices belong to the same colour class in the parents, while the particular colour assigned to a vertex is not important. This is intuitively clear, since the validity of a k-colouring is unaffected by permuting the colours.

MA-GH has been tested on large $G_{n,p}$, Leighton and flat graphs (these are available from Johnson et al. [2003b]), and its performance has been compared to that of long runs of TS-GH-GCP [Galinier and Hao, 1999]. Empirical results indicate that MA-GH achieves significantly better performance than TS-GH-GCP, especially on large $G_{n,p}$ graphs; for almost all instances (and number of colours) tested, it found better k-colourings than TS-GH-GCP or produced the same k-colourings substantially faster. The overall excellent performance of MA-GH is confirmed by the fact that for four 1 000 vertex graphs (three randomly generated $G_{n,p}$ graphs and one flat graph) it was able to improve on the previously best known colourings; however, substantial fine-tuning of the algorithm was required to achieve these results. Overall, MA-GH currently appears to be one of the best SLS algorithms known for finding high-quality solutions of hard GCP instances.

Related Problems

There are a number of extensions of the GCP that arise in the context of various application problems, mainly in scheduling and telecommunications (see, for example, Eisenblätter et al. [2002]). One of these is the *List Colouring Problem*, in which a graph G is given along with lists of colours $L(v)$ for each vertex v, and the objective is to find a minimal k-colouring that assigns to each vertex a colour from its associated list $L(v)$ [Erdõs et al., 1979].

A second extension is the *T-Colouring Problem*, where each edge (u, v) has an associated set of weights D_{uv} and a colouring has to satisfy additional separation constraints $|a(u) - a(v)| \notin D_{uv}$. This problem was introduced in the context of the Frequency Assignment Problem (FAP) [Hale, 1980] to model situations in which specific differences between frequencies (colours) assigned to transmitters (vertices) have to be maintained in order to avoid interference (see also Section 7.3, page 343*ff.*). Note that the GCP is a special case of the T-Colouring Problem with $D_{uv} = \{0\}$ for all edges (u, v). The special case of the T-Colouring Problem, where $D_{uv} = \{0, 1, \ldots, d_{uv}\}$, is called the Bandwidth Colouring Problem; in this case, the separation constraint becomes $|a(u) - a(v)| > d_{uv}$. The combination of the List Colouring and the T-Colouring Problems is known as the *List T-Colouring Problem* [Tesman, 1993a].

In another extension of the GCP, a number of colours can be simultaneously assigned to each vertex; this is the *Set Colouring Problem* (also known as the *Multi-Colouring Problem*) [Roberts, 1979]. Here, for each vertex v a weight $k_v < k$ is given, and the goal is to find a set colouring for the given graph, that is, an assignment of colour sets $C(v) \subseteq \{1, \ldots, k\}$ to all vertices $v \in V$ such that each vertex v is assigned a set of exactly k_v colours, and for each edge (u, v) the respective colour sets have no elements in common (i.e., $C(u) \cap C(v) = \emptyset$). The Set Colouring Problem has also been extended with the constraints of the T-Colouring Problem to the *Set T-Colouring Problem* [Tesman, 1993b].

In general, most SLS algorithms for the GCP can be extended in a rather straightforward way to these related problems. An example for such an extension can be found in the work of Dorne and Hao, who developed a tabu search algorithm for the T-Colouring and Set T-Colouring Problems that is closely related to TS-GH-GCP [Dorne and Hao, 1999].

10.2 The Quadratic Assignment Problem

The *Quadratic Assignment Problem (QAP)* has been the subject of an enormous amount of research efforts; besides the Travelling Salesman Problem, the QAP is one of the most studied combinatorial optimisation problems [Çela, 1998]. It is of particular interest in the context of SLS algorithms, which typically outperform all other types of QAP algorithms by a large margin. The QAP can be best described as the problem of assigning objects to a set of locations with given distances between the locations and given flows between the objects; the flows correspond, for example, to the amount of materials or products to be exchanged between machines in a production environment. The goal is to assign the objects to locations in such a way that the sum of the product between flows and distances is minimal.

Formally, a QAP instance is specified by n objects and n locations, where both the objects and the locations can be represented as integers from the set $I := \{1, \ldots, n\}$, and two positive real-valued $n \times n$ matrices A and B, where a_{ij} is the *distance* between locations i and j, and b_{rs} is the *flow* between objects r and s — A and B are called the *distance matrix* and the *flow matrix*, respectively. The objective in the QAP is to find an optimal *assignment* of objects to locations, that is, a mapping $\psi : I \mapsto I$ such that every object is assigned to exactly one location, no location is assigned more than one object, and the function

$$f(\psi) := \sum_{i=1}^{n} \sum_{j=1}^{n} b_{ij} a_{\psi(i)\psi(j)} \qquad (10.1)$$

is minimised. Here, $\psi(i)$ denotes the location of object i under assignment ψ, and the term $b_{ij} a_{\psi(i)\psi(j)}$ intuitively represents the cost contribution of simultaneously assigning object i to location $\psi(i)$ and object j to location $\psi(j)$. In other words, given n, A and B, the objective in the QAP is to find $\psi^* \in \operatorname{argmin}_{\psi \in \Psi(I)} f(\psi)$, where $\Psi(I)$ is the set of all assignments of objects to locations. Note that every assignment ψ corresponds to a permutation of the index set $\{1, \ldots, n\}$. (Equivalently, the QAP can be stated as the problem of assigning locations to objects.)

EXAMPLE 10.3 **Keyboard Layout as a Quadratic Assignment Problem**

As an example for a QAP instance, we consider the problem of optimising the layout of a computer (or typewriter) keyboard. In this case, the objects correspond to the letters of the alphabet and the locations to the keys of a keyboard. The flow between two letters captures the empirical frequency of the corresponding combination; for example, b_{we} is the empirical frequency of the letter 'w' being directly followed by 'e' (these frequencies are obviously language dependent). The distance between two keys is defined as the empirically measured time required for pressing them in sequence, for example, a_{ij} is the time needed to press key i followed directly by key j. Note that in this example, the flow and distance matrices are typically asymmetric. Based on these definitions, the optimisation objective measures the typing efficiency in terms of the total time required to type a typical text. (Obviously, this is an abstraction of the real problem of optimising a keyboard layout; an optimisation approach to a more realistic keyboard layout problem has been described by Wagner et al. [2003].)

The definition of the QAP given above corresponds to the so-called *Koopman–Beckman formulation*. Note that (as is done in the original paper by Koopman and Beckman [1957]), it is straightforward to additionally include fixed assignment costs of objects to locations by adding a cost term $\sum_{i=1}^{n} c_{i\,\psi(i)}$ to Equation 10.1.

The QAP is an $\mathcal{NP}$-hard optimisation problem, and no polynomial-time deterministic algorithm exists that is guaranteed to achieve any constant approximation ratio [Sahni and Gonzalez, 1976]; it is also considered to be one of the hardest optimisation problems in practice, since the size of QAP instances for which provably optimal solutions can be found using complete algorithms is limited to around $n = 30$ [Hahn et al., 2001; Hahn and Krarup, 2001; Anstreicher et al., 2002].

At the time of this writing, the largest non-trivial QAP instance from QAPLIB, a benchmark library for the QAP [Hahn, 2003], that has been solved provably optimally is the instance `ste36a` with only 36 locations [Brixius and Anstreicher, 2001; Nyström, 1999]; this is in stark contrast with the state of the art in complete algorithms for the TSP, which have been used for solving instances with up to 15 112 cities. (It may be noted that TSP instances can be directly encoded into QAP instances of the same size.) Despite the small size of this instance, solving it optimally using a state-of-the-art systematic search algorithm for the QAP required approximately 180 hours of CPU time on an 800 MHz Pentium III processor [Brixius and Anstreicher, 2001]; several smaller instances take substantially longer to solve. For only few QAP instance classes, larger instances with a size of up to 75 objects have been solved provably optimally, which again, required substantial computation times [Drezner et al., to appear].

In contrast, the best-performing SLS algorithms for the QAP typically require only a few seconds on a comparable machine to find the optimal solution for instance `ste36a` and other instances of similar size. Hence, SLS methods currently represent the only feasible approach for solving large QAP instances.

Applications and Benchmark Instances

The QAP can be seen as an abstract model of a variety of practical layout and location problems. Examples of problems that have been formulated as a QAP include backboard wiring [Steinberg, 1961], where computer components have to be placed such that the total amount of wiring required to connect them is minimised; hospital layout [Elshafei, 1977; Krarup and Pruzan, 1978], where the goal is to place facilities of a hospital in buildings such that the total amount of communication times distance is minimised; typewriter keyboard design [Burkard and Offermann, 1977], the problem described in Example 10.3; half-tone rendering of specific shades of grey [Taillard, 1995]; and many others [Burkard et al., 1998a; Çela, 1998].

In the context of the numerous existing research studies on the QAP and QAP algorithms, a large number of benchmark instances have been used. Many of these instances are available through QAPLIB [Burkard et al., 1997; Hahn, 2003], an online resource for the QAP that comprises benchmark instances, several QAP solver implementations and further information on the QAP. The QAPLIB benchmarks stem from various classes of instances with strongly varying characteristics. It is well known that the particular instance class and their corresponding characteristics have a considerable influence on the performance of SLS methods [Taillard, 1995; Gambardella et al., 1999; Stützle and Hoos, 1999]. Generally, most QAPLIB instances can be classified into the following four categories [Taillard, 1995]:

- **Unstructured, randomly generated instances.** In these instances, the distance and flow matrix entries are generated randomly according to a uniform distribution over a given range of values. These instances are among the hardest to solve optimally. However, most SLS algorithms can reach solution qualities within 1–3% of the optimum relatively quickly compared to complete QAP algorithms.

- **Instances with grid-based distance matrices.** In this class of instances, the distance matrix stems from a $n_1 \times n_2$ grid and the distances are defined as the Manhattan distance between grid points. These instances have multiple global optima (at least four, if $n_1 \neq n_2$ and at least eight, if $n_1 = n_2$) due to the symmetries of the distance matrices. The flow matrices of these instances are generated according to various types of probability distributions.

- **'Real-life' instances.** Instances from this class stem from practical applications of the QAP, such as those mentioned above. The matrix entries of real-life QAP instances exhibit a clear structure; in particular, the flow matrices have many zero entries and the remaining entries are not uniformly distributed.

- **Random 'real-life like' instances.** This class of instances comprises randomly generated instances whose distance and flow matrix entries have been generated according to non-uniform probability distributions in order to obtain problem characteristics that resemble those of real-life instances [Taillard, 1995]. These instances complement the real-life instances, which are often relatively small, in a useful way.

QAPLIB contains a number of additional instances, for example, from a generator that produces instances with known optimal assignments [Li and Pardalos, 1992] and from encodings of weighted tree problems [Christofides and Benavent, 1989]. Unfortunately, many of the QAPLIB instances are of relatively small size ($n \leq 50$) and do not appear to pose serious challenges to high-performance SLS algorithms. In addition, QAPLIB does not provide sets of instances with systematically varied characteristics and hence does not support comprehensive studies of algorithm behaviour in dependence of instance features.

To overcome these limitations, recent studies have proposed and used a large number of additional instances. The largest instance collection has been compiled by Stützle in the context of research performed in the Metaheuristics Network (see also Dorigo [2003]). It comprises instances of size 50 to 500 with systematically varied characteristics of the underlying distance and flow matrices. Large, randomly generated real-life like instances with up to 768 objects have been proposed by Taillard [2003].

Several measures have been used to characterise QAP instances. One of these is the *flow dominance* (*fd*), which is defined as the coefficient of variation of entries of the flow matrix B multiplied by 100, that is, $fd(B) := 100 \cdot \hat{\sigma}_B / \hat{\mu}_B$, where

$$\hat{\mu}_B = \frac{1}{n^2} \cdot \sum_{i=1}^{n} \sum_{j=1}^{n} b_{ij}$$

and

$$\hat{\sigma}_B = \sqrt{\frac{1}{n^2 - 1} \cdot \sum_{i=1}^{n} \sum_{j=1}^{n} (b_{ij} - \hat{\mu}_B)}.$$

A high flow dominance indicates that a large part of the overall flow is exchanged among relatively few objects. Hence, randomly generated problems according to a uniform distribution will have a rather low flow dominance, whereas real-life problems, in general, have much higher flow dominance values. The distance dominance can be defined analogously. Another feature that has been used to characterise QAP instances is the sparsity of the flow or distance matrix, defined as $sp := n_0/n^2$, where n_0 is the number of zero-entries in the given flow or distance matrix, respectively.

Analyses of the search spaces of given QAP instances have shown that the different problem characteristics also translate into different search space characteristics (cf. Chapter 5). In general, real-life and real-life like instances have a much higher fitness-distance correlation than unstructured, randomly generated instances [Merz and Freisleben, 2000a; Stützle and Hoos, 2000]; in fact, FDC values for instances of the latter type are close to zero. Similarily, systematic variations in autocorrelation lengths have been reported between instances from the different classes [Merz and Freisleben, 2000a].

Reactive Tabu Search for the QAP

Most 'simple' SLS algorithms for the QAP are based on a 2-exchange neighbourhood relation that contains all assignments that can be obtained by swapping the locations of two objects, that is, $N(\psi, \psi') :\Leftrightarrow \exists r, s : (r \neq s \wedge \psi'(r) = \psi(s) \wedge \psi'(s) = \psi(r) \wedge \forall i \notin \{r, s\} : \psi'(i) = \psi(i))$; this neighbourhood relation also provides the basis for the well-known *Reactive Tabu Search algorithm for the QAP (RTS-QAP)* [Battiti and Tecchiolli, 1994].

To initialise the search process, RTS-QAP generates an assignment uniformly at random. RTS-QAP can be seen as an extension of a simple best-improvement search algorithm that uses the objective function of the QAP for evaluating candidate solutions. (It may be noted that by using various speed-up techniques, including efficient caching and incremental updating of the evaluation function contributions of individual 2-exchange steps, each best-improvement search step can be performed in time $\Theta(n^2)$ after the search has been initialised in time $\Theta(n^3)$.) Tabu status is associated with atomic assignments of individual objects to locations. A search step is tabu if, and only if, both objects r and s involved in the respective exchange would become assigned to a location that they occupied in the most recent tt iterations. At any time, a search step that leads to an improvement in the incumbent candidate solution can be performed regardless of its tabu status (aspiration mechanism).

During the search process, RTS-QAP dynamically adjusts the most critical parameter of this basic tabu search procedure, the tabu tenure tt. Furthermore, an escape mechanism is triggered when severe search stagnation is detected. The mechanism used for dynamically adjusting tt is based on the search history. More specifically, RTS-QAP stores all the candidate solutions encountered in the search trajectory since the last escape phase (or search initialisation) together with some additional information, such as the iteration number or how often a given candidate solution has been encountered (stored in variable ψ_{rep}) in a hash table (a hash table is used for efficiency reasons; we refer to Chapter 11 in Cormen et al. [2001] for details on hash tables) to check whether the search process is cycling (i.e., whether candidate solutions are being frequently revisited).

For the reaction mechanism, RTS-QAP uses the variables ma, the moving average of recurrences of candidate solutions in the search trajectory; $sttc$, the number of iterations since the last change of the tabu tenure; and nr, which counts the number of frequently repeated candidate solutions. These variables

```
procedure RTS-QAP(n, A, B)
    input: dimension n ∈ ℕ, distance matrix A, flow matrix B
    output: assignment ψ̂

    ψ := init; ψ̂ := ψ; tt := 1;
    while not terminate(ψ, maxIter) do
        ψ' := stepTS(ψ, tt);
        if not checkRepetitions(ψ, ψ', tt) then
            ψ := ψ';
            if f(ψ) < f(ψ̂) then
                ψ̂ = ψ;
            end
        else
            ψ := escape(ψ);
        end
    end
    return ψ̂
end RTS-QAP
```

Figure 10.4 Outline of the RTS-QAP algorithm; the tabu tenure, tt, is modified within the functions *checkRepetitions* and *escape*. (For further details, see text.)

are initialised to zero at the start of the algorithm and after each application of the escape mechanism.

Figure 10.4 shows a high-level outline of the RTS-QAP algorithm. The essential function for the reaction mechanism for parameter tt is *checkRepetitions*, which works as follows. First the new candidate solution ψ' is searched in the hash table. If this search is successful, the fact that with ψ' a previously visited candidate solution is encountered indicates that the search process may be trapped. There are two possible ways to react to such a situation. As an immediate solution, increasing the tabu tenure tt helps to avoid revisiting candidate solutions. This is done by increasing tt by a factor α_{inc}; multiple such reactions lead to a geometric increase of tt and, hence, help to avoid revisiting candidate solutions.

However, it still may happen that a number of candidate solutions, which are stored in a set of frequently revisited candidate solutions, are encountered many times; this is taken as a sign that the search trajectory is trapped in a limited area of the search space and the escape mechanism is triggered. In particular, the escape mechanism is triggered if more than nr candidate solutions are frequently revisited, where nr is a parameter of the algorithm set to three. Technically, a candidate solution is considered 'frequently visited' if it has been encountered more than a user-specified number of times since the last escape (or search initialisation).

If tt is only increased, this may lead to an overly strongly confined search trajectory. To counteract this problem, the tabu tenure is occasionally reduced by a factor α_{dec}; this is done when the tabu tenure has not changed for $sttc$ iterations. Additionally, the tabu tenure is decreased whenever a situation is encountered in which all possible search steps are tabu.

The escape mechanism first clears the hash table and the tabu status of all tabu attributes, sets $tt := 1$, and then applies a large random modification to the current candidate solution. In particular, it performs $1 + (1 + r) \cdot ma/2$ random search steps, where r is a random number drawn according to a uniform distribution over the interval $[0, 1]$. In each of these steps, the pair of objects to be exchanged is selected uniformly at random, ignoring the tabu status of any such exchange step.

Experimental results indicate that on randomly generated, unstructured instances, RTS-QAP performs better than a simple tabu search algorithm and Robust Tabu Search [Taillard, 1991]. Further experiments have shown that the performance of RTS-QAP is typically much more robust w.r.t. the additional parameters controlling the reaction mechanism than the underlying tabu search algorithm is w.r.t. the tabu tenure. In fact, experiments with various parameter settings have shown that (i) the escape mechanism is important for achieving optimal performance and that (ii) the parameter settings of α_{dec} and α_{inc} have only a minor influence on the performance of the algorithm, as long as they are chosen within reasonable limits (Battiti and Tecchiolli recommend values of 0.9 and 1.1, respectively). In general, RTS-QAP appears to be particularly well-suited for solving unstructured QAP instances, where it achieves state-of-the-art performance. However, on more structured instances, it is often surpassed by hybrid SLS algorithms such as the one presented next.

Population-Based Iterated Local Search

One of the best-performing SLS algorithms for the QAP is a particular population-based iterated local search method [Stützle, 1999]. In the following, we first describe the underlying ILS algorithm, *ILS-S-QAP*, and then explain the population-based extension.

ILS-S-QAP starts from a random assignment and applies a first-improvement 2-opt local search procedure based on the same neighbourhood that is used in RTS-QAP [Stützle, 1999]. To speed up the local search, don't look bits are associated with each object (see also Section 8.2, page 375f.). If during the neighbourhood scan for an object no improving step is found, its don't look bit is turned on (set to 1) and the object is not considered as a starting object for a neighbourhood scan in the next iteration. Whenever an object is involved in a 2-exchange step and changes its location, its don't look bit is turned off again.

Additionally, the local search procedure in ILS-S-QAP has the peculiarity that it randomly changes the order in which the neighbourhood is scanned between different applications of the procedure; this is done by generating a random permutation ϕ_r each time the local search is initialised, which is subsequently used to determine the order in which the neighbours of the current candidate solution are evaluated in each search step. Hence, even when initialised with the same assignment, the local search procedure may produce different locally optimal candidate solutions. In the context of the ILS algorithm, this has the advantage that even after applying a relatively weak perturbation to a locally optimal assignment ϕ', a subsequent local search is rather unlikely to return to the same ϕ'. The local search procedure terminates when all don't look bits are set to one.

Each perturbation phase of ILS-S-QAP consists of one random k-exchange move. Like in basic Variable Neighbourhood Search (VNS) [Hansen and Mladenović, 1999; 2001], the value of k is dynamically modified during the search. In particular, k varies between two values k_{min} and k_{max}. At the beginning of the search process, k is set to k_{max}; after each iteration of the algorithm (consisting of a perturbation and subsequent local search phase), k is decreased by one until it reaches the value k_{min}. Subsequently, k is increased by one after each iteration in which no improvement in the incumbent solution has been achieved, otherwise it is set to k_{min}. Whenever k reaches the value k_{max}, it is reset to k_{min}. This mechanism for dynamically modifying k makes the algorithm's performance somewhat more robust compared to using a fixed value of k; this is particularly important since there appears to be no single fixed value for k that achieves peak performance across a diverse set of instances.

As outlined in Chapter 8, page 375*f.*, the don't look bit technique is integrated into the perturbation by resetting to zero only the don't look bits of those objects that change their location in the respective k-exchange move. This resetting strategy results in an additional significant speed up and, if k is not too large, it allows to substantially increase the number of local search phases that can be performed within a given amount of CPU time compared to the variant in which after each perturbation phase all don't look bits are set to zero.

The acceptance criterion of ILS-S-QAP selects the new assignment if, and only if, its quality is equal to or better than the previous locally optimal assignment; hence, the ILS algorithm performs an iterated descent in the space of 2-exchange local optima.

The population-based ILS extension of this algorithm, *PBILS-S-QAP*, follows — at least in principle — the same ideas as earlier population-based ILS algorithms for the TSP (see Section 8.4, page 400*ff.*). In each iteration, PBILS-S-QAP performs one iteration of ILS-S-QAP on each assignment in the current population. Then, a new population is determined by a variant of the

$(\mu + \lambda)$-selection strategy that is frequently used in Evolution Strategies [Schwefel, 1981]. In the original $(\mu + \lambda)$-selection mechanism, μ is the population size and λ the number of offspring; the new population is comprised of the μ best out of $\mu + \lambda$ candidate solutions (in PBILS-S-QAP, $\lambda := \mu$).

This selection mechanism strongly favours the best candidate solutions and can hence easily lead to premature convergence of the population and subsequent search stagnation. In order to avoid this problem, two additional diversification techniques are applied.

Firstly, when selecting the assignments that will comprise the next generation sp', the $\mu + \lambda$ given assignments (current population + offspring) are considered in the order of their solution quality. Before inserting an assignment ψ into sp', we first measure its distance to each of the assignments already included in sp'. Here, we define the distance between two assignments ψ and ψ' to be the number of objects that are placed on different locations in ψ and ψ', that is, $d(\psi, \psi') := \#\{i \mid \psi(i) \neq \psi'(i)\}$. (This is a direct extension of the well-known Hamming distance for bit strings.) An assignment is inserted into the new population sp' if, and only if, the minimum distance to any of the assignments already included in sp' is larger than a threshold distance d_θ; the value of d_θ is gradually decreased during the search process, until a given minimum value d_{min} is reached. This ensures that a higher degree of diversification is achieved in earlier stages of the search process.

Secondly, PBILS-S-QAP uses a restart mechanism that applies a strong perturbation to all candidate solutions in the current population; this mechanism is invoked when no improving assignment has been found for a number of iterations or when the average distance between the elements in the population falls below some given threshold value.

PBILS-S-QAP generally shows better performance than QAP variants of several of the population-based ILS algorithms described in Chapter 8, Section 8.4, as well as ILS-S-QAP and several other ILS algorithms that use only a single candidate solution [Stützle, 1999]. Experimental results indicate that PBILS-S-QAP is a state-of-the-art algorithm for real-life and real-life like QAP instances [Stützle, 1999]. PBILS-S-QAP has also been shown to achieve the best performance across a large set of QAP instances in a recent extensive experimental evaluation of several SLS algorithms for the QAP that has been performed in the context of the Metaheuristics Network (see Dorigo [2003]).

Generalisations and Related Problems

Several variants and generalisations of the QAP can be found in the literature. Lawler introduced a slightly more general variant of the QAP, in which a four-dimensional array of coefficients $C := (c_{ijkl})$ is given and the objective is to minimise $f(\psi) := \sum_{i=1}^{n} \sum_{j=1}^{n} c_{ij\psi(i)\psi(j)}$ [Lawler, 1963]. Clearly, every QAP according to our definition (i.e., the Koopman–Beckman formulation) can be seen as a special case of Lawler's variant in which $c_{ijkl} = b_{ij} \cdot a_{kl}$.

The *Bottleneck QAP* is a variant of the standard QAP in which the goal is to minimise the objective function

$$f_b(\psi) := \max\{b_{ij}a_{\psi(i)\psi(j)} \mid 1 \le i, j \le n\}. \tag{10.2}$$

That is, in the Bottleneck QAP the goal is to minimise the maximum cost instead of the sum of the costs. The Bottleneck QAP first arose in the backboard wiring application of Steinberg [1961] in cases where the goal is to minimise the maximum wire length.

In the *Quadratic Semi-Assignment Problem (QSAP)*, $m < n$ locations are given and the goal is to assign all objects to locations such that (i) every location is assigned at least one object and (ii) the sum of the flows between the objects times the distances between the locations of the objects is minimised. The QSAP is $\mathcal{NP}$-hard and difficult to solve in practice [Milis and Magirou, 1995]; a tabu search approach to the QSAP has been proposed by Domschke et al. [1992].

The *Biquadratic Assignment Problem (BiQAP)* is a generalisation of the QAP to a quartic assignment problem, where two four-dimensional $n \times n \times n \times n$ matrices $A := (a_{ijkl})$ and $B := (b_{rsuv})$ are given and, under the straightforward extension of the assignment constraints, the function

$$f(\psi) := \sum_{i=1}^{n} \sum_{j=1}^{n} \sum_{k=1}^{n} \sum_{l=1}^{n} b_{ijkl}a_{\psi(i)\psi(j)\psi(k)\psi(l)} \tag{10.3}$$

is to be minimised. The BiQAP arises in the design of Very Large Scale Integrated (VLSI) sequential circuits. Burkard et al. [1994] have given a detailed description of the VLSI design problem that leads to the BiQAP. In addition, they investigated lower bounds for the BiQAP, but found that these still left a large gap to the optimal solution and concluded that complete algorithms are likely to perform extremely poorly on the BiQAP [Burkard et al., 1994]. Therefore, several SLS algorithms have been proposed for the BiQAP, including algorithms based on Iterative Improvement, Simulated Annealing, Tabu Search (all of which were implemented by Burkard and Çela [1995]) as well as GRASP [Mavridou et al., 1998]; these algorithms have been tested on randomly generated instances with known optimal solutions obtained from a generator described in Burkard et al. [1994]. Of the SLS algorithms tested, the GRASP algorithm appears to achieve the highest quality solutions for long run-times.

10.3   **Set Covering**

The *Set Covering Problem (SCP)* is a well-known combinatorial optimisation problem. Its importance stems from the large number of real-world applications, ranging from crew scheduling in airlines or railway companies [Hoffmann and Padberg, 1993; Caprara et al., 1997; Housos and Elmoth, 1997] and driver scheduling in public transportation [Lourenço et al., 2001] to scheduling and production planning in various industries [Vasko and Wolf, 1987].

The SCP can be formalised as follows. Given a finite set $A := \{a_1, \ldots, a_m\}$ and a family $F := \{A_1, \ldots, A_n\}$ of subsets $A_i \subseteq A$ that covers A, that is, every element of A appears in at least one set in F, in the *Minimum SCP* the goal is to find a minimum size subset $C^* \subseteq F$ that covers A, that is, to find $C^* \in \text{argmin}_{C' \in Covers(A,F)} \#C'$, where $Covers(A, F) := \{C \mid C \subseteq F \land \bigcup_{A' \in C} A' = A\}$. In the *Weighted SCP*, additionally a weight function $w : F \mapsto \mathbb{R}^+$ is given, which assigns a positive weight (or cost) to each element of F, and the objective is to find a set cover C^* with minimal total weight, that is, $C^* \in \text{argmin}_{C' \in Covers(A,F)} w(C')$, where the weight of C' is defined as $w(C') := \sum_{A' \in C'} w(A')$. Clearly, the Minimum SCP can be seen as a special case of the Weighted SCP: Every Minimum SCP instance is equivalent to a Weighted SCP instance in which all elements of F have the same weight. For that reason, the Minimum SCP is also know as the *Unicost SCP*. In the remainder of this section, we will mainly focus on the more general Weighted SCP, to which we will refer simply as the SCP.

Frequently, the SCP is formalised as an integer programming (IP) problem. In this case, a binary variable x_i is associated with each subset A_i and $x_i = 1$ indicates that subset A_i is chosen to be in the current candidate cover C, while $x_i = 0$ indicates that $A_i \notin C$. Each set A_i, $i = 1, \ldots, n$, is represented by a column in a $m \times n$ matrix B, with $b_{ji} := 1$ if $a_j \in A_i$ and $b_{ji} := 0$ otherwise. Intuitively, each row of B corresponds to an element of A, and we say that a column i covers a row j if, and only if, element a_i is contained in subset A_i (i.e., $b_{ji} = 1$). Furthermore, if $w(A_i)$ is the weight of A_i, let $c_i := w(A_i)$ be the cost associated with including column i in a solution. The SCP can then be represented by the following integer programming problem:

$$\text{minimise} \qquad f(x_1, \ldots, x_n) := \sum_{i=1}^{n} c_i \cdot x_i \qquad (10.4)$$

$$\text{subject to} \qquad \sum_{i=1}^{n} b_{ji} \cdot x_i \geq 1 \quad j = 1, \ldots, m \qquad (10.5)$$

$$x_i \in \{0, 1\} \qquad i = 1, \ldots, n \qquad (10.6)$$

The constraints 10.5 enforce that each element of A (i.e., each row of B) is covered by at least one element of F (i.e., by a column i of B such that $x_i = 1$), and the integrality constraints 10.6 specify the domains of the variables x_i.

Most of the literature on SLS algorithms for the SCP uses the IP formulation; we will follow the same convention and refer to the elements of A as *rows* (of the matrix B), to the subsets A_i as *columns* and to the weights $w(A_i)$ as *column costs*.

EXAMPLE 10.4 **A Simple Instance of the Set Covering Problem**

Consider the instance of the Minimum Set Covering Problem that is defined by $A := \{a, b, c, d, e, f, g\}$ and $F := \{A_1, \ldots, A_6, A_7\}$ with

$$
\begin{aligned}
A_1 &:= \{a, b, f, g\} \\
A_2 &:= \{a, b, g\} \\
A_3 &:= \{a, b, c\} \\
A_4 &:= \{e, f, g\} \\
A_5 &:= \{f, g\} \\
A_6 &:= \{d, f\} \\
A_7 &:= \{d\}
\end{aligned}
$$

Notice that A_2 and A_5 are proper subsets of A_1, which is also the largest of the sets A_i. However, the only two possible solutions that include A_1, $C_1 := \{A_1, A_3, A_4, A_6\}$ and $C_2 := \{A_1, A_3, A_4, A_7\}$, have a cardinality of four, while the two optimal solutions ($C_1^* := \{A_3, A_4, A_6\}$, $C_2^* := \{A_3, A_4, A_7\}$) require only three subsets.

In the integer programming formulation, this SCP instance is represented by the matrix

$$
B := \begin{pmatrix}
1 & 1 & \mathbf{1} & 0 & 0 & 0 & 0 \\
1 & 1 & \mathbf{1} & 0 & 0 & 0 & 0 \\
0 & 0 & \mathbf{1} & 0 & 0 & 0 & 0 \\
0 & 0 & \mathbf{0} & 0 & 0 & 1 & 1 \\
0 & 0 & \mathbf{0} & 1 & 0 & 0 & 0 \\
1 & 0 & \mathbf{0} & 1 & 1 & 1 & 0 \\
1 & 1 & \mathbf{0} & 1 & 1 & 0 & 0
\end{pmatrix}
$$

where the columns corresponding to the A_i contained in the optimal cover C_1^* are shown in bold. Note that the first column of B contains entries $b_{ji} = 1$

for the four elements of A_1, a, b, f and g. This leads to the following integer program (the variables x_i that have value 1 in the optimal solution C_1^* are shown in bold):

$$\text{minimise} \qquad f(x_1, \ldots, x_7) := \sum_{i=1}^{7} x_i$$

subject to

$$
\begin{array}{rcccccccccc}
x_1 & + & x_2 & + & \mathbf{x_3} & & & & & & \geq 1 \\
x_1 & + & x_2 & + & \mathbf{x_3} & & & & & & \geq 1 \\
& & & & \mathbf{x_3} & & & & & & \geq 1 \\
& & & & & & & \mathbf{x_6} & + & x_7 & \geq 1 \qquad (10.7) \\
& & & & & \mathbf{x_4} & & & & & \geq 1 \\
x_1 & + & & & & \mathbf{x_4} & + & x_5 & + & \mathbf{x_6} & \geq 1 \\
x_1 & + & x_2 & + & & \mathbf{x_4} & + & x_5 & & & \geq 1
\end{array}
$$

$$x_i \in \{0, 1\} \qquad i = 1, \ldots, 7$$

It may be noted that although C_1^* covers one element twice and C_2^* covers each element of A exactly once, both have the same objective function value (number of subsets A_i) and are, hence, equally optimal.

The SCP is an $\mathcal{NP}$-hard combinatorial optimisation problem. In the case of the Minimum SCP, there exists a polynomial-time approximation algorithm that is guaranteed to return a solution that is worse than the minimum number of sets required to cover A by a factor of at most $1 + \log m$ [Johnson, 1974]; the same bound also holds for the Weighted SCP [Chvátal, 1979]. However, for the Minimum SCP it has been shown that reaching an approximation ratio of $c \cdot \log m$ is $\mathcal{NP}$-hard [Raz and Safra, 1997]; in particular, there is no polynomial-time deterministic algorithm that achieves a constant approximation ratio.

Applications and Benchmark Instances

The SCP has many important real-world applications, such as airline crew scheduling. How aspects of this application problem are mapped onto the SCP is illustrated in the following example. (For more details, see Wedelin [1995].)

EXAMPLE 10.5 **SCP Formulation of Crew Scheduling Problems**

In this example, we outline how a particular problem that arises in the context of scheduling crews in the airline industry can be formulated as a weighted SCP. Given a timetable of flight legs, the problem is to assign a crew to each flight leg so that the overall cost of a solution, which is a function of the individual assignments, is as low as possible. Each crew has a home base, and a schedule for a crew comprises a series of flight legs that start and end at its home base. A possible schedule has to obey legal and contractual rules, such as prescribed rest times or maximum working times. In this example, the flight legs correspond to the elements of A, each feasible schedule for a crew is represented by a set A_i, and the weight of set A_i, $w(A_i)$, is the cost of a specific crew schedule.

To solve the problem of finding optimal schedules that cover every flight leg, first a large number of possible schedules are generated for each crew and their costs are computed. Each crew schedule (in this context also called a *pairing*) represents a sequence of flight legs that can be performed by a single crew. Then, a subset of the generated schedules is selected such that every flight leg is included in at least one pairing and the total cost is minimised. Obviously, every flight leg needs to be covered, since there must be at least one crew for each flight leg; however, it is possible that a flight leg is covered several times and one crew is simply travelling on a flight as passengers. In real applications, instances with up to a few thousand flight legs and hundred thousands of possible schedules may be obtained [Wedelin, 1995].

As it can be expected, in this application the cardinality of each of the sets A_i, that is, the number of non-zero entries in each column, is typically rather small. In the IP formulation, this leads to matrices B with very low density, that is, matrices that contain a very small fraction of non-zero entries.

The application relevance of the SCP is witnessed by the fact that many commonly used benchmark instances stem from real-world SCP applications, including the benchmark set introduced by Balas and Carrera [1996], which is derived from applications in the airline industry (aa*) or bus companies (bus*). The instances range in size from 105 to 681 rows (items in A) and from 2 241 to 9 524 columns (subsets $A_i \in F$). In these 'real-world' SCP applications, the density of the matrices is low, ranging from 0.51% to 4.11%, which reflects the fact that the sets A_i are small compared to A.

Another prominent set of benchmark instances stems from the FASTER (Ferrovie Airo Set covering TendER) competition organised by the Italian railway company; these instances are derived from set covering problems arising in railway crew scheduling applications. These instances range from 507 rows and 63 009 columns to 4 872 rows and 968 672 columns; they have the peculiarity that all column costs are either one or two, and each column covers at most 12 rows.

Several SCP instances from crew-scheduling applications in airline companies have been proposed by Wedelin [1995]. The original instances have been reduced by a preprocessing stage that eliminates certain columns and rows. The resulting instances range from as few as 29 rows and 157 columns to 1 585 rows and 105 804 columns. The matrix density drops from 8.2% for the smallest instance to 0.3% for the largest.

Algorithms for the SCP are also often tested on randomly generated instances. The most widely used set of such instances is available from OR-Library [Beasley, 2003]; it comprises instances from 200 rows and 1 000 columns up to 1 000 rows and 10 000 columns. These instances are randomly generated in such a way that every column covers at least one row (i.e., none of the A_i is empty), and every row is covered by at least two columns. The column costs are randomly generated integers from the set $\{1, \ldots, 100\}$. The densities of these instances vary between 2% and 20%. In addition, one set of relatively small unicost instances with 50 rows and 500 columns is available. Interestingly, results from a recent study indicate that these randomly generated instances differ from instances derived from real-world SCP applications in terms of their search space characteristics [Finger et al., 2002].

Iterated Greedy Algorithms for the SCP

Greedy construction heuristics were among the first heuristic approaches to solving the SCP [Chvátal, 1979; Balas and Ho, 1980]. These iteratively add columns (i.e., subsets) to a partial candidate solution until all rows (i.e., elements of A) of a given SCP instance are covered. Greedy construction heuristics are also underlying some of the currently best-performing SLS algorithms. These so-called *iterated greedy (IG)* algorithms can be seen as a variant of Iterated Local Search, in which local search and perturbation phases are replaced by construction and destruction phases [Pranzo and Stützle, 2003]. During a destruction phase, solution components are removed from a complete candidate solution. In a construction phase, on the other hand, solution components are added to a partial candidate solution until a complete candidate solution has been obtained. As in ILS, an acceptance criterion is used to decide whether the search continues from this new complete candidate solution or from the one that served as the starting point of the most recent destruction phase.

The SCP algorithm by Jacobs and Brusco, IG-JB, which will be discussed in the following, can be seen as an IG algorithm that uses an acceptance criterion based on the Metropolis condition from Simulated Annealing [Jacobs and

```
procedure IG-JB((B, c))
    input: 0/1 matrix B, weight vector c
    output: cover Ĉ
    C := constructBH((B, c));
    C := removeRedundantColumns((B, c), C);
    Ĉ := C;
    T := T₀;
    while not terminate((B, c), C) do
        C' := destruct((B, c), C);
        C' := construct((B, c), C');
        C' := removeRedundantColumns((B, c), C');
        if (f(C') < f(Ĉ)) then
            Ĉ := C';
        end
        C := acceptSA((B, c), T, C, C');
        T := α · T;
    end
    return Ĉ
end IG-JB
```

Figure 10.5 Outline of the IG-JB algorithm for Weighted SCP. (For details, see text.)

Brusco, 1995]. It is based on a representation of candidate solutions as sets of columns of the matrix B from the IP formulation of the given Weighted SCP instance. An outline of the IG-JB algorithm is shown in Figure 10.5.

The initial solution in IG-JB is generated by a greedy construction heuristic that is based on work by Balas and Ho [1980]. The respective procedure *constructBH* assumes that all columns are ordered according to their cost values in non-decreasing order. It then iterates through the following two steps until all rows are covered:

1. Select a currently uncovered row j uniformly at random.

2. Add the lowest cost column i that covers row j.

Once a complete cover C is constructed, redundant columns, that is, columns that only cover rows that are also covered by other columns in C, are eliminated. This is done by the procedure *removeRedundantColumns*, which examines the columns of the solution in non-increasing order of their cost and removes columns that are found to be redundant.

The procedure *destruct* iteratively removes a fixed number of $k_1 \cdot \#C$ columns from the current cover, where k_1 is a parameter with $0 < k_1 < 1$ and $\#C$ is the number of columns in C. During this process, the columns to be removed are chosen independently and uniformly at random from the remaining elements of C.

Next, in procedure *construct* a complete candidate solution is recovered as follows. First, a candidate set is built that consists of all columns with cost less than $k_2 \cdot \max\{c_i \mid i \in C\}$, where $k_2 > 0$ is a parameter, c_i is the cost of column i and C is the cover (i.e., a set of columns) before invoking the destruction phase. The parameter k_2 has a direct influence on the size of the candidate set. Note that if the candidate set is too small, there is a risk to never find an optimal cover. Then, for each of the columns in the candidate set the *cover value* $\gamma_i := c_i/b_i$ is computed, where b_i is the number of additional rows covered when adding column i to the current partial cover. In other words, the cover value gives the unit cost of covering one additional row. Finally, a column with minimum cover value $\gamma_{min} := \min\{\gamma_i \mid i \in \{1, \ldots, n\}\}$ is added to the partial candidate solution; if there is more than one column with minimum cover value, one of them is chosen uniformly at random. These steps are iterated until a complete cover is obtained, whose redundant columns are then eliminated using the previously described *removeRedundantColumns* procedure.

The acceptance criterion in IG-JB is taken from Simulated Annealing: Improving covers C' are always accepted, while worsening covers C' are accepted with probability $e^{(f(C)-f(C'))/T}$, where the temperature parameter T is subject to a standard geometric cooling schedule. (See Chapter 2, page 76*ff.* for details on Simulated Annealing.)

IG-JB was tested on a set of large, randomly generated instances from OR-Library [Beasley, 2003] and on an additional set of instances where the cost of the columns is positively correlated with the number of ones in a column (i.e., with the number of rows covered by a column). Computational results show that IG-JB outperforms an earlier Lagrangian heuristic for the SCP by Beasley [1990]. However, nowadays, IG-JB appears to be outperformed by more recent SLS algorithms for the SCP [Brusco et al., 1999; Caprara et al., 1999; Yagiura et al., 2001], including the algorithm presented next.

The IG Algorithm by Marchiori and Steenbeck

Another iterated greedy algorithm for the SCP, *IG-MS*, has been proposed by Marchiori and Steenbeck [2000a]. In addition to the construction and destruction procedures that are common to all IG algorithms, IG-MS uses (i) a procedure *recomputeCore* that occasionally computes a new core instance, that is, a smaller

SCP instance containing only a subset of the columns that are most likely to appear in optimal solutions, and (ii) an additional local optimisation procedure that tries to improve upon the candidate solutions generated in the construction process.

Before explaining the components of the IG-MS algorithm in detail, we introduce the heuristic values that control the construction and destruction phases of IG-MS; these can be seen as a refinement of the cover values described for IG-JB. Let C be a current (partial) cover and $cov(C)$ the set of rows that are covered by C; $cov(i, C)$ is the set of rows that are covered by column i, but are not covered by any columns in $C \setminus \{A_i\}$. (Recall that column i in B corresponds to the subset A_i in the set theoretic formulation of the SCP.) We denote by $c_{min}(j)$ the minimum cost of a column that covers row j. Then, for IG-MS the cover utility $cu(i, C)$ that evaluates the usefulness of column i with respect to a cover C is defined as

$$cu(i, C) := \sum_{j \in cov(i,C)} c_{min}(j)$$

Note that if $cu(i, C) = 0$, then column i is redundant with respect to C. The values $cu(i, C)$ are used to define selection costs $sc(i, C)$ as follows

$$sc(i, C) := \begin{cases} \infty & \text{if } i \text{ is redundant w.r.t. } C; \\ c_i / cu(i, C) & \text{otherwise.} \end{cases} \qquad (10.8)$$

The construction procedure of IG-MS iteratively adds columns that are not in a partial cover C and have minimum selection cost. Additionally, each time a column is added, columns may be removed again from C, if a predicate *removeColumns* is true. This is the case if the partial or full cover C contains at least one redundant column. If C does not contain any redundant columns, *removeColumns* is set to true with probability w_r and to false in the remaining cases. If $C = \emptyset$ then *removeColumns* is always false. The procedure *selectRemove* selects a column of C with maximum selection cost. An algorithm outline of this construction procedure is shown in Figure 10.6 (next page).

The construction procedure generates a complete cover C that does not contain any redundant columns. In a next step, C is locally optimised. The local optimisation tentatively adds a column i to C trying to render at least two of the columns in C other than i redundant in such a way that the sum of the costs of the columns that can be removed after adding i is larger than c_i; such a column i is called a *superior column*. The local optimisation procedure first generates a list of superior columns that is ordered according to non-increasing gains, and then iteratively adds these columns to C and removes redundant columns, whenever an improvement can be obtained in this way. This latter condition has to be tested, because after each such step, the subsequent columns in the list may no longer be superior.

procedure *construct*((B,c),C)
    input: $0/1$ *matrix B, weight vector c, partial cover C*
    output: *complete cover C*
    while C is not a complete cover of B **do**
        $j := selectAdd((B,c),C)$;
        $C := C \cup \{A_j\}$;
        while *removeColumns* **do**
            $j := selectRemove((B,c),C)$;
            $C := C \backslash \{A_j\}$;
        end
    end
    return C
end *construct*

Figure 10.6 Algorithm outline for the construction procedure of IG-MS. (For details, see text.)

The acceptance criterion used in IG-MS always selects the better of the two given covers; as a consequence, it always returns the incumbent cover, (i.e., the best cover encountered up to that point in the search process). The destruction phase builds a partial cover from a set E of *elite columns*, that is, columns from the given cover C for which $cu(i, C) > c_i$. For each of the columns in E the number of times they occurred in an incumbent cover $\widehat{C}$ is counted. The destruction procedure selects from E the columns that occurred rarely in an incumbent cover, while the other columns are selected independently with a probability that is set to a random value from the interval $[0.1, 0.9]$. In general, because of the acceptance criterion that is used to select the cover to which *destruct* is applied, this destruction procedure returns a partial cover that contains a subset of the columns of the incumbent cover.

Finally, a procedure *recomputeCore* is used to determine the SCP core, that is, a candidate list of columns that are considered for inclusion in a candidate solution. The core instance is iteratively built as follows. The columns of the given SCP instance are considered in non-decreasing order of their selection cost. If a given column i is an elite column, it is added to the core with probability close to one; otherwise, if a row j exists that is covered by i and for which $c_i < k_0 \cdot c_{min}(j)$, i is also added to the core; finally, i is added to the core if it covers at least one row j that is covered by less than k_1 columns in the current core (the values $k_0, k_0 > 1$, and k_1 are parameters of the algorithm). The core is computed at the beginning of the search process and re-computed every 100 iterations of the algorithm.

The construction and destruction procedures both only consider columns that are included in the current core, which significantly increases their efficiency, particularly for large SCP instances.

IG-MS has been empirically compared against the CFT heuristic [Caprara et al., 1999], one of the currently best performing algorithms for the SCP, as well as against two genetic algorithms [Beasley and Chu, 1996; Eremeev, 1999]. In general, IG-MS was found to reach the same solution qualities as the ones obtained by CFT on all instances tested, including Wedelin's and Balas-Carrera's real-world instances, as well as the random OR-Library instances; the only exceptions are three instances of Wedelin, for one of which IG-MS obtained a slightly worse solution, while for the two others it achieved slightly better solution qualities. For the two genetic algorithms, performance comparisons were only conducted on the random instances from OR-Library; here, IG-MS found better solutions for two of the largest SCP instances, while on the other instances, all algorithms obtained the same solution qualities. The run-time required by IG-MS for obtaining these solution qualities appears to be competitive with that of CFT and lower than that of the genetic algorithms; however, because of differences in the experimental protocols used for evaluating the algorithms by their respective authors and the different hardware used, further empirical analysis is required to establish their relative performance more precisely. Overall, these results suggest that IG-MS can currently be considered a state-of-the-art algorithm for the SCP.

Related Problems

There are a number of subset problems related to the Set Covering Problem. The *Minimum k-Set Covering Problem* is a special case of the SCP in which the cardinality of all subsets A_i in A is bounded from above by a constant k; this problem is efficiently approximable with approximation ratio $\sum_{i=1}^{k} \frac{1}{i} - 1/2$ [Duh and Fürer, 1997].

Another related problem is the *Set Partitioning Problem* which can be seen as a variant of the SCP in which the objective is to find a collection C of subsets A_i that covers every element of A exactly once, that is, for any $A_i, A_j \in C$ with $A_i \neq A_j$ we have that $A_i \cap A_j = \emptyset$. An integer programming formulation of this problem is obtained from that for the SCP (cf. page 488) by changing the inequalities in Equation 10.5 into equalities. Similarly to the SCP, the Set Partitioning Problem arises in a wide variety of applications, such as crew scheduling and vehicle routing [Balas and Padberg, 1976]. Unlike the situation for the SCP, only few SLS algorithms for the Set Partitioning Problem have been proposed in the literature [Chu and Beasley, 1998; Maniezzo and Milandri, 2002].

In the *Set Packing Problem (SPP)* we are given a set $A := \{1, \ldots, m\}$ of m elements and n subsets $A_i, i = 1, \ldots, n$, but unlike in the SCP, the objective is to find the maximum number of sets such that all chosen subsets A_i are mutually disjoint, that is, if s is a candidate solution of a given SPP instance, then for all $A_i, A_j \in s$ we require $A_i \cap A_j = \emptyset$. Like for the SCP, a weighted variant of this problem can be easily defined by assigning to each set A_i a positive real weight $w(A_i)$; in this *Weighted Set Packing Problem* the objective is to determine a set of mutually disjoint subsets A_i with maximum total weight. Note that the integer programming formulation of this problem is analogous to that of the SCP except that we have to replace the '$\geq$' in the inequalities in Equation 10.5 by '$\leq$'. The Set Packing Problem arises in many practical resource allocation problems; in particular, the Weighted SPP has recently received considerable attention in the form of the Combinatorial Auctions Winner Determination Problem discussed in the next section.

10.4 Combinatorial Auctions

Auctions play an important role in economics as well as in multi-agent systems, where auction mechanisms are used for resource allocation and task distribution. The items that are auctioned range from household goods to radio frequencies, network bandwidth and pollution rights. In a *combinatorial auction*, bids can be placed on bundles of items. In situations where a complete bundle of goods is required for a certain purpose or task, the ability to bid directly for a bundle instead of bidding individually on all respective items allows bidders to minimise their risk of getting stuck with incomplete bundles. At the same time, overlaps between such bundle bids make it difficult to determine an assignment of items to bids that maximises the revenue of the auctioneer. SLS algorithms are amongst the best-performing methods for finding optimal or very high-quality solutions to this *Combinatorial Auctions Winner Determination Problem*.

Winner Determination in Combinatorial Auctions

In a combinatorial auction, a seller has a *set of items* $I := \{i_1, \ldots, i_m\}$ to be auctioned. Potential buyers value different subsets or *bundles* of items, $S \subseteq I$, and submit *bids* of the form $b := (S, p)$, where p, the *price of b*, is a positive real number that represents the amount the buyer is willing to pay for bundle S. An instance of a combinatorial auction is then given by a set of items I and a set of bids $B = \{b_1, \ldots, b_n\}$.

An *allocation* is a set of bids $A \subseteq B$ that are considered winning. For a given allocation A, all bids $b \in A$ are called *satisfied* or *winning*, and all $b \notin A$ are called *unsatisfied*. An allocation $A \subseteq B$ is called *feasible* if it does not contain any pair of bids that require the same item, that is, if $\nexists i \in I : \#\{(S, p) \in A \mid i \in S\} > 1$; the set of all feasible allocations for given sets of items and bids is denoted $FAlloc(I, B)$. The value of a feasible allocation A, denoted $val(I, B, A)$, is defined as the sum of the prices of the bids satisfied under A.

The *Combinatorial Auctions Winner Determination Problem (CAWDP)* is defined as follows: Given a set of items I and a set of bids B, find a feasible allocation A^* with maximal value, that is, find $A^* \in \text{argmax}_{A \in FAlloc(I,B)} val(I, B, A)$ or, equivalently, $A^* \in \text{argmin}_{A \in FAlloc(I,B)}(-val(I, B, A))$. (Although it may be more intuitive to think of the CAWDP as a maximisation problem, following our general convention, we present it here as a minimisation problem.) Typically, the output desired from a winner determination algorithm is a set of winning bids (i.e., a feasible allocation) as well as an assignment of items to winning bids; the latter is easily and efficiently computed from the former. Here and in the following, it is assumed that any items that are not assigned to a winning bid do not cause any direct cost to the auctioneer; if this so-called *free disposal assumption* is not met, the winner-determination problem can become much harder to solve [Sandholm et al., 2002].

EXAMPLE 10.6 **Simple CAWDP Instance**

Consider a combinatorial auction with items $I := \{a, b, c, d, e\}$ and the following set of bids, B:

$$
\begin{aligned}
b_1 &:= (\{a, c\}, 5.5) \\
b_2 &:= (\{a, c, d\}, 15) \\
b_3 &:= (\{b\}, 1) \\
b_4 &:= (\{b, d\}, 12) \\
b_5 &:= (\{d\}, 8) \\
b_6 &:= (\{d, e\}, 10)
\end{aligned}
$$

Note that the combined value of the two bids for the individual items b and d is lower than the value of the bundle bid for both (b_4), which reflects the complementarity of these items.

Let us consider the allocations $A_1 := \{b_2, b_4\}$ and $A_2 := \{b_1, b_4\}$. While A_2 is feasible, A_1 is infeasible because b_2 and b_4 both require item d. The value of A_2, $val(I, B, A_2)$, is 17.5, which is the maximal value over all possible feasible allocations for this problem instance. Under the optimal allocation A_2, bids b_1 and b_4 win, with items a and c assigned to b_1 and b, d assigned to b_4. Note that e remains unassigned under this allocation; there is a feasible allocation that assigns all items to bids ($A_3 := \{b_1, b_3, b_6\}$), but its value is lower than 17.5.

The CAWDP can be represented as a Boolean linear programming (BLP) problem. In this formulation, allocations are represented using a binary variable x_j for each bid $b_j \in B$ that indicates whether b_j is part of the current allocation ($x_j = 1$) or not ($x_j = 0$). Optimal allocations correspond to solutions of

$$\text{minimise} \quad f(x_1, \ldots, x_n) := -\sum_{j=1}^{n} x_j \cdot p_j \tag{10.9}$$

$$\text{subject to} \quad \sum_{j=1}^{n} c_{ij} \cdot x_j \leq 1 \quad i = 1, \ldots, m \tag{10.10}$$

$$x_j \in \{0, 1\} \quad j = 1, \ldots, n \tag{10.11}$$

where $c_{ij} = 1$ if item i is an element of S_j for some $(S_j, p_j) \in B$, and 0 otherwise. Using this BLP formulation, standard solvers for BLPs or more general integer programming (IP) problems can be applied to the Combinatorial Auction Winner Determination Problem. It may be noticed that this formulation is similar to the one for the Set Covering Problem (see Equations 10.4 to 10.6 on page 488); the only difference is that the '$\geq$' in Equation 10.5 is replaced by a '$\leq$' in Equation 10.10. This latter formulation corresponds exactly to that of the Weighted Set Packing Problem mentioned in the previous section, which illustrates the fact that both problems are equivalent. Consequently, any algorithm or result for CAWDP is directly applicable to Weighted Set Packing and vice versa.

Winner determination in combinatorial auctions is an $\mathcal{NP}$-hard problem. It has been shown that no polynomial-time deterministic algorithm can achieve approximation ratios better than or equal to $n^{1-\epsilon}$, where n is the number of bids, for any $\epsilon > 0$ (unless $\mathcal{NP} = \mathcal{ZPP}$; $\mathcal{ZPP}$ is the class of problems that can be solved in expected polynomial time by a probabilistic algorithm with zero error probability) [Sandholm, 1999; 2002]. The best known approximation algorithm for the CAWDP achieves a worst-case approximation ratio of $O(n/log(n))$ [Halldórsson, 2000].

Benchmark Instances for the CAWDP

Along with the development of CAWDP algorithms, various classes of random instance distributions have been proposed and used for empirical performance evaluations. These instance distributions are based on generative probabilistic models of varying complexity. The simplest instance generators are based on simple distributions for bundle sizes and compositions as well as prices. The design of complex generators, on the other hand, is based on features of various types of real-world application domains of combinatorial auctions.

Given a number of items and bids, simple random instance distributions are obtained by using a combination of probability distributions for determining for each bid $b := (S, p)$ independently the bid length (i.e., the size of S), the items in S and the price p. In one of the simplest cases, the so-called 'uniform' (or constant) model, the same fixed number of items is used in each bid, and the bid price is chosen uniformly at random from the interval $[0, 1]$ [Sandholm, 1999]. Slightly more complex models use uniform or parameterised normal, binomial or exponential bid-length distributions in combination with various price distributions, including uniform distributions over an interval with a size linear in the length of the respective bid, and normal distributions [Boutilier et al., 1999; Fujisima et al., 1999; Sandholm, 1999; Andersson et al., 2000]. In all of these cases, the items in S are drawn uniformly at random from the set of all items (without replacement).

The CATS (Combinatorial Auctions Test Suite) benchmark collection contains a number of probability distributions of problem instances (in the form of random instance generators) that are modelled based on (potential) real-world applications of combinatorial auctions as well as generators for the artificial random distributions mentioned above [Leyton-Brown et al., 2000a]. The CATS distributions model problems involving paths in space, such as combinatorial auctions for truck routes or network bandwidth; spatial proximity problems that correspond to situations arising in real estate or drilling rights auctions; temporal matching problems, where corresponding time-slices must be secured on multiple resources, for example, machines; and temporal scheduling problems, which essentially correspond to distributed job shop scheduling problems with one resource (see also Chapter 9, Section 9.4). These parameterised distributions are defined based on random generation mechanisms for bundles and bid prices that are based on simplified models of realistic combinatorial auction applications.

Since their introduction, CATS benchmark instances have been used for a number of empirical studies on CAWDP algorithms. There is some evidence, however, that most instances from the 'real-world' CATS distributions (particularly those from the matching and scheduling models) tend to be relatively easy compared to similarly sized instances obtained from some of the artificial random distributions, including the uniform model mentioned above; at the same time, other artificial distributions have been shown to produce mostly easy instances [Leyton-Brown et al., 2002]. Recent research uses regression learning techniques to investigate features that render CAWDP instances computationally hard for state-of-the-art algorithms [Leyton-Brown et al., 2002] and rejection sampling in combination with empirical hardness models to obtain hard instances from existing instance distributions, such as the CATS distributions [Leyton-Brown, 2003; Leyton-Brown et al., 2003].

The Casanova Algorithm

There are many ways in which SLS methods can be applied to the Combinatorial Auctions Winner Determination Problem. One approach is to search the space of feasible allocations in such a way that in each step an unsatisfied bid $b \notin A$ is selected and added to the current allocation A, after all satisfied bids in A that overlap with b have been removed from A. Formally, this approach uses the neighbourhood relation defined by $N(A, A') :\Leftrightarrow \exists (S', p') \notin A : A' = A \setminus \{(S, p) \in A \mid S \cap S' \neq \emptyset\} \cup \{(S', p')\}$. It may be noted that this neighbourhood relation is not symmetric: in general, $N(A, A')$ does not imply $N(A', A)$. Consequently, the search steps in SLS algorithms based on this neighbourhood relation are typically not directly reversible.

The class of SLS algorithms for the CAWDP that are based on this neighbourhood relation is known as CASLS. This class contains one of the currently best-performing SLS algorithms for the CAWDP, *Casanova* [Hoos and Boutilier, 2000]; this algorithm is conceptually very similar to Novelty$^+$, one of the best-performing SLS algorithms for SAT (see also Chapter 6, page 276*ff.*). Like Novelty$^+$, Casanova is based on a randomised best-improvement method with a limited form of memory; the evaluation function used in Casanova measures the sum of the revenue per item over all bids in the currently satisfied bids, that is, $g(I, B, A) := \sum_{(S,p) \in A} p/\#S$; the value $p/\#S$ is also called the *score* of bid (S, p).

Casanova works as follows (a GLSM for Casanova is shown in Figure 10.7): The search process starts from an empty allocation. Then, in each step, with probability wp (walk probability), a currently unsatisfied bid is selected uniformly at random (random walk step); with probability $1 - wp$, a bid is selected greedily based on the score and age of each bid, where the *age* of a bid b, $age(b)$, is defined as the number of search steps since b last became satisfied, or, if b has never been satisfied since the last search initialisation, the number of steps since the search was last initialised. In a greedy selection step, first all bids are ranked according to their score. Then, either the highest ranked bid b_1 or the second-highest bid b_2 is selected as follows: if $age(b_1) \geq age(b_2)$, select b_1; otherwise select b_2 with probability np (novelty probability) and b_1 with probability $1 - np$ (see Figure 10.8 for a decision tree representation of the greedy Casanova step). After *maxSteps* steps, the search is reinitialised with an empty allocation. After a total of *maxTries* such independent tries, the search is terminated and the best

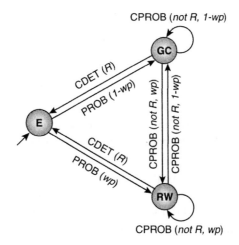

Figure **10.7** GLSM model for Casanova; the restart predicate R is equal to countm($maxSteps$), GSLM state E initialises the search at the empty allocation, GC performs a greedy Casanova step and RW performs a random walk step (see text for details).

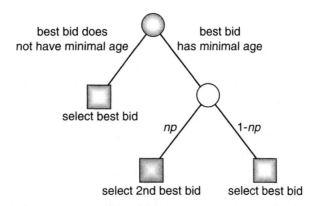

Figure **10.8** Decision tree representation of a greedy Casanova step. Deterministic and probabilistic choices are represented by grey and white circles, respectively; edges are labelled with the respective conditions and probabilities. Grey boxes indicate bid selection actions.

allocation in terms of total revenue found throughout the entire search process is returned as the search result.

Empirical studies have shown that compared to CASS [Fujisima et al., 1999], a state-of-the-art systematic search algorithm for the CAWDP at the time when Casanova was developed, Casanova achieves superior performance on a broad range of benchmark instances, both in terms of the CPU time required for obtaining optimal solutions (without proving optimality) as well as in terms of the solution qualities obtained for fixed run-time. Casanova's performance advantage is particularly pronounced for certain types of structured randomised instances that are obtained from encoding generalised types of bids, such as so-called *CNF* and *k-of* bids that can directly express substitutabilities between bundles of items, into standard CAWDP instances [Hoos and Boutilier, 2000] (see also page 506).

There is also limited empirical evidence that Casanova outperforms CPLEX, a state-of-the-art commercial integer programming package, on various types of CAWDP instances from the CATS benchmark suite [Schuurmans et al., 2001]. While the relative performance of Casanova compared to the most recent systematic search algorithms for the CAWDP (e.g., the CABOB algorithm by Sandholm et al. [2001]) is currently unclear, it is reasonable to assume that there is considerable room for improving Casanova's performance by optimising its implementation as well as the underlying SLS method.

The Exponentiated Subgradient Algorithm for the CAWDP

Like Casanova, the *Exponentiated Subgradient Algorithm* for the Combinatorial Auction Winner Determination Problem, *ESG-CAWDP*, is closely related to a high-performance SAT algorithm, ESG-SAT (see Chapter 6, page 289*ff*.). ESG-CAWDP is a dynamic local search algorithm that associates a penalty weight $itp(i)$ with each given item i and performs an iterative best-improvement search in the space of all allocations. It uses a neighbourhood relation under which two allocations A and A' are neighbours if, and only if, they differ in exactly one bid, that is, $N(A, A') :\Leftrightarrow \exists b \in B : (A' = A \cup \{b\}$ or $A = A' \cup \{b\})$, and a modified evaluation function of the form $g'(I, B, A) := \sum_{(S,p) \in A} p - \sum_{i \in I} (itp(i) \cdot ovd(i, A))$, where $ovd(i, A) := \#\{(S,p) \in A \mid i \in S\} - 1$ is the 'overdemand' for item i under allocation A, and $itp(i)$ is the penalty for item i, which is adjusted during the search. Any item i with $ovd(i, A) > 0$ for a given allocation A constitutes a *conflict* in A; feasible allocations are those that have no conflicts.

ESG-CAWDP works as follows. The search is initialised by selecting an allocation A uniformly at random such that for any given bid the probability that it is contained in A is $1/2$; furthermore, all item penalties are set to one. Then, a series of best improvement steps is performed. In each of these steps, one of the allocations A' that has a maximal evaluation function value $g'(I, B, A')$ amongst all the neighbours of A, is selected uniformly at random. When this best improvement search reaches a local maximum w.r.t. g', with probability p_r, the search is continued by selecting a neighbour of A uniformly at random; otherwise, the local search phase is terminated.

After each local search phase, the item penalties are updated in a two-stage process: First, each item penalty $itp(i)$ is multiplied by a factor $\alpha^{\theta(i)}$, where $\theta(i) := -1/2$ if i is not involved in a conflict in the current allocation A, and $\theta(i) := ovd(i, A) - 1/2$ otherwise (scaling stage); α is a parameter of the algorithm. Then, all item penalties are smoothed using the formula $itp(i) := itp(i) \cdot \rho + (1 - \rho) \cdot \overline{itp}$, where $\overline{itp}$ is the average over all item penalties after scaling and ρ is a parameter between zero and one. The algorithm terminates after a given number of local search phases and subsequent penalty updates have been performed or a specified solution quality has been reached.

The ESG-CAWDP algorithm can be easily generalised to arbitrary Boolean linear programming problems; an efficient implementation of this more general version, *ESG-BLP*, has been used for a performance evaluation on various sets of CAWDP instances [Schuurmans et al., 2001]. Empirical results indicate that on a range of CAWDP instances from the CATS benchmark suite, ESG-BLP does not quite reach the performance of Casanova in terms of the CPU time or number of search steps required for finding optimal solutions. ESG-BLP typically also falls short of the performance of CPLEX, when comparing the CPU time required for finding optimal quality solutions (without proving optimality). However, in many cases, its performance appears to be reasonably close to that of CPLEX and Casanova. Interestingly, ESG-CAWDP has been shown to outperform both Casanova and CPLEX on sets of CAWDP-encoded SAT problems, which suggests that for certain types of structured CAWDP instances, ESG-CAWDP is the best-performing algorithm currently known. It is unclear, however, whether similar performance advantages can be obtained for more practically relevant CAWDP instances.

Generalisations and Related Problems

Combinatorial Auctions can be generalised in various ways. One of the most straightforward generalisations allows multiple units of the same item to be offered and bid on (see, for example, Leyton-Brown et al. [2000b]). In principle, the corresponding *Multi-Unit Combinatorial Auctions Winner Determination Problem* can be easily cast as a (single unit) CAWDP instance according to our definition, by listing each unit of each item separately (after disambiguating their representation), making CAWDP algorithms applicable. A natural and more compact representation can be obtained by associating quantities with all items, both in the set of available items, as well as for the items requested within any bid. This representation can be easily transformed into a BLP formulation very similar to the one for the single-unit case, and the resulting problem instances can in principle be solved using ESG-BLP. Generally, one would expect algorithms that work directly on the compact representation of multi-unit combinatorial auctions to solve these problems more efficiently than conventional CAWDP algorithms applied to the respective CAWDP-encoded instances; however, the extent to which this is the case is presently unclear.

Other variants and generalisations of combinatorial auction problems are obtained by considering different *bidding languages*, which allow bidders to express various types of preferences and valuations over bundles of items. One example for such a generalised bidding language supports the submission of sets of bids that are connected by an XOR-constraint; feasible allocations can then include no more than one bid from any such set [Sandholm, 1999; Nisan, 2000]. Such XOR-bids allow bidders to express valuations under which the value of a bundle is less than the sum of the values of its components or subsets (substitutability). Combinatorial auctions with XOR-bids can be easily and efficiently encoded into standard combinatorial auctions using so-called *dummy goods* that are included in each set of XOR-bids [Fujisima et al., 1999; Sandholm, 1999; 2002].

Another way of allowing bidders to express substitutabilities and other forms of complex valuations is to consider bids in the form of propositional formulae, where propositional variables corresponds to items [Hoos and Boutilier, 2000; Boutilier and Hoos, 2001]. In such *logical bidding languages*, prices can be attached to the formulae comprising the bids as well as to subformulae, and the value of a bid is determined based on the satisfaction status of its respective formula and subformulae, where each item is satisfied if, and only if, it is assigned to the respective bid (for details, see Boutilier and Hoos [2001]). Standard combinatorial auctions correspond to a case where each bid is a conjunction of items, and prices are only attached to entire bids. A slightly generalised logical bidding language allows so-called CNF bids, that is, conjunctions of disjunctions of items, which provide another mechanism for expressing certain substitutabilities [Hoos and Boutilier, 2000]. Logical bidding languages allow certain types of complex valuations to be expressed substantially more concisely than standard combinatorial auctions or combinatorial auctions with XOR-bids [Boutilier and Hoos, 2001].

Finally, there are various problems that are closely related to combinatorial auctions. In *combinatorial reverse auctions*, a buyer wants to obtain certain items at the lowest possible cost, and various sellers submit offers (also called *asks*) for bundles of these items; the objective is to find a cost-minimal set of offers that covers the need of the buyer [Sandholm et al., 2002]. Reverse combinatorial auctions have important applications, for example, in procurement.

A *combinatorial exchange* is a market mechanism that allows multiple users (who may each buy, sell, or both) to submit bids and asks for bundles of items; here, the objective is to label the bids and asks as winning and losing such that the supply does not exceed the demand, and the surplus, that is, the difference between the total value of the winning bids and the total value of the winning offers, is maximised [Sandholm et al., 2002]. Combinatorial exchanges can be seen as a direct generalisation of both, combinatorial auctions and reverse auctions; hence, the same hardness and inapproximability results as for the CAWDP apply to the problem of determining winners in combinatorial exchanges.

10.5 DNA Code Design

DNA is one of the most important classes of biomolecules, because the genetic information of all organisms is stored in the form of long strands of DNA. At an abstract level, a DNA strand can be represented as a string over the four-letter alphabet {A, C, T, G}. DNA strands can in principle encode any kind of information, which can then be processed using established laboratory techniques. This provides the basis for various approaches to biomolecular computation, nanostructure design and molecular tagging (see, for example, Frutos et al. [1997], Seeman [1990], and Brenner and Lerner [1992]). DNA molecules also play a crucial role in many biochemical techniques, including polymerase chain reaction (PCR) and DNA microarray technology, both of which have countless applications in the biological and biomedical sciences. Many of these applications require sets of short DNA strands that satisfy certain combinatorial constraints. The problem of designing such sets in many ways resembles well-known problems from coding theory, particularly problems in designing error-correcting codes, and is therefore known as *DNA Code Design Problem*.

DNA Code Design Problems

A DNA (deoxyribonucleic acid) strand is a linear biopolymer that consists of nucleotide subunits each of which is formed by a section of a sugar-phosphate backbone and a nitrogenous base. In DNA, four different nucleotides occur, labelled A, C, G and T according to the bases they contain (adenine, cytosine, guanine and thymine). Each DNA strand has two chemically distinct ends, called the $5'$- and the $3'$-end. In the context of this section, DNA strands can be represented as strings over the four-letter alphabet {A, C, G, T}, where the left end of such a string corresponds to the $5'$-end and the right end to the $3'$-end of the strand. We call such strings *DNA words*.

Hydrogen bonds can form between A and T as well as between C and G; the base pairs A•T and C•G (as well as G•C and T•A) are called *complementary*. The complementarity extends to entire strands: two DNA strands are complementary if, and only if, one can be obtained from the other by replacing every base by its complement and reversing the orientation of the strand. Formally, this is captured in the following definition:

DEFINITION 10.1   **DNA Words and Complementarity**

> *A* DNA word *is a string over the four letter alphabet* {A, C, G, T}. *The length of a DNA word w is denoted $|w|$. The* complement *of a DNA word $w :=$ $b_1 b_2 \cdots b_{k-1} b_k$ is defined as $compl(w) := \overline{b_k}\, \overline{b_{k-1}} \,\cdots\, \overline{b_2}\, \overline{b_1}$, where $\overline{A} := T$, $\overline{T} := A$, $\overline{C} := G$ and $\overline{G} := C$.*

For example, for $w :=$ GATTACCA, $|w| = 8$ and $compl(w) =$ TGGTAATC. Note that the complement of a complement of a word w is always w itself (i.e., $compl(compl(w)) = w$); furthermore, complementarity is symmetric, that is, $w = compl(w')$ if, and only if, $w' = compl(w)$.

Two complementary DNA strands can bond to each other; this process is called *hybridisation* and leads to the formation of the well-known double-helix structure. The complementarity between bases or entire DNA strands also underlies many important biological and technological processes involving DNA, such as protein biosynthesis, DNA replication and genetic recombination. Hybridisation can also occur between strands that are not perfect complements of each other, that is, in cases where one of the two strands differs in some positions from the complement of the other.

A *DNA code* is a set of (typically short) DNA words that satisfies certain constraints. The constraints reflect properties of the DNA strands that minimise the potential for undesired hybridisation interactions and maximise the efficiency of desired interactions. In many applications of DNA codes, one set of code words is used for representing information, while the set of the complements of these words is used for performing certain operations on the information-carrying strands, such as marking occurrences of a certain piece of information i by means of hybridisation between all strands containing i and strands containing the complement of the representation of i. The following constraints have been considered in a number of DNA code design applications, particularly in DNA computing [Frutos et al., 1997], where DNA words are used for representing the data being processed.

The Hamming Distance Constraint, HD(d)
Any two words w_1, w_2 from the set have Hamming distance at least d, that is, disagree in at least d positions. This constraint intuitively ensures that any two words are sufficiently different to not be 'confused' in any hybridisation-based operation on specific words.

The Complement Hamming Distance Constraint, CHD(d)

For any two words w_1, w_2 from the set (where w_2 can be identical to w_1), the Hamming distance between w_1 and $compl(w_2)$ is at least d. This constraint intuitively ensures that hybridisations between different code words or between multiple copies of a code word (in the case $w_1 = w_2$) do not occur; such hybridisations can, for example, greatly decrease the efficiency of a DNA-based computation.

The GC Content Constraint, GCC(p)

Any word w in the set contains G or C in exactly $\lceil p \cdot |w| \rceil$ positions, where p, $0 \leq p \leq 1$, is a parameter. Since base pairings involving G and C are stronger (i.e., thermodynamically more stable) than those involving A and T, fixing the GC content of all code words ensures that the desired hybridisation between any word and its complement is of approximately equal strength.

The problem of designing a DNA code for a given combination of constraints and word length can now be defined as a combinatorial optimisation problem as follows:

DEFINITION 10.2 **The DNA Code Design Problem**

An instance of the DNA Code Design Problem (DNA-CDP) *for a given set of constraints* C *on sets of code words is given by a word length* n*; the objective is to find a maximum size set* S *of DNA words of length* n *that satisfies all constraints in* C*. By using the objective function* $f(S) := -\#S$*, this can be formulated as a minimisation problem.*

In DNA-CDP, the candidate solutions are *sets of DNA words*, and these sets are feasible if they satisfy *all* given constraints. Many applications of DNA-CDP, particularly in DNA computing, require DNA codes of a certain size – this obviously corresponds to the decision variants of the DNA-CDP as defined here.

EXAMPLE 10.7 **A Simple DNA Code Design Problem**

Consider the problem of designing a maximum size DNA code with word length $n = 8$ given the constraints HD(6), CHD(6) and GCC(0.5). The set S_1 comprising the DNA words

    AGCTCTGT   GACGTTTG   TTACGCGT   GGTAGGAT
    TGTCATCG   GCTTCCTA   CCCAAAAG

is a solution to this problem instance. Note that the CHD constraint ensures that every word w from this set and its complement, $compl(w)$, have Hamming distance at least 6.

If the first word, $w_1 := $ AGCTCTGT, is replaced by $w_1' := $ AGCTCTGA, a DNA code is obtained that still satisfies the given GCC constraint, but there are conflicts between the words w_1' and $w_6 := $ GCTTCCTA (the Hamming distance between these words is 5) and between w_1' and $w_3 := $ TTACGCGT (the Hamming distance between w_1' and $compl(w_3)$ is 5).

In fact, the seven word set given above is the best solution (i.e., the largest DNA code) for this problem instance currently known. When increasing the word length to $n = 10$ and keeping the same constraints, codes of size up to 41 are known. For $n = 8$, when relaxing the HD(d) and CHD(d) constraints to $d = 4$, solutions with up to 112 words can be obtained. Similarly, when only using the constraints HD(6) and CHD(6), DNA codes of size 10 are known.

Although code design problems that are related to the DNA-CDP have been studied extensively in coding theory, not much is known regarding the theoretical complexity of constructing or approximating maximum size codes for the combinatorial constraints considered here. In our formulation, the space of candidate solutions for a DNA code design problem is doubly exponential in the given word length, n. While codes up to certain sizes can be obtained efficiently using construction methods such as the Gilbert-Varshamov algorithm (see, e.g., [MacWilliams and Sloane, 1977]), empirical evidence indicates that finding optimal or close to optimal solutions to DNA code design problems requires at least exponential time in the length of the code words [Tulpan et al., 2003].

Applications and Benchmark Instances

Currently, there are four main application areas in which DNA code design problems arise: DNA computing, DNA nanostructure design, DNA tagging in chemical libraries and DNA microarray design.

In *DNA computing*, DNA words are used for encoding information in the form of DNA strands. There are two main approaches for using DNA-based computation for solving suitably encoded combinatorial problems, such as the Hamiltonian Path Problem or SAT. In the solution-based approach, most of the crucial steps of a computation occur in solution, that is, in a situation where all DNA strands can move relatively freely in a liquid (see, for example, Adleman [1994]). In the surface-based approach, on the other hand, some of the information-carrying DNA strands are affixed to a surface, and crucial computation steps involve exposing this surface and the affixed DNA strands to solutions containing other DNA strands or reactive agents, in particular specific enzymes that replicate or degrade DNA strands under certain conditions [Frutos et al., 1997]. In both approaches, established biomolecular methods such as sequence-specific enzymatic digestion (i.e., strand cleavage), PCR (for strand replication), gel electrophoresis (for length-specific separation of DNA strands) and DNA microarray assays are used to perform the steps of these computations.

In *DNA nanostructure design*, DNA codes are used for creating DNA molecules that assemble into larger structures with well–defined properties, such as 2d-crystals, cubes, cages and simple nanomechanical devices. Such DNA nanostructures can be constructed from building blocks consisting of multibranched complexes formed by partial hybridisation between several strands of DNA; these building blocks attach to each other via 'sticky ends', that is, free ends of single-stranded DNA. In this context, DNA codes are used for designing DNA strands that will assemble correctly and efficiently into the multibranched complexes required for building larger structures with specific properties, such as a specific 2d-crystal [Seeman, 1990].

Specific DNA nanostructures can be used for performing self-assembly computations. The underlying idea is to create 'DNA tiles', that is, multi-strand DNA complexes with sticky ends, for encoding data as well as computation rules. The tiles are designed in such a way that specific tiles can assemble into larger structures via hybridisation between their respective sticky ends. It is known that the self-assembly of such tiles into larger structures can simulate a Turing machine and hence perform arbitrary universal computations. Similar as in nanostructure design, DNA codes are used for designing tiles that combine in such a way that the correctness of the self-assembly computation is ensured [Winfree et al., 1999].

Sequence-specific hybridisations between complementary strands of DNA can also be used in the context of *DNA tagging*, a technique that uses DNA strands as 'molecular barcodes' for identifying and accessing specific elements from chemical libraries that can encompass thousands of distinct chemical compounds, such as protein sequences. Such encoded combinatorial chemical libraries can be used, for example, for identifying proteins that show specific interactions with a target molecule — a key step in many approaches to drug design. Combinatorial chemistry methods can be used to create resin beads to which multiple copies of a candidate protein as well as of a unique DNA tag identifying that protein are bonded. In this context, DNA codes are used for designing DNA tags that can be effectively detected and isolated based on correct hybridisation with their respective complementary strands [Brenner and Lerner, 1992].

Finally, DNA codes can be applied in the design of DNA tags used in *universal DNA microarrays*. Here, DNA tags are used in specifically designed adaptors that allow the reliable and massively parallel detection of arbitrary genomic sequences using the same universal DNA microarray. Designing and synthesising these adaptors is substantially easier and cheaper than creating a customised DNA microarray, and through the use of carefully designed DNA codes, the universal microarray system can be designed in such a way that the potential for errors due to mishybridisations is typically much smaller than when using customly designed DNA microarrays [Gerry et al., 1999].

So far, the sets of DNA words used for these 'real-world' applications have been mostly constructed manually using ad-hoc methods that in some cases are based on results from coding theory.

The SLS-THC Algorithm

The DNA word design algorithm by Tulpan, Hoos and Condon, SLS-THC [Tulpan et al., 2003], is based on a randomised iterative improvement search method similar to WalkSAT, one of the best known algorithms for the propositional satisfiability problem [Selman et al., 1994; McAllester et al., 1997] (see Chapter 6, page 273*ff.*) and Casanova, one of the best-performing algorithms for the Combinatorial Auctions Winner Determination Problem [Hoos and Boutilier, 2000] (see Section 10.4, page 502*ff.*). The search space used in this approach consists of the DNA word sets with the target number of words. Two candidate solutions (i.e., DNA word sets) are direct neighbours if, and only if, one can be obtained from the other by changing exactly one position in one word — this is called the *1-mutation neighbourhood relation*.

The SLS-THC algorithm is based on an evaluation function g that counts the number of word pairs in a given candidate solution S that violate the given binary constraints; here, this can be any combination of the HD and CHD constraints (a variant of the algorithm that supports the GCC constraint will be discussed later). The main idea behind the algorithm is to iteratively pick a pair of words that are in conflict w.r.t. a given constraint and to modify one of them such that the number of conflicts (i.e., the number of word pairs that violate at least one constraint) is maximally reduced.

An outline of the SLS-THC algorithm is shown in Figure 10.9. The search process is initialised at a set of words that is determined by a simple randomised process that generates any DNA word of length n with equal probability. Note that the initial word set may contain multiple copies of the same word.

In each step of the search process (i.e., one execution of the inner for-loop from Figure 10.9), first, a pair of words violating one of the Hamming distance constraints is selected uniformly at random. Then, for each of these words, all possible single-base modifications are considered. As an example of single-base

procedure *SLS-THC*($k,n,C,maxTries,maxSteps, \theta$)
    input: *number of code words* $k \in N$, *word length* $n \in N$, *set of*
            constraints C, *maxTries* $\in \mathbb{N}$ **and** *maxSteps* $\in \mathbb{N}, \theta \in [0,1]$
    output: *set* S *of m words that fully or partially satisfies* C

    for $i := 1$ **to** *maxTries* **do**
        $S :=$ initial set of words;
        $\widehat{S} := S$;
        for $j := 1$ **to** *maxSteps* **do**
            if $g(S) = 0$ **then**
                return S
            end
            Randomly select words $w_1, w_2 \in S$ that violate one of the
                constraints in C;
            $M_1 :=$ all set of words obtained from w_1 by substituting one base;
            $M_2 :=$ all set of words obtained from w_2 by substituting one base;
            with probability θ **do**
                select word w' from $M_1 \cup M_2$ uniformly at random;
            otherwise
                select word w' from $M_1 \cup M_2$ such that the number of
                    constraint violations is maximally decreased;
            end
            if $w' \in M_1$ **then**
                replace w_1 by w' in S;
            else
                replace w_2 by w' in S;
            end
            if $g(S) < g(\widehat{S})$ **then**
                $\widehat{S} := S$;
            end
         end
        end
        return $\widehat{S}$
    end *SLS-THC*

Figure **10.9** Outline of the SLS-THC algorithm for DNA word design; $g(S)$ denotes the number of constraint violations in word set S (see text for details).

modifications, consider the code word ACTT. When replacing, for example, the first base with the three possible alternatives, the new code words GCTT, CCTT and TCTT are obtained. For a pair of words of length n this yields $6n$ new words, some of which may be identical.

With a fixed probability θ, one of these modifications is accepted uniformly at random, regardless of the number of constraint violations that will result from it. In the remaining cases, a modification that leads to a maximal decrease in the number of constraint violations is accepted. (If there are multiple such modifications, one of them is chosen uniformly at random.) The noise parameter θ controls the greediness of the search process: for high values of θ, constraint violations are not resolved efficiently, while for low values of θ, the search is more directed but it has more difficulties to escape from local optima of the underlying search space. Note that in each step of the algorithm, exactly one base in one word is modified. When counting the number of constraint violations, the degree of violation is not considered, that is, when considering, for example, the HDD(4) constraint, one violation is counted for any two words with Hamming distance $k < 4$, regardless of whether their actual Hamming distance k is 3 or 0.

Throughout the run of the algorithm, the best candidate solution encountered so far, that is, the DNA word set with the fewest constraint violations, is memorised. Note that even if the algorithm terminates without finding a valid set of size k, a valid subset can always be obtained by iteratively selecting pairs of words that violate a Hamming distance constraint and removing one of the two words involved in that conflict from the set. Hence, a word set of size k with v constraint violations can always be reduced to a valid set of at least size $k - v$.

To obtain good performance, it is crucial to implement this algorithm in such a way that Hamming distances between words and/or their reverse complements are not recomputed in each iteration of the algorithm; instead, these are computed once after generating the initial set, and updated after each search step. This can be done efficiently, since any modification of a single word can only affect the Hamming distances between this word and the $k - 1$ remaining words in the set, while a naïve implementation would potentially recompute all $\Theta(k^2)$ Hamming distances between words and/or their reverse complements. (Note that similar considerations arise for many SLS algorithms for other problems, such as SAT or CSP; see also the in-depth section on page 271.)

This basic SLS algorithm can be easily extended to also accomodate the GCC constraint, by restricting the candidate solutions to sets of words that all satisfy a given GCC constraint [Tulpan et al., 2003]. Furthermore, a variant of the SLS-THC algorithm that initialises some of the DNA words to elements of a previously determined word set, while the remaining words are generated randomly as described above, can be used for extending known DNA codes.

Empirical analyses of run-time distributions have shown that on hard word design problems, SLS-THC often suffers from search stagnation that severely compromises its performance. This can be overcome by extending the algorithm with a mechanism for diversifying the search by occasional random replacement of a small fraction of the current set of DNA words [Tulpan et al., 2003]. It has been empirically demonstrated that SLS-THC in many cases reaches or exceeds the code sizes obtained from the best known theoretical construction methods. Particularly impressive results have been reported for word lengths ranging from 4 to 12 under combinations of HD and CHD constraints, where for 56 out of 57 problem instances previously best theoretical results could be empirically matched or improved [Marathe et al., 2001; Tulpan et al., 2003], and practically relevant problem instances with combinations of HD, CHD and GCC constraints [Frutos et al., 1997; Tulpan et al., 2003].

There is some indication that by using different neighbourhood relations that allow more than one base to be modified in each search step, the performance of the SLS-THC algorithm can be further improved, not only in terms of the number of search steps, but also in terms of the CPU time required for solving given instances of the DNA-CDP problem [Tulpan et al., 2003; Tulpan and Hoos, 2003].

Generalisations and Related Problems

There are many variants of the DNA Code Design Problem that use additional or alternative constraints to the HD, CHD and GCC constraints discussed here. As an example, consider constraints on the Hamming distance between misaligned strands and constraints on overlapped pairings involving three or more strands (see Figure 10.10 on the next page); both types of constraints play an important role in applications where information is encoded into DNA by concatenating individual code words, similar to the character-by-character encoding of text strings into bit vectors. The SLS algorithm presented earlier in this section can be extended in a relatively straightforward way to accomodate these additional constraints.

As previously motivated, the Hamming distance constraints considered here are designed to optimise desired hybridisation pairings between code words while reducing the occurrence of erroneous pairings as much as possible. In reality, however, the number of complementary base pairs between two strands is only a very crude measure for their probability of bonding and for the strength of the pairing. It is therefore often useful to consider more accurate models of DNA hybridisation that, for instance, reflect the different strength of the bonds underlying G•C and A•T pairings or the distribution of mismatched base pairs over an imperfectly hybridised pair of DNA strands (see Figure 10.10c). Based on thermodynamic models of DNA hybridisation, Gibbs free energy values can be calculated for pairs of DNA strands, which can then be used for predicting the temperature at which a given hybridised pair of strands tends to separate into single strands under specific reaction conditions (melting temperature of

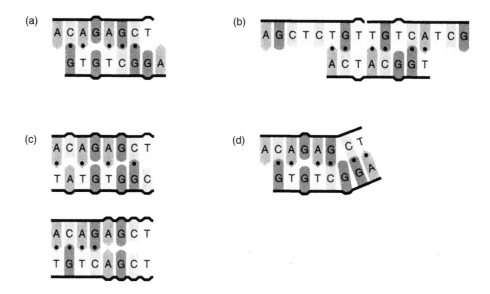

Figure **10.10** Illustration of strand interactions that give rise to additional constraints for DNA code design: (a) Slide misalignment leads to 5 complementary base pairings compared to only 2 pairs for the same two strands under perfect alignment; (b) overlap between three strands can lead to additional correct pairings; (c) a given number of base pairs can be distributed differently, leading to differences in the stability of the duplex; (d) DNA secondary structure formation (here a single bulge loop) can increase thermodynamic stability.

the duplex). Thermodynamic constraints, that is, constraints on the free energies or melting temperatures of pairs of DNA strands, can be used for designing DNA codes with more accurately controlled physical properties. While results from coding theory can be exploited to some extent for DNA code design using Hamming distance constraints, the use of thermodynamic constraints requires different approaches. Interestingly, the SLS algorithm presented above can be extended to deal with thermodynamic constraints in a rather natural way.

Additional complications in the design and application of DNA codes arise from the fact that DNA strands can pair in ways that do not correspond to a double helix structure where each base in one strand is juxtaposed to exactly one base in the other strand, to which it may or may not be complementary. Instead, strands may pair in ways that give rise to asymmetric bulges or loop structures (see Figure 10.10d). Furthermore, under certain conditions, a single strand can partially hybridise with itself. In general, the structures that arise from such types of base pairings are known as DNA secondary structures. (Primary structure refers to the base sequence of a DNA strand, and tertiary structure to its three-dimensional conformation.) Constraints that reflect certain aspects of DNA secondary structure can be used in the design of DNA codes for applications where the formation of secondary structure, for example, of DNA strands corresponding to individual code words or concatenations of multiple code words, interferes with desired hybridisation pairings. Such constraints are typically based on thermodynamic models of DNA secondary structure.

Finally, RNA (ribonucleic acid) molecules share many of the salient properties of DNA and play a similarly crucial role in many biological processes. Most of the considerations and techniques for the design of DNA codes discussed in this section apply analogously to closely related RNA code design problems.

10.6 **Further Readings and Related Work**

For each of the problems discussed in this chapter there is a large and steadily growing body of relevant literature. In the following, we mention a selection of the most relevant work along with some relatively recent studies.

The Graph Colouring Problem has been widely studied, and a large number of heuristic GCP algorithms have been proposed in the literature. Constructive search methods include (i) simple greedy algorithms that, given some predefined order of the vertices, iteratively assign colours according to some heuristic function, (ii) the DSATUR algorithm [Brélaz, 1979], which in each step chooses a vertex with maximal saturation degree, that is, a vertex that is adjacent to a maximal number of distinctly coloured vertices (ties are broken according to the maximal degree of a vertex in the uncoloured subgraph), and (iii) the Recursive Largest First (RLF) algorithm [Leighton, 1979], which builds colour classes successively according to the results of a heuristic search for large independent sets that leave the number of edges in the uncoloured subgraph minimal. Good performance has been reported for a randomised extension of RLF, XRLF, that uses limited enumeration once a subgraph becomes sufficiently small [Johnson et al., 1991]. Note that the order in which vertices are assigned colours is fixed in the case of the simple greedy heuristics, while DSATUR and RLF are examples of dynamic construction heuristics, where the order in which vertices are assigned colours depends on the partial solution constructed so far.

A wide range of SLS methods have been applied to the GCP. These include several tabu search algorithms [Hertz and de Werra, 1987; Fleurent and Ferland, 1996; Dorne and Hao, 1999], the most effective of which are similar to the TSGH algorithm by Hao and Galinier for binary CSPs (see Chapter 6, Section 6.6), GRASP [Laguna and Martí, 2001]; Iterated Local Search [Chiarandini and Stützle, 2002; Paquete and Stützle, 2002]; Distributed Local Search [Morgenstern, 1996]; various pure and hybrid evolutionary algorithms [Davis, 1991;

Fleurent and Ferland, 1996; Eiben et al., 1998]; iterated greedy algorithms [Culberson and Luo, 1996]; and ACO algorithms [Costa and Hertz, 1997]. A few studies have also considered SLS algorithms based on neighbourhood relations other than the standard 1-exchange neighbourhood, such as Kempe Chains [Johnson et al., 1991], Variable Neighbourhood Search [Avanthay et al., 2003] and cyclic k–exchange neighbourhoods [Chiarandini et al., 2003]. However, it is currently not clear whether these can contribute to enhancing the state-of-the-art in GCP solving.

Although the performance of these and other SLS algorithms for the GCP strongly varies across different classes of graphs, one common trend appears to be that hybrid algorithms such as the one presented in Section 10.1 are among the top performers when very high-quality colourings are required [Galinier and Hao, 1999]. Search space analysis methods have yielded some insights into the factors underlying the performance of SLS algorithms for the GCP [Hamiez and Hao, 2003]; however, further significant research efforts appear to be necessary to gain a better understanding of the behaviour of these algorithms.

The Quadratic Assignment Problem has been the subject of a large number of studies and continues to be of significant interest to many researchers. For general overviews with a special emphasis on more mathematical background, we refer to the overview papers by Pardalos et al. [1994] and Burkard et al. [1998a], as well as to the book by Çela [1998]; these also provide more details on extensions and special cases of the QAP, as well as asymptotic results on the complexity of specific classes of QAP instances.

The literature on SLS algorithms for the QAP is vast, primarily because the QAP is extremely difficult to solve for complete algorithms, and large instances can only be solved by SLS methods. Additionally, similar to the TSP, the QAP plays a central role as a benchmark problem for computational approaches to $\mathcal{NP}$-hard optimisation problems. While pure construction heuristics do not play a very important role in QAP solving, most of the research on SLS algorithms is concentrated on 'simple' SLS algorithms and hybrid SLS techniques. These include applications of Simulated Annealing [Burkard and Rendl, 1984; Connolly, 1990], Threshold Accepting [Nissen and Paul, 1995], Tabu Search [Battiti and Tecchiolli, 1994; Skorin-Kapov, 1990; Taillard, 1991; Misevicius, 2002], Memetic Algorithms [Drezner, 2003; Fleurent and Ferland, 1994; Merz and Freisleben, 2000a; Vazquez and Whitley, 2000], Evolution Strategies [Nissen, 1994], GRASP [Li et al., 1994], ACO Algorithms [Gambardella et al., 1999; Maniezzo, 1999; Maniezzo et al., 1994; Stützle and Dorigo, 1999a; Stützle, 1997; Stützle and Hoos, 2000] and Scatter Search [Cung et al., 1997]. As in the case of the GCP, the performance of these SLS algorithms depends strongly on characteristics of the given problem instance. While for unstructured, randomly generated instances algorithms, such as the reactive tabu search algorithm described in Section 10.2, perform best, state-of-the-art performance on structured, real-life QAP instances and large randomly generated real-life like instances is achieved by hybrid SLS algorithms such as the population-based ILS algorithm introduced in Section 10.2, the memetic algorithms of Merz and Freisleben [2000a] or that of Drezner [2003], or by ant colony optimisation algorithms [Stützle and Dorigo, 1999a; Dorigo and Stützle, 2004].

The Set Covering Problem is of enormous practical relevance and therefore a large number of solution approaches, including complete and incomplete algorithms, have been proposed for this problem. Complete algorithms can solve instances with up to a few hundred rows and several thousand columns; a comparison of various complete algorithms can be found in Caprara et al. [2000]. It is worth noting that among the complete SCP algorithms, general-purpose integer programming software such as CPLEX performs extremely well, outperforming several other complete algorithms that were specifically developed for the SCP. However, complete algorithms are restricted to rather small sized SCP instances and SLS algorithms are important for obtaining good solutions for large instances.

Many SLS algorithms for the SCP are based on the iterative application of greedy construction heuristics. Several of the best-performing heuristic algorithms of this type are based on combinations of Lagrangian relaxation and subgradient optimisation techniques. The essential idea is to compute surrogate costs (Lagrangian costs) that give an indication of the utility of selecting a specific column; these approaches have a rather long history [Balas and Ho, 1980; Beasley, 1990; Balas and Carrera, 1996; Ceria et al., 1998; Caprara et al., 1999].

In recent years, an increasing number of general-purpose SLS techniques has been applied to the SCP. These include Genetic Algorithms [Beasley and Chu, 1996; Eremeev, 1999] and especially SLS algorithms that are based on the principles of Iterated Greedy Search [Brusco et al., 1999; Jacobs and Brusco, 1995; Marchiori and Steenbeek, 2000a; Yagiura et al., 2001]. In general, the best performance appears to be reached by the algorithm of Marchiori and Steenbeck [2000a] (presented earlier in this chapter) and the SLS algorithm by Yagiura et al. [2001] that exploits a novel 3-exchange neighbourhood for the SCP. Along with the CFT algorithm by Caprara et al. [1999] (which is based on Lagrangian relaxation), the latter two hybrid SLS algorithms are currently among the best-performing algorithms for the SCP.

Although the Combinatorial Auctions Winner Determination Problem has not received as much attention as the GCP, QAP or SCP, there is a substantial (and steadily growing) body of literature on the CAWDP and related problems. The article by Sandholm [2002] provides a good and detailed introduction to combinatorial auctions and the winner determination problem, complexity results, and various earlier approaches for solving the CAWDP. Recent high-performance systematic search algorithms include CASS [Fujisima et al., 1999] and its multi-unit generalisation CAMUS [Leyton-Brown et al., 2000b], as well as CABOB [Sandholm et al., 2001]. Currently, the two algorithms described in this chapter are the only SLS algorithms for the CAWDP.

However, the CAWDP is equivalent to the Maximum Weighted Set Packing Problem, a well-studied problem that is conceptually closely related to the SCP, and winner determination in multi-unit combinatorial auctions is equivalent to the multi-dimensional knapsack problem; for both problems, various SLS algorithms and construction heuristics have been proposed and studied (see, e.g., de Vries and Vohra [2003], Holte [2001], Arkin and Hassin [1997], Chandra and Halldörsson [2000], and Vasquez and Hao [2001], but there is currently no indication that these algorithms reach or exceed the performance of Casanova or ESG-CAWDP. Finally, for an overview of generalisations of the CAWDP as well as related problems we refer the interested reader to the article by Sandholm et al. [2002] or the article by de Vries and Vohra [2003].

The design of DNA codes subject to various combinatorial constraints is a relatively young area of study that has already attracted a considerable number of researchers. The survey paper by Brennemann and Condon [2002] provides a good introduction to DNA code design problems and their applications. Besides the SLS-THC algorithm described here, several evolutionary algorithms have been used for solving DNA-CDP instances with various constraints in the context of particular applications in DNA computing [Deaton et al., 1996; 1999; Zhang and Shin, 1998]; however, these algorithms are not described in sufficient detail to assess their performance compared to other DNA-CDP algorithms.

Generally, DNA code design is strongly related to the problem of designing other types of codes, particularly constant weight binary codes, which has been studied extensively in coding theory. A good introduction to code design problems and local search algorithms for these problems can be found in a book chapter by Honkala and Östergård [1997]. In fact, various SLS methods, including Simulated Annealing and Evolutionary Algorithms, have been successfully applied to binary code design problems [Gamal et al., 1987; Comellas and Roca, 1993].

10.7   Summary

In this chapter, we gave an overview of SLS applications to five combinatorial problems. While these problems may not be the most representative for the types of problems that have been solved successfully using SLS algorithms, they illustrate a range of computational problems for which SLS methods achieve state-of-the-art performance. We introduced the Graph Colouring Problem, the Quadratic Assignment Problem, the Set Covering Problem, the Combinatorial Auctions Winner Determination Problem and the DNA Code Design Problem; we briefly discussed their respective computational complexity, applications and commonly used benchmark instances; we presented selected SLS algorithms for each of the problems; and we gave a brief overview of generalisations and related problems.

The *Graph Colouring Problem (GCP)* is a well-studied combinatorial problem in graph theory that involves assigning colours to the vertices of a given graph such that there is no edge between any two vertices of the same colour. This problem can be seen as a special case of the Constraint Satisfaction Problem (see Chapter 6, Section 6.5), but is typically solved with specialised algorithms. We first presented three simulated annealing algorithms that use a standard geometric cooling schedule. They mainly differ in the underlying evaluation function and neighbourhood relation. Next, we described the Memetic Algorithm by Galinier and Hao (MA-GH), which combines a genetic algorithm that is based on a partition-based crossover operator with an effective tabu search procedure in order to improve solutions. MA-GH is currently one of the best-performing SLS algorithms for the GCP.

In the *Quadratic Assignment Problem (QAP)*, the objective is to assign a number of objects to locations such that the product of the distances between the locations and the flow between the corresponding objects is minimised. The QAP serves as a model for a number of practical layout and location problems; as such, it is one of the most widely studied combinatorial optimisation problems. We presented a reactive tabu search algorithm that is particularly successful for randomly generated, unstructured QAP instances. The central goal of the reaction mechanism used in this algorithm is to adapt the tabu list length during the search process. The second algorithm we discussed, PBILS-S-QAP, is a population-based extension of a simple iterated local search algorithm that maintains diversity in its search process by enforcing a minimum distance between pairs of candidate solutions in the population; this distance is varied during the search process. Additionally, a restart mechanism is used to overcome search stagnation. PBILS-S-QAP is currently a state-of-the-art algorithm for real-world and large, randomly generated real-world like QAP instances.

The *Minimum (Weighted) Set Covering Problem (SCP)* is another hard combinatorial optimisation problem with many real-world applications. Given a set A and a number of weighted subsets of A, the objective is to find a combination of subsets with minimal total weight that covers every element of A. We described two *iterated greedy (IG) algorithms* for the SCP; IG is an SLS method that can be seen as a variant of Iterated Local Search. While the first of these two IG algorithms is a rather straightforward adaptation of the general IG method to the

SCP, the second uses a more complex construction mechanism that considers occasional removals of solution components; it also uses an additional local search procedure to post-optimise the candidate solutions generated by the construction procedure, as well as a dynamically updated candidate list of subsets considered during the search process. While the former of these two IG algorithms is mainly of historical interest, the latter is considered a state-of-the-art SCP solver.

The *Combinatorial Auctions Winner Determination Problem (CAWDP)* is a very application-relevant problem that recently has received a steadily increasing amount of attention in the artificial intelligence and operations research communities. This problem models an auction mechanism in which bids are submitted for bundles of items; the objective is to determine a feasible, that is, non-overlapping set of winning bids such that the revenue of the auctioneer is maximised. We presented the first SLS algorithm devised for this problem, Casanova. This algorithm, whose performance is still considered state-of-the-art, is based on a relatively simple randomised best-improvement method with a limited form of memory similar to the Novelty$^+$ algorithm for SAT. We also covered ESG-CAWDP, another recent SLS algorithm for the CAWDP based on Dynamic Local Search. Overall, very little research has been done on SLS algorithms for CAWDP, which suggests that compared to GCP, QAP or SCP, there is considerable potential for the development of improved algorithms.

The final problem, the *DNA Code Design Problem (DNA-CDP)*, arises from applications in biomolecular computing and biotechnology. In this hard optimisation problem, the goal is to find maximum size sets of DNA sequences (words) that satisfy given combinatorial constraints, for example, constraints on the Hamming distance between pairs of words. We described SLS-THC, a recent SLS algorithm for this problem that has been successfully used to obtain improvements over the best known solutions for a number of prominent DNA Code Design Problems. Similar to Casanova for the CAWDP, SLS-THC is a relatively simple randomised best-improvement method motivated by well-known SLS algorithms, such as the WalkSAT algorithm family for SAT and the Min Conflicts Heuristic for CSP.

The algorithms presented in this chapter prominently illustrate a number of crucial issues in the development of high-performance SLS algorithms. On one side, problem-specific knowledge and techniques play an important role; this is exemplified by the Kempe chain neighbourhood and the GPX recombination operator for the GCP, by the 1-mutation neighbourhood for the DNA-CDP, as well as by the use of the candidate list ('core instance') in the IG-MS algorithm for the SCP. On the other side, the use of appropriately chosen general SLS methods in combination with particular implementation techniques is of equally central importance; these methods and techniques are often similar between largely different problems, such as SAT and Combinatorial Auctions Winner

Determination. One of the most crucial factors in this context is the combination of SLS mechanisms that achieve a balance between an efficient exploitation of suitable heuristic information and a sufficient exploration of the search space. In this context, diversification mechanisms such as the distance-based selection in the PBILS-S-QAP algorithm for the QAP are of particular interest.

In order to design state-of-the-art SLS algorithms for combinatorial problems such as the ones presented in this chapter these major issues need to be addressed in an appropriate way. While good knowledge of SLS methods in general and the problem to be solved in particular goes a long way towards success in this endeavour, typically this needs to be complemented by thorough and principled experimentation.

Exercises

10.1 [*Medium*] Benchmark instances for several of the problems treated here are obtained from encodings of other problems. Common examples include:

(a) encodings of the Travelling Salesman Problem and the Graph Partitioning Problem as Quadratic Assignment Problems;

(b) encodings of SAT as a Combinatorial Auctions Winner Determination Problem.

Give the corresponding encoding for each of these examples. Do you expect any advantages in solving the respective problems by using these encodings?

10.2 [*Medium*] Prove that in the Penalty Function SA algorithm for the Graph Colouring Problem, described in Section 10.1 (page 471*ff.*), all local minima correspond to legal colourings of the given graph G.

10.3 [*Hard*] In many cases, the results obtained from search space analysis techniques, such as fitness-distance analysis, strongly depend on the underlying definition of distance between candidate solutions (see also Chapter 5). Define an appropriate distance measure for the Graph Colouring Problem and determine whether it can be computed efficiently, that is, deterministically in polynomial time.

10.4 [*Medium; Hands-On*] Implement a simple iterated local search algorithm for the QAP along the lines of the ILS-S-QAP algorithm from Section 10.2 (page 484*ff.*). Accept only better or at least equally good candidate solutions in the acceptance criterion. Study the run-time behaviour of this ILS algorithm, especially taking into account possible stagnation behaviour, along the lines discussed in Chapter 4, Section 4.4.

Does your run-time analysis give an indication why a population-based ILS extension may be a viable way to improve performance?

Study other possibilities of improving performance by (i) allowing larger perturbations or (ii) accepting also worse candidate solutions in the acceptance criterion. Which of the two possibilities do you find preferable?

10.5 [*Medium*] To efficiently implement iterative improvement algorithms, evaluation function values are typically incrementally updated after every search step (delta evaluations; see also Chapter 1, Section 1.5).

Derive a numerical formula for the delta evaluation for the general Quadratic Assignment Problem, where the flow and distance matrices may both be asymmetric. Once you have obtained this formula, simplify it for the case where the two matrices are symmetric and have a zero diagonal.

10.6 [*Easy*] Consider the SCP instance from Example 10.4 (page 489); formulate and solve the corresponding instances of the Set Partitioning Problem and the Set Packing Problem.

10.7 [*Medium*] Design at least two 'simple' neighbourhood relations for (perturbative) local search procedure algorithms for the Set Covering Problem.

10.8 [*Easy*] Conceptually compare the Casanova algorithm for the Combinatorial Auctions Winner Determination Problem (CAWD) with the Novelty$^+$ algorithm for SAT. Point out similarities and differences of the underlying search techniques and relate these to the common features and differences of the CAWDP and SAT.

10.9 [*Medium*] Specify an extension of the SLS-THC algorithm for the DNA Code Design Problem that finds DNA codes satisfying the HD(c) constraint and an additional SHD(c') constraint that ensures a minimal Hamming distance of c' between any two DNA words under any possible slide misalignment (see Figure 10.10, page 516). Discuss implementation and algorithmic efficiency issues of your extended algorithm.

10.10 [*Medium*] Consider the well-known *k-Clustering Problem*.

(a) Give a concise definition of the problem, as found in the standard literature, and show how it fits the definition of a combinatorial optimisation problem as stated in Chapter 1, Section 1.1.

(b) Based on a basic literature search, what can you say about the computational complexity of this problem? (Give at least one reference to support your statements. Your answer should be brief, but as precise and informative as possible. Do not give references to papers that you have not seen.)

(c) Briefly describe at least two stochastic local search algorithms that have been applied to the k-Clustering Problem (with references); explain the neighbourhood relations used in these algorithms.

Note: There are many different clustering problems other than the k-Clustering Problem, some of which you will likely encounter during your literature search.

E PILOGUE

Let us return to Augsburg and the problem of visiting all of its 127 Biergärten in a single day as well as to the logic puzzle, solving which would pay for all the beer along the way. It is easy to see that the former problem can be formalised as a TSP instance (see Chapters 1 and 8), while the latter can be modelled as an instance of SAT or CSP (see Chapters 1 and 6). Consequently, they could be solved using a variety of SLS algorithms for these problems (see Chapters 2, 6 and 8). Given the relatively small instance sizes, even relatively simple SLS algorithms would solve these problems easily, at least when run on a modern computer. (Clearly, many other search methods could be used to solve these problems similarly efficiently.)

Generally, Moore's Law, that is, the exponential increase over time in processing power and memory size (or more precisely: memory density) that occurred since the beginning of modern computing, plays a considerable role in pushing the limits of practical solvability of hard combinatorial problems. More importantly, at least in the case of SLS methods, it facilitates the rapid evaluation of algorithmic behaviour and makes possible new forms of empirical analysis, which provide the basis for the development of improved algorithms. Both factors combined, hardware and algorithmic improvements, have led to dramatic progress in our ability to solve large instances of hard combinatorial problems; and in many cases, SLS methods provide the algorithmic basis for these developments.

Nevertheless, many challenges remain in solving hard combinatorial problems in practice. There are problems, such as the QAP (see Chapter 10, Section 10.2), for which even relatively small instances are beyond the reach of state-of-the-art SLS methods (and any other algorithmic technique). Tight time

527

constraints arising in real-time applications pose formidable challenges in many cases. And, as we will discuss later in this section, many application problems have aspects that are not well captured when formalising them as static combinatorial problems, such as the problems covered in this book.

SLS methods are amongst the most promising algorithmic approaches for meeting these challenges, both as a result of inherent advantages of the paradigm and the large amount of research and development in this highly active area. Still, the study and application of SLS algorithms (not unlike other heuristic approaches) is sometimes regarded sceptically. From a scientific point of view, some view the use of SLS methods as a crutch. Common criticisms include a lack of theoretical underpinnings or provable performance guarantees; the notion that developing and applying SLS algorithms involves mainly tweaking algorithmic details or parameters, and is neither based on nor leads to deeper understanding of the problem to be solved; and a certain irreproducability or even arbitrariness of SLS performance and behaviour.

Craft, Art or Science?

Much of the work on designing and applying SLS algorithms in many ways resembles a craft rather than a science. This is, for instance, reflected in the types of questions that are raised and investigated in many, if not most, papers on SLS methods and their applications: 'How can SLS be applied to a given problem? How can a given algorithm be implemented efficiently? How can the performance or robustness of a given SLS algorithm be improved?' These questions typically have no simple answers, and experience, rather than intellectual understanding, is often the key to achieving the underlying goals.

One might even argue that when true mastery of the subtleties arising in the context of SLS methods and their successful applications is achieved, the craft can turn into an art. At that level, intuitions that are extremely hard to communicate or rationalise appear to play a crucial role; creativity and ingenuity complement good knowledge and skilful application of techniques; and aesthetic considerations such as formal elegance and a notion of style become relevant.

Clearly, very similar considerations apply to other areas of computing science, particularly to algorithm, software and system design: While some algorithms, pieces of software or systems (software and/or hardware) are merely fulfilling a certain purpose more or less reliably or correctly, others are well-designed, elegant or even beautiful — in the same way in which these concepts would apply to a bridge, an airplane or the proof of a mathematical theorem. At the same time, much of the criticism mentioned above in the context of SLS methods, similarly applies to areas such as software and system design.

In many ways, it appears that a more scientific approach to the design, analysis and application of SLS methods (and, more generally: of algorithms or systems) would be highly desirable (see also Hooker [1994]). Above and beyond the 'how?' questions mentioned above, such an approach is focussed on 'why?' questions, such as 'Why do (or don't) certain SLS techniques work well for a given problem or instance type? Why does an algorithm suffer from stagnation in a given situation? Why does Algorithm 1 perform better than Algorithm 2 on Problem A, but worse on Problem B?' Clearly, the answers to such questions contribute to a deeper understanding of SLS methods and their behaviour. It may also be noted that this approach is much more based on intellectual curiosity than solely on performance results as a driving force.

The classical scientific approach is based on the idea of understanding phenomena by means of models or theories obtained from an iterated process of observing, hypothesising and experimenting; models and theories are assessed by their explanatory and predictive power (scope and accuracy), as well as by their conceptual simplicity (Occam's Razor). Other aspects of the scientific approach are the crucial role of logical and precise arguments and the use of formalism and conceptual frameworks for achieving increased clarity and conciseness. As stated in the Prologue, one of the major goals of this book is to present important components and concepts for a scientific approach to the design and application of SLS methods.

It should be noted that, certainly in the case of SLS methods, the scientific approach and the more traditional, 'craft-like' approach are not incompatible; quite on the contrary, at the present stage, both approaches complement each other well. Most scientific studies of SLS methods require a substantial amount of experience, intuition and mastery of the nature outlined above to yield meaningful results in terms of a deeper understanding of the observed phenomena and behaviour. At the same time, the design and successful application of SLS algorithms often benefits from the deeper understanding and insights obtained from such scientific studies.

Crucial Aspects of Successful SLS Applications

Although our current understanding of SLS algorithms is still rather limited, many important insights into SLS behaviour and performance have been obtained and there are a number of factors that are now known to be critical for the development and application of most successful SLS algorithms.

As discussed in Chapters 1, 2 and 5, the neighourhood relation and the evaluation function play an important role for any SLS algorithm. Intuitively, they determine the mobility and the heuristic guidance of the search process;

together, they induce important characteristics of the search landscape and of the behaviour of any SLS algorithm based on these two components. A poor choice of the neighbourhood relation or evaluation function can make it hard or impossible to find any SLS strategy that performs well on the given problem. Conversely, the choice or design of a good SLS strategy depends on the underlying neighbourhood relation and evaluation function.

For example, one neighbourhood relation may lead to a large number of local minima (or closed plateau regions), requiring more powerful diversification and local minima escape mechanisms in order to avoid stagnation behaviour; but the same mechanisms may impede search progress in conjunction with larger local neighbourhoods that induce search landscapes with fewer local minima. For many problems (such as SAT or CSP), there are few natural neighbourhood relations and evaluation functions, while in other cases, such as for most scheduling problems, the range of choices for these components is much broader. Especially in these latter cases, good problem knowledge in conjunction with search space analysis techniques are often crucial for selecting a neighbourhood relation and evaluation function that provides the basis for a high-performance SLS algorithm.

Obviously, the performance and behaviour of any SLS algorithm critically depends on the search strategy, which probabilistically determines the trajectory of an SLS algorithm within a given search landscape. Choosing or designing a suitable search strategy generally requires good knowledge of standard SLS methods and their typical properties (cf. Chapter 2). Furthermore, important high-level issues need to be considered in this context, in particular the balance between intensification and diversification mechanisms within a given SLS strategy, which plays a crucial role for the behaviour and performance of hybrid SLS methods, such as Iterated Local Search.

Overall, particularly in the case of complex or hybrid SLS methods, the design process itself can have an important impact on the quality of the resulting algorithm. While an explorative design approach that incrementally extends an algorithm in which components and features are added in a step-by-step fashion can yield good results, it often leads to very complex, but poorly understood algorithms with a large number of parameters.

A more principled approach to the design of SLS algorithms starts from a unified formal framework, such as the GLSM model introduced in Chapter 3, and employs thorough empirical analyses at each major step of the design process. This helps to eliminate components or mechanisms that have been rendered ineffective or unnecessary by later modifications of the algorithm under design, and typically leads to simpler, more elegant SLS algorithms with fewer parameters whose behaviour is better understood.

Clearly, appropriate methods for empirically analysing the behaviour and performance of SLS algorithms play a crucial role not only in the context of such a more principled design approach, but also for the critical assessment of any newly designed SLS algorithm and for the informed comparison of various algorithms for a given problem (cf. Chapter 4).

Finally, most successful SLS applications rely heavily on efficient implementations of the underlying SLS algorithms. Key components of many efficient implementations are caching and incremental updating schemes for evaluation function values, standard speed-up techniques such as 'don't look bits', and advanced data structures that facilitate efficient access to and manipulation of candidate solutions (see, for example, Chapter 8).

Chance and Necessity: Is Stochasticity Needed?

It may appear that the notions of stochasticity and the role of probabilistic choice for the search process are notably absent from our discussion of the general issues arising in the context of developing and applying successful SLS algorithms. This raises a fundamental question: Is stochasticity needed? Or could it be the case that, as our knowledge and expertise in solving specific combinatorial problems increases, at some point we will be able to design deterministic search algorithms whose performance generally matches or exceeds that of the best SLS methods?

One approach to answering this question is followed in complexity theory, where randomised models of computation and complexity classes are studied along with their deterministic and non-deterministic counterparts. (Note that, as explained in Chapter 1, the theoretical notion of non-deterministic computation is very different from the notion of stochastic computation.) Unfortunately, the precise relationship between randomised and (non-)deterministic complexity classes is in many cases unknown, and the question whether randomised algorithms can solve certain problems more efficiently remains open. However, there are good reasons to believe that, from a theoretical point of view, randomisation may not substantially reduce the worst-case time complexity of algorithms for solving $\mathcal{NP}$-hard problems; more precisely, it is strongly believed that $\mathcal{P} = \mathcal{BPP}$, where $\mathcal{BPP}$ is the class of problems that can be solved in polynomial time in the worst case by a probabilistic algorithm with bounded error probability $1/2 - \epsilon$ for some $\epsilon > 0$ (see, for example, Kabenets [2002]).

This belief is partially based on the observation that many stochastic algorithms can be derandomised without significant changes in performance or behaviour. A general method for derandomising a given stochastic algorithm is to base all its probabilistic decisions on information obtained from a deterministic procedure, such as a pseudo-random number generator. Basically all programming environments include such generators, and almost all implementations of randomised algorithms are based on these. Interestingly, when empirically analysing such deterministic implementations of theoretically well-understood stochastic algorithms, the observed behaviour is very similar to and often indistinguishable from the behaviour that can be derived analytically under the assumption of truly random decisions.

Hence, although sources of true randomness could be realised in a reasonably straightforward way (under the assumption that certain physical processes, such as atomic decay, are truly random), all current evidence appears to suggest that using such devices would not lead to any significant algorithmic or performance advantages. This is certainly the case for SLS algorithms, where — different from other situations (such as the randomised generation of problem instances) — typically even the use of relatively low-quality pseudo-random number generators does not appear to have a significant effect on empirically observable performance and behaviour.

However, this type of derandomisation remains somewhat unsatisfying, since it is essentially based on the idea of using information that is independent of the problem or problem instance to be solved to replace truly stochastic decisions. Sometimes it is felt that this kind of decision making, like true random choice, is merely a crutch: When you don't know the right decision, choose randomly — this way, at least with a certain probability you will make a good choice.

A slightly different view of this issue reveals the main reasons underlying the effectiveness of stochasticity (or pseudo-random decision making) in the context of heuristic search: Because imperfect heuristics can lead to poor choices, a certain degree of stochasticity can reduce the chance of paying a high price, for example in terms of computation time, as a result of such misguidance. This certainly holds for local search algorithms, which are mostly based on relatively simple heuristics for selecting initial candidate solutions and determining search steps (consider, e.g., Iterative Best Improvement). But randomising a heuristically guided search process comes at a price: In situations where the heuristic guidance would be beneficial or even optimal, randomisation may introduce a significant chance of making a worse decision, and hence lead to undesirable variability and unpredictability in the performance and behaviour of a search algorithm.

This raises the question of whether by using better heuristics, the need for stochasticity (true or simulated) can be reduced. There is some evidence indicating that in general, this is not the case, particularly when considering recent developments in state-of-the-art systematic search algorithms for SAT, where it has been shown that the use of powerful heuristics and inference techniques often leads to algorithms whose performance varies extremely when applying simple equivalence transformations (such as clause or variable reordering) to a given input formula. It has been shown that by randomising the heuristic choices that drive the search process, the performance of these algorithms can be substantially improved, both in terms of expected solution time as well as in terms of performance robustness [Gomes et al., 1998]; consequently, many high-performance systematic search algorithms incorporate randomised choice, and overall, it appears that when using stronger heuristics within a search algorithm, there is often an increased need for mechanisms that alleviate the cost of misguidance.

In the case of local search approaches, there are various types of such mechanisms: They include the use of stochastic (or pseudo-stochastic) choices for search initialisation and for determining search steps as well as the application of random (or pseudo-random) perturbations to the input problem instance. In some cases, such as for certain tabu search or dynamic local search algorithms, only a relatively small number of actual (pseudo-)stochastic choices is required for achieving the desired robustness. In general, in order to achieve the best possible performance and robustness, the heuristic guidance and inference components of a search algorithm need to be balanced against stochastic and other mechanisms that diversify its behaviour in order to minimise the impact of heuristic misguidance. Achieving this balance is one of the most important issues in the design and successful application of SLS algorithms, and the SLS methods presented and discussed in this book merely illustrate well known areas in the broad spectrum of possible solutions to that problem.

The Future of SLS: Thoughts, Challenges, Opportunities

Combinatorial problems are ubiquitous in computing science, operations research and an extremely wide and diverse range of application areas; the theoretical, practical and economic significance of these problems and the methods for solving them can hardly be overestimated. Historically, stochastic local search has been one of the most widely and successfully used approaches for solving hard combinatorial problems, and especially in the recent past, the interest in SLS methods has increased steadily and significantly. As a consequence, to date, stochastic local search is an extremely active and dynamic research area, which is not only stimulated and driven by ideas, techniques and results from a number of disciplines, but also has a significant impact on these disciplines themselves.

While it is difficult to anticipate the developments within the area of stochastic local search even in the near future, there are numerous challenges and opportunities that, in our view, define at least some very likely directions and goals. Many of these have been known for a while, others can be extrapolated from current developments, and some represent areas of budding interest. In the following, we briefly discuss challenges, opportunities and directions in five main areas.

Empirical analysis of SLS behaviour. As motivated in Chapter 4, computational experiments and empirical analysis methods play a crucial role in the development, characterisation and application of SLS algorithms. Yet, compared to mature empirical sciences, such as physics, chemistry, or biology, the area of algorithmics and, more generally, computing science as a discipline, is still in the early stages of adopting and exploiting a well-founded empirical approach. While solid foundations for the empirical analysis of SLS behaviour are now available and widely accepted, their widespread use and application by researchers and practitioners needs yet to be achieved.

This development can be substantially aided by software environments that facilitate and automate crucial aspects of experimental design, data collection and statistical data analysis (in particular, by means of appropriate statistical tests). There are tremendous opportunities in this area, as the results obtained from advanced empirical analyses of SLS behaviour are likely to have a large impact on our understanding of SLS methods as well as on the design of future SLS algorithms and their successful application for solving practically relevant problems.

Problem characteristics, search space structure and SLS behaviour.   The connections between problem or instance characteristics, features of the landscape searched by a given family of SLS algorithms, and the behaviour of particular SLS algorithms are of considerable interest in the context of understanding SLS behaviour and problem hardness. Existing work in this area has yielded some interesting insights (cf. Chapter 5), but many questions currently remain unanswered. Future work, particularly based on relatively recent models of search space structure, such as plateau connection graphs and basin partition trees, is likely to identify, for example, the factors responsible for the large, unexplained variability in SLS behaviour across sets of problem instances that do not differ significantly in the features known to have an impact on SLS performance. Such results can provide the basis for the principled design of SLS algorithms for particular types of problems or instance classes, and for the development of self-tuning SLS algorithms that show robust performance over a wide range of instances of a given problem.

Theoretical results and foundations.   One of the biggest challenges in the theoretical study of SLS algorithms is the analytical characterisation of the performance and behaviour of practically relevant SLS methods. It is interesting to note that some of the tightest lower bounds on the complexity of $\mathcal{NP}$-complete problems, such as SAT, have been proven based on relatively simple SLS algorithms. The fact that empirically, the performance of these methods falls far behind that of state-of-the-art SLS algorithms suggests that there may be room for substantial improvements on these complexity results, and that by analysing more sophisticated SLS algorithms, it may be possible to achieve these improvements.

We also see enormous potential for improving our theoretical understanding of SLS behaviour based on further pursuing the connection between SLS algorithms and the theory of Markov processes, and in particular, Markov chain Monte Carlo (MCMC) methods. While it is easy to directly model the semantics of an SLS algorithm by a Markov chain, these models typically appear to be far too complex to be useful, for instance, for the purpose of theoretical analysis. It may be the case, however, that such models can be simplified drastically while retaining important aspects of their behaviour, which would make them much more accessible to further analysis. Overall, improved theoretical foundations of stochastic local search will likely provide insights and ideas for new SLS methods with theoretical performance guarantees.

New SLS methods. As previously explained, the traditional approach to the design of SLS algorithms is largely based on intuition, trial and error. This typically leads to combinations or variations of known techniques that are often conceptually complex and poorly understood. While completely new SLS methods, inspired by theoretical insights or by models from other disciplines, will eventually arise, it is likely that the development of new SLS methods will mostly be based on an increasingly principled iterative design process involving human designers with a good knowledge of existing SLS techniques and their properties as well as machine learning methods and automated empirical analysis techniques. In this context, we anticipate that unifying conceptual frameworks (cf. Chapter 3) as well as software environments that provide automated support for the typical development stages of an SLS algorithm (including conceptual design, implementation, evaluation, parameter tuning and realization of production versions) will play an increasingly important role.

We see substantial opportunities in the principled design and study of hybrid SLS methods as well as of combinations of SLS and systematic search methods for solving hard combinatorial problems. We also anticipate an increasing emphasis on developing SLS algorithms with provable performance guarantees; one particular challenge in this context is the development of competitive complete SLS algorithms for problems such as SAT (cf. Selman et al. [1997]).

Practical impact and new applications. From the beginning, much of the work on SLS algorithms has been directly inspired by combinatorial problems from various areas of practical interest. Nowadays, hard combinatorial problems arise wherever computing is applied, and in many cases, these problems are solved using SLS methods. Still, many combinatorial problems cannot be solved in a satisfactory manner, and new application problems arise frequently. Hence, particularly in areas such as bioinformatics, telecommunications, networking, system design and machine learning, we see substantial opportunities for the successful application of known or newly designed SLS methods.

Furthermore, while in this book we have solely covered static combinatorial decision and single-objective optimisation problems, many problems from practice are dynamic or have multiple optimisation objectives. Consider, for example, the problem of finding a feasible roundtrip through Augsburg's 127 Biergärten.

Assuming that you found a good trip, part-way through completing your circuit, you find that an important bus connection does not run this evening — can you (efficiently) recover a solution to the modified problem?

Instances of dynamic problems may change during or after a solution attempt, and the objective is often to minimise the delay until the modified instance has been solved. In the case of stochastic problems, additional knowledge is given on the statistical nature of the changes (usually in the form of probability distributions). Finally, in multi-objective optimisation, multiple optimisation criteria are given and need to be considered simultaneously; for such problems, one common goal is to characterise the trade-offs between these various measures of solution quality by identifying sets of Pareto-optimal solutions [Steuer, 1986].

Different from many other approaches to combinatorial problem solving, SLS methods can be quite naturally extended to these more general problems. We see tremendous opportunities in developing and using SLS methods in this context, and although many challenges need to be addressed (e.g., with respect to empirical performance analysis), we believe that SLS will become one of the standard approaches for solving dynamic, stochastic and multi-objective problems in practice.

Overall, given the sustained interest SLS methods have received since the early days of modern computing, and considering the opportunities in this area, both in terms of fundamental scientific problems and exciting new applications, it is difficult to imagine anything other than an exciting future for stochastic local search. It is likely that, based on the synthesis of insights from various disciplines in which SLS methods are studied, including algorithmics and theory of computation, artificial intelligence, operations research and statistics, as well as on increasing cross-fertilisation of theoretical and empirical approaches, the next generation of SLS methods will have even more impact in many application areas.

Finally, we believe that the area of stochastic local search plays a pivotal role in a paradigm shift that concerns many areas of computing science: While the roots of computing lie mainly in discrete mathematics and engineering, the empirical approach that underlies most other sciences is becoming increasingly important as the academic discipline concerned with computation and information evolves into a mature science. The study of SLS methods relies heavily on this new, third foundation, and offers a rich ground for its development and exploration, a fascinating and relatively easily accessible microcosm brimming with intriguing concepts, open problems and exciting applications. We hope that this book served you, the reader, as a suitable introduction and guide to this microcosm, and inspires you to explore beyond its limits.

GLOSSARY

Underlined terms within entries are also defined in the glossary.

2-Opt: Simple iterative improvement algorithm for the Travelling Salesman Problem that is based on the 2-exchange neighbourhood on edges of the given graph. The candidate tours obtained by 2-opt local search are called 2-*optimal* (or 2-*opt*) tours.

3-Opt: Well-known iterative improvement algorithm for the Travelling Salesman Problem (TSP) that is based on the 3-exchange neighbourhood on edges of the given graph. 3-opt local search provides the basis for a number of high-performance stochastic local search algorithms for the TSP. The candidate tours obtained by 3-opt local search are called 3-*optimal* (or 3-*opt*) tours.

k-Exchange Neighbourhood: Neighbourhood in which a candidate solution differs from any of its direct neighbours in up to k solution components. Prominent examples include the 2- and 3-*exchange neighbourhoods* for the Travelling Salesman Problem; 1-*exchange neighbourhoods* are widely used in stochastic local search algorithms for many combinatorial problems, including the Satisfiability Problem and the finite discrete Constraint Satisfaction Problem.

k-Flip Neighbourhood: k-exchange neighbourhood for the Satisfiability Problem (SAT) or the Maximum Satisfiability Problem (MAX-SAT), under which the direct neighbours of a candidate solution are precisely those truth assignments that are obtained by flipping (i.e., changing) the value of up to k propositional variables. Most stochastic local search algorithms for SAT and MAX-SAT are based on the 1-*flip neighbourhood*.

Active Schedule: A schedule in which no operation can be completed earlier by any change in the processing sequence of any machine without delaying other operations.

Adaptive Iterative Construction Search (AICS): Stochastic local search method that is based on multiple iterations of a constructive search procedure; the construction is guided by heuristic information and by experience gained from past iterations. The latter is represented in the form of *weights* associated with the elementrary decisions that can be made during the construction process; these weights are adapted during the search. AICS is conceptually closely related to Ant Colony

Optimisation (ACO), and the pheromone trails in ACO correspond to the weights in AICS.

Algorithm Portfolio: Collection of algorithms that are simultaneously or selectively applied to a given problem instance. Compared to their constituting algorithms, portfolios can often solve combinatorial problems more robustly, particularly in situations where different algorithms are likely to perform best on various types of problem instances, depending on partially unknown features of the instance under consideration. Because of the additional uncertainty in run-time, portfolios of Las Vegas algorithms are particularly interesting. Portfolios of stochastic local search algorithms can be adequately represented using co-operative GLSM models.

Annealing Schedule: Often also called *cooling schedule*; a mapping used in Simulated Annealing (SA) that determines for each run-time t the value of the temperature parameter, $T(t)$. Annealing schedules are commonly specified by an initial temperature, $T(0)$, a temperature update scheme, a number of iterations to be performed at each temperature and a termination condition. Although the name suggests a monotonic decrease of temperature over time, annealing schedules can be arbitrary functions and may contain segments of increasing or constant temperature.

Ant Colony Optimisation (ACO): Stochastic local search method that is inspired by the pheromone trail following behaviour of some ant species. In analogy to the biological example, ACO is based on the indirect communication of a colony of simple agents, called (artificial) *ants*, mediated by (artificial) *pheromone trails*. The pheromone trails in ACO serve as a distributed, numerical memory that the ants use to probabilistically construct candidate solutions to the given problem instance; like the weights in Adaptive Iterated Construction Search, the pheromone trails are adapted during the search process and represent the collective experience of the ants. A large number of different ACO algorithms has been proposed, including $\mathcal{MAX}\text{-}\mathcal{MIN}$ Ant System and Ant Colony System. Most of the best-performing ACO algorithms for combinatorial problems use subsidiary local search to improve the candidate solutions constructed by the ants.

Ant Colony System (ACS): Ant Colony Optimisation algorithm that uses an aggressive construction procedure in which, with high probability, deterministic choices are made instead of the usual stochastically biased choice. Furthermore, unlike many other ACO algorithms, ACS performs additional updates of the pheromone trails during the construction of candidate solutions.

Approximate (Optimisation) Algorithm: Incomplete search algorithm for an optimisation problem, that is, an algorithm for solving an optimisation problem that is not guaranteed to find an optimal solution, even if run for an arbitrarily long (finite) amount of time.

Approximation Algorithm: Algorithm for solving an optimisation problem that has provable performance guarantees, typically in the form of constant bounds on the worst-case approximation ratio. (The term is sometimes used incorrectly to refer to optimisation algorithms that may return suboptimal candidate solutions or solutions but do not have performance guarantees.)

Approximation Ratio: Performance criterion for optimisation algorithms; for an algorithm A, the approximation ratio r on a given instance π' of an optimisation problem Π' with objective function f is defined as $r := \max\{\hat{q}/q^*, q^*/\hat{q}\}$, where $\hat{q}$ is the best solution quality achieved by A on π', and q^* is the optimal solution quality of π'. The (worst-case) approximation ratio of A on problem Π', that is, the maximum approximation ratio of A over all instances $\pi' \in \Pi'$, plays an important role in computational complexity theory.

Arc Consistency: Property of an instance of the Constraint Satisfaction Problem (CSP); a CSP instance is arc consistent if, and only if, there is no variable x whose domain includes a value d such that x occurs in a constraint C with no satisfying tuple for which x has value d. Arc consistency can always be enforced in polynomial time w.r.t. to the number and maximal domain size of the variables in a given CSP instance. Procedures for enforcing arc consistency play an important role in the context of systematic search algorithms and stochastic local search algorithms for the CSP.

Aspiration Criterion: Condition used in a tabu search algorithm to override the tabu status of candidate solutions or solution components, for example, when the respective search step leads to an improvement in the current incumbent candidate solution.

Autocorrelation Coefficient: Measure of ruggedness of a search landscape; empirically determined from the *autocorrelation function* of an uninformed random walk. Intuitively, larger autocorrelation coefficients indicate smoother landscapes. Closely related to the autocorrelation coefficient is the correlation length, which has a similar intuitive interpretation.

Backtracking: Algorithmic technique in which a search algorithm, upon encountering a 'dead end' from which further search is impossible or unpromising, reverts to an earlier point in its search trajectory. Many systematic search algorithms are based on combinations of constructive search methods with backtracking.

Basin Partition Tree (BPT): Represents the basin structure of a search landscape; a BPT T forms a complete partition of a given landscape S, that is, the vertices of T represent disjoint subsets of S, and every position in S is represented by exactly one vertex in T. BPTs can be seen as abstractions of plateau connection graphs and are closely related to basin trees.

Basin Tree (BT): Represents the basin structure of a search landscape; the vertices of a BT represent the *barrier-level basins* of a given landscape, and the edges indicate immediate containment between the respective basins. BTs are closely related to basin partition trees.

Benchmark Instance: Problem instance used for the empirical evaluation and characterisation of algorithmic behaviour. Sets of benchmark instances (also referred to as *benchmark suites*) are often made available via on-line benchmark libraries.

Biergarten: Open air pub where people enjoy a refreshing *Maß of Weißbier* (a litre of a specific type of Bavarian beer), (light) supper and good company, typically whilst sitting under large, old chestnut trees on a mild summer evening; Biergärten are an essential part of Southern German culture. Many of the ideas and concepts described in this book were at some point discussed between its authors during their many pleasant visits to a local Biergarten.

Bivariate Run-Time Distribution: Joint probability distribution of run-time and solution quality that characterises the behaviour of an optimisation Las Vegas algorithm.

Branch & Bound: Systematic search method for combinatorial optimisation problems that exploits *upper and lower bounding techniques* for effectively pruning the search tree of a given problem instance. In particular, the search can be pruned at any partial solution whose lower bound exceeds the current upper bound on the quality of an optimal solution for the given problem instance (in the case of a minimisation problem). In the context of branch & bound algorithms for minimisation problems, stochastic local search algorithms can be used for obtaining upper bounds on the optimal solution quality of a given problem instance. Branch & bound algorithms are amongst the best-performing systematic search methods for many hard combinatorial optimisation problems, such as the Maximum Satisfiability Problem.

Branch & Cut: Method for solving integer programming problems that works by solving a series of linear programming relaxations. At each stage, *cuts*, that is, linear inequalities that are satisfied by all integer feasible solutions but not by the non-integer optimal solution to the current relaxation, are added to the linear optimisation problem; this makes the relaxation more closely approximate the optimal solution of the original integer programming problem. This process is iterated until finding 'good' cuts becomes hard, at which point the current problem is split into two subproblems, to which branch & cut is applied recursively. For many problems, such as the Travelling Salesman Problem, branch & cut algorithms are currently among the best-performing complete search algorithms.

Candidate Solution: Element of the search space of a given combinatorial problem; typically a grouping, ordering or assignment of solution components. For many combinatorial problems, there is a natural distinction between *partial candidate solutions*, which may be further extended with additional solution components (without violating the fundamental structure of a candidate solution for the given problem), and *complete candidate solutions*, for which this is not the case. For example, in the case of the Satisfiability Problem, where natural solution components are assignments of truth values to individual propositional variables, partial candidate solutions may assign values to only some of the variables in the given formula, while complete candidate solutions specify truth values for all variables. Candidate solutions are also often called search positions; other terms that are sometimes used in the literature are *(search) state* and *(search) configuration*.

Casanova: High-performance stochastic local search algorithm for the Combinatorial Auction Winner Determination Problem; shares features with Novelty$^+$, a WalkSAT algorithm for the Satisfiability Problem, and with the Min-Conflicts Heuristic for the Constraint Satisfaction Problem.

Chained Lin-Kernighan (CLK): Iterated local search algorithm for the Travelling Salesman Problem.

Chained Local Optimisation: This term is sometimes used to refer to Iterated Local Search.

Clause: Disjunction of *literals* (i.e., of propositional variables or their negations) in a CNF formula. A clause is satisfied if, and only if, at least one of its literals is evaluated to true.

Clause Penalty: Numerical value associated with a clause in an instance of the Satisfiability Problem or the Maximum Satisfiability Problem. Clause penalties are used by dynamic local search algorithms and intuitively reflect the relative importance of ensuring the satisfaction of the respective clauses; they are often also referred to as a *penalty weights* or *clause weights*.

Clause Weight: Numerical value associated with a clause in an instance of the Weighted Maximum Satisfiability Problem (MAX-SAT) that reflects the relative importance of satisfying the respective clause. The term is also sometimes used to refer to clause penalties; but unlike the clause penalties used in dynamic local search algorithms for Weighted MAX-SAT, clause weights are an integral part of a given problem instance and are not modified during the search.

CNF Formula: Propositional formula in conjunctive normal form, that is, a disjunction of CNF clauses, where each CNF clause is a disjunction of *literals* (a literal is a propositional variables or its negation).

Combinatorial Auction: Auction mechanism that allows bidders to place bids for bundles of items. Combinatorial auctions have many practical applications. In contrast to conventional auctions, many basic problems associated with combinatorial auctions, such as the Combinatorial Auctions Winner Determination Problem, are computationally hard.

Combinatorial Auction Winner Determination Problem (CAWDP): Combinatorial problem in which the objective is to determine a feasible set of winning bids in a combinatorial auction such that the auctioneer's revenue is maximised. The CAWDP is an $\mathcal{NP}$-hard combinatorial optimisation problem that does not have an efficient approximation algorithm. Stochastic local search algorithms such as Casanova are among the state-of-the-art methods for solving this problem.

Combinatorial Problem: Problem in which given a set of *solution components*, the objective is to find a combination of these components with certain properties. For *combinatorial decision problems*, such as the Satisfiability Problem (SAT), the desired properties are stated in the form of logical conditions, while in the case of *combinatorial optimisation problems*, such as the Travelling Salesman Problem (TSP), the desired properties mainly consist of optimisation objectives which may be accompanied by additional logical conditions. Many (but not all) combinatorial problems are computationally hard and are solved by searching exponentially large spaces of candidate solutions.

Complete Search Algorithm: Search algorithm that, given sufficiently high run-time, is guaranteed to either find a solution of a given problem instance, or, if the instance is insoluble, to determine this fact with certainty. Complete optimisation algorithms, if run sufficiently long, are guaranteed to find solutions with provably optimal objective function values for any given problem instance. All systematic search algorithms are complete, while most stochastic local search algorithms are incomplete.

Completion Time: Time at which an operation or a job finishes processing in a scheduling problem.

Constraint: Relation between the values certain variables are allowed to take in a given instance of the Constraint Satisfaction Problem. The term is also used to refer to a logical condition on the properties of a candidate solution or solution component for a given combinatorial problem. *Hard constraints* represent conditions that any candidate solution needs to satisfy in order to be considered useful in a given application context, while *soft constraints* capture optimisation goals, not all of which may be satisfiable simultaneously.

Constraint Satisfaction Problem (CSP): Prominent combinatorial decision problem in artificial intelligence; given a set of variables and a set of relations (*constraints*) between these variables, the objective of the decision variant is to decide whether there exists an assignment of values to variables such that all constraints are simultaneously satisfied — such an assignment is called a *satisfying tuple*. *Finite discrete CSP*, the special case in which all variables have finite, discrete domains, is a widely studied $\mathcal{NP}$-complete problem of high theoretical and practical interest.

Constructive Search: Search paradigm in which the search process starts from an empty candidate solution and iteratively adds solution components until a complete candidate solution has been obtained. Constructive search algorithms are also known as *construction heuristics*. Constructive search can be seen as a special form of local search in a space of partial candidate solutions. A typical example for a constructive search algorithm is the Nearest Neighbour Heuristic for the Travelling Salesman Problem.

Co-operative GLSM Model: Extension of the Generalised Local Search Machine (GLSM) model that captures the simultaneous application of multiple search processes to solving a given problem instance; each of these search processes is represented by an individual GLSM. Co-operative GLSMs can be classified into *homogeneous co-operative GLSMs*, in which all individual GLSMs are identical, and *heterogeneous co-operative GLSMs*, which are comprised of different types of individual GLSMs. Furthermore, a distinction is made between *co-operative GLSMs without communication* and *co-operative GLSMs with communication*; while in the former case, the individual GLSMs do not communicate during the search, in the latter case, they exchange information, for instance, by means of a shared memory or a message passing mechanism. Co-operative GLSMs can be used to adequately represent various types of stochastic local search methods, including population-based SLS methods and algorithm portfolios.

Correlation Length: Measure of the ruggedness of a search landscape; typically determined empirically based on the autocorrelation function of an uninformed random walk in the landscape. Intuitively, smoother landscapes have larger correlation lengths. The correlation length is closely related to the autocorrelation coefficient and has a similar intuitive interpretation.

CPU Time: Measure for the run-time of an algorithm that is based on the actual computation time used by the respective implementation (process) on a given processor, typically measured in seconds, where one CPU second corresponds to one second of wall-clock time during which the CPU executes the given process only. CPU time is (mostly) independent of other processes running at the same time on a multi-tasking operating system, but depends on the processor type (model, speed, cache size) and may be affected by other aspects of the execution environment, such as main memory, compiler and operating system characteristics. When reporting run-times in CPU seconds, at least the processor model and speed, the amount of RAM, as well as the operating system type and version should be specified.

Critical Block: Sequence of critical operations that are assigned to the same machine in a candidate solution of a given instance of a scheduling problem.

Critical Operation: Operation in a candidate solution s of a scheduling problem instance that cannot be delayed without increasing the makespan of s.

Critical Path: Maximum length sequence of critical operations in a candidate solution of a scheduling problem instance; a critical path can contain jobs that are assigned to different machines.

Crossover: Type of recombination mechanism that is based on assembling pieces or fragments from a linear representation of two or more candidate solutions (*parents*) into one or more new candidate solutions (*offspring*). Crossover is used in many Evolutionary Algorithms; sometimes, the term is used broadly to refer to any type of recombination mechanism.

Cutoff Time: Fixed time limit after which a run of a given algorithm is terminated, typically specified in terms of CPU time; in the context of restart mechanisms, the term is also used to refer to the (maximal) time interval after which the respective search process is restarted.

Decision Problem: Computational problem in which given a problem instance, the objective is to decide whether it satisfies a certain property. A prominent example is the Satisfiability Problem — the problem of deciding whether there exists an assignment of truth values to the variables in a given propositional formula such that the formula evaluates to true.

Degenerate Distribution: Fundamental type of probability distribution that characterises a random variable with only a single possible value. The cumulative distribution function of a degenerate distribution is a step function of the form $\theta[v](x) = 0$ if $x < v$ and 1 otherwise, where v is the single value of the corresponding random variable. (This type of distribution is sometimes also called 'Dirac delta distribution'.)

Discrete Lagrangian Method (DLM): Family of dynamic local search algorithms that comprises several high-performance algorithms for the Satisfiability Problem and the Maximum Satisfiability Problem. DLM was originally motivated by Lagrangian methods for solving continuous optimisation problems.

Dispatching Rule: Heuristic function used in the context of a constructive search algorithm for scheduling problems for choosing in each construction step the operation or job to be scheduled next; often synonymously referred to as *priority rule*. An example for a well-known dispatching rule is the *Earliest Due Date (EDD) rule*, which always selects the unscheduled job with the earliest due date.

Diversification: Important property of a search process; *diversification mechanisms* help to ensure sufficiently strong exploration of the search space in order to avoid search stagnation and entrapment of the search process in regions of the search space that do not contain (sufficiently high-quality) candidate solutions. An important issue in the design of stochastic local search algorithms is to achieve a good (problem-specific) balance between diversification and intensification of the search process.

DNA Code Design Problem (DNA-CDP): Combinatorial optimisation problem with applications in biomolecular computation as well as in the design of DNA microarrays. The objective is to find a set of DNA words, that is, of strings over the alphabet {A, C, G, T}, that satisfy given constraints. Such a set is called a *DNA code*; typical constraints involve the Hamming distance between words or between words and their reverse complements, or the GC content of DNA words. The DNA Code Design Problem is related to well-known problems in coding theory. Although its theoretical complexity is still unknown, the DNA Code Design Problem appears to be computationally hard in practice.

DNA Word: Representation of a short strand of deoxyribonucleic acid (DNA) in the form of a string over the alphabet {A, C, G, T}. The DNA Code Design Problem involves finding sets of DNA words with certain properties.

Domain: In the sense of *variable domain*: set of values a given variable can take (e.g., in the Constraint Satisfaction Problem); in the sense of *problem/application domain*: particular type of problem or area of application.

Double-Bridge Move: Specific type of search step in the 4-exchange neighbourhood on edges in the Travelling Salesman Problem (TSP) that cannot be easily reversed by a short sequence of steps in the 2-exchange neighbourhood as performed, for example, by the Lin-Kernighan Algorithm; because of this latter property, double-bridge moves are often used in the perturbation phase of iterated local search algorithms for the TSP.

Due Date: Deadline for the completion of a job in a scheduling problem. Scheduling problems with due dates often use the (weighted) number of the jobs that are finished after their respective due dates or the sum of the weighted delays as an objective function.

Dynamic Local Search (DLS): Stochastic local search method that modifies the evaluation function of a subsidiary local search algorithm during the search. This is typically realised by associating *penalty weights* with components of a candidate solution; these weights are updated whenever the subsidiary local search procedure encounters a local minimum of the current evaluation function. Specific DLS algorithms differ primarily in their subsidiary local search and weight update procedures.

Dynasearch: Iterative improvement method that builds complex search steps based on optimal combinations of simple, independent search steps; search steps are considered *independent* if, and only if, they do not interfere with each other with respect to their effect on the evaluation function value and the feasibility of candidate solutions. Optimal combinations of simple, independent search steps are identified exploiting an algorithmic technique called *dynamic programming* (hence the name *Dyna*search).

Essentially Incomplete: Property of a Las Vegas algorithm; an algorithm is essentially incomplete if, and only if, it is not probabilistically approximately complete. Essentially incomplete SLS algorithms can get permanently trapped in non-solution regions of a given search space.

Evaluation Function: Function that assigns numerical values, typically real numbers greater than or equal to zero, to the candidate solutions of a given problem instance; most stochastic local search algorithms use an evaluation function for guiding the search process. In the case of SLS algorithms for optimisation problems, the evaluation function is often, but not always, identical to the objective function of the respective problem instance.

Evolutionary Algorithms (EAs): Large and diverse class of population-based SLS algorithms that is inspired by the process of biological evolution through selection, mutation and recombination. Evolutionary Algorithms are iterative algorithms that start with an initial population of candidate solutions and then repeatedly apply a series of the *genetic operators* selection, mutation and recombination. Using these operators, in each iteration of an EA, the current population is (completely or partially) replaced by a new set of *individuals*, that is, candidate solutions. Historically, several types of EAs can be distinguished: *genetic algorithms, evolutionary programming methods* and *evolution strategies*; but more recently, the differences between these types of EAs are becoming increasingly blurred. Many high-performance EAs for combinatorial problems use a subsidiary local search procedure for improving candidate solutions in each iteration of the search process; in many cases, these hybrid algorithms are based on genetic algorithms and are hence referred to as *genetic local search methods*. Slightly more generally, EAs that use subsidiary local search procedures are also known as *memetic search methods* or *memetic algorithms*.

Exact (Optimisation) Algorithm: Complete search algorithm for an optimisation problem, that is, an algorithm for solving an optimisation problem that is guaranteed to find an optimal solution for any given problem instance in time bounded by a function of the instance size.

Exit: Search position on the border of a given plateau that has a direct neighbour at a lower level; plateaus without exits, so-called *closed plateaus*, tend to make a problem instance difficult to solve for stochastic local search algorithms.

Exponential Distribution: Fundamental type of probability distribution with a cumulative distribution function of the form $ed[m](x) := 1 - 2^{-x/m}$ (where m is the median of the distribution) or, equivalently, $Exp[\lambda](x) := 1 - e^{-\lambda \cdot x}$. Some prominent stochastic local search algorithms, such as GSAT with Random Walk or Novelty$^+$, show run-time distributions that closely resemble exponential distributions; SLS algorithms with exponential RTDs are not affected by static restart mechanisms and achieve optimal parallelisation speedup under multiple independent runs parallelisation.

Exponentiated Subgradient Algorithm (ESG): Dynamic local search method that uses *multiplicative penalty updates* whenever its underlying iterative best improvement algorithm encounters a local minimum of the given evaluation function; at the same time, a *smoothing mechanism* is applied to all penalty values. ESG has been motivated by subgradient optimisation methods for Lagrangian relaxation. The general framework has given rise to high-performance stochastic local search algorithms for the Satisfiability Problem and the Combinatorial Auctions Winner Determination Problem.

Feasible Candidate Solution: Alternative term for solution, that is, for a candidate solution that is an element of the solution set of a given stochastic local search algorithm and problem instance.

Fitness: Alternative term for the evaluation function value of a given candidate solution; originating from and often used in the context of Evolutionary Algorithms. Similarly, the term *fitness landscape* is often used synonymously with search landscape.

Fitness-Distance Analysis: A type of search space analysis in which the correlation between the fitness (i.e., evaluation function value) of candidate solutions and their distance to the next (optimal) solution is studied; this correlation is also known as *fitness-distance correlation (FDC)*. FDC can be measured by the *fitness-distance correlation coefficient* and graphically illustrated or analysed by means of *fitness-distance plots*.

Flow Shop Scheduling Problem (FSP): Scheduling problem in which each job consists of a number of atomic operations that are to be performed by different machines. In the FSP, all jobs have to pass through the machines in the same machine order. In the *Permutation FSP*, all jobs are processed in the same order on all machines, that is, a *candidate solution* is uniquely defined by a permutation of the jobs. The FSP is a special case of the Group Shop Scheduling Problem (GSP).

Gantt Chart: A graphical representation of a schedule in which the horizontal axis represents time, while different machines are distinguished along the vertical axis. Each operation is represented by a box that indicates when and where it is scheduled to be processed; the start and finish time of the operation correspond to the left and right boundaries of the respective box. Colours, shadings or labels are often used to indicate the grouping of operations into jobs or other properties, such as completion of a given operation after the due date of the respective job.

Generalised Local Search Machine (GLSM): Formal model of a hybrid SLS method that explicitly represents the search control in the form of a finite state machine whose *states* correspond to simple search strategies, while *transitions* between the states correspond to switches between search strategies. GLSM representations are often useful for designing, conceptualising and analysing complex stochastic local search algorithms.

Genetic Algorithm (GA): A type of Evolutionary Algorithm in which candidate solutions are represented typically by vectors of integers. While historically, candidate solutions were represented as bit strings, that is, by vectors of binary values 0 and 1, it is nowadays acknowledged that specialised representations (such as permutations of the integers $1, \dots, n$ — which, for instance, are used to represent tours for the Travelling Salesman Problem) can be essential for the performance of GAs.

Graph: Mathematical structure consisting of a set of *vertices* and a set of *edges*, where each edge connects two vertices. An edge e is said to be *incident* to a vertex v if, and only if, v is one of the vertices connected by e. In a *directed graph*, the edges are oriented, such that an edge from v to v', represented by (v, v'), is different from an edge from v' to v, represented by (v', v). In an *undirected graph*, the edges have no orientation, such that an edge between two vertices v, v' connects v with v' as well as v' with v; these undirected edges are often represented as sets $\{v, v'\}$. In an *edge-weighted graph*, a numerical value called an *edge weight* is associated with every edge; these weights can, for example, represent the length or cost of the respective edges.

Graph Colouring Problem (GCP): Well-known combinatorial problem in which the objective is to colour the vertices of a given graph in such a way that any two vertices connected by an edge are assigned different colours. Such an assignment of colours to vertices is called a *k-colouring* if, and only if, it uses at most k different colours. The GCP is an $\mathcal{NP}$-hard problem with many applications, and stochastic local search algorithms are among the state-of-the-art methods for solving hard GCP instances.

Greedy Randomised Adaptive Search Procedures (GRASP): Class of stochastic local search algorithms that uses randomised greedy constructive search heuristics to generate a large number of different high-quality candidate solutions that are then further improved by a subsidiary local search procedure. These cycles of construction and local search phases are iterated until a termination condition is satisfied. The construction process in GRASP is called 'adaptive' because the heuristic value of a solution component in the constructive search typically depends on the components that are already present in the current partial candidate solution.

Group Shop Scheduling Problem (GSP): Scheduling problem that can be seen as a generalisation of the Job Shop, Open Shop and Flow Shop Scheduling Problems. In the GSP, each job consists of a number of atomic operations that are to be performed by different machines. The operations of each job are partitioned into a number of *groups*, and a total ordering of the groups of each job is given, while within each group the operations can be processed in any order.

GSAT: Simple stochastic local search algorithm for the Satisfiability Problem (SAT); essentially an iterative best improvement algorithm which in each search step flips (i.e., changes) the truth value assigned to one propositional variable. GSAT is also sometimes used to refer to the broader family of GSAT algorithms, which also includes variants of the basic GSAT algorithm, such as GSAT/Tabu or GSAT with Random Walk.

GSAT/Tabu: Simple stochastic local search algorithm for the Satisfiability Problem (SAT); essentially a simple tabu search variant of GSAT, in which whenever a variable is flipped it is declared tabu for a constant number of subsequent search steps.

GSAT with Random Walk (GWSAT): Prominent stochastic local search algorithm for the Satisfiability Problem (SAT); essentially a randomised iterative improvement algorithm. In each search step, the variable to be flipped is either selected as in GSAT or uniformly at random from the set of variables that occur in currently unsatisfied clauses. GWSAT can be seen as a simple hybrid combination of basic GSAT and a simple *conflict directed random walk* algorithm.

Heuristic: Function used for making certain decisions within an algorithm; in the context of search algorithms, typically used for guiding the search process, for example, by selecting solution components to be added to or modified in the current candidate solution. The term 'heuristic' is also commonly used to refer to algorithms that use heuristic functions and do not have certain performance guarantees, in particular w.r.t. finding solutions or reaching specific solution quality bounds.

Hybrid SLS Method: Stochastic local search method that combines different subsidiary SLS strategies with the goal to improve search performance or robustness. One of the simplest examples is Randomised Iterative Improvement, which probabilistically combines standard Iterative Improvement and Uninformed Random Walk. Hybrid SLS methods can often be modelled in an intuitive and representationally adequate way as Generalised Local Search Machines.

Incomplete Search Algorithm: Search algorithm that is not complete; incomplete algorithms may find (optimal) solutions to a given soluble problem instance, but generally cannot be guaranteed to do so, even if arbitrary (finite) amounts of run-time are allowed. Furthermore, incomplete algorithms generally cannot determine the insolubility of a given problem instance.

Incumbent Candidate Solution: Highest quality candidate solution encountered during the run of a stochastic local search algorithm for an optimisation problem, where solution quality is measured by the given objective function; often also referred to as *incumbent solution*. The term 'incumbent solution' is sometimes also used to refer to candidate solutions with the best evaluation function value encountered during a run of an SLS algorithm for a decision problem.

Initialisation: Elementary operation of a stochastic local search algorithm that is performed at the beginning of the search process and involves selecting a candidate solution from which the search is started. In SLS methods that use restart mechanisms, the same operation is also used for restarts during the search process.

Initialising State: State of a Generalised Local Search Machine that is used for search initialisation or restart; the search position after one search step performed in an initialising state is probabilistically independent of the position before the step.

Insertion Neighbourhood: Neighbourhood relation on permutations (or orderings) of solution components under which two permutations are direct neighbours if, and only if, one can be obtained from the other by removing an element from one position and inserting it at another position. The insertion neighbourhood is often used in scheduling problems; it is sometimes also called *shift neighbourhood*.

Insoluble: Property of a problem instance; a problem instance is called insoluble if, and only if, it has an empty solution set, and soluble otherwise. The term is typically applied to instances of decision problems, but also applies to optimisation problems whose instances may have empty solution sets.

Integer Programming: A set of methods for solving integer programming problems which comprises techniques such as branch & bound, branch & cut and dynamic programming.

Integer Linear Programming (ILP) Problem: Special case of the Integer Programming Problem in which the objective function as well as all feasibility constraints are linear functions of the decision variables. The set of techniques for solving ILP problems is called *Integer Linear Programming*. The special case in which the decision variables are restricted to the domain $\{0, 1\}$ is known as *0-1 Integer Linear Programming Problem* or *(Overconstrained) Pseudo-Boolean CSP*.

Integer Programming (IP) Problem: Combinatorial problem in which, given a set of numerical *decision variables*, a set of *feasibility constraints* and an optimisation objective, the goal is to find an optimal assignment of integer values to the variables such that all feasibility constraints are satisfied. Many combinatorial problems, such as the Set Covering Problem, can be represented as IP problems in an intuitive and straightforward way. The generalisation of IP in which some variables are not restricted to integer values is known as the *Mixed Integer Programming (MIP) Problem*. IP problems in which the variables are restricted to binary values 0 and 1 are often called *0-1 integer programming problems* or *Boolean programming problems*.

Intensification: Important property of a search process; *intensification mechanisms* aim to carefully examine a specific area of the given search space in order to find a solution or higher-quality candidate solutions. Intensification strategies are often strongly based on greedy heuristic guidance mechanisms. In high-performance stochastic local search algorithms, intensification strategies are often complemented with suitable diversification mechanisms.

Interchange Neighbourhood: Neighbourhood relation on permutations (or orderings) of solution components under which two permutations are direct neighbours if, and only if, one can be obtained from the other by interchanging the elements at two different positions. This neighbourhood relation is often used in scheduling problems but also in assignment problems, such as the Quadratic Assignment Problem; it is occasionally also called *swap neighbourhood*.

Iterated Greedy (IG): Stochastic local search method that can be seen as a variant of Iterated Local Search (ILS) in which *construction* and *destruction phases* are used instead of local search and perturbation phases. During each *construction phase*, a constructive search procedure is used to extend a partial candidate solution into a complete candidate solution. During the *destruction phase*, solution components are removed from the given complete candidate solution. Like ILS, IG uses an *acceptance criterion* to decide whether the search is continued from the new candidate solution s' obtained from a given candidate solution s by destruction and subsequent construction, or whether s' is abandoned and the search is continued from s. IG algorithms are among the state-of-the-art methods for solving the Set Covering Problem.

Iterated Lin-Kernighan (ILK): Iterated local search algorithm for the Travelling Salesman Problem that uses the Lin-Kernighan (LK) algorithm or a variant thereof as its subsidiary local search procedure; the perturbation procedure of ILK is based on random double-bridge moves, and its acceptance criterion only accepts new locally optimal candidate solutions that represent an improvement in the incumbent candidate solution.

Iterated Local Search (ILS): Stochastic local search method that is based on the repeated application of alternating *local search* and *perturbation phases*. Starting from an initial candidate solution, a *subsidiary local search procedure* is applied until a local minimum w.r.t. a given evaluation function is reached. Then, the following sequence of steps is iterated until a termination condition is satisfied. First, a *perturbation mechanism* is applied to the current candidate solution s, resulting in an intermediate candidate solution s', which is typically not locally optimal. Next, *local search* is applied to s', which leads to a locally optimal candidate solution s''. Finally, an *acceptance criterion* is used to decide from which candidate solution (typically either s'' or s) the search is continued. ILS algorithms can be seen as performing a biased walk in the space of local minima w.r.t. to a given evaluation function. Iterated local search algorithms are also known under various other names, including *Chained Local Optimisation* and *Large Step Markov Chains*.

Iterated Robust Tabu Search (IRoTS): Iterated local search algorithm for the Maximum Satisfiability Problem (MAX-SAT) whose subsidiary local search and perturbation phases are both based on Robust Tabu Search; furthermore, IRoTS uses a randomised acceptance criterion that is biased towards better-quality candidate solutions.

Iterative Improvement: Stochastic local search method that in each search step achieves an improvement in the evaluation function value of the current candidate solution. Depending on the pivoting rule that is used for selecting the search steps, different types of iterative improvement algorithms can be distinguished: *Iterative best improvement* algorithms always select a candidate solution with best function value within the current neighbourhood, while *iterative first improvement* algorithms check the neighbourhood in a given order and perform the search step that corresponds to the first improving candidate solution encountered during this check. Iterative improvement algorithms terminate when they encounter a local minimum of the given evaluation function.

Job Shop Scheduling Problem (JSP): Scheduling problem in which each job consists of a number of atomic operations that are to be performed by different machines; there is a total ordering of all the operations within each job, that is, the operations have to be performed in a given, fixed order. The JSP is a special case of the Group Shop Scheduling Problem (GSP).

Large Step Markov Chains: Iterated local search method that typically uses the Metropolis acceptance criterion. The term is occasionally used more broadly to refer to the general SLS method of Iterated Local Search.

Las Vegas algorithm (LVA): Stochastic algorithm that is guaranteed to only return correct solutions, but may run arbitrarily long without finding a solution for a given, soluble problem instance. The run-time of a Las Vegas algorithm on a given problem instance is characterised by a random variable. Stochastic local search algorithms form an important subclass of Las Vegas algorithms.

Lateness: Difference between the completion time of a job in a scheduling problem and its due date under a given schedule. Note that jobs that are completed before their due date have negative lateness.

Level: Alternative term for the evaluation function value of a candidate solution; mostly used in the context of search space analysis.

Lin-Kernighan (LK) Algorithm: Variable depth search algorithm for the Travelling Salesman Problem (TSP), also known as *Lin-Kernighan Heuristic*. Variants of this local search algorithm form the basis of most current state-of-the-art algorithms for the TSP. Historically, LK was one of the first variable depth search algorithms.

Local Minima Density: Relative frequency of local minima within a given search space; analogous to the solution density, the local minima density of a search space of size $\#S$ with l local minima is defined as $l/\#S$. Problem instances with higher local minima density tend to be easier to solve for stochastic local search algorithms. Local minima density is closely related to the ruggedness of a given search landscape and can sometimes be estimated using ruggedness measures, such as correlation length.

Local Minimum: Candidate solution whose evaluation function value is smaller than or equal to the evaluation function values of any of the candidate solutions in its neighbourhood; a candidate solution is a *strict local minimum* if, and only if, all of its neighbours have strictly larger evaluation function values.

Local Search: Algorithmic method for searching a given space of candidate solutions that starts from an initial candidate solution and then iteratively moves from one candidate solution to a candidate solution from its direct neighbourhood, based on local information, until a termination condition is satisfied.

Log-Log Plot: Graphical representation in which both axes of a given coordinate system are shown in logarithmic scale. Log-log plots are particularly useful in the context of scaling and correlation analyses, since the graphs of polynomial functions appear as straight lines in such plots; they are also often used when analysing the tails of distributions, such as run-time distributions.

Makespan: Commonly used objective function for scheduling problems that measures the maximum completion time of any job under a given candidate schedule.

Many-Valued Satisfiability Problem (MV-SAT): Generalisation of the Satisfiability Problem (SAT), in which variables can have domains with more than two values. An instance of MV-SAT is given by a generalisation of a CNF formula in which each literal specifies or rules out one or more values of a many-valued variable. MV-SAT is a special case of the Constraint Satisfaction Problem (CSP). Like CSP and SAT, MV-SAT is $\mathcal{NP}$-complete. MV-SAT is also known as *Multi-Valued SAT*.

Maximum Constraint Satisfaction Problem (MAX-CSP): Generalisation of the Constraint Satisfaction Problem (CSP) in which, given a CSP instance, the objective is to find a variable assignment that maximises the number of satisfied constraints. In *Weighted MAX-CSP*, weights associated with each constraint can be used to prioritise the satisfaction of constraints, and the objective is to find an assignment that maximises the total weight of the satisfied constraints.

Maximum Satisfiability Problem (MAX-SAT): Generalisation of the Satisfiability Problem in which, given a CNF formula, the objective is to find a variable assignment that maximises the number of satisfied clauses. In *Weighted MAX-SAT*, weights associated with each clause can be used to prioritise the satisfaction of clauses, and the objective is to find an assignment that maximises the total weight of the satisfied clauses.

$\mathcal{MAX}$–$\mathcal{MIN}$ Ant System ($\mathcal{MMAS}$): Class of Ant Colony Optimisation algorithms that uses limits on the feasible values of the pheromone trails as well as additional intensification and diversification mechanisms. $\mathcal{MMAS}$ forms the basis for many successful applications of ACO algorithms to $\mathcal{NP}$-hard optimisation problems. Unlike many other ACO algorithms, as a result of its use of limits on the pheromone trails, $\mathcal{MMAS}$ is provably probabilistically approximately complete (PAC).

Memetic Algorithms (MAs): Class of population-based SLS algorithms that combines the evolutionary operators mutation, recombination and selection with a subsidiary local search procedure or, more generally, problem-specific heuristic information. MAs that use subsidiary local search can be seen as evolutionary algorithms that search a space of locally optimal candidate solutions.

Metaheuristic: Generic technique or approach that is used to guide or control an underlying problem-specific heuristic method (for example a local search algorithm or a constructive search algorithm) in order to improve its performance or robustness. The term is also widely used to refer to simple, hybrid and population-based SLS methods, such as Simulated Annealing, Tabu Search, Iterated Local Search, Evolutionary Algorithms and Ant Colony Optimisation.

Metric TSP: Special case of the Travelling Salesman Problem (TSP), in which the vertices of the given graph correspond to points in a metric space, and the edge weights correspond to metric distances between pairs of points. A prominent special case is the *Euclidean TSP*, which is based on the standard Euclidean metric.

Metropolis Acceptance Criterion: Widely used acceptance criterion that deterministically accepts a new candidate solution s' if it has a better evaluation function value than the current candidate solution s; otherwise, s' is accepted with probability $e^{-|g(s)-g(s')|/T}$, where $g(s)$ and $g(s')$ are the evaluation function values of s and s', respectively, and T is a parameter called *temperature*. This acceptance criterion is frequently used in simulated annealing algorithms and can also be found in a number of iterated local search algorithms.

Min-Conflicts Heuristic (MCH): Well-known stochastic local search algorithm for the finite discrete Constraint Satisfaction Problem. MCH is basically an iterative improvement method, which in each search step first selects a variable that appears in a currently unsatisfied constraint, and then assigns a value to this variable such that the number of unsatisfied constraints is maximally reduced, that is, minimised within the local neighbourhood.

Multiple Independent Runs Parallelisation: Conceptually simple form of parallelisation, in which multiple runs of a given stochastic algorithm applied to the same problem instance are executed in parallel and independently from each other. Multiple independent runs parallelisation typically only involves minimal communication between the respective processes (or no communication at all) and is very easy to implement. Most stochastic local search algorithms achieve high parallelisation speedup under this form of parallelisation.

Mutation: Basic operation used in Evolutionary Algorithms to introduce modifications to a member of the population, that is, to a candidate solution.

Nearest Neighbour Heuristic: Constructive search method for the Travelling Salesman Problem (TSP) which, starting at a randomly chosen vertex in the given graph, in each step follows an edge with minimal weight connecting the current vertex to one of the neighbouring vertices that have not yet been visited. The resulting *nearest neighbour tours* are typically of substantially lower quality than those obtained by commonly used stochastic local search algorithms for the TSP.

Neighbourhood: The set of candidate solutions that are direct neighbours of a given candidate solution (e.g., the current candidate solution) under a given neighbourhood relation. (The term is sometimes also used to refer to a neighbourhood relation.)

Neighbourhood Graph: Graph whose vertices are the candidate solutions of a given problem instance and whose edges connect candidate solutions that are direct neighbours of each other under the given neighbourhood relation. The search trajectory of any local search algorithm can be seen as a walk in the respective neighbourhood graph. The *diameter of the neighbourhood graph*, defined as the maximal distance between any pair of vertices s and s', where distance is measured as the minimal number of edges that need to be traversed to reach s from s', is an important concept in search space analysis.

Neighbourhood Relation: Important component of a local search algorithm; a binary relation that determines the direct neighbours of any candidate solution of a given problem instance, that is, the set of candidate solutions that can potentially be reached in a single search step.

NK-Landscapes: Statistical model of search landscapes with controllable ruggedness. The model has two main parameters: N specifies a number of binary variables and K indicates the number of other variables on which the evaluation function contribution of each variable depends. For fixed N, the ruggedness of an

NK-landscape increases with K. The problem of finding a global minimum in an NK-landscape is $\mathcal{NP}$-hard.

Noise Parameter: Parameter that controls the degree of randomness used within the search steps of a given stochastic local search algorithm. Examples for noise parameters are the walk probability in GSAT with Random Walk, or the tabu tenure in WalkSAT/Tabu.

Non-Boolean Satisfiability Problem (NB-SAT): Generalisation of the Satisfiability Problem (SAT). An instance of NB-SAT is given by a generalisation of a CNF formula in which each literal specifies or rules out precisely one value of the corresponding non-Boolean variable. NB-SAT is a special case of the Constraint Satisfaction Problem. Like SAT and CSP, NB-SAT is $\mathcal{NP}$-complete.

Non-Delay Schedule: A schedule under which no machine in a given scheduling problem is ever kept idle as long as there is an operation that can be processed.

Non-Oblivious SLS Method: Stochastic local search method that uses a *non-oblivious evaluation function*, that is, an evaluation function that captures the degree to which constraints of a given problem are satisfied or violated. In the case of the Satisfiability or Maximum Satisfiability Problem, the constraints are the clauses of the given CNF formula, and the degree of satisfaction of a clause c under a given assignment a is the number of literals in c that are satisfied under a.

Novelty$^+$: Prominent high-performance stochastic local search algorithm for the Satisfiability Problem (SAT). Novelty$^+$ is a WalkSAT algorithm that uses a history-based, randomised greedy mechanism for selecting the propositional variable to be flipped in each search step. Through the use of an unconditional random walk mechanism, Novelty$^+$ is provably probabilistically approximately complete (PAC).

$\mathcal{NP}$: Computational complexity class; a decision problem is in $\mathcal{NP}$ if, and only if, it can be solved in polynomial time w.r.t. the size of a given problem instance by a *nondeterministic Turing machine* — an idealised, theoretical model of computation. Intuitively, a problem is in $\mathcal{NP}$ if there is a polynomial-time algorithm for checking the correctness of a solution to any given problem instance.

$\mathcal{NP}$-complete: Computational complexity property; a decision problem is $\mathcal{NP}$-complete if, and only if, it is in the complexity class $\mathcal{NP}$ and it is $\mathcal{NP}$-hard. Intuitively, $\mathcal{NP}$-complete problems are the hardest problems that can be solved in polynomial time w.r.t. the size of a given problem instance by a *nondeterministic Turing machine* — an idealised, theoretical machine model. It is strongly suspected (though yet unproven) that solving $\mathcal{NP}$-complete problems with any conventional, implementable algorithm (i.e., using a computational model equivalent to a deterministic Turing machine) takes time exponential in instance size.

$\mathcal{NP}$-hard: Computational complexity property; a problem is $\mathcal{NP}$-hard if, and only if, any algorithm that would solve it in polynomial time w.r.t. the size of a given instance could also be used to solve (suitably encoded instances of) any other problem in the complexity class $\mathcal{NP}$ in polynomial time. Unlike the notion of $\mathcal{NP}$-completeness, $\mathcal{NP}$-hardness applies to optimisation problems as well as to decision problems.

Objective Function: Important component of any combinatorial optimisation problem; a function that assigns a numerical value called solution quality (typically a real number greater than or equal to zero) to each candidate solution of a given problem instance. The objective function is the measure to be minimised or maximised in a combinatorial optimisation problem and typically models the cost or quality of a candidate solution.

Oblivious SLS Method: Stochastic local search method that uses an *oblivious evaluation function*, that is, an evaluation function that ignores the degree to which constraints of a given problem are satisfied or violated; most widely known, high-performance stochastic local search algorithms are of this type. (See also non-oblivious SLS methods.)

Open Shop Scheduling Problem (OSP): A scheduling problem in which each job consists of a number of atomic operations that are to be performed by different machines. In the OSP, there are no precedence constraints among the operations of each job, that is, the operations of each job can be processed in arbitrary order. The OSP is a special case of the Group Shop Scheduling Problem.

Operation: Basic action to be performed by a machine in a scheduling problem on a given job; depending on the type of scheduling problem, jobs may consist of one or more operations (single- *vs* multi-stage scheduling problems).

Optimal Solution: Solution of an optimisation problem whose objective function value is provably minimal within the entire search space of the given problem instance in case of a minimisation problem, or provably maximal in case of a maximisation problem. For many hard and large instances of optimisation problems, (provably) optimal solutions are not known, and empirically best or quasi-optimal solutions have to be used instead, for example, in the context of assessing the performance of stochastic local search algorithms.

Optimisation Las Vegas Algorithm (OLVA): Las Vegas algorithm for an optimisation problem for which, when applied to a given problem instance, the solution quality obtained at run-time t is characterised by a random variable $SQ(t)$. The performance of an optimisation Las Vegas algorithm on a given problem instance is characterised by a bivariate run-time distribution. Stochastic local search algorithms for optimisation problems form an important subclass of the optimisation Las Vegas algorithms.

Optimisation Problem: Computational problem in which given a problem instance π' and objective function f, the goal is to find a candidate solution of π' that minimises (or maximises) f. A prominent example is the Travelling Salesman Problem — the problem of finding a round trip of minimal length in a given edge-weighted graph that visits every vertex exactly once.

Parallel Machine Scheduling Problem: Type of scheduling problem in which several machines are available for processing the given jobs; each job consists of a single operation that is performed by one machine. In *identical parallel machine problems*, the processing time of a job is independent of the machine on which it is processed; in *uniform parallel machine problems*, each machine has a speed that uniformly affects the processing times of all jobs assigned to it; and in *unrelated parallel machine problems*, the processing times of jobs may depend in a non-uniform way on the machines to which they are assigned.

Parallelisation Speedup: Speedup achieved by a parallel algorithm A_p as compared to a functionally equivalent sequential algorithm A_s. Formally, parallelisation speedup is defined as the ratio of the run-time of A_s (running on a single processor) and the parallel run-time of A_p running on multiple processors, where one unit of parallel run-time consists of one unit of sequential run-time on each of the processors that are involved in the execution of the parallel algorithm A_p at the given time. Parallelisation speedup depends on the given algorithms A_p and A_s as well as on the given input (i.e., problem instance).

Path: Sequence p of vertices in a given graph G such that any pair of successive vertices in p are connected by an edge in G. In a *cyclic path* or *cycle in G*, the first and the last vertex in p are identical. A *Hamiltonian path in G* contains every vertex in G exactly once; a *Hamiltonian cycle in G* is a cyclic path p that contains every vertex in G exactly once with the exception of the first and last vertex in p. If G is an edge-weighted graph, the *weight of a path p in G* is the total weight of the edges in p.

Peak Performance: Maximal performance achieved by a given parameterised algorithm applied to a given problem instance (or set of problem instances) when using optimised parameter settings. Even though in many cases peak performance is only obtained for manually tuned, instance-specific parameter settings, it is a useful measure for assessing the *performance potential* of an algorithm.

Permutation: Ordering of a set of objects; formally, a permutation over a set S with n elements can be defined as a mapping of the integers $1, \ldots, n$ to elements of S such that each integer is mapped to a unique element of S.

Perturbative Search: Search paradigm in which candidate solutions are iteratively perturbed by modifying one or more solution components in each search step. Typically, the candidate solutions used in a perturbative search algorithm are complete candidate solutions. Iterative improvement algorithms, such as 2-opt local search for the Travelling Salesman Problem, are typical examples of perturbative search methods.

Pivoting Rule: Rule that defines the mechanism for determining search steps in an iterative improvement algorithm. The most widely used pivoting rules are the so-called *best improvement* and *first improvement* strategies. *Best improvement* always selects the search step that leads to the neighbouring candidate solution with the best evaluation function value. *First improvement* searches the neighbourhood of the current candidate solution s in a given order and selects the search step leading to the first neighbour with a strictly better evaluation function value than s encountered during this process.

Plateau: Maximally connected set of search positions at the same level of a given search landscape; search spaces with extensive plateaus, as can be found for typical instances of the Satisfiability Problem (SAT), can be challenging for stochastic local search (SLS) algorithms, since plateaus are by definition regions in which the algorithm does not have any heuristic guidance. Plateaus with exits are called *open plateaus*, while plateaus without exits are referred to as *closed plateaus*; the latter consists entirely of local minima or strict local minima positions and tend to render problem instances difficult to solve for SLS algorithms.

Plateau Connection Graph (PCG): Representation of the plateau structure of a search landscape; the vertices of a PCG represent plateaus in the given landscape, and the edges correspond to exits connecting the respective plateaus. A PCG forms a complete partition of a given landscape, which can be seen as a refinement of the respective basin partition tree.

Population-Based SLS Method: Class of stochastic local search algorithms that maintain a *population*, that is, a set of candidate solutions of the given problem instance. In each search step, one or more elements of the population (i.e., individual candidate solutions) may be modified. The use of a population of candidate solutions often helps to achieve adequate diversification of the search process. Examples for population-based SLS algorithms include Evolutionary Algorithms, Memetic Algorithms and Ant Colony Optimisation.

Precedence Constraint: Constraint that indicates that a job (or operation) in a scheduling problem has to be executed before (but not necessarily immediately before) another job (or operation). Precedence constraints occur in many real-world scheduling problems.

Probabilistically Approximately Complete (PAC): Property of a stochastic search algorithm; an algorithm is PAC if, and only if, for increasingly long run-times the probability of finding a solution (or optimal solution) of any soluble problem instance gets arbitrarily close to one. Algorithms that are PAC can never get trapped in non-solution regions of the search space. However, they may still require a very long time to escape from such regions.

Probabilistic Domination: Performance relationship between two Las Vegas algorithms; LVA A probabilistically dominates LVA B if, and only if, for any given run-time, A achieves at least as high a success probability as B, and there is at least one run-time value for which the success probability of A exceeds that of B. For optimisation Las Vegas algorithms, probabilistic domination is defined analogously, based on the probability of reaching a given solution quality bound q within the given time, and A probabilistically dominates B if, and only if, it probabilistically dominates B for any given q.

Probabilistic Iterative Improvement (PII): SLS method that probabilistically accepts a neighbouring candidate solution s' based on an acceptance criterion that takes into account the difference in evaluation function value of s' and the current candidate solution, s. Iterative Improvement as well as Randomised Iterative Improvement can be seen as special cases of PII.

Pseudo-Boolean CSP (PB-CSP): Special case of the finite discrete Constraint Satisfaction Problem, also known as *Linear Pseudo-Boolean Programming*. In Pseudo-Boolean CSP, all variables have domains $\{0, 1\}$, and the constraints have the form of linear inequalities, such as $x_1 - 2x_2 + 3x_3 \geq 0$ (or other relations, including equalities and strict inequalities). *Overconstrained Pseudo-Boolean CSP (OPB-CSP)* is an optimisation variant of PB-CSP, in which in addition to a set of constraints as in PB-CSP (*hard constraints*) a set of *soft constraints* of the same general form is given, and the objective is to find an assignment that leaves a minimal number of soft constraints unsatisfied while satisfying all hard constraints. OPB-CSP can be seen as a special case of the 0-1 Integer Linear Programming (ILP) Problem.

Quadratic Assignment Problem (QAP): Combinatorial optimisation problem in which, given a number of *objects* and *locations*, *flow values* between the objects and *distances* between the locations, the goal is to assign all objects to different locations such that the sum of the products of the given flow values between pairs of objects and the distances between the respective locations is minimised. The QAP has various practical applications, such as facility layout and halftone rendering problems. The QAP is $\mathcal{NP}$-hard and is considered to be one of the empirically hardest combinatorial optimisation problems. Stochastic local search algorithms currently represent the only feasible approach for solving large QAP instances.

Qualified Run-Time Distribution (QRTD): Probability distribution of the time required by an optimisation Las Vegas algorithm to reach or exceed a given solution quality bound b for a given problem instance; corresponds precisely to the run-time distribution required by a variant of the same algorithm that terminates as soon as the specified solution quality bound has been reached or exceeded, and hence solves the decision variant for bound b associated with the given problem instance.

Quantile: Statistical measure; given a numerical random variable X, the α-quantile of the respective probability distribution, denoted $q_\alpha(X)$, is defined as the minimal value x' such that $P(X \leq x') \geq \alpha$. The 0.5-quantile is the *median* of the distribution, and the $p/100$-quantiles are also known as the *p-th percentiles*. The medians of empirical distributions are often statistically more stable than the respective means and can be easily and intuitively read off a graphical representation of the cumulative distribution function. *Quantile ratios*, such as $q_{0.9}(X)/q_{0.1}(X)$, are measures of variation which, like the variation coefficient, are invariant w.r.t. multiplicative scaling of the underlying random variable. Quantiles and quantile ratios are often used in the empirical analysis of the performance and behaviour of Las Vegas algorithms.

Quantile-Quantile Plot (QQ Plot): 2-dimensional graphical representation of the correlation between the quantiles of two probability distributions, in which the two coordinates of each data point correspond to the same quantile of two given distributions. Quantile-quantile plots are often used to visually illustrate the differences and similarities between two distributions. They are also useful for informally testing whether a given empirical distribution, such as a run-time distribution or solution quality distribution, can be modelled by a theoretical distribution, such as an exponential or normal distribution. (Formally, this type of hypothesis can be tested using statistical hypothesis tests, such as the *Kolmogorov-Smirnov* or χ^2 *goodness-of-fit tests*.)

Quasi-Optimal Solution: Solution of an optimisation problem that is believed or suspected to be an optimal solution, but whose optimality has not been proven. Quasi-optimal solutions are often obtained by long runs of high-performance stochastic local search (SLS) algorithms. (Such algorithms need to be able to find provably optimal solutions for instances that are small enough to allow the application of state-of-the-art complete search algorithms; ideally, they should also be probabilistically approximately complete.) Often, as substantial improvements are made in state-of-the-art complete algorithms for a problem, quasi-optimal solutions for well-known benchmark instances are verified to be optimal, while sometimes, they turn out to be suboptimal.

Randomised Iterative Improvement (RII): Hybrid SLS method in which with a probability wp, called walk probability or noise parameter, a neighbouring candidate solution is picked at random, while with a probability of $1 - wp$, a standard iterative improvement step is performed. Note that this mechanism allows for arbitrary long sequences of uninformed random walk steps, which renders RII provably probabilistically approximately complete.

Reactive Tabu Search (RTS): Tabu search method that dynamically adapts the tabu tenure during the search based on a limited amount of memory on previously visited search positions. Additionally, an escape mechanism is triggered whenever sufficient evidence for search stagnation has been gathered; this escape mechanism typically consists of a number of uninformed random walk steps that are executed starting from the current candidate solution.

Recombination: Fundamental operation in Evolutionary Algorithms or Memetic Algorithms that generates one or more new candidate solutions (*offspring*) by combining parts (solution components) of two or more existing candidate solutions (*parents*). In many cases, a special type of recombination called crossover is used.

Release Date: Earliest time at which a job in a scheduling problem is available for processing.

Restart Mechanism: Mechanism that restarts a search process from a new initial candidate solution under certain conditions. When using a *static restart mechanism*, re-initialisation occurs regularly after a fixed number of search steps, while *dynamic restart mechanisms* may restart the search if no improvement in the incumbent candidate solution has been achieved for a certain number of steps, or if the search process is trapped in a local minimum of the given evaluation function.

Robust Tabu Search (RoTS): Tabu search method that repeatedly chooses the value of the tabu tenure parameter uniformly at random from a given integer interval during the search; this often results in increased robustness of performance compared to standard Tabu Search, which uses a constant tabu tenure value. Many high-performance RoTS algorithms also use additional diversification mechanisms to prevent or overcome search stagnation.

RTQ (Run-Time Over Solution Quality): Development of a statistical measure of the run-time in dependence of the solution quality achieved by an optimisation Las Vegas algorithm; statistical measures commonly used in this context include the mean, median and quantiles of the qualified run-time distribution for the respective solution quality bound. Like SQTs, RTQs summarise the underlying bivariate run-time distributions and reflect the tradeoff between run-time and solution quality for the given algorithm on a specific problem instance. RTQs are closely related to SQTs but somewhat less intuitive; consequently, they are rarely used in the analysis of the behaviour of optimisation LVAs.

Ruggedness: Property of a search landscape that is closely related to its local minima density; intuitively, rugged landscapes have many local minima, and the respective problem instances tend to be harder to solve for stochastic local search algorithms. Landscape ruggedness is often measured by means of the autocorrelation coefficient or correlation length.

Run-Time Distribution (RTD): Probability distribution of the time required by a Las Vegas algorithm to solve a given problem instance. The RTDs for optimisation Las Vegas algorithms are bivariate distributions of run-time and solution quality. RTDs that are based on measurements of run-time in terms of elementary operations of an algorithm (such as search steps in the case of stochastic local search algorithms) are called *run-length distributions (RLDs)*. RTDs and RLDs play an important role in the empirical analysis of the behaviour and performance of SLS algorithms.

Satisfiability Problem (SAT): Prototypical combinatorial decision problem in which given a propositional formula F, the objective is to decide whether there is an assignment of truth values to the variables in F such that F becomes satisfied. Commonly, SAT is used to refer to a version of the general satisfiability problem that is restricted to CNF formulae. SAT (in general and for CNF formulae) is $\mathcal{NP}$-complete. Stochastic local search algorithms are among the best methods for solving certain types of hard SAT instances.

Scaling: Dependence of a property of an algorithm, such as run-time, on the size of the input, that is, the problem instance to be solved. Often, scaling is analysed for a specific family or distribution of problem instances. The term is also used occasionally to refer to the dependence of certain properties of the instances themselves on instance size. Finally, the term scaling is often applied to the operation of data renormalisation, for example, multiplication with a constant. *Finite size scaling* is a technique originally developed in statistical physics that is used for extrapolating ensemble properties of very large problem instances, such as the occurrence of a solubility phase transition, from empirical data obtained for small instance sizes.

Scaling and Probabilistic Smoothing (SAPS): Dynamic local search method that is closely related to the Exponentiated Subgradient Algorithm (ESG). Different from ESG, SAPS performs penalty smoothing probabilistically, which results in substantial performance improvements. *Reactive SAPS (RSAPS)* adaptively modifies the smoothing probability to achieve increased performance robustness. SAPS and RSAPS have been shown to reach state-of-the-art performance for certain types of instances of the Satisfiability Problem and the Maximum Satisfiability Problem.

Scheduling Problems: Large and important class of combinatorial problems in which given a set of *jobs* that have to be processed by a set of *machines*, the goal is to find an optimal *schedule*, that is, a mapping of jobs to machines and times at which they are processed, under a given *objective function* and subject to *feasibility constraints*. In *single-stage scheduling problems*, each job consists of one atomic operation that is executed by exactly one machine in a single processing stage, while in *multi-stage scheduling problems*, a job can consist of multiple operations that may have to be performed by different machines in multiple stages. Similarly, in a *single-machine scheduling problem*, only one machine is available to process the jobs, while in a *multi-machine scheduling problem*, several machines are available. Scheduling problems arise in many application areas; in many, but not all cases, they are $\mathcal{NP}$-hard. Stochastic local search algorithms are among the best methods for solving hard scheduling problems.

Search Cost Distribution (SCD): Probability distribution of *search cost* across a set of problem instances, where search cost is a measure of the run-time (or other resources) required by a given algorithm for solving a specific problem instance. For stochastic search algorithms, search cost is typically defined as the mean or a quantile of the run-time distribution for a given problem instance. In this case, the run-time distribution characterises the variation of run-time over multiple runs of the algorithm on the same problem instance, while a search cost distribution reflects the variation of run-time over a given set or distribution of problem instances. In the literature, SCDs are sometimes referred to as 'hardness distributions'.

Search Landscape: Mathematical structure comprised of the search space, the neighbourhood relation and an evaluation function for a given problem instance. Characteristics of this landscape, such as its solution density, local minima density or ruggedness, have a crucial impact on the behaviour of stochastic local search algorithms.

Search Position: Alternative term for a candidate solution of a given problem instance; often used in the context of search space analysis.

Search Space: Set of all candidate solutions of a given problem instance. The size of the search space, that is, the number of candidate solutions, typically scales at least exponentially w.r.t. instance size within a given family of problem instances, and it can strongly depend on the representation of candidate solutions. For example, for an instance of the Travelling Salesman Problem with n vertices and $\Theta(n^2)$ edges, candidate solutions may be represented as cyclic permutations of the n vertices or by $\Theta(n^2)$ binary variables, each of which indicates whether or not a specific edge is part of a given candidate solution; these two different representations lead to search spaces of size $\Theta(n!)$ and $\Theta(2^{(n^2)})$, respectively. The term *search space* is occasionally used more broadly to refer to a search landscape.

Search Space Analysis: Investigation of features and characteristics of the search space, or — more generally — the search landscape of a given problem instance. Search space analysis plays an important role in explaining, understanding and improving the behaviour of stochastic local search algorithms.

Search Step: Elementary operation of a stochastic local search algorithm, in which the search moves from a candidate solution s to a candidate solution s' in the direct neighbourhood of s. A search step typically involves the modification, addition or removal of one or more solution components.

Search Trajectory: Finite sequence of search positions (i.e., candidate solutions) as (possibly) visited in successive search steps of a given stochastic local search algorithm.

Selection: Fundamental operation in Evolutionary Algorithms or Memetic Algorithms that is used for selecting the individuals (i.e., candidate solutions) from the current population to be retained for the next iteration of the search process — these surviving individuals form the next generation population. More generally, selection mechanisms may also be used for choosing the individuals that undergo specific operations in a population-based SLS algorithm, such as mutation or recombination in the context of Evolutionary Algorithms, or local search in the context of Memetic Algorithms or Ant Colony Optimisation.

Semi-Log Plot: Graphical representation in which one of the two axes of a coordinate system is shown in logarithmic scale. Semi-log plots with a logarithmic x-axis are particularly useful in the context of scaling and correlation analyses, since the graphs of exponential functions appear as straight lines in such plots; they are also often used when analysing distributions, such as run-time distributions, whose shape in a semi-log plot with logarithmic x-axis appears invariant under multiplication by a constant factor.

Sequencing Problems: Class of scheduling problems in which the goal is to determine an optimal ordering for processing a given set of jobs. Different from a schedule, such an ordering or sequence does not assign time-slots to jobs (or operations); however, in many cases, a schedule can be easily derived from the solution of a sequencing problem. The term is also often used to refer to scheduling problems whose candidate solutions naturally correspond to sequences of jobs or operations. Examples of sequencing problems include the Permutation Flow Shop Scheduling Problem and the Single-Machine Total Weighted Tardiness Scheduling Problem.

Set Covering Problem (SCP): Well-known combinatorial optimisation problem in which, given a set A of objects and a family F of subsets of A, the goal is to find the minimum number of subsets from F such that every element of A occurs in at least one of the chosen subsets, that is, the set A is *covered* by the selected sets. This problem is also known as the *Minimum SCP* or *Unicost SCP*. In the *Weighted SCP*, a weight is associated with each subset in F, and the objective is to find a selection of subsets from F covering A with minimum total weight. The SCP arises in various practical applications, such as crew scheduling in airline, railway and bus companies; stochastic local search algorithms are among the best methods for solving this $\mathcal{NP}$-hard optimisation problem.

Simulated Annealing (SA): Stochastic local search method that is inspired by the physical process of the annealing of solids. In each search step, a neighbour s' of a candidate solution s is generated by a *proposal mechanism*, and an *acceptance criterion* is used to probabilistically decide whether the search will be continued from s or s'; this decision is made based on the evaluation function values of s and s' as well as on the value of a parameter T called *temperature*, which is modified during the search process according to an annealing schedule. Many simulated annealing algorithms use the Metropolis acceptance criterion.

Single-Machine Total Weighted Tardiness Scheduling Problem (SMTWTP): Single-machine scheduling problem in which for each job a processing time, a due date and an importance weight are given; all jobs are available for processing at time zero, and the objective is to find a schedule that minimises the sum of the weighted tardiness values of the jobs. The SMTWTP is an $\mathcal{NP}$-hard optimisation problem. The special case in which all jobs have the same weight is known as the *Single-Machine Total Tardiness Scheduling Problem (SMTTP)*.

Solubility Phase Transition: Transition from one to zero in the probability that a randomly drawn instance from a parameterised family of problem instance distributions is soluble as some parameter is varied. A classical example is

Uniform Random k-SAT for a fixed number of variables and variable number of clauses. In many cases, a solubility phase transition is accompanied by qualitative changes in other properties of the instance distribution, in particular, the expected time for solving a given problem instance.

Soluble: Property of a problem instance; a problem instance is called *soluble* if, and only if, it has at least one solution, and *insoluble* otherwise. The term is typically applied to instances of decision problems; many optimisation problems are defined in such a way that their instances are always soluble in the sense that they have a non-empty set of (feasible) solutions.

Solution: Element of the solution set, S', of a given stochastic local search (SLS) algorithm for a given problem instance. A solution of a combinatorial decision problem is typically defined as a candidate solution that satisfies certain logical conditions. For example, the solutions of an instance of the Satisfiability Problem (SAT) are the models of the given propositional formula. For combinatorial optimisation problems, the solution set can be defined in various ways, for example, it may comprise only candidate solutions with optimal solution quality, or candidate solutions that satisfy certain logical conditions or even all candidate solutions. Note that SLS algorithms for optimisation problems do not necessarily terminate as soon as a solution is found, but they return only candidate solutions that are guaranteed to be an element of the given solution set. Solutions as defined here are often also referred to as *valid* or *feasible candidate solutions*.

Solution Density: Relative frequency of solutions within the search space of a given problem instance; the solution density of a search space of size $\#S$ with k solutions is defined as $k/\#S$. Problem instances with higher solution density tend to be easier to solve for stochastic local search algorithms.

Solution Probability: Probability with which a Las Vegas algorithm finds a solution of a given problem instance within a given time; also often referred to as *success probability*.

Solution Quality: Value of the objective function for a given optimisation problem, typically as applied to a candidate solution. The *relative solution quality* of a candidate solution with solution quality q is defined as $q/q^* - 1$ for minimisation problems if q^*, the quality of an optimal (or quasi-optimal) solution, is not equal to zero, and as $(1 + q)/(1 + q^*)$ otherwise. Relative solution quality is useful in empirically evaluating the performance of optimisation algorithms across sets of instances with different optimal solution qualities; small relative solution quality values are often specified in percent, where a value of 1% is equivalent to a relative solution quality of 0.01. Relative solution quality is often simply referred to as solution quality.

Solution Quality Distribution (SQD): Probability distribution of the solution quality achieved by an optimisation Las Vegas algorithm within a fixed run-time. An *asymptotic SQD* is the solution quality distribution obtained in the limit as run-time approaches infinity; asymptotic SQDs are useful for the analysis of optimisation LVAs that are not probabilistically asymptotically complete, such as simple iterative improvement algorithms or constructive search methods.

SQT (Solution Quality over Time): Development of a statistical measure of the solution quality over the run-time of an optimisation Las Vegas algorithm; statistical measures commonly used in this context include the mean, median and quantiles of the solution quality distribution for the respective run-time. SQTs summarise the underlying bivariate run-time distributions and reflect the tradeoff between run-time and solution quality for the given algorithm on a specific problem instance. When reporting SQTs, care should be taken to reflect the variability of solution quality for each run-time, for example, by showing SQT curves for several quantiles; it is often advisable to additionally study solution quality distributions and qualified run-time distributions, which reflect different aspects of the underlying bivariate RTD.

Stagnation: Situation in which the *efficiency* of a search process decreases, where efficiency is formally defined based on the rate of increase over time in the solution probability of a given Las Vegas algorithm for a given problem instance. In the case of stochastic local search algorithms, stagnation occurs when the search process gets trapped (permanently or for a substantial amount of time) in a non-solution area of the given search space. Stagnation can typically be detected based on the run-time distributions of a given SLS algorithm; it can often be avoided or overcome by means of additional diversification mechanisms.

Statistical Hypothesis Test: Formal method for assessing the validity of statements about relations between or properties of sets of statistical data. For example, the *Mann-Whitney U-test* can be used to determine whether two stochastic local search algorithms applied to the same problem instance show statistically significant differences in median run-time. The statement to be tested (or its negation) is called the *null hypothesis* of the test. The *significance level* determines the maximum allowable probability of incorrectly rejecting the null hypothesis, while the probability of correct acceptance is bounded from below by the *power* of the test; for a given test, the power determines the required *sample size*. When applying statistical tests, often a significance level of 0.05 and a power of at least 0.8 is used. The application of a test to a given data set results in a *p-value*, which represents the probability that the null hypothesis is incorrectly rejected. Statistical hypothesis tests that are useful in the empirical analysis of the performance and behaviour of SLS algorithms include the *Mann-Whitney U-test*, the *binomial sign test*, the *Wilcoxon matched pairs signed-rank test*, *Spearman's rank order test*, the χ^2- and *Kolmogorov-Smirnov goodness-of-fit tests* and the *Shapiro-Wilk normality test*.

Steepest Ascent Mildest Descent (SAMD): An early stochastic local search algorithm for the Maximum Satisfiability Problem that is conceptually closely related to Tabu Search. SAMD is based on a simple iterative best improvement algorithm, but unlike in Simple Tabu Search, only variables flipped in non-improving search steps are declared tabu.

Stochastic Local Search (SLS) Algorithm: Local search algorithm that makes use of randomised choices in generating or selecting candidate solutions for a given instance of a combinatorial problem. SLS algorithms may use random choices for search initialisation and/or the computation of search steps.

Stochastic Search: Search paradigm that makes use of randomised decisions. Stochastic local search algorithms as well as randomised systematic search algorithms fall into this general class of search methods.

Success Probability: Probability with which a Las Vegas algorithm finds a solution of a given problem instance within a given time; also often referred to as *solution probability*.

Systematic Search: Search paradigm under which the entire search space of a problem instance is traversed (or pruned) in a systematic manner, which renders the respective algorithms complete (unless the search process is terminated prematurely). Many systematic search algorithms perform some form of *tree search*, where each node n_s in the search tree corresponds to a partial candidate solution s, and the children of n_s represent all (feasible) partial candidate solutions that can be obtained by adding one solution component to s. When performed in a depth-first manner, this type of search can be seen as a constructive search process with backtracking.

Tabu Search: SLS method that strongly exploits memory of the search history for guiding the search process. The most basic form of tabu search, *Simple Tabu Search*, is obtained by enhancing an iterative improvement algorithm with a form of short-term memory that enables it to escape from local optima. In Simple Tabu Search, typically the solution components modified in a search step are declared *tabu* for a fixed number of subsequent search steps (*tabu tenure*); during their tabu tenure, solution components may not be added to or removed from the current candidate solution. An aspiration criterion can be used to override the tabu status of solution components. Simple tabu search algorithms are often enhanced by different types of intensification and diversification strategies. Tabu search algorithms are among the best-performing stochastic local search algorithms for many combinatorial problems.

Tabu Tenure: Important parameter of tabu search algorithms; the tabu tenure corresponds to the number of search steps for which the tabu status of a solution component c is maintained after c has been declared tabu and thereby determines the amount of time for which c cannot be modified.

Tardiness: Amount of time by which a job in a scheduling problem is completed after its due date. In practical applications, tardiness often incurs additional costs, such as contractual penalties due to late deliveries.

Termination Condition: Component of a stochastic local search algorithm that determines when the search is ended; formally specified in the form of a predicate of the current candidate solution as well as partial historical or statistical information on the search trajectory. Typical termination conditions are satisfied if a solution of a decision problem is reached or a given solution quality of an optimisation problem has been achieved, if a certain bound on run-time or number of search steps has been exceeded, or if no improvement in evaluation function value has been observed for a certain amount of time.

TMCH: Stochastic local search algorithm for the Constraint Satisfaction Problem (CSP); a simple tabu search algorithm based on the Min-Conflicts Heuristic, in which after each search step, the respective variable/value pair is declared tabu for a fixed number of steps.

Transition Action: Action that is performed when a specific transition between two states of a Generalised Local Search Machine (GLSM) occurs. Transition actions may be used for modifying global parameters, such as the temperature parameter in simulated annealing algorithms, for communication between individual GLSMs in co-operative GLSM models or for input/output functionality.

Transpose Neighbourhood: Neighbourhood relation on permutations (or orderings) of solution components under which two permutations are direct neighbours if, and only if, one can be obtained from the other by interchanging two elements at adjacent positions. This neighbourhood relation is sometimes used in scheduling problems; however, local search algorithms based on the transpose neighbourhood perform often substantially worse than those based on the insertion or interchange neighbourhood.

Travelling Salesman Problem (TSP): Also known as *Travelling Salesperson Problem*; a prototypical combinatorial optimisation problem in which given an edge-weighted graph G, the objective is to find a cyclic path (also called *tour*) that visits every node in G exactly once and whose total edge-weight is minimal. Often, the TSP is restricted to the special case where the vertices in G (which are also called *cities*) correspond to points in the Euclidean plane, G is fully connected, and the weight of any edge is defined as the Euclidean distance between the two points that correspond to its incident vertices (*Euclidean TSP*). The TSP (in general and in the Euclidean case) is $\mathcal{NP}$-hard. Stochastic local search algorithms are amongst the best known methods for solving hard and large TSP instances.

TS-GH: State-of-the-art tabu search algorithm for the finite discrete Constraint Satisfaction and Maximum Constraint Satisfaction Problems; TS-GH is based on an iterative best improvement procedure that in each step changes the value assigned to a variable appearing in a currently unsatisfied constraint in order to minimise the number (or total weight) of unsatisfied constraints.

TS-NS-JSP: State-of-the-art tabu search algorithm for the Job Shop Scheduling Problem. TS-NS-JSP uses a strongly reduced neighbourhood relation; conceptually, this algorithm is quite similar to TS-NS-PFSP.

TS-NS-PFSP: State-of-the-art tabu search algorithm for the Permutation Flow Shop Scheduling Problem (PFSP), which strongly exploits neighbourhood pruning for the PFSP and occasionally restarts the local search process at the incumbent candidate solution.

TSPLIB: A library of widely used benchmark instances for the Travelling Salesman Problem (TSP) and several related problems. As of September 2003, the largest TSP benchmark instance from TSPLIB solved to optimality has 15 112 vertices.

Uniform Random k-SAT: Prominent class of benchmark instances for the Satisfiability Problem (SAT); every clause in a Uniform Random k-SAT instance consists of exactly k literals that are chosen uniformly at random from a given set of n propositional variables and their negations, where no clause may contain multiple copies of the same literal or a variable and its negation. The mean empirical hardness of Uniform Random k-SAT instances strongly depends on the ratio between the number of clauses and variables.

Uninformed Random Picking: Simple stochastic search method that repeatedly samples the given search space by selecting a candidate solution uniformly at random. Many stochastic local search algorithms use Uninformed Random Picking for search initialisation.

Uninformed Random Walk: Simple SLS method that in each search step selects a candidate solution s' uniformly at random from the neighbourhood of the current candidate solution s. Because it lacks heuristic guidance, Uninformed Random Walk alone performs very poorly, but this simple SLS method and its variants are not only of considerable theoretical interest, but are also sometimes used as diversification mechanisms in high-performance stochastic local search algorithms.

Unit Propagation: Polynomial preprocessing method for the Satisfiability Problem (SAT); eliminates *unit clauses*, that is, clauses that contain only one literal, from a given CNF formula and propagates the effects to all other clauses containing the same variable. Unit propagation plays a crucial role in high-performance systematic search algorithms for SAT.

Valid Candidate Solution: Alternative term for solution, that is, for a candidate solution that is an element of the solution set of a given stochastic local search algorithm and problem instance.

Variable Depth Search (VDS): Iterative improvement method that is based on *complex search steps*, which are obtained by concatenating a variable number of *simple search steps*. VDS forms the basis of many of the best-performing iterative improvement algorithms, such as the well-known Lin-Kernighan Algorithm for the Travelling Salesman Problem.

Variation Coefficient: Statistical measure of the variation of a given probability distribution. The variation coefficient vc is defined as the ratio of the standard deviation and the mean of the given distribution. Compared to the standard deviation, the variation coefficient vc has the advantage that it is invariant w.r.t. multiplicative scaling of the underlying distribution.

Variable Neighbourhood Search (VNS): Stochastic local search method based on the general idea of changing the neighbourhood relation during search. Examples of VNS algorithms include *Variable Neighbourhood Descent (VND)*, an iterative improvement method that systematically switches between different neighbourhood relations, and *Basic VNS*, which can be seen as an iterated local search method that systematically modifies the strength of its perturbation procedure.

Walk Probability: Important parameter of randomised iterative improvement (RII) algorithms; the walk probability determines the probability with which at any given step of an RII algorithm a random walk step is performed. This parameter also appears in various high-performance hybrid stochastic local search algorithms for the Satisfiability Problem and the finite discrete Constraint Satisfaction Problem.

WalkSAT: Prominent family of high-performance stochastic local search algorithms for the Satisfiability Problem (for CNF formulae). WalkSAT is based on the 1-flip neighbourhood; in each search step, first a currently unsatisfied clause c is selected uniformly at random, and then a heuristically chosen variable from c is flipped, rendering c satisfied. WalkSAT algorithms differ in the selection mechanisms used for choosing the variable from c to be flipped. The term is also used to refer to one specific WalkSAT algorithm, *WalkSAT/SKC*, which is based on a randomised best improvement variable selection mechanism.

WMCH: Well-known stochastic local search algorithm for the finite discrete Constraint Satisfaction Problem; a variant of the Min-Conflicts Heuristic (MCH) that can be seen as a randomised iterative improvement method, which in each search step first selects a variable that appears in a currently unsatisfied constraint, and then with a certain probability sets this variable to a randomly chosen value, and otherwise chooses a value that minimises the number of violated constraints (as in MCH).

BIBLIOGRAPHY

A [Aardal et al., 2003] K. I. Aardal, C. P. M. van Hoesel, A. M. C. A. Koster, C. Mannino, and A. Sassano. Models and solution techniques for the frequency assignment problem. *4OR*, 1(4):261–317, 2003.

[Aarts and Korst, 1989] E. H. L. Aarts and J. H. M. Korst. *Simulated Annealing and Boltzman Machines – A Stochastic Approach to Combinatorial Optimization and Neural Computation*. John Wiley & Sons, Chichester, UK, 1989.

[Aarts and Lenstra, 1997] E. H. L. Aarts and J. K. Lenstra, editors. *Local Search in Combinatorial Optimization*. John Wiley & Sons, Chichester, UK, 1997.

[Aarts et al., 1997] E. H. L. Aarts, J. H. M. Korst, and P. J. M. van Laarhoven. Simulated annealing. In E. H. L. Aarts and J. K. Lenstra, editors, *Local Search in Combinatorial Optimization*, pages 91–120. John Wiley & Sons, Chichester, UK, 1997.

[Abramson and Randall, 1999] D. Abramson and M. Randall. A simulated annealing code for general integer linear programs. *Annals of Operations Research*, 86:3–21, 1999.

[Abramson et al., 1996] D. Abramson, H. Dang, and M. Krishnamoorthy. A comparison of two methods for solving 0–1 integer programs using a general purpose simulated annealing algorithm. *Annals of Operations Research*, 63:129–150, 1996.

[Achugbue and Chin, 1982] J. O. Achugbue and F. Y. Chin. Scheduling the open shop to minimize mean flow time. *SIAM Journal on Computing*, 11(4):709–720, 1982.

[Adams et al., 1988] J. Adams, E. Balas, and D. Zawack. The shifting bottleneck procedure for job shop scheduling. *Management Science*, 34(3):391–401, 1988.

[Adleman, 1994] L. M. Adleman. Molecular computation of solutions to combinatorial problems. *Science*, 266:1021–1024, 1994.

[Ahuja and Orlin, 1996] R. K. Ahuja and J. B. Orlin. Use of representative operation counts in computational testing of algorithms. *INFORMS Journal on Computing*, 8(3):318–330, 1996.

[Ahuja et al., 2002] R. K. Ahuja, O. Ergun, J. B. Orlin, and A. P. Punnen. A survey of very large-scale neighborhood search techniques. *Discrete Applied Mathematics*, 123(1–3):75–102, 2002.

[Aiex et al., 2002] R. M. Aiex, M. G. C. Resende, and C. C. Ribeiro. Probability distribution of solution time in GRASP: An experimental investigation. *Journal of Heuristics*, 8(3):343–373, 2002.

[Alimonti, 1994] P. Alimonti. New local search approximation techniques for maximum generalized satisfiability problems. In M. A. Bonuccelli, P. Crescenzi, and R. Petreschi, editors, *Algorithms and Complexity, Second Italian Conference, CIAC '94*, volume 778 of *Lecture Notes in Computer Science*, pages 40–53. Springer Verlag, Berlin, Germany, 1994.

[Alimonti, 1996] P. Alimonti. New local search approximation techniques for maximum generalized satisfiability problems. *Information Processing Letters*, 57(3):151–158, 1996.

[Alsinet et al., 2003] T. Alsinet, F. Manya, and J. Planes. Improved branch and bound algorithms for Max-SAT. In *Proceedings of the Sixth International Conference on Theory and Applications of Satisfiability Testing (SAT 2003)*, pages 408–415, Santa Margherita Ligure, Italy, 2003.

[Altenberg, 1997a] L. Altenberg. Fitness distance correlation analysis: An instructive counterexample. In T. Bäck, editor, *Proceedings of the Seventh International Conference on Genetic Algorithms*, pages 57–64. Morgan Kaufmann Publishers, San Mateo, CA, USA, 1997.

[Altenberg, 1997b] L. Altenberg. NK fitness landscapes. In T. Bäck, D. Fogel, and Z. Michalewicz, editors, *Handbook of Evolutionary Computation*. Oxford University Press, New York, NY, USA, 1997.

[Alur et al., 1993] R. Alur, C. Courcoubetis, T. A. Henzinger, and P.-H. Ho. Hybrid automata: An algorithmic approach to the specification and verification of hybrid systems. In R. L. Grossman, A. Nerode, A. P. Ravn, and H. Rischel, editors, *Hybrid Systems*, volume 736 of *Lecture Notes in Computer Science*, pages 209–229. Springer Verlag, Berlin, Germany, 1993.

[Anderson et al., 1997] E. J. Anderson, C. A. Glass, and C. N. Potts. Machine scheduling. In E. H. L. Aarts and J. K. Lenstra, editors, *Local Search in Combinatorial Optimization*, pages 361–414. John Wiley & Sons, Chichester, UK, 1997.

[Andersson et al., 2000] A. Andersson, M. Tenhunen, and F. Ygge. Integer programming for combinatorial auction winner determination. In *Fourth International Conference on Multi-Agent Systems*, pages 39–46. IEEE Computer Society Press, Los Alamitos, CA, USA, 2000.

[Angel and Zissimopoulos, 1998] E. Angel and V. Zissimopoulos. Autocorrelation coefficient for the graph bipartitioning problem. *Theoretical Computer Science*, 191(1–2): 229–243, 1998.

[Angel and Zissimopoulos, 2000] E. Angel and V. Zissimopoulos. On the classification of NP-complete problems in terms of their correlation coefficient. *Discrete Applied Mathematics*, 99(1–3):261–277, 2000.

[Angel and Zissimopoulos, 2001] E. Angel and V. Zissimopoulos. On the landscape ruggedness of the quadratic assignment problem. *Theoretical Computer Science*, 263(1–2):159–172, 2001.

[Anstreicher et al., 2002] K. M. Anstreicher, N. W. Brixius, J.-P. Goux, and J. Linderoth. Solving large quadratic assignment problems on computational grids. *Mathematical Programming*, 91(3):563–588, 2002.

[Applegate and Cook, 1991] D. Applegate and W. Cook. A computational study of the job-shop scheduling problem. *ORSA Journal on Computing*, 3(2):149–156, 1991.

[Applegate et al., 1998] D. Applegate, R. Bixby, V. Chvátal, and W. Cook. On the solution of traveling salesman problems. *Documenta Mathematica*, Extra Volume ICM III:645–656, 1998.

[Applegate et al., 1999] D. Applegate, R. Bixby, V. Chvátal, and W. Cook. Finding tours in the TSP. Technical Report 99885, Forschungsinstitut für Diskrete Mathematik, University of Bonn, Germany, 1999.

[Applegate et al., 2003a] D. Applegate, R. Bixby, V. Chvátal, and W. Cook. Implementing the Dantzig-Fulkerson-Johnson algorithm for large traveling salesman problems. *Mathematical Programming Series B*, 97(1–2):91–153, 2003.

[Applegate et al., 2003b] D. Applegate, R. Bixby, V. Chvátal, and W. Cook. Solving traveling salesman problems. http://www.math.princeton.edu/tsp, 2003. Version visited last on 15 September 2003.

[Applegate et al., 2003c] D. Applegate, W. Cook, and A. Rohe. Chained Lin-Kernighan for large traveling salesman problems. *INFORMS Journal on Computing*, 15(1):82–92, 2003.

[Arkin and Hassin, 1997] E. M. Arkin and R. Hassin. On local search for weighted k-set packing. In R. E. Burkard and G. J. Woeginger, editors, *Algorithms – ESA '97, 5th Annual European Symposium*, volume 1284 of *Lecture Notes in Computer Science*, pages 13–22. Springer Verlag, Berlin, Germany, 1997.

[Arora, 1998] S. Arora. Polynomial time approximation schemes for Euclidean traveling salesman and other geometric problems. *Journal of the ACM*, 45(5):753–782, 1998.

[Asahiro et al., 1996] Y. Asahiro, K. Iwama, and E. Miyano. Random generation of test instances with controlled attributes. In D. S. Johnson and M. A. Trick, editors, *Cliques, Coloring, and Satisfiability: The Second DIMACS Implementation Challenge*, volume 26 of *DIMACS Series on Discrete Mathematics and Theoretical Computer Science*, pages 377–393. American Mathematical Society, Providence, RI, USA, 1996.

[Asano and Williamson, 2000] T. Asano and D. P. Williamson. Improved approximation algorithms for MAX SAT. In D. B. Shmoys, editor, *Proceedings of the Eleventh Annual ACM-SIAM Symposium on Discrete Algorithms*, pages 96–105. Society for Industrial and Applied Mathematics, Philadelphia, PA, USA, 2000.

[Asano, 1997] T. Asano. Approximation algorithms for MAX SAT: Yannakakis vs. Goemans-Williamson. In Y. Mansour and B. Schieber, editors, *Proceedings of the Fifth Israel Symposium on Theory of Computing and Systems*, pages 24–37. IEEE Computer Society Press, Los Alamitos, CA, USA, 1997.

[Ausiello et al. , 1999] G. Ausiello, P. Crescenzi, G. Gambosi, V. Kann, A. Marchetti-Spaccamela, and M. Protasi. *Complexity and Approximation: Combinatorial Optimization Problems and Their Approximability Properties*. Springer Verlag, Berlin, Germany, 1999.

[Avanthay et al., 2003] C. Avanthay, A. Hertz, and N. Zufferey. A variable neighborhood search for graph coloring. *European Journal of Operational Research*, 151(2):379–388, 2003.

B [Babai, 1979] L. Babai. Monte Carlo algorithms in graph isomorphism testing. Technical Report DMS 79-10, Université de Montréal, Montréal, Canada, 1979.

[Bacchus et al., 2002] F. Bacchus, X. Chen, P. van Beek, and T. Walsh. Binary vs non-binary constraints. *Artificial Intelligence*, 140(1–2):1–37, 2002.

[Bäck, 1996] T. Bäck. *Evolutionary Algorithms in Theory and Practice*. Oxford University Press, New York, NJ, USA, 1996.

[Bailey et al., 2001] D. D. Bailey, V. Dalmau, and P. G. Kolaitis. Phase transitions of PP-complete satisfiability problems. In B. Nebel, editor, *Proceedings of the Seventeenth*

International Joint Conference on Artificial Intelligence, pages 183–192. Morgan Kaufmann Publishers, San Francisco, CA, USA, 2001.

[Baker and Bertrand, 1982] K. R. Baker and J. W. M. Bertrand. A dynamic priority rule for scheduling against due-dates. *Journal of Operations Management*, 1(1):37–42, 1982.

[Balas and Carrera, 1996] E. Balas and M. C. Carrera. A dynamic subgradient-based branch and bound procedure for set covering. *Operations Research*, 44(6):875–890, 1996.

[Balas and Ho, 1980] E. Balas and A. Ho. Set covering algorithms using cutting planes, heuristics, and subgradient optimization: A computational study. *Mathematical Programming Study*, 12:37–60, 1980.

[Balas and Padberg, 1976] E. Balas and M. W. Padberg. Set partitioning: A survey. *SIAM Review*, 18:710–760, 1976.

[Balas and Simonetti, 2001] E. Balas and N. Simonetti. Linear time dynamic programming algorithms for new classes of restricted TSPs: A computational study. *INFORMS Journal on Computing*, 13(1):56–75, 2001.

[Balas and Vazacopoulos, 1998] E. Balas and A. Vazacopoulos. Guided local search with shifting bottleneck for job shop scheduling. *Management Science*, 44(2):262–275, 1998.

[Baluja and Caruana, 1995] S. Baluja and R. Caruana. Removing the genetics from the standard genetic algorithm. In A. Prieditis and S. Russell, editors, *Proceedings of the Twelfth International Conference on Machine Learning (ML-95)*, pages 38–46. Morgan Kaufmann Publishers, Palo Alto, CA, USA, 1995.

[Barr et al., 1995] R. S. Barr, B. L. Golden, J. P. Kelly, M. G. C. Resende, and W. R. Stewart. Designing and reporting on computational experiments with heuristic methods. *Journal of Heuristics*, 1(1):9–32, 1995.

[Battiti and Protasi, 1997a] R. Battiti and M. Protasi. Reactive search, a history-based heuristic for MAX-SAT. *ACM Journal of Experimental Algorithmics*, 2, 1997.

[Battiti and Protasi, 1997b] R. Battiti and M. Protasi. Solving MAX-SAT with non-oblivious functions and history-based heuristics. In D. Du, J. Gu, and P. M. Pardalos, editors, *Satisfiability problem: Theory and Applications*, volume 35 of *DIMACS Series on Discrete Mathematics and Theoretical Computer Science*, pages 649–667. American Mathematical Society, Providence, RI, USA, 1997.

[Battiti and Protasi, 1998] R. Battiti and M. Protasi. Approximate algorithms and heuristics for MAX-SAT. In D.-Z. Du and P. M. Pardalos, editors, Handbook of Combinatorial Optimization, volume 1, pages 77–148. Kluwer Academic Publishers, Dordrecht, The Netherlands, 1998.

[Battiti and Protasi, 1999] R. Battiti and M. Protasi. Reactive local search techniques for the maximum k-conjunctive constraint satisfaction problem. *Discrete Applied Mathematics*, 96–97:3–27, 1999.

[Battiti and Protasi, 2001] R. Battiti and M. Protasi. Reactive local search for the maximum clique problem. *Algorithmica*, 29(4):610–637, 2001.

[Battiti and Tecchiolli, 1992] R. Battiti and G. Tecchiolli. Parallel biased search for combinatorial optimization: Genetic algorithms and TABU. *Microprocessors and Microsystems*, 16(7):351–367, 1992.

[Battiti and Tecchiolli, 1994] R. Battiti and G. Tecchiolli. The reactive tabu search. *ORSA Journal on Computing*, 6(2):126–140, 1994.

[Bauer et al., 1999] A. Bauer, B. Bullnheimer, R. F. Hartl, and C. Strauss. An ant colony optimization approach for the single machine total tardiness problem. In P. J. Angeline, Z. Michalewicz, M. Schoenauer, X. Yao, and A. Zalzala, editors, *Proceedings of the 1999 Congress on Evolutionary Computation (CEC'99)*, pages 1445–1450. IEEE Press, Piscataway, NJ, USA, 1999.

[Baum, 1986a] E. B. Baum. Iterated descent: A better algorithm for local search in combinatorial optimization problems. Manuscript, 1986.

[Baum, 1986b] E. B. Baum. Towards Practical "Neural" Computation for Combinatorial Optimization Problems. In *Neural Networks for Computing, AIP Conference Proceedings*, pages 53–64, 1986.

[Baxter, 1981] J. Baxter. Local optima avoidance in depot location. *Journal of the Operational Research Society*, 32(9):815–819, 1981.

[Bayardo Jr. and Pehoushek, 2000] R. J. Bayardo Jr. and J. D. Pehoushek. Counting models using connected components. In *Proceedings of the Seventeenth National Conference on Artificial Intelligence*, pages 157–162. AAAI Press/The MIT Press, Menlo Park, CA, USA, 2000.

[Beardwood et al., 1959] J. Beardwood, J. H. Halton, and J. M. Hammersley. The shortest path through many points. *Proceedings of the Cambridge Philosophical Society*, 55:299–327, 1959.

[Beasley and Chu, 1996] J. E. Beasley and P. C. Chu. A genetic algorithm for the set covering problem. *European Journal of Operational Research*, 94(2):392–404, 1996.

[Beasley, 1990] J. E. Beasley. A Lagrangian heuristic for set covering problems. *Naval Research Logistics*, 37(1):151–164, 1990.

[Beasley, 2003] J. E. Beasley. OR-Library. `http://mscmga.ms.ic.ac.uk/info.html`, 2003. Version visited last on 15 September 2003.

[Béjar and Manyà, 1999] R. Béjar and F. Manyà. Solving combinatorial problems with regular local search algorithms. In H. Ganzinger and D. McAllester, editors, *Proceedings of the 6th International Conference on Logic for Programming and Automated Reasoning*, volume 1705 of *Lecture Notes in Artificial Intelligence*, pages 33–43. Springer Verlag, Berlin, Germany, 1999.

[Bentley, 1992] J. L. Bentley. Fast algorithms for geometric traveling salesman problems. *ORSA Journal on Computing*, 4(4):387–411, 1992.

[Beringer et al., 1994] A. Beringer, G. Aschemann, H. H. Hoos, M. Metzger, and A. Weiß. GSAT versus simulated annealing. In A. G. Cohn, editor, *Proceedings of the 11th European Conference on Artificial Intelligence*, pages 130–134. John Wiley & Sons, Chichester, UK, 1994.

[Bertoni et al., 2000] A. Bertoni, P. Campadelli, M. Carpentieri, and G. Grossi. A genetic model: Analysis and application to MAXSAT. *Evolutionary Computation*, 8(3):291–309, 2000.

[Bertsekas, 1995] D. Bertsekas. *Dynamic Programming and Optimal Control*. Athena Scientific, Belmont, MA, USA, 1995.

[Bessière et al., 1999] C. Bessière, E. C. Freuder, and J. C. Régin. Using constraint meta-knowledge to reduce arc consistency computation. *Artificial Intelligence*, 107(1):125–148, 1999.

[Biere et al., 1999a] A. Biere, A. Cimatti, E. M. Clarke, and Y. Zhu. Symbolic model checking without BDDs. In R. Cleaveland, editor, *Tools and Algorithms for Construction and Analysis of Systems (TACAS '99)*, volume 1579 of *Lecture Notes in Computer Science*, pages 193–207. Springer Verlag, Berlin, Germany, 1999.

[Biere et al., 1999b] A. Biere, A. Cimatti, E. M. Clarke, M. Fujita, and Y. Zhu. Symbolic model checking using SAT procedures instead of BDDs. In M. J. Irwin, editor, *Proceedings of the 36th ACM/IEEE Conference on Design Automation*, pages 317–320. ACM press, New York, NY, USA, 1999.

[Bierwirth and Stöppler, 1992] C. Bierwirth and S. Stöppler. The application of a parallel genetic algorithm to the $n/m/P/C_{\max}$ flowshop problem. In G. Fandel, T. Gulledge, and A. Jones, editors, *New Directions for Operations Research in Manufacturing*, pages 161–179. Springer Verlag, Berlin, Germany, 1992.

[Biggs, 1993] N. Biggs. *Algebraic Graph Theory*. Cambridge University Press, Cambridge, UK, 1993.

[Birattari et al., 2002] M. Birattari, T. Stützle, L. Paquete, and K. Varrentrapp. A racing algorithm for configuring metaheuristics. In W. B. Langdon et al., editors, *Proceedings of the Genetic and Evolutionary Computation Conference (GECCO-2002)*, pages 11–18. Morgan Kaufmann Publishers, San Francisco, CA, USA, 2002.

[Birnbaum and Lozinskii, 1999] E. Birnbaum and E. L. Lozinskii. The good old Davis-Putnam procedure helps counting models. *Journal of Artificial Intelligence Research*, 10:457–477, 1999.

[Bistarelli et al., 1997] S. Bistarelli, U. Montanari, and F. Rossi. Semiring-based constraint solving and optimization. *Journal of the ACM*, 44(2):201–236, 1997.

[Blazewicz et al., 1993] J. Blazewicz, K. Ecker, G. Schmidt, and J. Weglarz. *Scheduling in Computer and Manufacturing Systems*. Springer Verlag, Berlin, Germany, 1993.

[Blum, 2003a] C. Blum. An ACO algorithm to tackle shop scheduling problems — software and results. `http://iridia.ulb.ac.be/~cblum/gss`, 2003. Version visited last on 15 September 2003.

[Blum, 2003b] C. Blum. An ant colony optimization algorithm to tackle shop scheduling problems. Technical Report TR/IRIDIA/2003-1, IRIDIA, Université Libre de Bruxelles, Belgium, 2003.

[Blum, to appear] C. Blum. Beam-ACO—hybridizing ant colony optimization with beam search: An application to open shop scheduling. *Computers & Operations Research*, to appear.

[Boese et al., 1994] K. D. Boese, A. B. Kahng, and S. Muddu. A new adaptive multi-start technique for combinatorial global optimization. *Operations Research Letters*, 16(2):101–113, 1994.

[Boese, 1996] K. D. Boese. *Models for Iterative Global Optimization*. PhD thesis, University of California, Computer Science Department, Los Angeles, CA, USA, 1996.

[Boettcher and Percus, 2000] S. Boettcher and A. G. Percus. Nature's way of optimizing. *Artificial Intelligence*, 119(1–2):275–286, 2000.

[Boettcher, 2000] S. Boettcher. Extremal optimization: Heuristics via co-evolutionary avalanches. *Computing in Science and Engineering*, 2(6):75–82, 2000.

[Bonabeau et al., 1999] E. Bonabeau, M. Dorigo, and G. Theraulaz. *Swarm Intelligence: From Natural to Artificial Systems*. Oxford University Press, New York, NY, USA, 1999.

[Borchers and Furman, 1999] B. Borchers and J. Furman. A two-phase exact algorithm for MAX-SAT and weighted MAX-SAT problems. *Journal of Combinatorial Optimization*, 2(4):299–306, 1999.

[Borůvka, 1926] O. Borůvka. On a certain minimal problem (in Czech). *Práce Moravské Přírodovědecké Společnosti*, 3:37–58, 1926. English translation available at http://citeseer.nj.nec.com/nesetril00otakar.html.

[Boutilier and Hoos, 2001] C. Boutilier and H. H. Hoos. Bidding languages for combinatorial auctions. In B. Nebel, editor, *Proceedings of the Seventeenth International Joint Conference on Artificial Intelligence*, pages 1211–1217. Morgan Kaufmann Publishers, San Francisco, CA, USA, 2001.

[Boutilier et al., 1999] C. Boutilier, M. Goldszmidt, and B. Sabata. Sequential auctions for allocation of resources with complementarities. In T. Dean, editor, *Proceedings of the Sixteenth International Joint Conference on Artificial Intelligence*, pages 527–534. Morgan Kaufmann Publishers, San Francisco, CA, USA, 1999.

[Box and Jenkins, 1970] G. E. P. Box and G. M. Jenkins. *Time Series Analysis, Forecasting and Control*. Prentice Hall, Englewood Cliffs, NJ, USA, 1970.

[Box and Muller, 1958] G. E. P. Box and M. E. Muller. A note on the generation of random normal deviates. *Annals of Mathematical Statistics*, 29:610–611, 1958.

[Boyan and Moore, 1998] J. A. Boyan and A. W. Moore. Learning evaluation functions for global optimization and Boolean satisfiability. In *Proceedings of the Fifteenth National Conference on Artificial Intelligence*, pages 3–10. AAAI Press/The MIT Press, Menlo Park, CA, USA, 1998.

[Brady, 1985] R. M. Brady. Optimization strategies gleaned from biological evolution. *Nature*, 317:804–806, 1985.

[Brafman and Hoos, 1999] R. I. Brafman and H. H. Hoos. To encode or not to encode – I: Linear planning. In T. Dean, editor, *Proceedings of the Sixteenth International Joint Conference on Artificial Intelligence*, pages 988–993. Morgan Kaufmann Publishers, San Francisco, CA, USA, 1999.

[Bray and Moore, 1980] A. J. Bray and M. A. Moore. Metastable states in spin glasses. *Journal of Physics C*, 13(19):L469–L476, 1980.

[Brélaz, 1979] D. Brélaz. New methods to color the vertices of a graph. *Communications of the ACM*, 22(4):251–256, 1979.

[Brenneman and Condon, 2002] A. Brenneman and A. Condon. Strand design for bio-molecular computation. Theoretical Computer Science, 287(1):39–58, 2002.

[Brenner and Lerner, 1992] S. Brenner and R. A. Lerner. Encoded combinatorial chemistry. *Proceedings of the National Academy of Sciences of the United States of America*, 89(12):5381–5383, 1992.

[Bresina, 1996] J. L. Bresina. Heuristic-biased stochastic sampling. In *Proceedings of the Thirteenth National Conference on Artificial Intelligence*, pages 271–278. AAAI Press/The MIT Press, Menlo Park, CA, USA, 1996.

[Briggs et al., 1994] P. Briggs, K. D. Cooper, and L. Torczon. Improvements to graph coloring register allocation. *ACM Transactions on Programming Languages and Systems*, 16(3):428–455, 1994.

[Brixius and Anstreicher, 2001] N. W. Brixius and K. M. Anstreicher. The Steinberg wiring problem. Technical report, College of Business Administration, University of Iowa, Iowa City, USA, October 2001.

[Brucker and Knust, 2003] P. Brucker and S. Knust. Complexity results for scheduling problems. `http://www.mathematik.uni-osnabrueck.de/research/OR/class`, 2003. Version visited last on 15 September 2003.

[Brucker et al., 1997] P. Brucker, J. Hurink, B. Jurisch, and B. Wöstmann. A branch & bound algorithm for the open-shop problem. *Discrete Applied Mathematics*, 76(1–3): 43–59, 1997.

[Brucker, 1998] P. Brucker. *Scheduling Algorithms*. Springer Verlag, Heidelberg, Germany, 1998.

[Brusco et al., 1999] M. J. Brusco, L. W. Jacobs, and G. M. Thompson. A morphing procedure to supplement a simulated annealing heuristic for cost- and coverage-correlated set covering problems. *Annals of Operations Research*, 86:611–627, 1999.

[Bryant, 1986] R. E. Bryant. Graph-based algorithms for Boolean function manipulation. *IEEE Transactions on Computers*, C-35(8):677–691, 1986.

[Bullnheimer et al., 1999] B. Bullnheimer, R. F. Hartl, and C. Strauss. A new rank-based version of the Ant System: A computational study. *Central European Journal for Operations Research*, 7(1):25–38, 1999.

[Buriol et al., 2004] L. S. Buriol, P. M. França, and P. Moscato. A new memetic algorithm for the asymmetric traveling salesman problem. Submitted for publication, 2004.

[Burkard and Çela, 1995] R. E. Burkard and E. Çela. Heuristics for biquadratic assignment problems and their computational comparison. *European Journal of Operational Research*, 83(2):283–300, 1995.

[Burkard and Offermann, 1977] R. E. Burkard and J. Offermann. Entwurf von Schreibmaschinentastaturen mittels quadratischer Zuordnungsprobleme. *Zeitschrift für Operations Research, Serie B: Praxis*, 21:B121–B132, 1977.

[Burkard and Rendl, 1984] R. E. Burkard and F. Rendl. A thermodynamically motivated simulation procedure for combinatorial optimization problems. *European Journal of Operational Research*, 17(2):169–174, 1984.

[Burkard et al., 1994] R. E. Burkard, E. Çela, and B. Klinz. On the biquadratic assignment problem. In P. M. Pardalos and H. Wolkowicz, editors, *Quadratic Assignment and Related Problems*, volume 16 of *DIMACS Series on Discrete Mathematics and Theoretical Computer Science*, pages 117–146. American Mathematical Society, Providence, RI, USA, 1994.

[Burkard et al., 1997] R. E. Burkard, S. E. Karisch, and F. Rendl. QAPLIB – a quadratic assignment problem library. *Journal of Global Optimization*, 10(4):391–403, 1997.

[Burkard et al., 1998a] R. E. Burkard, E. Çela, P. M. Pardalos, and L. S. Pitsoulis. The quadratic assignment problem. In P. M. Pardalos and D.-Z. Du, editors, *Handbook of Combinatorial Optimization*, volume 2, pages 241–338. Kluwer Academic Publishers, Dordrecht, The Netherlands, 1998.

[Burkard et al., 1998b] R. E. Burkard, V. G. Deineko, R. van Dal, J. A. A. van der Veen, and G. J. Woeginger. Well-solvable special cases of the traveling salesman problem: A survey. *SIAM Review*, 40(3):496–546, 1998.

[Burke et al., 1996] E. K. Burke, D. Elliman, P. Ford, and R.Weare. Examination timetabling in British universities — a survey. In E. K. Burke and P. Ross, editors, *Practice and Theory of Automated Timetabling*, volume 1153 of *Lecture Notes in Computer Science*, pages 76–90. Springer Verlag, Berlin, Germany, 1996.

[Burke et al., 2001] E. K. Burke, P. I. Cowling, and R. Keuthen. Effective local and guided variable neighbourhood search methods for the asymmetric travelling salesman problem. In E. J. W. Boers, J. Gottlieb, P. L. Lanzi, R. E. Smith, S. Cagnoni, E. Hart, G. R. Raidl, and H. Tijink, editors, *Applications of Evolutionary Computing: Proceedings of EvoWorkshops 2001*, volume 2037 of *Lecture Notes in Computer Science*, pages 203–212. Springer Verlag, Berlin, Germany, 2001.

C   [Cadoli et al., 2002] M. Cadoli, M. Schaerf, A. Giovanardi, and M. Giovanardi. An algorithm to evaluate quantified Boolean formulae and its experimental evaluation. *Journal of Automated Reasoning*, 28(2):101–142, 2002.

[Caprara et al., 1997] A. Caprara, M. Fischetti, P. Toth, D. Vigo, and P. L. Guida. Algorithms for railway crew management. *Mathematical Programming*, 79:125–141, 1997.

[Caprara et al., 1999] A. Caprara, M. Fischetti, and P. Toth. A heuristic method for the set covering problem. *Operations Research*, 47(5):730–743, 1999.

[Caprara et al., 2000] A. Caprara, M. Fischetti, and P. Toth. Algorithms for the set covering problem. *Annals of Operations Research*, 98:353–371, 2000.

[Carlier and Pinson, 1989] J. Carlier and E. Pinson. An algorithm for solving the job-shop problem. *Management Science*, 35(2):164–176, 1989.

[Carter et al., 1996] M. W. Carter, G. Laporte, and S. Lee. Examination timetabling: Algorithmic strategies and applications. *Journal of the Operational Research Society*, 47(3):373–383, 1996.

[Carter, 1986] M. W. Carter. A survey of practical applications of examination timetabling algorithms. *Operations Research*, 34(2):193–202, 1986.

[Çela, 1998] E. Çela. *The Quadratic Assignment Problem: Theory and Algorithms*. Kluwer Academic Publishers, Dordrecht, The Netherlands, 1998.

[Ceria et al., 1998] S. Ceria, P. Nobili, and A. Sassano. A Lagrangian-based heuristic for large-scale set covering problems. *Mathematical Programming Series B*, 81(2):215–228, 1998.

[Cerný, 1985] V. Cerný. A thermodynamical approach to the traveling salesman problem. *Journal of Optimization Theory and Applications*, 45(1):41–51, 1985.

[Cha and Iwama, 1995] B. Cha and K. Iwama. Performance test of local search algorithms using new types of random CNF formulas. In C. S. Mellish, editor, *Proceedings of the Fourteenth International Joint Conference on Artificial Intelligence*, pages 304–310. Morgan Kaufmann Publishers, San Francisco, CA, USA, 1995.

[Cha and Iwama, 1996] B. Cha and K. Iwama. Adding new clauses for faster local search. In *Proceedings of the Thirteenth National Conference on Artificial Intelligence*, pages 332–337. AAAI Press/The MIT Press, Menlo Park, CA, USA, 1996.

[Chaitin et al., 1981] G. J. Chaitin, M. Auslander, A. K. Chandra, J. Cocke, M. E. Hopkins, and P. Markstein. Register allocation via coloring. *Computer Languages*, 6(1):47–57, 1981.

[Chaitin, 1982] G. J. Chaitin. Register allocation and spilling via graph coloring. In *Proceedings of the 1982 SIGPLAN Symposium on Compiler Construction*, pages 98–101. ACM press, New York, NY, USA, 1982.

[Chams et al., 1987] M. Chams, A. Hertz, and D. De Werra. Some experiments with simulated annealing for coloring graphs. *European Journal of Operational Research*, 32(2):260–266, 1987.

[Chandra and Halldórsson, 2000] B. Chandra and M. M. Halldórsson. Greedy local improvement and weighted set packing approximation. *Journal of Algorithms*, 39(2):223–240, 2000.

[Chandra et al., 1999] B. Chandra, H. Karloff, and C. Tovey. New results on the old k-opt algorithm for the TSP. *SIAM Journal on Computing*, 28(6):1998–2029, 1999.

[Charon and Hudry, 2000] I. Charon and O. Hudry. Application of the noising method to the travelling salesman problem. *European Journal of Operational Research*, 125(2):266–277, 2000.

[Chazelle, 2000] B. Chazelle. A minimum spanning tree algorithm with inverse-Ackermann type complexity. *Journal of the ACM*, 47(6):1028–1047, 2000.

[Cheeseman et al., 1991] P. Cheeseman, B. Kanefsky, and W. M. Taylor. Where the really hard problems are. In J. Mylopoulos and R. Reiter, editors, *Proceedings of the 12th International Joint Conference on Artificial Intelligence*, pages 331–337. Morgan Kaufmann Publishers, San Francisco, CA, USA, 1991.

[Chen et al., 1997] J. Chen, D. K. Friesen, and H. Zheng. Tight bound on Johnson's algorithm for Max-SAT. In *Proceedings of the 12th Annual IEEE Conference on Computational Complexity*, pages 274–281. IEEE Computer Society Press, Los Alamitos, CA, USA, 1997.

[Chiarandini and Stützle, 2002] M. Chiarandini and T. Stützle. An application of iterated local search to the graph coloring problem. In D. S. Johnson, A. Mehrotra, and M. Trick, editors, *Proceedings of the Computational Symposium on Graph Coloring and its Generalizations*, pages 112–125, Ithaca, New York, USA, 2002.

[Chiarandini et al., 2003] M. Chiarandini, I. Dumitrescu, and T. Stützle. Local search for the graph colouring problem—a computational study. Technical Report AIDA-03-07, FG Intellektik, FB Informatik, TU Darmstadt, Germany, 2003.

[Chow and Hennessy, 1990] F. C. Chow and J. L. Hennessy. The priority-based coloring approach to register allocation. *ACM Transactions on Programming Languages and Systems*, 12(4):501–536, 1990.

[Christofides and Benavent, 1989] N. Christofides and E. Benavent. An exact algorithm for the quadratic assignment problem. *Operations Research*, 37(5):760–768, 1989.

[Christofides, 1976] N. Christofides. Worst-case analysis of a new heuristic for the travelling salesman problem. Technical Report 388, Graduate School of Industrial Administration, Carnegie-Mellon University, Pittsburgh, PA, USA, 1976.

[Chrobak et al., 1990] M. Chrobak, T. Szymacha, and A. Krawczyk. A data structure useful for finding Hamiltonian cycles. *Theoretical Computer Science*, 71(3):419–424, 1990.

[Chu and Beasley, 1998] P. C. Chu and J. E. Beasley. Constraint handling in genetic algorithms: The set partitioning problem. *Journal of Heuristics*, 4(4):323–357, 1998.

[Chvátal, 1979] V. Chvátal. A greedy heuristic for the set covering problem. *Mathematics of Operations Research*, 4:233–235, 1979.

[Cirasella et al., 2001] J. Cirasella, D. S. Johnson, L. A. McGeoch, and W. Zhang. The asymmetric traveling salesman problem: Algorithms, instance generators, and tests. In A. L. Buchsbaum and J. Snoeyink, editors, *Algorithm Engineering and Experimentation, Third International Workshop, ALENEX 2001*, volume 2153 of *Lecture Notes in Computer Science*, pages 32–59. Springer Verlag, Berlin, Germany, 2001.

[Clark et al., 1996] D. Clark, J. Frank, I. P. Gent, E. MacIntyre, N. Tomov, and T. Walsh. Local search and the number of solutions. In E. C. Freuder, editor, *Proceedings of the Second International Conference on Principles and Practice of Constraint Programming*, volume 1118 of *Lecture Notes in Computer Science*, pages 119–133. Springer Verlag, Berlin, Germany, 1996.

[Clarke and Wright, 1964] G. Clarke and J. W. Wright. Scheduling of vehicles from a central depot to a number of delivery points. *Operations Research*, 12(4):568–581, 1964.

[Clearwater et al., 1991] S. H. Clearwater, B. A. Huberman, and T. Hogg. Cooperative solution of constraint satisfaction problems. *Science*, 254:1181–1183, 1991.

[Clearwater et al., 1992] S. H. Clearwater, B. A. Huberman, and T. Hogg. Cooperative problem solving. In B. A. Huberman, editor, *Computation: The Micro and the Macro View*, pages 33–70. World Scientific, Singapore, 1992.

[Codenotti et al., 1996] B. Codenotti, G. Manzini, L. Margara, and G. Resta. Perturbation: An efficient technique for the solution of very large instances of the Euclidean TSP. *INFORMS Journal on Computing*, 8(2):125–133, 1996.

[Cohen, 1995] P. R. Cohen. *Empirical Methods for Artificial Intelligence*. MIT Press, Cambridge, MA, USA, 1995.

[Colbourn, 1984] C. Colbourn. The complexity of completing latin squares. *Discrete Applied Mathematics*, 8(1):25–30, 1984.

[Coleman and Moré, 1983] T. F. Coleman and J. J. Moré. Estimation of sparse Jacobian matrices and graph coloring problems. *SIAM Journal on Numerical Analysis*, 20(1):187–209, 1983.

[Collins et al., 1988] N. E. Collins, R. W. Eglese, and B. L. Golden. Simulated annealing — an annotated bibliography. *American Journal of Mathematical and Management Sciences*, 8(3–4):209–307, 1988.

[Comellas and Roca, 1993] F. Comellas and R. Roca. Using genetic algorithms to design constant weight codes. In *Applications of Neural Networks to Telecommunications*, pages 119–124. Lawrence Erlbaum Associates, Mahwah, NJ, USA, 1993.

[Congram et al., 2002] R. K. Congram, C. N. Potts, and S. van de Velde. An iterated dynasearch algorithm for the single-machine total weighted tardiness scheduling problem. *INFORMS Journal on Computing*, 14(1):52–67, 2002.

[Congram, 2000] R. K. Congram. *Polynomially Searchable Exponential Neighbourhoods for Sequencing Problems in Combinatorial Optimization*. PhD thesis, Southampton University, Faculty of Mathematical Studies, Southampton, UK, 2000.

[Connolly, 1990] D. T. Connolly. An improved annealing scheme for the QAP. *European Journal of Operational Research*, 46(1):93–100, 1990.

[Connolly, 1992] D. T. Connolly. General purpose simulated annealing. *Journal of the Operational Research Society*, 43(5):495–505, 1992.

[Conover, 1999] W. J. Conover. *Practical Nonparametric Statistics*, third edition. John Wiley & Sons, New York, NY, USA, 1999.

[Conway et al., 1967] R. W. Conway, W. L. Maxwell, and L. W. Miller. *Theory of Scheduling*. Addison-Wesley, Reading, MA, USA, 1967.

[Cook and Mitchell, 1997] S. A. Cook and D. G. Mitchell. Finding hard instances of the satisfiability problem: A survey. In D. Du, J. Gu, and P. M. Pardalos, editors, *Satisfiability Problem: Theory and Applications*, volume 35 of *DIMACS Series on Discrete Mathematics and Theoretical Computer Science*, pages 1–18. American Mathematical Society, Providence, RI, USA, 1997.

[Cook and Seymour, 2003] W. Cook and P. Seymour. Tour merging via branch-decomposition. *INFORMS Journal on Computing*, 15(3):233–248, 2003.

[Cook et al., 1997] W. Cook, W. H. Cunningham, W. R. Pulleybank, and A. Schrijver. *Combinatorial Optimization*. John Wiley & Sons, New York, NY, USA, 1997.

[Cook, 1971] S. A. Cook. The complexity of theorem proving procedures. In M. A. Harrison, R. B. Banerji, and J. D. Ullman, editors, *Proceedings of the Third Annual ACM Symposium on Theory of Computing*, pages 151–158. ACM press, New York, NY, USA, 1971.

[Cordón et al., 2002] O. Cordón, I. Fernández de Viana, and F. Herrera. Analysis of the best-worst Ant System and its variants on the TSP. *Mathware & Soft Computing*, 9(2–3):177–192, 2002.

[Cormen et al., 2001] T. H. Cormen, C. E. Leiserson, R. L. Rivest, and C. Stein. *Introduction to Algorithms*, second edition. MIT Press, Cambridge, MA, USA, 2001.

[Corne et al., 1999] D. Corne, M. Dorigo, and F. Glover. *New Ideas in Optimization*. McGraw Hill, London, UK, 1999.

[Costa and Hertz, 1997] D. Costa and A. Hertz. Ants can colour graphs. *Journal of the Operational Research Society*, 48(3):295–305, 1997.

[Coy et al., 2001] S. P. Coy, B. L. Golden, G. C. Runger, and E. A. Wasil. Using experimental design to find effective parameter settings for heuristics. *Journal of Heuristics*, 7(1):77–97, 2001.

[Craenen et al., 2000] B. Craenen, A. E. Eiben, and E. Marchiori. Solving constraint satisfaction problems with heuristic-based evolutionary algorithms. In *Proceedings of the 2000 Congress on Evolutionary Computation*, pages 1571–1577. IEEE Press, Piscataway, NJ, USA, 2000.

[Crauwels et al., 1998] H. A. J. Crauwels, C. N. Potts, and L. N. Van Wassenhove. Local search heuristics for the single machine total weighted tardiness scheduling problem. *INFORMS Journal on Computing*, 10(3):341–350, 1998.

[Crawford and Auton, 1996] J. M. Crawford and L. D. Auton. Experimental results on the crossover point in random 3SAT. *Artificial Intelligence*, 81(1–2):31–57, 1996.

[Croes, 1958] G. A. Croes. A method for solving traveling salesman problems. *Operations Research*, 6(6):791–812, 1958.

[Crowder et al., 1979] H. Crowder, R. Dembo, and J. Mulvey. On reporting computational experiments with mathematical software. *ACM Transactions on Mathematical Software*, 5(2):193–203, 1979.

[Crowston et al., 1963] W. B. Crowston, F. Glover, G. L. Thompson, and J. D. Trawick. Probabilistic and parametric learning combinations of local job shop scheduling rules. ONR Research Memorandum No. 117, GSIA, Carnegie-Mellon University, Pittsburgh, PA, USA, 1963.

[Culberson and Luo, 1996] J. C. Culberson and F. Luo. Exploring the *k*-colorable landscape with iterated greedy. In D. S. Johnson and M. A. Trick, editors, *Cliques, Coloring, and Satisfiability: Second DIMACS Implementation Challenge*, volume 26 of *DIMACS Series on Discrete Mathematics and Theoretical Computer Science*, pages 245–284. American Mathematical Society, Providence, RI, USA, 1996.

[Culberson et al., 1995] J. Culberson, A. Beacham, and D. Papp. Hiding our colors. In *Proceedings of the CP'95 Workshop on Studying and Solving Really Hard Problems*, pages 31–42, Cassis, France, September 1995.

[Culberson et al., 2000] J. Culberson, I. P. Gent, and H. H. Hoos. On the probabilistic approximate completeness of WalkSAT for 2-SAT. Technical Report APES-15a-2000, APES Research Group, 2000.

[Culberson, 2003] J. Culberson. Graph coloring page. `http://www.cs.ualberta.ca/~joe/Coloring/`, 2003. Version visited last on 15 September 2003.

[Cung et al., 1997] V.-D. Cung, T. Mautor, P. Michelon, and A. Tavares. A scatter search based approach for the quadratic assignment problem. In T. Baeck, Z. Michalewicz, and X. Yao, editors, *Proceedings of the 1997 IEEE International Conference on Evolutionary Computation (ICEC'97)*, pages 165–170. IEEE Press, Piscataway, NJ, USA, 1997.

D   [Dannenbring, 1977] D. G. Dannenbring. Procedures for estimating optimal solution values for large combinatorial problems. *Management Science*, 23(12):1273–1283, 1977.

[Dantsin et al., 2001] E. Dantsin, M. Gavrilovich, E. Hirsch, and B. Konev. MAX SAT approximation beyond the limits of polynomial-time approximation. *Annals of Pure and Applied Logic*, 113(1–3):81–94, 2001.

[Dantzig et al., 1954] G. Dantzig, R. Fulkerson, and S. Johnson. Solution of a large-scale traveling salesman problem. *Operations Research*, 2(4):393–410, 1954.

[Davenport et al., 1994] A. Davenport, E. Tsang, C. J. Wang, and K. Zhu. GENET: A connectionist architecture for solving constraint satisfaction problems by iterative improvement. In *Proceedings of the 11th National Conference on Artificial Intelligence*, pages 325–330. AAAI Press/The MIT Press, Menlo Park, CA, USA, 1994.

[Davidor, 1991] Y. Davidor. Epistasis variance: A viewpoint on GA-hardness. In G. J. E. Rawlins, editor, *Foundations of Genetic Algorithms*, volume 1, pages 23–35. Morgan Kaufmann Publishers, San Mateo, CA, USA, 1991.

[Davis et al., 1962] M. Davis, G. Logemann, and D. W. Loveland. A machine program for theorem proving. *Communications of the ACM*, 5(7):394–397, 1962.

[Davis, 1991] L. Davis. Order-based genetic algorithms and the graph coloring problem. In L. Davis, editor, *Handbook of Genetic Algorithms*, pages 72–90. Van Nostrand Reinhold, New York, USA, 1991.

[de Kleer, 1989] J. de Kleer. A comparison of ATMS and CSP techniques. In N. S. Sridharan, editor, *Proceedings of the 11th International Joint Conference on Artificial Intelligence*, pages 290–296. Morgan Kaufmann Publishers, San Francisco, CA, USA, 1989.

[de Leeuw et al., 1955] K. de Leeuw, E. F. Moore, C. E. Shannon, and N. Shapirio. Computability by probabilistic machines. In C. E. Shannon and J. McCarthy, editors, *Automata Studies*, pages 183–212. Princeton University Press, Princeton, NJ, USA, 1955.

[de Vries and Vohra, 2003] S. de Vries and R. Vohra. Combinatorial auctions: A survey. *INFORMS Journal on Computing*, 15(3):284–309, 2003.

[de Werra, 1985] D. de Werra. An introduction to timetabling. *European Journal of Operational Research*, 19(2):151–162, 1985.

[Dean and Voss, 2000] A. Dean and D. Voss. *Design and Analysis of Experiments*. Springer Verlag, Berlin, Germany, 2000.

[Deaton et al., 1996] R. Deaton, M. Garzon, R. C. Murphy, J. A. Rose, D. R. Franceschetti, and S. E. Stevens Jr. Genetic search of reliable encodings for DNA-based computation. In *Late Breaking Papers at the Genetic Programming 1996 Conference*, pages 9–15, Stanford, CA, USA, July 28–31, 1996.

[Deaton et al., 1999] R. Deaton, R. C. Murphy, M. Garzon, D. R. Franceschetti, and S. E. Stevens Jr. Good encodings for DNA-based solutions to combinatorial problems. In *DNA Based Computers II*, volume 44 of *DIMACS Series on Discrete Mathematics and Theoretical Computer Science*, pages 247–258. American Mathematical Society, Providence, RI, USA, 1999.

[Debruyne and Bessière, 2001] R. Debruyne and C. Bessière. Domain filtering consistencies. *Journal of Artificial Intelligence Research*, 14:205–230, 2001.

[Dechter and Pearl, 1989] R. Dechter and J. Pearl. Tree clustering schemes for constraint-processing. *Artificial Intelligence*, 38(3):353–366, 1989.

[Dechter, 2003] R. Dechter. *Constraint Processing*. Morgan Kaufmann Publishers, San Francisco, CA, USA, 2003.

[del Val, 2000] A. del Val. On 2-SAT and renamable Horn. In *Proceedings of the Seventeenth National Conference on Artificial Intelligence*, pages 279–284. AAAI Press/The MIT Press, Menlo Park, CA, USA, 2000.

[Dell'Amico and Trubian, 1993] M. Dell'Amico and M. Trubian. Applying tabu search to the job shop scheduling problem. *Annals of Operations Research*, 41:231–252, 1993.

[Dempster et al., 1977] A. P. Dempster, N. M. Laird, and D. B. Rubin. Maximum likelihood from incomplete data via the EM algorithm. *Journal of the Royal Statistical Society, Series B*, 39(1):1–38, 1977.

[den Besten et al., 2000] M. L. den Besten, T. Stützle, and M. Dorigo. Ant colony optimization for the total weighted tardiness problem. In M. Schoenauer, K. Deb, G. Rudolph, X. Yao, E. Lutton, J. J. Merelo, and H.-P. Schwefel, editors, *Proceedings of PPSN-VI, Sixth International Conference on Parallel Problem Solving from Nature*, volume 1917 of *Lecture Notes in Computer Science*, pages 611–620. Springer Verlag, Berlin, Germany, 2000.

[den Besten et al., 2001] M. L. den Besten, T. Stützle, and M. Dorigo. Design of iterated local search algorithms: An example application to the single machine total weighted tardiness problem. In E. J. W. Boers, J. Gottlieb, P. L. Lanzi, R. E. Smith, S. Cagnoni, E. Hart, G. R. Raidl, and H. Tijink, editors, *Applications of Evolutionary Computing: Proceedings of EvoWorkshops 2001*, volume 2037 of *Lecture Notes in Computer Science*, pages 441–452. Springer Verlag, Berlin, Germany, 2001.

[Di Caro and Dorigo, 1998] G. Di Caro and M. Dorigo. AntNet: Distributed stigmergetic control for communications networks. *Journal of Artificial Intelligence Research*, 9:317–365, 1998.

[Di Gaspero and Schaerf, 2002] L. Di Gaspero and A. Schaerf. Writing local search algorithms using EASYLOCAL++. In S. Voß and D. L. Woodruff, editors, *Optimization Software Class Libraries*, pages 155–175. Kluwer Academic Publishers, Boston, MA, USA, 2002.

[Dijkstra, 1959] E. W. Dijkstra. A note on two problems in connection with graphs. *Numerical Mathematics*, 1:269–271, 1959.

[Domschke et al., 1992] W. Domschke, P. Forst, and S. Voß. Tabu search techniques for the quadratic semi-assignment problem. In G. Fandel, T. Gulledge, and A. Jones, editors, *New Directions for Operations Research in Manufacturing*, pages 389–405. Springer Verlag, Berlin, Germany, 1992.

[Dorigo and Di Caro, 1999] M. Dorigo and G. Di Caro. The ant colony optimization meta-heuristic. In D. Corne, M. Dorigo, and F. Glover, editors, *New Ideas in Optimization*, pages 11–32. McGraw Hill, London, UK, 1999.

[Dorigo and Gambardella, 1997] M. Dorigo and L. M. Gambardella. Ant colony system: A cooperative learning approach to the traveling salesman problem. *IEEE Transactions on Evolutionary Computation*, 1(1):53–66, 1997.

[Dorigo and Stützle, 2004] M. Dorigo and T. Stützle. *Ant Colony Optimization*. MIT Press, Cambridge, MA, USA, 2004.

[Dorigo et al., 1991] M. Dorigo, V. Maniezzo, and A. Colorni. Positive feedback as a search strategy. Technical Report 91-016, Dipartimento di Elettronica, Politecnico di Milano, Italy, 1991.

[Dorigo et al., 1996] M. Dorigo, V. Maniezzo, and A. Colorni. Ant System: Optimization by a colony of cooperating agents. *IEEE Transactions on Systems, Man, and Cybernetics – Part B*, 26(1):29–41, 1996.

[Dorigo et al., 1999] M. Dorigo, G. Di Caro, and L. M. Gambardella. Ant algorithms for discrete optimization. *Artificial Life*, 5(2):137–172, 1999.

[Dorigo, 1992] M. Dorigo. *Optimization, Learning and Natural Algorithms* (in Italian). PhD thesis, Dipartimento di Elettronica, Politecnico di Milano, Italy, 1992. 140 pages.

[Dorigo, 2003] M. Dorigo. Metaheuristics network. http://www.metaheuristics. net, 2003. Version visited last on 15 September 2003.

[Dorne and Hao, 1999] R. Dorne and J. K. Hao. Tabu search for graph coloring, T-colorings and set T-colorings. In S. Voss, S. Martello, I. H. Osman, and C. Roucairol, editors, *Meta-heuristics: Advances and Trends in Local Search Paradigms for Optimization*, pages 77–92. Kluwer Academic Publishers, Boston, MA, USA, 1999.

[Dowling and Gallier, 1984] W. F. Dowling and J. H. Gallier. Linear time algorithms for testing the satisfiability of propositional Horn formulae. *Journal of Logic Programming*, 1(3):267–284, 1984.

[Dowsland, 1993] K. A. Dowsland. Simulated annealing. In C. R. Reeves, editor, *Modern Heuristic Techniques for Combinatorial Problems*. Blackwell Scientific Publications, Oxford, UK, 1993.

[Dozier et al., 1998] G. Dozier, J. Bowen, and A. Homaifar. Solving constraint satisfaction problems using hybrid evolutionary search. *IEEE Transactions on Evolutionary Computation*, 2(1):23–33, 1998.

[Drezner et al., to appear] Z. Drezner, P. Hahn, and É. D. Taillard. A study of quadratic assignment problem instances that are difficult for meta-heuristic methods. *Annals of Operations Research*, to appear.

[Drezner, 2003] Z. Drezner. A new genetic algorithm for the quadratic assignment problem. *INFORMS Journal on Computing*, 15(3):320–330, 2003.

[Du and Leung, 1990] J. Du and J. Y.-T. Leung. Minimizing total tardiness on one machine is NP–hard. *Mathematics of Operations Research*, 15(3):483–495, 1990.

[Dubois and Dequen, 2001] O. Dubois and G. Dequen. A backbone-search heuristic for efficient solving of hard 3-SAT formulae. In B. Nebel, editor, *Proceedings of the Seventeenth International Joint Conference on Artificial Intelligence*, pages 248–253. Morgan Kaufmann Publishers, San Francisco, CA, USA, 2001.

[Dudek et al., 1992] R. Dudek, S. Panwalker, and M. Smith. The lessons of flowshop scheduling research. *Operations Research*, 40(1):7–13, 1992.

[Dueck and Scheuer, 1990] G. Dueck and T. Scheuer. Threshold accepting: A general purpose optimization algorithm appearing superior to simulated annealing. *Journal of Computational Physics*, 90(1):161–175, 1990.

[Duh and Fürer, 1997] R.-C. Duh and M. Fürer. Approximation of k-set cover by semi-local optimization. In F. T. Leighton and P. Shor, editors, *Proceedings of the Twenty-ninth Annual ACM Symposium on Theory of Computing*, pages 256–264. ACM Press, New York, NY, USA, 1997.

E   [Eiben et al., 1998] A. E. Eiben, J. K. Hauw, and J. I. Van Hemert. Graph coloring with adaptive evolutionary algorithms. *Journal of Heuristics*, 4(1):25–46, 1998.

[Eiben, 2001] A. E. Eiben. Evolutionary algorithms and constraint satisfaction: Definitions, survey, methodology, and research directions. In L. Kallel, B. Naudts, and A. Rogers, editors, *Theoretical Aspects of Evolutionary Computing*, pages 13–58. Springer Verlag, Berlin, Germany, 2001.

[Eisenblätter and Koster, 2003] A. Eisenblätter and A. M. C. A. Koster. FAP web — a website about frequency assignment problems. http://fap.zib.de, 2003. Version visited last on 15 September 2003.

[Eisenblätter et al., 2002] A. Eisenblätter, M. Grötschel, and A. M. C. A. Koster. Frequency planning and ramifications of coloring. *Discussiones Mathematicae Graph Theory*, 22(1):51–88, 2002.

[Elshafei, 1977] A. N. Elshafei. Hospital layout as a quadratic assignment problem. *Operations Research Quarterly*, 28(1):167–179, 1977.

[Engebretsen, 2000] L. Engebretsen. *Approximate Constraint Satisfaction*. PhD thesis, Royal Institute of Technology, Stockholm, Sweden, 2000.

[Erdós et al., 1979] P. Erdós, A. L. Rubin, and H. Taylor. Choosability in graphs. *Congressus Numerantium*, 26:125–157, 1979.

[Eremeev, 1999] A. V. Eremeev. A genetic algorithm with a non-binary representation for the set covering problem. In P. Kall and H.-J. Lüthi, editors, *Operations Research Proceedings 1998*, pages 175–181. Springer Verlag, Berlin, Germany, 1999.

[Ernst et al., 1997] M. D. Ernst, T. D. Millstein, and D. S. Weld. Automatic SAT-compilation of planning problems. In M. E. Pollack, editor, *Proceedings of the Fifteenth International Joint Conference on Artificial Intelligence*, pages 1169–1177. Morgan Kaufmann Publishers, San Francisco, CA, USA, 1997.

[Ertel and Luby, 1994] W. Ertel and M. Luby. Optimal parallelization of Las Vegas algorithms. In P. Enjalbert, E. W. Mayr, and K. W. Wagner, editors, *STACS 94, 11th Annual Symposium on Theoretical Aspects of Computer Science*, volume 775 of *Lecture Notes in Computer Science*, pages 463–475. Springer Verlag, Berlin, Germany, 1994.

[Everitt et al., 2001] B. S. Everitt, S. Landau, and M. Leese. *Cluster Analysis*. Oxford University Press, New York, NY, USA, 2001.

F

[Faigle and Kern, 1992] U. Faigle and W. Kern. Some convergence results for probabilistic tabu search. *ORSA Journal on Computing*, 4(1):32–37, 1992.

[Fang et al., 1994] H.-L. Fang, P. Ross, and D. Corne. A promising genetic algorithm approach to job-shop scheduling, rescheduling, and open-shop scheduling problems. In A. G. Cohn, editor, *Proceedings of the 11th European Conference on Artificial Intelligence*, pages 590–594. John Wiley & Sons, Chichester, UK, 1994.

[Feige and Goemans, 1995] U. Feige and M. X. Goemans. Approximating the value of two proper proof systems, with applications to MAX-2SAT and MAX-DICUT. In *Proceedings of the 3rd Israel Symposium on Theory and Computing Systems*, pages 182–189. IEEE Computer Society Press, Los Alamitos, CA, USA, 1995.

[Feige and Kilian, 1998] U. Feige and J. Kilian. Zero knowledge and the chromatic number. *Journal of Computer and System Science*, 57(2):187–199, 1998.

[Feo and Resende, 1989] T. A. Feo and M. G. C. Resende. A probabilistic heuristic for a computationally difficult set covering problem. *Operations Research Letters*, 8(2):67–71, 1989.

[Feo and Resende, 1995] T. A. Feo and M. G. C. Resende. Greedy randomized adaptive search procedures. *Journal of Global Optimization*, 6:109–133, 1995.

[Ferreira et al., 2000] F. F. Ferreira, J. F. Fontanari, and P. F Stadler. Landscape statistics of the low autocorrelated binary string problem. *Journal of Physics A*, 33(48):8635–8647, 2000.

[Festa and Resende, 2001] P. Festa and M. G. C. Resende. GRASP: An annotated bibliography. In P. Hansen and C. C. Ribeiro, editors, *Essays and Surveys on Metaheuristics*, pages 325–367. Kluwer Academic Publishers, Boston, MA, USA, 2001.

[Fiechter, 1994] C.-N. Fiechter. A parallel tabu search algorithm for large traveling salesman problems. *Discrete Applied Mathematics*, 51(3):243–267, 1994.

[Fielding, 2000] M. Fielding. Simulated annealing with an optimal fixed temperature. *SIAM Journal on Optimization*, 11(2):289–307, 2000.

[Finger et al., 2002] M. Finger, T. Stützle, and H. R. Lourenço. Exploiting fitness distance correlation of set covering problems. In S. Cagnoni, J. Gottlieb, E. Hart, M. Middendorf, and G. R. Raidl, editors, *Applications of Evolutionary Computing*, volume 2279 of *Lecture Notes in Computer Science*, pages 61–71. Springer Verlag, Berlin, Germany, 2002.

[Fink and Voß, 2002] A. Fink and S. Voß. HotFrame: A heuristic optimization framework. In S. Voß and D. L. Woodruff, editors, *Optimization Software Class Libraries*, pages 81–154. Kluwer Academic Publishers, Boston, MA, USA, 2002.

[Fischetti et al., 1994] M. Fischetti, P. Toth, and D. Vigo. A branch and bound algorithm for the capacitated vehicle routing problem on directed graphs. *Operations Research*, 42(5):846–859, 1994.

[Fisher and Thompson, 1963] H. Fisher and G. L. Thompson. Probabilistic learning combinations of local job-shop scheduling rules. In J. F. Muth and G. L. Thompson, editors, *Industrial Scheduling*, pages 225–251. Prentice Hall, Englewood Cliffs, New Jersey, 1963.

[Fisher, 1981] M. L. Fisher. The Lagrangean relaxation method for solving integer programming problems. *Management Science*, 27(1):1–17, 1981.

[Fisher, 1985] M. L. Fisher. An applications oriented guide to Lagrangian relaxation. *Interfaces*, 15(2):10–21, 1985.

[Flamm et al., 2000] C. Flamm, W. Fontana, I. L. Hofacker, and P. Schuster. RNA folding at elementary step resolution. *RNA*, 6:325–338, 2000.

[Flamm et al., 2002] C. Flamm, I. L. Hofacker, P. F. Stadler, and M. T. Wolfinger. Barrier trees of degenerate landscapes. *Zeitschrift für Physikalische Chemie*, 216(2):155–173, 2002.

[Fleurent and Ferland, 1994] C. Fleurent and J. A. Ferland. Genetic hybrids for the quadratic assignment problem. In P. M. Pardalos and H. Wolkowicz, editors, *Quadratic Assignment and Related Problems*, volume 16 of *DIMACS Series on Discrete Mathematics and Theoretical Computer Science*, pages 173–187. American Mathematical Society, Providence, RI, USA, 1994.

[Fleurent and Ferland, 1996] C. Fleurent and J. A. Ferland. Genetic and hybrid algorithms for graph coloring. *Annals of Operations Research*, 63:437–464, 1996.

[Flood, 1956] M. M. Flood. The travelling salesman problem. *Operations Research*, 4(1):61–75, 1956.

[Fogel et al., 1966] L. J. Fogel, A. J. Owens, and M. J. Walsh. *Artificial Intelligence Through Simulated Evolution*. John Wiley & Sons, New York, NY, USA, 1966.

[Fonlupt et al., 1998] C. Fonlupt, D. Robilliard, and P. Preux. A bit-wise epistasis measure for binary search spaces. In A. E. Eiben, T. Bäck, M. Schoenauer, and H.-P. Schwefel, editors, *Proceedings of PPSN-V, Fifth International Conference on Parallel Problem Solving from Nature*, volume 1498 of *Lecture Notes in Computer Science*, pages 47–56. Springer Verlag, Berlin, Germany, 1998.

[Fontana et al., 1993] W. Fontana, P. F. Stadler, E. G. Bornberg-Bauer, T. Griesmacher, I. L. Hofacker, M. Tacker, P. Tarazona, E. D. Weinberger, and P. Schuster. RNA folding and combinatory landscapes. *Physical Review E*, 47(3):2083–2099, 1993.

[Franco and Paull, 1983] J. Franco and M. Paull. Probabilistic analysis of the Davis-Putnam procedure for solving satisfiability. *Discrete Applied Mathematics*, 5(1):77–87, 1983.

[Franco and Swaminathan, 1997] J. Franco and R. Swaminathan. Average case results for satisfiability algorithms under the random clause model. *Annals of Mathematics and Artificial Intelligence*, 20:357–391, 1997.

[Frank et al., 1997] J. Frank, P. Cheeseman, and J. Stutz. When gravity fails: Local search topology. *Journal of Artificial Intelligence Research*, 7:249–281, 1997.

[Frank, 1996] J. Frank. Weighting for Godot: Learning heuristics for GSAT. In *Proceedings of the Thirteenth National Conference on Artificial Intelligence*, pages 776–783. AAAI Press/The MIT Press, Menlo Park, CA, USA, 1996.

[Frank, 1997] J. Frank. Learning short-term clause weights for GSAT. In M. E. Pollack, editor, *Proceedings of the 15th International Joint Conference on Artificial Intelligence*, pages 384–389. Morgan Kaufmann Publishers, San Francisco, CA, USA, 1997.

[Frauenfelder et al., 1997] H. Frauenfelder, A. R. Bishop, A. Garcia, A. Perelson, P. Schuster, D. Sherrington, and P. J. Swart. Landscape paradigms in physics and biology: Concepts, structure, and dynamics. *Physica D*, 107(2–4), 1997.

[Fredman et al., 1995] M. L. Fredman, D. S. Johnson, L. A. McGeoch, and G. Ostheimer. Data structures for traveling salesmen. *Journal of Algorithms*, 18(3):432–479, 1995.

[Freisleben and Merz, 1996] B. Freisleben and P. Merz. A genetic local search algorithm for solving symmetric and asymmetric traveling salesman problems. In T. Baeck, T. Fukuda, and Z. Michalewicz, editors, *Proceedings of the 1996 IEEE International Conference on Evolutionary Computation (ICEC'96)*, pages 616–621. IEEE Press, Piscataway, NJ, USA, 1996.

[French, 1982] S. French. *Sequencing and Scheduling: An Introduction to the Mathematics of the Job Shop*. Horwood, Chichester, UK, 1982.

[Freuder and Wallace, 1992] E. C. Freuder and R. J. Wallace. Partial constraint satisfaction. *Artificial Intelligence*, 58(1–3):21–70, 1992.

[Freuder, 1989] E. C. Freuder. Partial constraint satisfaction. In N. S. Sridharan, editor, *Proceedings of the 11th International Joint Conference on Artificial Intelligence*, pages 278–283. Morgan Kaufmann Publishers, San Francisco, CA, USA, 1989.

[Frisch and Peugniez, 2001] A. Frisch and T. J. Peugniez. Solving non-boolean satisfiability problems with stochastic local search. In B. Nebel, editor, *Proceedings of the 17th International Joint Conference on Artificial Intelligence*, pages 282–288. Morgan Kaufmann Publishers, San Francisco, CA, USA, 2001.

[Frost et al., 1997] D. Frost, I. Rish, and L.Vila. Summarizing CSP hardness with continuous probability distributions. In *Proceedings of the Fourteenth National Conference on Artificial Intelligence*, pages 327–333. AAAI Press/The MIT Press, Menlo Park, CA, USA, 1997.

[Frutos et al., 1997] A. G. Frutos, Q. Liu, A. J. Thiel, A. M. W. Sanner, A. E. Condon, L. M. Smith, and R. M. Corn. Demonstration of a word design strategy for DNA computing on surfaces. *Nucleic Acids Research*, 25(23):4748–4757, 1997.

[Fujisima et al., 1999] Y. Fujisima, K. Leyton-Brown, and Y. Shoham. Taming the computational complexity of combinatorial auctions. In T. Dean, editor, *Proceedings of the Sixteenth International Joint Conference on Artificial Intelligence*, pages 548–553. Morgan Kaufmann Publishers, San Francisco, CA, USA, 1999.

G   [Galinier and Hao, 1997] P. Galinier and J.-K. Hao. Tabu search for maximal constraint satisfaction problems. In G. Smolka, editor, *Principles and Practice of Constraint Programming—CP 1997*, volume 1330 of *Lecture Notes in Computer Science*, pages 196–208. Springer Verlag, Berlin, Germany, 1997.

[Galinier and Hao, 1999] P. Galinier and J. K. Hao. Hybrid evolutionary algorithms for graph coloring. *Journal of Combinatorial Optimization*, 3(4):379–397, 1999.

[Gamal et al., 1987] A. A. El Gamal, L. A. Hemachandra, I. Shperling, and V. K. Wei. Using simulated annealing to design good codes. *IEEE Transactions on Information Theory*, 33(1):116–123, 1987.

[Gambardella et al., 1999] L. M. Gambardella, É. D. Taillard, and M. Dorigo. Ant colonies for the quadratic assignment problem. *Journal of the Operational Research Society*, 50(2):167–176, 1999.

[Gamst, 1986] A. Gamst. Some lower bounds for a class of frequency assignment problems. *IEEE Transactions of Vehicular Technology*, 35(1):8–14, 1986.

[García-Pelayo and Stadler, 1997] R. García-Pelayo and P. F. Stadler. Correlation length, isotropy, and meta-stable states. *Physica D*, 107(2–4):240–254, 1997.

[Garey and Johnson, 1979] M. R. Garey and D. S. Johnson. *Computers and Intractability: A Guide to the Theory of $\mathcal{NP}$-Completeness*. Freeman, San Francisco, CA, USA, 1979.

[Garey et al., 1976a] M. R. Garey, D. S. Johnson, and R. Sethi. The complexity of flowshop and jobshop scheduling. *Mathematics of Operations Research*, 1:117–129, 1976.

[Garey et al., 1976b] M. R. Garey, D. S. Johnson, and H. C. So. An application of graph coloring to printed circuit testing. *IEEE Transactions on Circuits and Systems*, 23:591–599, 1976.

[Garnier and Kallel, 2002] J. Garnier and L. Kallel. Efficiency of local search with multiple local optima. *SIAM Journal Discrete Mathematics*, 15(1):122–141, 2002.

[Geman and Geman, 1984] S. Geman and D. Geman. Stochastic relaxation, Gibbs distribution, and the Bayesian restoration of images. *IEEE Transactions on Pattern Analysis and Machine Intelligence*, 6:721–741, 1984.

[Gendreau et al., 1992] M. Gendreau, A. Hertz, and G. Laporte. New insertion and post optimization procedures for the traveling salesman problem. *Operations Research*, 40(6):1086–1094, 1992.

[Gent and Walsh, 1993a] I. P. Gent and T. Walsh. An empirical analysis of search in GSAT. Journal of *Artificial Intelligence Research*, 1:47–59, 1993.

[Gent and Walsh, 1993b] I. P. Gent and T. Walsh. Towards an understanding of hill-climbing procedures for SAT. In *Proceedings of the 10th National Conference on Artificial Intelligence*, pages 28–33. AAAI Press/The MIT Press, Menlo Park, CA, USA, 1993.

[Gent and Walsh, 1995] I. P. Gent and T. Walsh. Unsatisfied variables in local search. In J. Hallam, editor, *Hybrid Problems, Hybrid Solutions*, pages 73–85. IOS Press, Amsterdam, The Netherlands, 1995.

[Gent et al., 1997] I. P. Gent, S. A. Grant, E. MacIntyre, P. Prosser, P. Shaw, B. M. Smith, and T. Walsh. How not to do it. Research Report 97.27, School of Computing, University of Leeds, UK, May 1997.

[Gent et al., 2000] I. P. Gent, H. van Maaren, and T. Walsh, editors. *SAT2000 — Highlights of Satisfiability Research in the Year 2000*. IOS Press, Amsterdam, The Netherlands, 2000.

[Gent et al., 2003] I. P. Gent, T. Walsh, and B. Selman. CSPLIB: A problem library for constraints. `http://csplib.org`, 2003. Version visited last on 15 September 2003.

[Gent, 2002] I. P. Gent. Arc consistency in SAT. In F. van Harmelen, editor, *Proceedings of the 15th European Conference on Artificial Intelligence*, pages 121–125. IOS Press, Amsterdam, The Netherlands, 2002.

[Gerry et al., 1999] N. P. Gerry, N. E. Witowski, J. Day, R. P. Hammer, G. Barany, and F. Barany. Universal DNA microarray method for multiplex detection of low abundance point mutations. *Journal of Molecular Biology*, 292(2):251–261, 1999.

[Giffler and Thompson, 1960] B. Giffler and G. L. Thompson. Algorithms for solving production scheduling problems. *Operations Research*, 8(4):487–503, 1960.

[Gill, 1977] J. Gill. Computational complexity of probabilistic turing machines. *SIAM Journal on Computing*, 6(4):675–695, 1977.

[Gilmore et al., 1985] P. C. Gilmore, E. L. Lawler, and D. B. Shmoys. Well-solved special cases. In E. L. Lawler, J. K. Lenstra, A. H. G. Rinnooy Kan, and D. B. Shmoys, editors, *The Traveling Salesman Problem*, pages 207–249. John Wiley & Sons, Chichester, UK, 1985.

[Glover and Hanafi, 2002] F. Glover and S. Hanafi. Tabu search and finite convergence. *Discrete Applied Mathematics*, 119(1–2):3–36, 2002.

[Glover and Kochenberger, 2002] F. Glover and G. Kochenberger, editors. *Handbook of Metaheuristics*. Kluwer Academic Publishers, Norwell, MA, USA, 2002.

[Glover and Laguna, 1997] F. Glover and M. Laguna. *Tabu Search*. Kluwer Academic Publishers, Boston, MA, USA, 1997.

[Glover et al., 2001] F. Glover, G. Gutin, A. Yeo, and A. Zverovich. Construction heuristics for the asymmetric TSP. *European Journal of Operational Research*, 129(3):555–568, 2001.

[Glover et al., 2002] F. Glover, M. Laguna, and R. Martí. Scatter search and path relinking: Advances and applications. In F. Glover and G. Kochenberger, editors, *Handbook of Metaheuristics*, pages 1–35. Kluwer Academic Publishers, Norwell, MA, USA, 2002.

[Glover, 1977] F. Glover. Heuristics for integer programming using surrogate constraints. *Decision Sciences*, 8:156–164, 1977.

[Glover, 1986] F. Glover. Future paths for integer programming and links to artificial intelligence. *Computers & Operations Research*, 13(5):533–549, 1986.

[Glover, 1989] F. Glover. Tabu search – part I. *ORSA Journal on Computing*, 1(3):190–206, 1989.

[Glover, 1990] F. Glover. Tabu search – part II. *ORSA Journal on Computing*, 2(1):4–32, 1990.

[Glover, 1996] F. Glover. Ejection chain, reference structures and alternating path methods for traveling salesman problems. *Discrete Applied Mathematics*, 65(1–3):223–253, 1996.

[Goemans and Williamson, 1994] M. X. Goemans and D. P. Williamson. A new 3/4 approximation algorithm for the maximum satisfiability problem. *SIAM Journal Discrete Mathematics*, 7(4):656–666, 1994.

[Goemans and Williamson, 1995] M. X. Goemans and D. P. Williamson. Improved approximation algorithms for maximum cut and satisfiability problems using semidefinite programming. *Journal of the ACM*, 42(6):1115–1145, 1995.

[Goldberg, 1979] A. Goldberg. On the complexity of the satisfiability problem. Courant Computer Science Report 16, New York University, NY, USA, 1979.

[Goldberg, 1989] D. E. Goldberg. *Genetic Algorithms in Search, Optimization, and Machine Learning*. Addison-Wesley, Reading, MA, USA, 1989.

[Golden and Steward, 1985] B. L. Golden and W. Steward. Empirical analysis of heuristics. In E. L. Lawler, J. K. Lenstra, A. H. G. Rinnooy Kan, and D. B. Shmoys, editors, *The Travelling Salesman Problem*, pages 207–249. John Wiley & Sons, Chichester, UK, 1985.

[Gomes and Selman, 1997a] C. P. Gomes and B. Selman. Algorithm portfolio design: Theory vs. practice. In D. Geiger and P. P. Shenoy, editors, *Proceedings of the Thirteenth Conference on Uncertainty in Artificial Intelligence*, pages 238–245. Morgan Kaufmann Publishers, San Francisco, CA, USA, 1997.

[Gomes and Selman, 1997b] C. P. Gomes and B. Selman. Problem structure in the presence of perturbations. In *Proceedings of the Fourteenth National Conference on Artificial Intelligence*, pages 221–226. AAAI Press/The MIT Press, Menlo Park, CA, USA, 1997.

[Gomes et al., 1997] C. P. Gomes, B. Selman, and N. Crato. Statistical analysis of backtracking on inconsistent CSPs. In G. Smolka, editor, *Principles and Practice of Constraint Programming — CP 1997*, volume 1330 of *Lecture Notes in Computer Science*, pages 121–135. Springer Verlag, Berlin, Germany, 1997.

[Gomes et al., 1998] C. P. Gomes, B. Selman, and H. Kautz. Boosting combinatorial search through randomization. In *Proceedings of the Fifteenth National Conference on Artificial Intelligence*, pages 431–437. AAAI Press/The MIT Press, Menlo Park, CA, USA, 1998.

[Gonzalez and Sahni, 1976] T. Gonzalez and S. Sahni. Open shop scheduling to minimize finish time. *Journal of the ACM*, 23(4):665–679, 1976.

[Gottlieb et al., 2002] J. Gottlieb, E. Marchiori, and C. Rossi. Evolutionary algorithms for the satisfiability problem. *Evolutionary Computation*, 10(1):35–50, 2002.

[Grabowski and Wodecki, 2001] J. Grabowski and M. Wodecki. A new very fast tabu search algorithm for the job shop problem. Technical Report 21/2001, Wroclaw University of Technology, Institute of Engineering Cybernetics, Wroclaw, Poland, 2001.

[Grabowski and Wodecki, 2002] J. Grabowski and M. Wodecki. A very fast tabu search algorithm for the flow shop problem with makespan criterion. Technical Report 63/2002, Wroclaw University of Technology, Institute of Engineering Cybernetics, Wroclaw, Poland, 2002.

[Graham et al., 1979] R. L. Graham, E. L. Lawler, J. K. Lenstra, and A. H. G. Rinnooy Kan. Optimisation and approximation in deterministic sequencing and scheduling: A survey. *Annals of Discrete Mathematics*, 5:287–326, 1979.

[Grant and Smith, 1996] S. A. Grant and B. M. Smith. The phase transition behavior of maintaining arc consistency. In W. Wahlster, editor, *Proceedings of the 12th European Conference on Artificial Intelligence*, pages 175–179. John Wiley & Sons, Chichester, UK, 1996.

[Grosso et al., 2004] A. Grosso, F. Della Croce, and R. Tadei. An enhanced dynasearch neighborhood for the single-machine total weighted tardiness scheduling problem. *Operations Research Letters*, 32(1):68–72, 2004.

[Grötschel and Holland, 1991] M. Grötschel and O. Holland. Solution of large-scale symmetric traveling salesman problems. *Mathematical Programming*, 51:141–202, 1991.

[Grötschel and Padberg, 1985] M. Grötschel and M. W. Padberg. Polyhedral theory. In E. L. Lawler, J. K. Lenstra, A. H. G. Rinnooy Kan, and D. B. Shmoys, editors, *The Traveling Salesman Problem*, pages 251–306. John Wiley & Sons, Chichester, UK, 1985.

[Grötschel and Padberg, 1993] M. Grötschel and M. W. Padberg. Ulysses 2000: In search of optimal solutions to hard combinatorial problems. SC 93-34, Konrad-Zuse-Zentrum für Informationstechnik, Berlin, Germany, 1993.

[Gu and Puri, 1995] J. Gu and R. Puri. Asynchronous Circuit Synthesis with Boolean Satisfiability. *IEEE Transactions of Computer-Aided Design of Integrated Circuits and Systems*, 14(8):961–973, 1995.

[Gu et al., 1997] J. Gu, P. Purdom, J. Franco, and B. W. Wah. Algorithms for the satisfiability (SAT) problem: A survey. In D. Du, J. Gu, and P. M. Pardalos, editors, *Satisfiability Problem: Theory and Applications*, volume 35 of *DIMACS Series on Discrete Mathematics and Theoretical Computer Science*, pages 19–151. American Mathematical Society, Providence, RI, USA, 1997.

[Gu, 1992] J. Gu. Efficient local search for very large-scale satisfiability problems. *SIGART Bulletin*, 3(1):8–12, 1992.

[Guéret and Prins, 1999] C. Guéret and C. Prins. A new lower bound for the open-shop problem. *Annals of Operations Research*, 92:165–183, 1999.

[Gutin and Punnen, 2002] G. Gutin and A. P. Punnen, editors. *The Traveling Salesman Problem and Its Variations*. Kluwer Academic Publishers, Dordrecht, The Netherlands, 2002.

[Gutjahr, 2002] W. J. Gutjahr. ACO algorithms with guaranteed convergence to the optimal solution. *Information Processing Letters*, 82(3):145–153, 2002.

H   [Haahr, 2003] M. Haahr. random.org, true random number service. `http://random.org`, 2003. Version visited last on 15 September 2003.

[Hahn and Krarup, 2001] P. Hahn and J. Krarup. A hospital facility layout problem finally solved. *Journal of Intelligent Manufacturing*, 12(5–6):487–496, 2001.

[Hahn et al., 2001] P. Hahn, W. L. Hightower, T. A. Johnson, M. Guignard-Spielberg, and C. Roucairol. Tree elaboration strategies in branch and bound algorithms for solving the quadratic assignment problem. *Yugoslavian Journal of Operational Research*, 11(1):41–60, 2001.

[Hahn, 2003] P. Hahn. QAPLIB — a quadratic assignment problem library. `http://www.seas.upenn.edu/qaplib`, 2003. Version visited last on 15 September 2003.

[Hajek, 1988] B. Hajek. Cooling schedules for optimal annealing. *Mathematics of Operations Research*, 13(2):311–329, 1988.

[Hale, 1980] W. K. Hale. Frequency assignment: Theory and applications. *Proceedings of the IEEE*, 68(12):1497–1513, 1980.

[Hall and Sriskandrajah, 1996] N. G. Hall and C. Sriskandrajah. A survey of machine scheduling problems with blocking and no-wait in process. *Operations Research*, 44(3):510–525, 1996.

[Halldórsson, 1993] M. M. Halldórsson. A still better performance guarantee for approximate graph coloring. *Information Processing Letters*, 45(1):19–23, 1993.

[Halldórsson, 2000] M. M. Halldórsson. Approximations of weighted independent set and hereditary subset problems. *Journal of Graphs Algorithms and Applications*, 4(1):1–16, 2000.

[Hamiez and Hao, 2003] J.-P. Hamiez and J.-K. Hao. An analysis of solution properties of the graph coloring problem. In M. G. C. Resende and J. P. de Sousa, editors, *Meta-heuristics: Computer Decision-Making*, pages 325–346. Kluwer Academic Publishers, Boston, MA, USA, 2003.

[Hanafi, 2001] S. Hanafi. On the convergence of tabu search. *Journal of Heuristics*, 7(1):47–58, 2001.

[Hansen and Jaumard, 1990] P. Hansen and B. Jaumard. Algorithms for the maximum satisfiability problem. *Computing*, 44:279–303, 1990.

[Hansen and Mladenović, 1999] P. Hansen and N. Mladenović. An introduction to variable neighborhood search. In S. Voss, S. Martello, I. H. Osman, and C. Roucairol, editors, *Meta-Heuristics: Advances and Trends in Local Search Paradigms for Optimization*, pages 433–458. Kluwer Academic Publishers, Boston, MA, USA, 1999.

[Hansen and Mladenovic, 2001a] P. Hansen and N. Mladenovic. Developments of variable neighborhood search. Les Cahiers du GERAD G-2001-24, Groupe d'études et de recherche en analyse des décisions (GERAD), Montreal, Canada, 2001.

[Hansen and Mladenović, 2001b] P. Hansen and N. Mladenović. Variable neighbourhood decomposition search. *Journal of Heuristics*, 7(4):335–350, 2001.

[Hansen and Mladenović, 2002] P. Hansen and N. Mladenović. Variable neighborhood search. In F. Glover and G. Kochenberger, editors, *Handbook of Metaheuristics*, pages 145–184. Kluwer Academic Publishers, Norwell, MA, USA, 2002.

[Hansen and Ribeiro, 2001] P. Hansen and C. C. Ribeiro, editors. *Essays and Surveys on Metaheuristics*. Kluwer Academic Publishers, Boston, MA, USA, 2001.

[Hansen et al., 2000] P. Hansen, B. Jaumard, N. Mladenović, and A. D. Parreira. Variable neighbourhood search for maximum weighted satisfiability problem. Les Cahiers du GERAD G-2000-62, Groupe d'études et de recherche en analyse des décisions (GERAD), Montreal, Canada, 2000.

[Hao and Pannier, 1998] J. K. Hao and J. Pannier. Simulated annealing and tabu search for constraint solving. In E. Boros and R. Greiner, editors, *Electronic Proceedings of the Fifth International Symposium on Artificial Intelligence and Mathematics*, Fort Lauderdale, Florida, USA, 1998. Available at http://rutcor.rutgers.edu/~amai/aimath98/Proceedings.html.

[Haralick and Elliot, 1980] R. Haralick and G. Elliot. Increasing tree search efficiency for constraint satisfaction problems. *Artificial Intelligence*, 14(3):263–313, 1980.

[Harrison, 1978] M. A. Harrison. *Introduction to Formal Language Theory*. Addison-Wesley, Reading, MA, USA, 1978.

[Hart and Shogan, 1987] J. P. Hart and A. W. Shogan. Semi-greedy heuristics: An empirical study. *Operations Research Letters*, 6(3):107–114, 1987.

[Håstad, 1997] J. Håstad. Some optimal inapproximability results. In F. T. Leighton and P. Shor, editors, *Proceedings of the Twenty-ninth Annual ACM Symposium on Theory of Computing*, pages 1–10. ACM Press, New York, NY, USA, 1997.

[Håstad, 2001] J. Håstad. Some optimal inapproximability results. *Journal of the ACM*, 48(4):798–859, 2001.

[Haupt, 1989] R. Haupt. A survey of priority rule-based scheduling. *OR Spektrum*, 11:3–16, 1989.

[Held and Karp, 1970] M. Held and R. M. Karp. The traveling salesman problem and minimum spanning trees. *Operations Research*, 18(6):1138–1162, 1970.

[Held and Karp, 1971] M. Held and R. M. Karp. The traveling salesman problem and minimum spanning trees: Part II. *Mathematical Programming*, 1(1):16–25, 1971.

[Held et al., 1974] M. Held, P. Wolfe, and H. P. Crowder. Validation of subgradient optimization. *Mathematical Programming*, 6:62–88, 1974.

[Helsgaun, 2000] K. Helsgaun. An effective implementation of the Lin-Kernighan traveling salesman heuristic. *European Journal of Operational Research*, 126(1):106–130, 2000.

[Helsgaun, 2003] K. Helsgaun. LKH. `http://www.dat.ruc.dk/~keld/research/LKH`, 2003. Version visited last on 15 September 2003.

[Hentenryck and Michel, 2003] P. Van Hentenryck and L. Michel. Control abstractions for local search. In F. Rossi, editor, *Principles and Practice of Constraint Programming – CP 2003*, volume 2833 of *Lecture Notes in Computer Science*, pages 65–80. Springer Verlag, Berlin, Germany, 2003.

[Henzinger, 1996] T. A. Henzinger. The theory of hybrid automata. In *Proceedings of the 11th Annual IEEE Symposium on Logic in Computer Science*, pages 278–292. IEEE Computer Society Press, Los Alamitos, CA, USA, 1996.

[Hertz and de Werra, 1987] A. Hertz and D. de Werra. Using tabu search techniques for graph coloring. *Computing*, 39:345–351, 1987.

[Hertz et al., 1997] A. Hertz, É. D. Taillard, and D. de Werra. A tutorial on tabu search. In E. H. L. Aarts and J. K. Lenstra, editors, *Local Search in Combinatorial Optimization*, pages 121–136. John Wiley & Sons, Chichester, UK, 1997.

[Hesse, 1943] H. Hesse. *Das Glasperlenspiel*. Fretz & Wasmuth, Zürich, Switzerland, 1943.

[Hirsch and Kojevnikov, 2001] E. Hirsch and A. Kojevnikov. Solving Boolean satisfiability using local search guided by unit clause elimination. In T. Walsh, editor, *Principles and Practice of Constraint Programming – CP 2001*, volume 2239 of *Lecture Notes in Computer Science*, pages 605–609. Springer Verlag, Berlin, Germany, 2001.

[Hoffman and Wolfe, 1985] A. Hoffman and P. Wolfe. History. In E. L. Lawler, J. K. Lenstra, A. H. G. Rinnooy Kan, and D. B. Shmoys, editors, *The Traveling Salesman Problem*, pages 1–15. John Wiley & Sons, Chichester, UK, 1985.

[Hoffmann and Padberg, 1993] K. L. Hoffmann and M. W. Padberg. Solving airline crew scheduling problems by branch-and-cut. *Management Science*, 39(6):657–682, 1993.

[Hofstadter, 1979] D. R. Hofstadter. *Gödel, Escher, Bach: An Eternal Golden Braid*. Basic Books, New York, NY, USA, 1979.

[Holland, 1975] J. H. Holland. *Adaption in Natural and Artificial Systems*. The University of Michigan Press, Ann Arbor, MI, 1975.

[Holte, 2001] R. C. Holte. Combinatorial auctions, knapsack problems, and hill-climbing search. In E. Stroulia and S. Matwin, editors, *Advances in Artificial Intelligence, 14th Conference of the Canadian Society for Computational Studies of Intelligence*, volume 2056 of *Lecture Notes in Computer Science*, pages 57–66. Springer Verlag, Berlin, Germany, 2001.

[Homaifar et al., 1993] A. Homaifar, S. Guan, and G. Liepins. A new approach on the traveling salesman problem by genetic algorithms. In S. Forrest, editor, *Proceedings of the Sixth International Conference on Genetic Algorithms (ICGA'95)*, pages 460–466. Morgan Kaufmann Publishers, San Mateo, CA, USA, 1993.

[Hong et al., 1997] I. Hong, A. B. Kahng, and B. R. Moon. Improved large-step Markov chain variants for the symmetric TSP. *Journal of Heuristics*, 3(1):63–81, 1997.

[Honkala and Östergård, 1997] I. S. Honkala and P. R. J. Östergård. Code design. In E. H. L. Aarts and J. K. Lenstra, editors, *Local Search in Combinatorial Optimization*, pages 415–456. John Wiley & Sons, Chichester, UK, 1997.

[Hooker, 1994] J. N. Hooker. Needed: An empirical science of algorithms. *Operations Research*, 42(2):201–212, 1994.

[Hooker, 1996] J. N. Hooker. Testing heuristics: We have it all wrong. *Journal of Heuristics*, 1(1):33–42, 1996.

[Hoos and Boutilier, 2000] H. H. Hoos and C. Boutilier. Solving combinatorial auctions using stochastic local search. In *Proceedings of the Seventeenth National Conference on Artificial Intelligence*, pages 22–29. AAAI Press/The MIT Press, Menlo Park, CA, USA, 2000.

[Hoos and O'Neill, 2000] H. H. Hoos and K. O'Neill. Stochastic local search methods for dynamic SAT — an initial investigation. In C. P. Gomes and H. H. Hoos, editors, *Proceedings of the AAAI-2002 Workshop on Leveraging Probability and Uncertainty in Computation*, pages 22–26, 2000.

[Hoos and Stützle, 1998] H. H. Hoos and T. Stützle. Evaluating Las Vegas algorithms — pitfalls and remedies. In G. F. Cooper and S. Moral, editors, *Proceedings of the Fourteenth Conference on Uncertainty in Artificial Intelligence*, pages 238–245. Morgan Kaufmann Publishers, San Francisco, CA, USA, 1998.

[Hoos and Stützle, 1999] H. H. Hoos and T. Stützle. Characterising the behaviour of stochastic local search. *Artificial Intelligence*, 112(1–2):213–232, 1999.

[Hoos and Stützle, 2000a] H. H. Hoos and T. Stützle. Local search algorithms for SAT: An empirical evaluation. *Journal of Automated Reasoning*, 24(4):421–481, 2000.

[Hoos and Stützle, 2000b] H. H. Hoos and T. Stützle. SATLIB: An Online Resource for Research on SAT. In H. van Maaren I. P. Gent and T. Walsh, editors, *SAT 2000*, pages 283–292. IOS Press, Amsterdam, The Netherlands, 2000.

[Hoos and Stützle, 2003] H. H. Hoos and T. Stützle. SATLIB — The Satisfiability Library. http://www.satlib.org, 2003. Version visited last on 15 September 2003.

[Hoos et al., 2002] H. H. Hoos, K. Smyth, and T. Stützle. Search space features underlying the performance of stochastic local search algorithms for MAX-SAT. Technical Report TR-2002-13, Computer Science Department, University of British Columbia, Vancouver, BC, Canada, 2002.

[Hoos et al., 2003] H. H. Hoos, K. Smyth, and T. Stützle. A new perspective on stochastic local search for MAX-SAT. Technical Report TR-2003-19, Computer Science Department, University of British Columbia, Vancouver, BC, Canada, 2003.

[Hoos, 1998] H. H. Hoos. *Stochastic Local Search — Methods, Models, Applications*. PhD thesis, TU Darmstadt, FB Informatik, Darmstadt, Germany, 1998.

[Hoos, 1999a] H. H. Hoos. On the run-time behaviour of stochastic local search algorithms for SAT. In *Proceedings of the Sixteenth National Conference on Artificial Intelligence*, pages 661–666. AAAI Press/The MIT Press, Menlo Park, CA, USA, 1999.

[Hoos, 1999b] H. H. Hoos. SAT-encodings, search space structure, and local search performance. In T. Dean, editor, *Proceedings of the Sixteenth International Joint Conference on Artificial Intelligence*, pages 296–302. Morgan Kaufmann Publishers, San Francisco, CA, USA, 1999.

[Hoos, 1999c] H. H. Hoos. *Stochastic Local Search — Methods, Models, Applications*. infix-Verlag, Sankt Augustin, Germany, 1999.

[Hoos, 2002a] H. H. Hoos. An adaptive noise mechanism for WalkSAT. In *Proceedings of the Eighteenth National Conference on Artificial Intelligence*, pages 655–660. AAAI Press/The MIT Press, Menlo Park, CA, USA, 2002.

[Hoos, 2002b] H. H. Hoos. A mixture-model for the behaviour of SLS algorithms for SAT. In *Proceedings of the Eighteenth National Conference on Artificial Intelligence*, pages 661–667. AAAI Press/The MIT Press, Menlo Park, CA, USA, 2002.

[Hopcroft et al., 2001] J. E. Hopcroft, R. Motwani, and J. D. Ullman. *Introduction to Automata Theory, Languages, and Computation*. Addison-Wesley, Reading, MA, USA, 2001.

[Hordijk et al., 2003] W. Hordijk, J. F. Fontanari, and P. F. Stadler. Shapes of tree representations of spin-glass landscapes. *Journal of Physics A*, 36(13):3103–3110, 2003.

[Hossain and Steihaug, 2002] S. Hossain and T. Steihaug. Graph coloring in the estimation of mathematical derivatives. In D. S. Johnson, A. Mehrotra, and M. Trick, editors, *Proceedings of the Computational Symposium on Graph Coloring and its Generalizations*, pages 9–16, Ithaca, NY, USA, 2002.

[Houdayer and Martin, 1999] J. Houdayer and O. C. Martin. Renormalization for discrete optimization. *Physical Review Letters*, 83(5):1030–1033, 1999.

[Housos and Elmoth, 1997] E. Housos and T. Elmoth. Automatic optimization of subproblems in scheduling airlines crews. *Interfaces*, 27(5):68–77, 1997.

[Hromkovič, 2003] J. Hromkovič. *Algorithmics for Hard Problems*, second edition. Springer Verlag, Berlin, Germany, 2003.

[Huberman et al., 1997] B. A. Huberman, R. Lukose, and T. Hogg. An economic approach to hard computational problems. *Science*, 275:51–54, 1997.

[Hutter et al., 2002] F. Hutter, D. A. D. Tompkins, and H. H. Hoos. Scaling and probabilistic smoothing: Efficient dynamic local search for SAT. In P. Van Hentenryck, editor, *Principles and Practice of Constraint Programming – CP 2002*, volume 2470 of *Lecture Notes in Computer Science*, pages 233–248. Springer Verlag, Berlin, Germany, 2002.

[Huynen et al., 1996] M. A. Huynen, P. F. Stadler, and W. Fontana. Smoothness within ruggedness: The role of neutrality in adaptation. *Proceedings of the National Academy of Sciences of the United States of America*, 93(1):397–401, 1996.

I   [Ibarra and Kim, 1975] O. H. Ibarra and C. E. Kim. Fast approximation for the knapsack and sum of subset problems. *Journal of the ACM*, 22(4):463–468, 1975.

[Ishibuchi et al., 1995] H. Ishibuchi, S. Misaki, and H. Tanaka. Modified simulated annealing algorithms for the flow shop sequencing problem. *European Journal of Operational Research*, 81(2):388–398, 1995.

[Iwama and Miyazaki, 1994] K. Iwama and S. Miyazaki. SAT-variable complexity of hard combinatorial problems. In B. Pehrson and I. Simon, editors, *Technology and*

Foundations–Information Processing '94, Proceedings of the IFIP 13th World Computer Congress, volume 1, pages 253–258. North-Holland, Amsterdam, The Netherlands, 1994.

J

[Jackson et al., 1990] R. Jackson, P. Boggs, S. Nash, and S. Powell. Report of the ad-hoc committee to revise the guidelines for reporting computational experiments in mathematical programming. *Mathematical Programming*, 49:413–425, 1990.

[Jackson, 1955] J. R. Jackson. Scheduling a production line to minimize maximum lateness. Technical Report 43, Management Science Research Project, University of California, CA, USA, 1955.

[Jacobs and Brusco, 1995] L. W. Jacobs and M. J. Brusco. A local search heuristic for large set-covering problems. *Naval Research Logistics*, 42(7):1129–1140, 1995.

[Jain and Meeran, 1999] A. S. Jain and S. Meeran. Deterministic job-shop scheduling: Past, present and future. *European Journal of Operational Research*, 113(2):390–434, 1999.

[Jain et al., 2000] A. S. Jain, B. Rangaswamy, and S. Meeran. New and "stronger" job-shop neighbourhoods: A focus on the method of Nowicki and Smutnicki. *Journal of Heuristics*, 6(4):457–480, 2000.

[Jájá, 1992] J. Jájá. *An Introduction to Parallel Algorithms*. Addison-Wesley, Reading, MA, USA, 1992.

[Jensen and Toft, 1994] T. R. Jensen and B. Toft. *Graph Coloring Problems*. John Wiley & Sons, New York, NY, USA, 1994.

[Jiang et al., 1995] Y. Jiang, H. Kautz, and B. Selman. Solving problems with hard and soft constraints using a stochastic algorithm for MAX-SAT, In *Proceedings of the 1st Workshop on Artificial Intelligence and Operations Research*, Timberline, Oregon, USA, June 6–10, 1995.

[Johnson and McGeoch, 1997] D. S. Johnson and L. A. McGeoch. The travelling salesman problem: A case study in local optimization. In E. H. L. Aarts and J. K. Lenstra, editors, *Local Search in Combinatorial Optimization*, pages 215–310. John Wiley & Sons, Chichester, UK, 1997.

[Johnson and McGeoch, 2002] D. S. Johnson and L. A. McGeoch. Experimental analysis of heuristics for the STSP. In G. Gutin and A. Punnen, editors, *The Traveling Salesman Problem and its Variations*, pages 369–443. Kluwer Academic Publishers, Dordrecht, The Netherlands, 2002.

[Johnson et al., 1988] D. S. Johnson, C. H. Papadimitriou, and M. Yannakakis. How easy is local search? *Journal of Computer and System Science*, 37(1):79–100, 1988.

[Johnson et al., 1989] D. S. Johnson, C. R. Aragon, L. A. McGeoch, and C. Schevon. Optimization by simulated annealing: An experimental evaluation: Part I, graph partitioning. *Operations Research*, 37(6):865–892, 1989.

[Johnson et al., 1991] D. S. Johnson, C. R. Aragon, L. A. McGeoch, and C. Schevon. Optimization by simulated annealing: An experimental evaluation: Part II, graph coloring and number partitioning. *Operations Research*, 39(3):378–406, 1991.

[Johnson et al., 1996] D. S. Johnson, L. A. McGeoch, and E. E. Rothberg. Asymptotic experimental analysis for the Held-Karp traveling salesman bound. In É. Tardos, editor, *Proceedings of the Seventh Annual ACM-SIAM Symposium on Discrete Algorithms*, pages 341–350. Society for Industrial and Applied Mathematics, Philadelphia, PA, USA, 1996.

[Johnson et al., 2002] D. S. Johnson, G. Gutin, L. A. McGeoch, A. Yeo, W. Zhang, and A. Zverovitch. Experimental analysis of heuristics for the ATSP. In G. Gutin and A. Punnen, editors, *The Traveling Salesman Problem and its Variations*, pages 445–487. Kluwer Academic Publishers, Dordrecht, The Netherlands, 2002.

[Johnson et al., 2003a] D. S. Johnson, L. A. McGeoch, F. Glover, and C. Rego. 8th DIMACS implementation challenge: The traveling salesman problem. `http://www.research.att.com/~dsj/chtsp`, 2003. Version visited last on 15 September 2003.

[Johnson et al., 2003b] D. S. Johnson, A. Mehrotra, and M. Trick. Color02/03: Graph coloring and its generalizations. `http://mat.gsia.cmu.edu/COLORING03`, 2003. Version visited last on 15 September 2003.

[Johnson, 1954] S. Johnson. Optimal two- and three-stage production scheduling with setup times included. *Naval Research Logistics Quarterly*, 1:61–68, 1954.

[Johnson, 1974] D. S. Johnson. Approximation algorithms for combinatorial problems. *Journal of Computer and System Science*, 9(3):256–278, 1974.

[Johnson, 1990] D. S. Johnson. Local optimization and the travelling salesman problem. In M. Paterson, editor, *Automata, Languages and Programming, 17th International Colloquium*, volume 443 of *Lecture Notes in Computer Science*, pages 446–461. Springer Verlag, Berlin, Germany, 1990.

[Johnson, 2002] D. S. Johnson. A theoretician's guide to the experimental analysis of algorithms. In M. H. Goldwasser, D. S. Johnson, and C. C. McGeoch, editors, *Data Structures, Near Neighbor Searches, and Methodology: Fifth and Sixth DIMACS Implementation Challenges*, DIMACS Series on Discrete Mathematics and Theoretical Computer Science, pages 215–250. American Mathematical Society, Providence, RI, USA, 2002.

[Jones and Forrest, 1995] T. Jones and S. Forrest. Fitness distance correlation as a measure of problem difficulty for genetic algorithms. In L. J. Eshelman, editor, *Proceedings of the Sixth International Conference on Genetic Algorithms*, pages 184–192. Morgan Kaufmann Publishers, San Mateo, CA, USA, 1995.

[Jonker and Volgenant, 1983] R. Jonker and T. Volgenant. Transforming asymmetric into symmetric traveling salesman problems. *Operations Research Letters*, 2(4):161–163, 1983.

[Joslin and Clements, 1999] D. E. Joslin and D. P. Clements. Squeaky wheel optimization. *Journal of Artificial Intelligence Research*, 10:353–373, 1999.

[Joy et al., 1997] S. Joy, J. Mitchell, and B. Borchers. A branch and cut algorithm for MAX-SAT and weighted MAX-SAT. In D. Du, J. Gu, and P. M. Pardalos, editors, *Satisfiability Problem: Theory and Applications*, volume 35 of *DIMACS Series on Discrete Mathematics and Theoretical Computer Science*, pages 519–536. American Mathematical Society, Providence, RI, USA, 1997.

K [Kabanets, 2002] V. Kabanets. Derandomization: A brief overview. *Bulletin of the European Association for Theoretical Computer Science*, 76:88–103, 2002.

[Kallel et al., 2001] L. Kallel, B. Naudts, and C. R. Reeves. Theoretical aspects of evolutionary computing. In L. Kallel, B. Naudts, and A. Rogers, editors, *Properties of Fitness*

Functions and Search Landscapes, pages 175–206. Springer Verlag, Berlin, Germany, 2001.

[Kanellakis and Papadimitriou, 1980] P. C. Kanellakis and C. H. Papadimitriou. Local search for the asymmetric traveling salesman problem. *Operations Research*, 28(5):1086–1099, 1980.

[Karloff and Zwick, 1997] H. J. Karloff and U. Zwick. A 7/8-approximation algorithm for MAX 3SAT? In *Proceedings of the 38th Annual IEEE Symposium on Foundations of Computer Science*, pages 406–415. IEEE Computer Society Press, Los Alamitos, CA, USA, 1997.

[Karp, 1979] R. M. Karp. A patching algorithm for the nonsymmetric traveling salesman problem. *SIAM Journal on Computing*, 8(4):561–573, 1979.

[Kasif, 1990] S. Kasif. On the parallel complexity of discrete relaxation in constraint satisfaction networks. *Artificial Intelligence*, 45(3):275–286, 1990.

[Kask and Dechter, 1995] K. Kask and R. Dechter. GSAT and local consistency. In C. S. Mellish, editor, *Proceedings of the Fourteenth International Joint Conference on Artificial Intelligence*, pages 616–623. Morgan Kaufmann Publishers, San Francisco, CA, USA, 1995.

[Kask and Dechter, 1996] K. Kask and R. Dechter. A graph-based method for improving GSAT. In *Proceedings of the Thirteenth National Conference on Artificial Intelligence*, pages 350–355. AAAI Press/The MIT Press, Menlo Park, CA, USA, 1996.

[Kask and Dechter, 2001] K. Kask and R. Dechter. A general scheme for automatic generation of search heuristics from specification dependencies. *Artificial Intelligence*, 129(1–2):91–131, 2001.

[Kask, 2000] K. Kask. New search heuristics for Max-CSP. In R. Dechter, editor, *Principles and Practice of Constraint Programming – CP 2000*, volume 1894 of *Lecture Notes in Computer Science*, pages 262–277. Springer Verlag, Berlin, Germany, 2000.

[Katayama and Narihisa, 1999] K. Katayama and H. Narihisa. Iterated local search approach using genetic transformation to the traveling salesman problem. In W. Banzhaf, J. Daida, A. E. Eiben, M. H. Garzon, V. Honavar, M. Jakiela, and R. E. Smith, editors, *Proceedings of the Genetic and Evolutionary Computation Conference (GECCO-1999)*, volume 1, pages 321–328. Morgan Kaufmann Publishers, San Francisco, CA, USA, 1999.

[Kauffman, 1993] S. A. Kauffman. *The Origins of Order*. Oxford University Press, New York, NY, USA, 1993.

[Kauffmann et al., 1988] S. Kauffmann, E. Weinberger, and A. S. Perelson. Maturation of the immune response via adaptive walks on affinity landscapes. In A. S. Perelson, editor, *Theoretical Immunology, Part I*. Addison-Wesley, Reading, MA, USA, 1988.

[Kautz and Selman, 1996] H. Kautz and B. Selman. Pushing the envelope: Planning, propositional logic, and stochastic search. In *Proceedings of the Thirteenth National Conference on Artificial Intelligence*, volume 2, pages 1194–1201. AAAI Press/The MIT Press, Menlo Park, CA, USA, 1996.

[Kautz and Walser, 1999] H. Kautz and J. P. Walser. State-space planning by integer optimization. In *Proceedings of the Sixteenth National Conference on Artificial Intelligence*, pages 526–533. AAAI Press/The MIT Press, Menlo Park, CA, USA, 1999.

[Kennedy et al., 2001] J. Kennedy, R. C. Eberhart, and Y. Shi. *Swarm Intelligence*. Morgan Kaufmann Publishers, San Francisco, CA, USA, 2001.

[Kernighan and Lin, 1970] B. W. Kernighan and S. Lin. An efficient heuristic procedure for partitioning graphs. *Bell Systems Technology Journal*, 49:213–219, 1970.

[Khanna et al., 1994] S. Khanna, R. Motwani, M. Sudan, and U. Vazirani. On syntactic versus computational views of approximability. In *Proceedings of the 35th Annual IEEE Symposium on Foundations of Computer Science*, pages 819–830. IEEE Computer Society Press, Los Alamitos, CA, USA, 1994.

[Kinnear Jr., 1994] K. E. Kinnear Jr. Fitness landscapes and difficulty in genetic programming. In *Proceedings of the 1994 IEEE World Conference on Computational Intelligence*, volume 1, pages 142–147. IEEE Press, Piscataway, NJ, USA, 1994.

[Kirkpatrick and Selman, 1994] S. Kirkpatrick and B. Selman. Critical behavior in the satisfiability of random Boolean expressions. *Science*, 264:1297–1301, 1994.

[Kirkpatrick and Toulouse, 1985] S. Kirkpatrick and G. Toulouse. Configuration space analysis of travelling salesman problems. *Journal de Physique*, 46(8):1277–1292, 1985.

[Kirkpatrick et al., 1983] S. Kirkpatrick, C. D. Gelatt Jr., and M. P. Vecchi. Optimization by simulated annealing. *Science*, 220:671–680, 1983.

[Klee and Minty, 1972] V. Klee and G. J. Minty. How good is the simplex algorithm? In O. Shisha, editor, *Inequalities, III*, pages 159–175. Academic Press, New York, NY, USA, 1972.

[Knox, 1994] J. Knox. Tabu search performance on the symmetric travelling salesman problem. *Computers & Operations Research*, 21(8):867–876, 1994.

[Knuth, 1997] D. E. Knuth. *The Art of Computer Programming, volume 2: Seminumerical Algorithms*, third edition. Addison-Wesley, Reading, MA, USA, 1997.

[Kolen and Pesch, 1994] A. Kolen and E. Pesch. Genetic local search in combinatorial optimization. *Discrete Applied Mathematics*, 48(3):273–284, 1994.

[Koopmans and Beckmann, 1957] T. C. Koopmans and M. J. Beckmann. Assignment problems and the location of economic activities. *Econometrica*, 25:53–76, 1957.

[Koulamas, 1998] C. Koulamas. A new constructive heuristic for the flowshop scheduling problem. *European Journal of Operational Research*, 105(1):66–71, 1998.

[Krakhofer and Stadler, 1996] B. Krakhofer and P. F. Stadler. Local minima in the graph bipartitioning problem. *Europhysics Letters*, 34(2):85–90, 1996.

[Krarup and Pruzan, 1978] J. Krarup and P. M. Pruzan. Computer-aided layout design. *Mathematical Programming Study*, 9:75–94, 1978.

[Krentel, 1989] M. W. Krentel. Structure in locally optimal solutions. In *Proceedings of the 30th Annual IEEE Symposium on Foundations of Computer Science*, pages 216–221. IEEE Computer Society Press, Los Alamitos, CA, USA, 1989.

[Krishnamurthy, 1989] E. V. Krishnamurthy. *Parallel Processing: Principles and Practice*. Addison-Wesley, Reading, MA, USA, 1989.

[Kumar et al., 1999] S. R. Kumar, A. Russell, and R. Sundaram. Approximating latin square extensions. *Algorithmica*, 24(2):128–138, 1999.

[Kwan, 1996] A. C. M. Kwan. Validity of normality assumption in CSP research. In N. Y. Foo and R. Goebel, editors, *PRICAI'96: Topics in Artificial Intelligence*, volume 1114 of *Lecture Notes in Computer Science*, pages 253–263. Springer Verlag, Berlin, Germany, 1996.

L [Laarhoven et al., 1992] P. J. M. Van Laarhoven, E. H. L. Aarts, and J. K. Lenstra. Job shop scheduling by simulated annealing. *Operations Research*, 40(1):113–125, 1992.

[Laburthe and Caseau, 1998] F. Laburthe and Y. Caseau. SALSA: A language for search algorithms. In M. Maher and J.-F. Puget, editors, *Principles and Practice of Constraint Programming — CP 1998*, volume 1520 of *Lecture Notes in Computer Science*, pages 310–324. Springer Verlag, Berlin, Germany, 1998.

[Laguna and Martí, 1999] M. Laguna and R. Martí. GRASP and path relinking for 2-layer straight line crossing minimization. *INFORMS Journal on Computing*, 11(1):44–52, 1999.

[Laguna and Martí, 2001] M. Laguna and R. Martí. A GRASP for coloring sparse graphs. *Computational Optimization and Applications*, 19(2):165–178, 2001.

[Laguna and Martí, 2003] M. Laguna and R. Martí. *Scatter Search: Methodology and Implementations in C*. Kluwer Academic Publishers, Boston, MA, USA, 2003.

[Larrañaga and Lozano, 2001] P. Larrañaga and J. A. Lozano. *Estimation of Distribution Algorithms. A New Tool for Evolutionary Computation*. Kluwer Academic Publishers, Dordrecht, The Netherlands, 2001.

[Larrosa and Dechter, 2002] J. Larrosa and R. Dechter. Boosting search with variable elimination in constraint optimization and constraint satisfaction problems. *Constraints*, 8(3):303–326, 2002.

[Larrosa and Meseguer, 2002] J. Larrosa and P. Meseguer. Partition-based lower bounds for Max-CSP. *Constraints*, 7(3–4):407–419, 2002.

[Larrosa et al., 1999] J. Larrosa, P. Meseguer, and T. Schiex. Maintaining reversible DAC for Max-CSP. *Artificial Intelligence*, 107(1):149–163, 1999.

[Lau and Watanabe, 1996] H. C. Lau and O. Watanabe. Randomized approximation of the constraint satisfaction problem. *Nordic Journal of Computing*, 3(4):405–424, 1996.

[Lau, 2002] H. C. Lau. A new approach for weighted constraint satisfaction. *Constraints*, 7(2):151–165, 2002.

[Lawler et al., 1985] E. L. Lawler, J. K. Lenstra, A. H. G. Rinnooy Kan, and D. B. Shmoys. *The Travelling Salesman Problem*. John Wiley & Sons, Chichester, UK, 1985.

[Lawler, 1963] E. L. Lawler. The quadratic assignment problem. *Management Science*, 9:586–599, 1963.

[Lawler, 1976] E. L. Lawler. *Combinatorial Optimization: Networks and Matroids*. Holt, Rinehart, and Winston, New York, NY, USA, 1976.

[Lawler, 1977] E. L. Lawler. A pseudopolynomial algorithm for sequencing jobs to minimize total tardiness. *Annals of Discrete Mathematics*, 1:331–342, 1977.

[Lawrence, 1984] S. Lawrence. Resource constrained project scheduling: an experimental investigation of heuristic scheduling techniques (supplement). Technical report, Graduate School of Industrial Administration, Carnegie-Mellon University, Pittsburgh, Pennsylvania, 1984.

[Laywine and Mullen, 1998] C. F. Laywine and G. L. Mullen. *Discrete Mathematics Using Latin Squares*. John Wiley & Sons, Chichester, UK, 1998.

[Leighton, 1979] F. T. Leighton. A graph coloring algorithm for large scheduling problems. *Journal of Research of the National Bureau of Standards*, 84(6):489–506, 1979.

[Lenstra and Rinnooy Kan, 1979] J. K. Lenstra and A. H. G. Rinnooy Kan. Computational complexity of discrete optimization problems. *Annals of Discrete Mathematics*, 4:121–140, 1979.

[Lenstra et al., 1977] J. K. Lenstra, A. H. G. Rinnooy Kan, and P. Bruckner. Complexity of machine scheduling problem. *Annals of Discrete Mathematics*, 1:343–362, 1977.

[Leyton-Brown et al., 2000a] K. Leyton-Brown, M. Pearson, and Y. Shoham. Towards a universal test suite for combinatorial auction algorithms. In *Proceedings of the 2nd ACM Conference on Electronic Commerce*, pages 66–76. ACM Press, New York, NY, USA, 2000.

[Leyton-Brown et al., 2000b] K. Leyton-Brown, Y. Shoham, and M. Tennenholtz. An algorithm for multiunit combinatorial auctions. In *Proceedings of the Seventeenth National Conference on Artificial Intelligence*, pages 56–61. AAAI Press/The MIT Press, Menlo Park, CA, USA, 2000.

[Leyton-Brown et al., 2002] K. Leyton-Brown, E. Nudelman, and Y. Shoham. Learning the empirical hardness of optimization problems: The case of combinatorial auctions. In P. Van Hentenryck, editor, *Principles and Practice of Constraint Programming – CP 2002*, volume 2470 of *Lecture Notes in Computer Science*, pages 556–572. Springer Verlag, Berlin, Germany, 2002.

[Leyton-Brown et al., 2003] K. Leyton-Brown, E. Nudelman, G. Andrew, J. McFadden, and Y. Shoham. A portfolio approach to algorithm selection. In F. Rossi, editor, *Principles and Practice of Constraint Programming – CP03*, volume 2833 of *Lecture Notes in Computer Science*, pages 899–903. Springer Verlag, Berlin, Germany, 2003.

[Leyton-Brown, 2003] K. Leyton-Brown. *Resource Allocation in Competitive Multiagent Systems*. PhD thesis, Stanford University, Computer Science Department, Stanford, CA, USA, 2003.

[Li and Pardalos, 1992] Y. Li and P. M. Pardalos. Generating quadratic assignment test problems with known optimal permutations. *Computational Optimization and Applications*, 1:163–184, 1992.

[Li et al., 1994] Y. Li, P. M. Pardalos, and M. G. C. Resende. A greedy randomized adaptive search procedure for the quadratic assignment problem. In P. M. Pardalos and H. Wolkowicz, editors, *Quadratic Assignment and Related Problems*, volume 16 of *DIMACS Series on Discrete Mathematics and Theoretical Computer Science*, pages 237–261. American Mathematical Society, Providence, RI, USA, 1994.

[Liaw, 1999] C.-F. Liaw. A tabu search algorithm for the open shop scheduling problem. *Computers & Operations Research*, 26(2):109–126, 1999.

[Liaw, 2000] C.-F. Liaw. A hybrid genetic algorithm for the open shop scheduling problem. *European Journal of Operational Research*, 124(1):28–42, 2000.

[Lin and Kernighan, 1973] S. Lin and B. W. Kernighan. An effective heuristic algorithm for the travelling salesman problem. *Operations Research*, 21(2):498–516, 1973.

[Lin, 1965] S. Lin. Computer solutions for the traveling salesman problem. *Bell Systems Technology Journal*, 44:2245–2269, 1965.

[Little et al., 1963] J. D. C. Little, K. G. Murty, D. W. Sweeney, and C. Karel. An algorithm for the traveling salesman problem. *Operations Research*, 11:972–989, 1963.

[Løkketangen, 2002] A. Løkketangen. Heuristics for 0/1 mixed integer programming. In P. M. Pardalos and M. G. C. Resende, editors, *Handbook of Applied Optimization*, pages 474–477. Oxford University Press, New York, NY, USA, 2002.

[Lourenço and Serra, 2002] H. R. Lourenço and D. Serra. Adaptive search heuristics for the generalized assignment problem. *Mathware & Soft Computing*, 9(2–3):209–234, 2002.

[Lourenço et al., 2001] H. R. Lourenço, R. Portugal, and J. P. Paixão. Multiobjective metaheuristics for the bus-driver scheduling problem. *Transportation Science*, 35(3):331–343, 2001.

[Lourenço et al., 2002] H. R. Lourenço, O. Martin, and T. Stützle. Iterated local search. In F. Glover and G. Kochenberger, editors, *Handbook of Metaheuristics*, pages 321–353. Kluwer Academic Publishers, Norwell, MA, USA, 2002.

[Lourenço, 1995] H. R. Lourenço. Job-shop scheduling: Computational study of local search and large-step optimization methods. *European Journal of Operational Research*, 83(2):347–364, 1995.

[Luby et al., 1993] M. Luby, A. Sinclair, and D. Zuckerman. Optimal speedup of Las Vegas algorithms. *Information Processing Letters*, 47(4):173–180, 1993.

[Lundy and Mees, 1986] M. Lundy and A. Mees. Convergence of an annealing algorithm. *Mathematical Programming*, 34:111–124, 1986.

M   [Mackworth, 1977] A. K. Mackworth. Consistency in networks of relations. *Artificial Intelligence*, 8(1):99–118, 1977.

[MacWilliams and Sloane, 1977] F. J. MacWilliams and N. J. A. Sloane. *The Theory of Error-Correcting Codes*. North-Holland, Amsterdam, The Netherlands, 1977.

[Mahajan and Ramesh, 1995] S. Mahajan and H. Ramesh. Derandomising semidefinite programming based approximation algorithms. In *Proceedings of the 36th Annual IEEE Symposium on Foundations of Computer Science*, pages 162–169. IEEE Computer Society Press, Los Alamitos, CA, USA, 1995.

[Malek et al., 1989] M. Malek, M. Guruswamy, and M. Pandya. Serial and parallel simulated annealing and tabu search algorithms for the traveling salesman problem. *Annals of Operations Research*, 21:59–84, 1989.

[Maler et al., 1992] O. Maler, Z. Manna, and A. Pnueli. From timed to hybrid systems. In J. W. de Bakker, C. Huizing, W. P. de Roever, and G. Rozenberg, editors, *Real-Time: Theory in Practice, REX Workshop*, volume 600 of *Lecture Notes in Computer Science*, pages 447–484. Springer Verlag, Berlin, Germany, 1992.

[Maniezzo and Milandri, 2002] V. Maniezzo and M. Milandri. An ant-based framework for very strongly constrained problems. In M. Dorigo, G. Di Caro, and M. Sampels, editors, *Proceedings of ANTS 2002 – From Ant Colonies to Artificial Ants: Third International Workshop on Ant Algorithms*, volume 2463 of *Lecture Notes in Computer Science*, pages 222–227. Springer Verlag, Berlin, Germany, 2002.

[Maniezzo et al., 1994] V. Maniezzo, A. Colorni, and M. Dorigo. The Ant System applied to the quadratic assignment problem. Technical Report IRIDIA/94-28, IRIDIA, Université Libre de Bruxelles, Belgium, 1994.

[Maniezzo, 1999] V. Maniezzo. Exact and approximate nondeterministic tree-search procedures for the quadratic assignment problem. *INFORMS Journal on Computing*, 11(4):358–369, 1999.

[Marathe et al., 2001] A. Marathe, A. Condon, and R. Corn. On combinatorial DNA word design. *Journal of Computational Biology*, 8(3):201–220, 2001.

[Marchiori and Steenbeek, 2000a] E. Marchiori and A. Steenbeek. An evolutionary algorithm for large scale set covering problems with application to airline crew scheduling. In S. Cagnoni, R. Poli, Y. Li, G. Smith, D. Corne, M. J. Oates, E. Hart, P. L. Lanzi, E. J. W. Boers, B. Paechter, and T. C. Fogarty, editors, *Real-World Applications of Evolutionary Computing, EvoWorkshops 2000*, volume 1803 of *Lecture Notes in Computer Science*, pages 367–381. Springer Verlag, Berlin, Germany, 2000.

[Marchiori and Steenbeek, 2000b] E. Marchiori and A. Steenbeek. A genetic local search algorithm for random binary constraint satisfaction problems. In J. Carroll, E. Damiani, H. Haddad, and D. Oppenheim, editors, *Proceedings of the 2000 ACM Symposium on Applied Computing*, pages 458–462. ACM Press, New York, NY, USA, 2000.

[Marquardt, 1963] D. W. Marquardt. An algorithm for least-squares estimation of nonlinear parameters. *Journal of the Society for Industrial and Applied Mathematics*, 11(2):431–441, 1963.

[Martin and Otto, 1996] O. C. Martin and S. W. Otto. Combining simulated annealing with local search heuristics. *Annals of Operations Research*, 63:57–75, 1996.

[Martin et al., 1991] O. C. Martin, S. W. Otto, and E. W. Felten. Large-step Markov chains for the traveling salesman problem. *Complex Systems*, 5(3):299–326, 1991.

[Martin et al., 1992] O. C. Martin, S. W. Otto, and E. W. Felten. Large-step Markov chains for the TSP incorporating local search heuristics. *Operations Research Letters*, 11(4):219–224, 1992.

[Matsumoto and Nishimura, 1998] M. Matsumoto and T. Nishimura. Mersenne twister: A 623-dimensionally equidistributed uniform pseudorandom number generator. *ACM Transactions on Modeling and Computer Simulation*, 8(1):3–30, 1998.

[Matsumoto, 2003] M. Matsumoto. Mersenne twister home page. `http://www.math.keio.ac.jp/~matumoto/emt.html`, 2003. Version visited last on 15 September 2003.

[Matsuo et al., 1987] H. Matsuo, C. J. Suh, and R. S. Sullivan. A controlled search simulated annealing method for the single machine weighted tardiness problem. Technical Report 87-12-2, Department of Management, University of Texas at Austin, Texas, USA, 1987.

[Matsuo et al., 1988] H. Matsuo, C. J. Suh, and R. S. Sullivan. A controlled search simulated annealing method for the general jobshop scheduling problem. Technical Report 03-04-88, Department of Management, University of Texas, at Austin, Texas, USA, 1988.

[Mavridou et al., 1998] T. Mavridou, P. M. Pardalos, L. S. Pitsoulis, and M. G. C. Resende. A GRASP for the biquadratic assignment problem. *European Journal of Operational Research*, 105(3):613–621, 1998.

[Mazure et al., 1997] B. Mazure, L. Sais, and E. Gregoire. TWSAT: A new local search algorithm for SAT – performance and analysis. In *Proceedings of the Fourteenth National Conference on Artificial Intelligence*, pages 281–285. AAAI Press/The MIT Press, Menlo Park, CA, USA, 1997.

[McAllester et al., 1997] D. McAllester, B. Selman, and H. Kautz. Evidence for invariants in local search. In *Proceedings of the Fourteenth National Conference on Artificial Intelligence*, pages 321–326. AAAI Press/The MIT Press, Menlo Park, CA, USA, 1997.

[McGeoch and Moret, 1999] C. C. McGeoch and B. M. E. Moret. How to present a paper on experimental work with algorithms. *SIGACT News*, 30(4):85–90, 1999.

[McGeoch, 1996] C. C. McGeoch. Toward an experimental method for algorithm simulation. *INFORMS Journal on Computing*, 8(1):1–15, 1996.

[McKay et al., 2002] K. N. McKay, M. Pinedo, and S. Webster. Practice-focused research issues for scheduling systems. *Production and Operations Management*, 11(2):249–258, 2002.

[Mehrotra and Trick, 1996] A. Mehrotra and M. A. Trick. A column generation approach for graph coloring. *INFORMS Journal on Computing*, 8(4):344–354, 1996.

[Merkle and Middendorf, 2003] D. Merkle and M. Middendorf. Ant colony optimization with global pheromone evaluation for scheduling a single machine. *Applied Intelligence*, 18(1):105–111, 2003.

[Merz and Freisleben, 1997] P. Merz and B. Freisleben. Genetic local search for the TSP: New results. In T. Bäck, Z. Michalewicz, and X. Yao, editors, *Proceedings of the 1997 IEEE International Conference on Evolutionary Computation (ICEC'97)*, pages 159–164. IEEE Press, Piscataway, NJ, USA, 1997.

[Merz and Freisleben, 1999] P. Merz and B. Freisleben. Fitness landscapes and memetic algorithm design. In D. Corne, M. Dorigo, and F. Glover, editors, *New Ideas in Optimization*, pages 244–260. McGraw Hill, London, UK, 1999.

[Merz and Freisleben, 2000a] P. Merz and B. Freisleben. Fitness landscape analysis and memetic algorithms for the quadratic assignment problem. *IEEE Transactions on Evolutionary Computation*, 4(4):337–352, 2000.

[Merz and Freisleben, 2000b] P. Merz and B. Freisleben. Fitness landscapes, memetic algorithms and greedy operators for graph bi-partitioning. *Evolutionary Computation*, 8(1):61–91, 2000.

[Merz and Freisleben, 2001] P. Merz and B. Freisleben. Memetic algorithms for the traveling salesman problem. *Complex Systems*, 13(4):297–345, 2001.

[Merz and Freisleben, 2002] P. Merz and B. Freisleben. Greedy and local search heuristics for the unconstrained binary quadratic programming problem. *Journal of Heuristics*, 8(2):197–213, 2002.

[Merz, 2000] P. Merz. *Memetic Algorithms for Combinatorial Optimization Problems: Fitness Landscapes and Effective Search Strategies*. PhD thesis, Department of Electrical Engineering and Computer Science, University of Siegen, Germany, 2000.

[Merz, 2002] P. Merz. Personal communication, 2002.

[Metropolis et al., 1953] N. Metropolis, A. W. Rosenbluth, M. N. Rosenbluth, A. Teller, and E. Teller. Equation of state calculations by fast computing machines. *Journal of Chemical Physics*, 21:1087–1092, 1953.

[Mezey, 1987] P. G. Mezey. *Potential Energy Hypersurfaces*. Elsevier, Amsterdam, The Netherlands, 1987.

[Michalewicz and Fogel, 2000] Z. Michalewicz and D. B. Fogel. *How to Solve It: Modern Heuristics*. Springer Verlag, Berlin, Germany, 2000.

[Michalewicz, 1994] Z. Michalewicz. *Genetic Algorithms + Data Structures = Evolution Programs*, second edition. Springer Verlag, Berlin, Germany, 1994.

[Michel and Hentenryck, 1999] L. Michel and P. Van Hentenryck. Localizer: A modeling language for local search. *INFORMS Journal on Computing*, 11(1):1–14, 1999.

[Michel and Hentenryck, 2000] L. Michel and P. Van Hentenryck. Localizer. *Constraints*, 5(1–2):43–84, 2000.

[Michel and Hentenryck, 2001] L. Michel and P. Van Hentenryck. Localizer++: An open library for local search. Technical Report CS-01-03, Department of Computer Science, Brown University, Providence, RI, USA, 2001.

[Milis and Magirou, 1995] I. Z. Milis and V. F. Magirou. A Lagrangean relaxation algorithm for sparse quadratic assignment problems. *Operations Research Letters*, 17(2):69–76, 1995.

[Mills and Tsang, 1999a] P. Mills and E. Tsang. Guided local search applied to the SAT problem. In *Proceedings of the 15th National Conference of the Australian Society for Operations Research (ASOR'99)*, pages 872–883, Queensland, Australia, 1999.

[Mills and Tsang, 1999b] P. Mills and E. Tsang. Solving the Max-SAT problem using guided local search. Technical Report CSM-327, University of Essex, Colchester, UK, 1999.

[Mills and Tsang, 2000] P. Mills and E. Tsang. Guided local search for solving SAT and weighted MAX-SAT problems. In I. P. Gent, H. van Maaren, and T. Walsh, editors, *SAT2000 — Highlights of Satisfiability Research in the Year 2000*, pages 89–106. IOS Press, Amsterdam, The Netherlands, 2000.

[Mills et al., 2003] P. Mills, E. Tsang, and J. Ford. Applying an extended guided local search on the quadratic assignment problem. *Annals of Operations Research*, 118:121–135, 2003.

[Minton et al., 1990] S. Minton, M. D. Johnston, A. B. Philips, and P. Laird. Solving large-scale constraint satisfaction and scheduling problems using a heuristic repair method. In *Proceedings of the 8th National Conference on Artificial Intelligence*, pages 17–24. AAAI Press/The MIT Press, Menlo Park, CA, USA, 1990.

[Minton et al., 1992] S. Minton, M. D. Johnston, A. B. Philips, and P. Laird. Minimizing conflicts: A heuristic repair method for constraint satisfaction and scheduling problems. *Artificial Intelligence*, 58(1–3):161–205, 1992.

[Minton, 1996] S. Minton. Automatically configuring constraint satisfaction programs: A case study. *Constraints*, 1(1–2):7–43, 1996.

[Misevicius, 2002] A. Misevicius. A tabu search algorithm for the quadratic assignment problem. Technical report, Kaunas University of Technology, Kaunas, Lithuania, 2002.

[Mitchell and Levesque, 1996] D. G. Mitchell and H. Levesque. Some pitfalls for experimenters with random SAT. *Artificial Intelligence*, 81(1–2):111–125, 1996.

[Mitchell et al., 1992] D. G. Mitchell, B. Selman, and H. Levesque. Hard and easy distributions of SAT problems. In *Proceedings of the 10th National Conference on Artificial Intelligence*, pages 459–465. AAAI Press/The MIT Press, Menlo Park, CA, USA, 1992.

[Mitchell, 1996] M. Mitchell. *An Introduction to Genetic Algorithms*. MIT Press, Cambridge, MA, USA, 1996.

[Mitchell, 2002] D. E. Mitchell. Branch-and-cut methods for combinatorial optimization problems. In P. M. Pardalos and M. G. C. Resende, editors, *Handbook of Applied Optimization*, pages 65–77. Oxford University Press, New York, NY, USA, 2002.

[Mladenović and Hansen, 1997] N. Mladenović and P. Hansen. Variable neighborhood search. *Computers & Operations Research*, 24(11):1097–1100, 1997.

[Möbius et al., 1999] A. Möbius, B. Freisleben, P. Merz, and M. Schreiber. Combinatorial optimization by iterative partial transcription. *Physical Review E*, 59(4):4667–4674, 1999.

[Monasson and Zecchina, 1996] R. Monasson and R. Zecchina. Entropy of the k-satisfiability problem. *Physical Review Letters*, 76(21):3881–3885, 1996.

[Monasson and Zecchina, 1997] R. Monasson and R. Zecchina. Statistical mechanics of the random K-SAT model. *Physical Review E*, 56(2):1357–1370, 1997.

[Montgomery, 2000] D. C. Montgomery. *Design and Analysis of Experiments*, fifth edition. John Wiley & Sons, New York, NY, 2000.

[Moret, 2002] B. M. E. Moret. Towards a discipline of experimental algorithmics. In M. H. Goldwasser, D. S. Johnson, and C. C. McGeoch, editors, *Data Structures, Near Neighbor Searches, and Methodology: Fifth and Sixth DIMACS Implementation Challenges*, DIMACS Series on Discrete Mathematics and Theoretical Computer Science, pages 197–214. American Mathematical Society, Providence, RI, USA, 2002.

[Morgenstern, 1996] C. Morgenstern. Distributed coloration neighbourhood search. In D. S. Johnson and M. A. Trick, editors, *Cliques, Coloring, and Satisfiability: Second DIMACS Implementation Challenge*, volume 26 of *DIMACS Series on Discrete Mathematics and Theoretical Computer Science*, pages 335–357. American Mathematical Society, Providence, RI, USA, 1996.

[Morris, 1993] P. Morris. The breakout method for escaping from local minima. In *Proceedings of the 11th National Conference on Artificial Intelligence*, pages 40–45. AAAI Press/The MIT Press, Menlo Park, CA, USA, 1993.

[Morton and Pentico, 1993] T. E. Morton and D. Pentico. *Heuristic Scheduling Systems*. John Wiley & Sons, New York, NY, USA, 1993.

[Morton et al., 1984] T. E. Morton, R. M. Rachamadugu, and A. Vepsalainen. Accurate myopic heuristics for tardiness scheduling. GSIA Working Paper 36-83-84, Carnegie–Mellon University, Pittsburgh, PA, USA, 1984.

[Moscato and Norman, 1992] P. Moscato and M. G. Norman. A 'memetic' approach for the traveling salesman problem. Implementation of a computational ecology for combinatorial optimization on message-passing systems. In M. Valero, E. Onate, M. Jane, J. L. Larriba, and B. Suarez, editors, *Parallel Computing and Transputer Applications*, pages 187–194. IOS Press, Amsterdam, The Netherlands, 1992.

[Moscato and Norman, 1998] P. Moscato and M. G. Norman. On the performance of heuristics on finite and infinite fractal instances of the Euclidean traveling salesman problem. *INFORMS Journal on Computing*, 10(2):121–132, 1998.

[Moscato, 1989] P. Moscato. On evolution, search, optimization, genetic algorithms and martial arts: Towards memetic algorithms. C3P Report 826, Caltech Concurrent Computation Program, 1989.

[Moscato, 1999] P. Moscato. Memetic algorithms: A short introduction. In D. Corne, M. Dorigo, and F. Glover, editors, *New Ideas in Optimization*, pages 219–234. McGraw Hill, London, UK, 1999.

[Mühlenbein and Paaß, 1996] H. Mühlenbein and G. Paaß. From recombination of genes to estimations of distributions. In H.-M. Voigt, W. Ebeling, I. Rechenberg, and H.-P. Schwefel, editors, *Proceedings of PPSN-IV, Fourth International Conference on Parallel Problem Solving from Nature*, volume 1141 of *Lecture Notes in Computer Science*, pages 178–187. Springer Verlag, Berlin, Germany, 1996.

[Mühlenbein et al., 1988] H. Mühlenbein, M. Gorges-Schleuter, and O. Krämer. Evolution algorithms in combinatorial optimization. *Parallel Computing*, 7:65–85, 1988.

[Mühlenbein, 1991] H. Mühlenbein. Evolution in time and space – the parallel genetic algorithm. In G. J. Rawlins, editor, *Foundations of Genetic Algorithms*, pages 316–337. Morgan Kaufmann Publishers, San Mateo, CA, USA, 1991.

[Muth and Thompson, 1963] J. F. Muth and G. L. Thompson, editors. *Industrial Scheduling*. Prentice Hall, Englewood Cliffs, NJ, USA, 1963.

N   [Nagata and Kobayashi, 1997] Y. Nagata and S. Kobayashi. Edge assembly crossover: A high-power genetic algorithm for the traveling salesman problem. In T. Bäck, editor, *Proceedings of the Seventh International Conference on Genetic Algorithms*, pages 450–457. Morgan Kaufmann Publishers, San Francisco, CA, USA, 1997.

[Narendra and Thathachar, 1989] K. S. Narendra and M. A. L. Thathachar. *Learning Automata: An Introduction*. Prentice-Hall, Englewood Cliffs, NJ, USA, 1989.

[Naudts and Kallel, 2000] B. Naudts and L. Kallel. A comparison of predictive measures of problem difficulty in evolutionary algorithms. *IEEE Transactions on Evolutionary Computation*, 4(1):1–16, 2000.

[Nawaz et al., 1983] M. Nawaz, E. Enscore Jr., and I. Ham. A heuristic algorithm for the m-machine, n-job flow-shop sequencing problem. *OMEGA*, 11(1):91–95, 1983.

[Nemhauser and Wolsey, 1988] G. L. Nemhauser and L. A. Wolsey. *Integer and Combinatorial Optimization*. John Wiley & Sons, New York, NY, USA, 1988.

[Neto and Pedroso, 2003] T. Neto and J. P. Pedroso. GRASP for linear integer programming. In M. G. C. Resende and J. P. de Sousa, editors, *Metaheuristics: Computer Decision-Making*, pages 545–574. Kluwer Academic Publishers, Dordrecht, The Netherlands, 2003.

[Neto, 1999] D. Neto. *Efficient Cluster Compensation for Lin-Kernighan Heuristics*. PhD thesis, University of Toronto, Department of Computer Science, Toronto, Canada, 1999.

[Neto, 2003] D. Neto. LK the program. `http://www.cs.toronto.edu/~neto/research/lk`, 2003. Version visited last on 15 September 2003.

[Nguyen et al., 2002] H. D. Nguyen, I. Yoshihara, K. Yamamori, and M. Yasunaga. Greedy genetic algorithms for symmetric and asymmetric TSPs. *IPSJ Transactions on Mathematical Modeling and Its Applications*, 43(10):165–175, 2002.

[Nisan, 2000] N. Nisan. Bidding and allocation in combinatorial auctions. In *Proceedings of the 2nd ACM Conference on Electronic Commerce*, pages 1–12. ACM Press, New York, NY, USA, 2000.

[Nissen and Paul, 1995] V. Nissen and H. Paul. A modification of threshold accepting and its application to the quadratic assignment problem. *OR Spektrum*, 17:205–210, 1995.

[Nissen, 1994] V. Nissen. Solving the quadratic assignment problem with clues from nature. *IEEE Transactions on Neural Networks*, 5(1):66–72, 1994.

[Nowicki and Smutnicki, 1996a] E. Nowicki and C. Smutnicki. A fast taboo search algorithm for the job-shop problem. *Management Science*, 42(2):797–813, 1996.

[Nowicki and Smutnicki, 1996b] E. Nowicki and C. Smutnicki. A fast tabu search algorithm for the permutation flow-shop problem. *European Journal of Operational Research*, 91(1):160–175, 1996.

[Nyström, 1999] M. Nyström. Solving certain large instances of the quadratic assignment problem: Steinberg's examples. Technical report, Department of Computer Science, California Institute of Technology, Pasadena, CA, 1999.

O [Ogbu and Smith, 1990] F. Ogbu and D. Smith. The application of the simulated annealing algorithm to the solution of the $n/m/C_{max}$ flowshop problem. *Computers & Operations Research*, 17(3):243–253, 1990.

[Ong and Moore, 1984] H. L. Ong and J. B. Moore. Worst-case analysis of two travelling salesman heuristics. *Operations Research Letters*, 2(6):273–277, 1984.

[Or, 1976] I. Or. *Traveling Salesman-type Problems and their Relation to the Logistics of Regional Blood Banking*. PhD thesis, Northwestern University, Department of Industrial Engineering and Management Science, Evanston, IL, USA, 1976.

[Osman and Kelly, 1996] I. H. Osman and J. P. Kelly, editors. *Meta-Heuristics: Theory & Applications*. Kluwer Academic Publishers, Boston, MA, USA, 1996.

[Osman and Laporte, 1996] I. H. Osman and G. Laporte. Metaheuristics: A bibliography. *Annals of Operations Research*, 63:513–628, 1996.

[Osman and Potts, 1989] I. H. Osman and C. N. Potts. Simulated annealing for permutation flow-shop scheduling. *OMEGA*, 17(6):551–557, 1989.

[Ow and Morton, 1988] P. S. Ow and T. E. Morton. Filtered beam search in scheduling. *International Journal of Production Research*, 26:297–307, 1988.

P [Pacciarelli and Pranzo, 2000] D. Pacciarelli and M. Pranzo. Production scheduling in a steelmaking-continuous casting plant. Technical Report 50, Università degli studi Roma Tre, Dipartimento di Informatica e Automazione, Rome, Italy, 2000.

[Padberg and Grötschel, 1985] M. W. Padberg and M. Grötschel. Polyhedral computations. In E. L. Lawler, J. K. Lenstra, A. H. G. Rinnooy Kan, and D. B. Shmoys, editors, *The Traveling Salesman Problem*, pages 307–360. John Wiley & Sons, Chichester, UK, 1985.

[Padberg and Rinaldi, 1991] M. Padberg and G. Rinaldi. A branch-and-cut algorithm for the resolution of large-scale symmetric travelling salesman problems. *SIAM Review*, 33(1):60–100, 1991.

[Palmer, 1991] R. Palmer. Optimization on rugged landscapes. In A. S. Perelson and S. A. Kauffman, editors, *Molecular Evolution on Rugged Landscapes: Proteins, RNA, and the Immune System*, pages 3–25. Addison Wesley, Redwood City, CA, 1991.

[Panwalker and Iskander, 1977] S. S. Panwalker and W. Iskander. A survey of scheduling rules. *Operations Research*, 25(1):45–61, 1977.

[Panwalker et al., 1973] S. S. Panwalker, R. A. Dudek, and M. L. Smith. Sequencing research and the industrial scheduling problem. In *Proceedings of the Symposium on Theory of Scheduling and its Applications*, pages 29–38. Springer Verlag, New York, NY, USA, 1973.

[Papadimitriou and Steiglitz, 1982] C. H. Papadimitriou and K. Steiglitz. *Combinatorial Optimization – Algorithms and Complexity*. Prentice Hall, Englewood Cliffs, NJ, USA, 1982.

[Papadimitriou, 1991] C. H. Papadimitriou. On selecting a satisfying truth assignment. In *Proceedings of the 32nd Annual IEEE Symposium on Foundations of Computer Science*, pages 163–169. IEEE Computer Society Press, Los Alamitos, CA, USA, 1991.

[Papadimitriou, 1992] C. H. Papadimitriou. The complexity of the Lin-Kernighan heuristic for the traveling salesman problem. *SIAM Journal on Computing*, 21(3):450–465, 1992.

[Papadimitriou, 1994] C. H. Papadimitriou. *Computational Complexity*. Addison-Wesley, Reading, MA, USA, 1994.

[Papoulis, 1991] A. Papoulis. *Probability, Random Variables and Stochastic Processes*, third edition. McGraw Hill, New York, NY, USA, 1991.

[Paquete and Stützle, 2002] L. Paquete and T. Stützle. An experimental investigation of iterated local search for coloring graphs. In S. Cagnoni, J. Gottlieb, E. Hart, M. Middendorf, and G. Raidl, editors, *Applications of Evolutionary Computing*, volume 2279 of *Lecture Notes in Computer Science*, pages 122–131. Springer Verlag, Berlin, Germany, 2002.

[Pardalos et al., 1994] P. M. Pardalos, F. Rendl, and H. Wolkowicz. The quadratic assignment problem: A survey and recent developments. In P. M. Pardalos and H. Wolkowicz, editors, *Quadratic Assignment and Related Problems*, DIMACS Series on Discrete Mathematics and Theoretical Computer Science, pages 1–42. American Mathematical Society, Providence, RI, USA, 1994.

[Park, 2002] J. Park. Using weighted MAX-SAT to approximate MPE. In *Proceedings of the Eighteenth National Conference on Artificial Intelligence*, pages 682–687. AAAI Press/The MIT Press, Menlo Park, CA, USA, 2002.

[Parkes and Walser, 1996] A. J. Parkes and J. P. Walser. Tuning local search for satisfiability testing. In *Proceedings of the Thirteenth National Conference on Artificial Intelligence*, volume 1, pages 356–362. AAAI Press/The MIT Press, Menlo Park, CA, USA, 1996.

[Paturi et al., 1997] R. Paturi, P. Pudlák, and F. Zane. Satisfiability coding lemma. In *Proceedings of the 38th Annual IEEE Symposium on Foundations of Computer Science*, pages 566–574. IEEE Computer Society Press, Los Alamitos, CA, USA, 1997.

[Paturi et al., 1998] R. Paturi, P. Pudlák, M. E. Saks, and F. Zane. An improved exponential-time algorithm for k-SAT. In *Proceedings of the 39th Annual IEEE Symposium on Foundations of Computer Science*, pages 628–637. IEEE Computer Society Press, Los Alamitos, CA, USA, 1998.

[Pearl, 1984] J. Pearl. *Heuristics: Intelligent Search Strategies for Computer Problem Solving*. Addison-Wesley, Reading, MA, USA, 1984.

[Pedroso, 1999] J. P. Pedroso. An evolutionary solver for pure integer linear programming. Technical Report 5/99, Centro de Investigação Operacional da Universidade de Lisboa, Lisbon, Portugal, 1999.

[Pekny and Miller, 1994] J. F. Pekny and D. L. Miller. A staged primal-dual algorithm for finding a minimum cost perfect two-matching in an undirected graph. *ORSA Journal on Computing*, 6(1):68–81, 1994.

[Percus and Martin, 1996] A. G. Percus and O. C. Martin. Finite size and dimensional dependence of the Euclidean traveling salesman problem. *Physical Review Letters*, 76(8):1188–1191, 1996.

[Pezzella and Merelli, 2000] F. Pezzella and E. Merelli. A tabu search method guided by shifting bottleneck for the job-shop scheduling problem. *European Journal of Operational Research*, 120(2):297–310, 2000.

[Pfahringer, 1996] B. Pfahringer. Multi-agent search for open shop scheduling: Adapting the ant-Q formalism. Technical Report TR-96-09, Austrian Research Institute for Artificial Intelligence, Vienna, Austria, 1996.

[Pinedo, 1995] M. Pinedo. *Scheduling: Theory, Algorithms, and Systems*. Prentice Hall, Englewood Cliffs, NJ, USA, 1995.

[Platzman and Bartholdi III, 1989] L. K. Platzman and J. J. Bartholdi III. Spacefilling curves and the planar travelling salesman problem. *Journal of the ACM*, 36(4):719–737, 1989.

[Poole et al., 1998] D. Poole, A. Mackworth, and R. Goebel. *Computational Intelligence: A Logical Approach*. Oxford University Press, New York, USA, 1998.

[Poole, 1984] D. Poole. Making 'clausal' theorem provers 'non-clausal'. In *Proceedings of the Canadian Society for Computational Studies of Intelligence National Conference (CSCSI-84)*, pages 124–125, London, Ontario, Canada, 1984.

[Potts and van de Velde, 1995] C. N. Potts and S. van de Velde. Dynasearch: Iterative local improvement by dynamic programming; part I, the traveling salesman problem. Technical Report LPOM–9511, Faculty of Mechanical Engineering, University of Twente, Enschede, The Netherlands, 1995.

[Potts and Wassenhove, 1985] C. N. Potts and L. N. van Wassenhove. A branch and bound algorithm for the total weighted tardiness problem. *Operations Research*, 33(2):363–377, 1985.

[Potts and Wassenhove, 1991] C. N. Potts and L. N. van Wassenhove. Single machine tardiness sequencing heuristics. *IIE Transactions*, 23:346–354, 1991.

[Potvin, 1996] J. Y. Potvin. Genetic algorithms for the traveling salesman problem. *Annals of Operations Research*, 63:339–370, 1996.

[Prais and Ribeiro, 2000] M. Prais and C. C. Ribeiro. Reactive GRASP: An application to a matrix decomposition problem in TDMA traffic assignment. *INFORMS Journal on Computing*, 12(3):164–176, 2000.

[Pranzo and Stützle, 2003] M. Pranzo and T. Stützle. Iterated greedy: Methodology, algorithms, and applications. Manuscript, 2003.

[Prestwich, 2003] S. Prestwich. Local search on SAT-encoded CSPs. In *Proceedings of the Sixth International Conference on Theory and Applications of Satisfiability Testing (SAT 2003)*, pages 388–399, Santa Margherita, Ligure, Italy, 2003.

[Prins, 2000] C. Prins. Competitive genetic algorithms for the open-shop scheduling problem. *Mathematical Methods of Operations Research*, 52(3):389–411, 2000.

Q [Quick et al., 1998] R. J. Quick, V. J. Rayward-Smith, and G. D. Smith. Fitness distance correlation and ridge functions. In A. E. Eiben, T. Bäck, M. Schoenauer, and H.-P. Schwefel, editors, *Proceedings of PPSN-V, Fifth International Conference on Parallel Problem Solving from Nature*, volume 1498 of *Lecture Notes in Computer Science*, pages 77–86. Springer Verlag, Berlin, Germany, 1998.

R [Rabin, 1976] M. O. Rabin. Probabilistic algorithms. In J. F. Traub, editor, *Algorithms and Complexity, Recent Results and New Directions*, pages 21–39. Academic Press, New York, NY, USA, 1976.

[Radcliffe and Surry, 1994] N. J. Radcliffe and P. D. Surry. Formal memetic algorithms. In T. Fogarty, editor, *Evolutionary Computing: AISB Workshop*, volume 865 of *Lecture Notes in Computer Science*, pages 1–16. Springer Verlag, Berlin, Germany, 1994.

[Rana and Whitley, 1998] S. Rana and D. Whitley. Genetic algorithm behavior in the MAXSAT domain. In A. E. Eiben, T. Bäck, M. Schoenauer, and H.-P. Schwefel, editors, *Proceedings of PPSN-V, Fifth International Conference on Parallel Problem Solving from Nature*, volume 1498 of *Lecture Notes in Computer Science*, pages 785–794. Springer Verlag, Berlin, Germany, 1998.

[Rana, 1999] S. Rana. *Examining the Role of Local Optima and Schema Processing in Genetic Search*. PhD thesis, Colorado State University, Department of Computer Science, Fort Collins, Colorado, USA, 1999.

[Randall and Abramson, 2001] M. Randall and D. Abramson. A general meta-heuristic based solver for combinatorial optimisation problems. *Computational Optimisation and Applications*, 20(2):185–210, 2001.

[Randerath et al., 2001] B. Randerath, E. Speckenmeyer, E. Boros, P. Hammer, A. Kogan, K. Makino, B. Simeone, and O. Cepek. A satisfiability formulation of problems on level graphs. In H. Kautz and B. Selman, editors, *LICS 2001 Workshop on Theory and Applications of Satisfiability Testing (SAT 2001)*, volume 9 of *Electronic Notes in Discrete Mathematics*. Elsevier, Amsterdam, The Netherlands, 2001.

[Rayward-Smith et al., 1996] V. J. Rayward-Smith, I. H. Osman, C. R. Reeves, and G. D. Smith, editors. *Modern Heuristic Search Methods*. John Wiley & Sons, Chichester, UK, 1996.

[Raz and Safra, 1997] R. Raz and S. Safra. A sub-constant error-probability low-degree test, and sub-constant error-probability PCP characterization of NP. In F. T. Leighton and P. Shor, editors, *Proceedings of the Twenty-ninth Annual ACM Symposium on Theory of Computing*, pages 475–484. ACM press, New York, NY, USA, 1997.

[Rechenberg, 1973] I. Rechenberg. *Evolutionsstrategie — Optimierung technischer Systeme nach Prinzipien der biologischen Information*. Fromman Verlag, Freiburg, Germany, 1973.

[Reeves and Rowe, 2003] C. R. Reeves and J. E. Rowe. *Genetic Algorithms — Principles and Perspectives*. Kluwer Academic Publishers, Dordrecht, The Netherlands, 2003.

[Reeves and Yamada, 1998] C. R. Reeves and T. Yamada. Genetic algorithms, path relinking, and the flowshop sequencing problem. *Evolutionary Computation*, 6(1):45–60, 1998.

[Reeves, 1993a] C. R. Reeves. Improving the efficiency of tabu search for machine sequencing problems. *Journal of the Operational Research Society*, 44(4):375–382, 1993.

[Reeves, 1993b] C. R. Reeves, editor. *Modern Heuristic Techniques for Combinatorial Problems*. Blackwell Scientific Publications, Oxford, UK, 1993.

[Reeves, 1995] C. R. Reeves. A genetic algorithm for flowshop sequencing. *Computers & Operations Research*, 22(1):5–13, 1995.

[Reeves, 1999] C. R. Reeves. Landscapes, operators and heuristic search. *Annals of Operations Research*, 86:473–490, 1999.

[Rego and Glover, 2002] C. Rego and F. Glover. Local search and metaheuristics for the traveling salesman problem. In G. Gutin and A. Punnen, editors, *The Traveling Salesman Problem and its Variations*, pages 309–368. Kluwer Academic Publishers, Dordrecht, The Netherlands, 2002.

[Reidys and Stadler, 2001] C. M. Reidys and P. F. Stadler. Neutrality in fitness landscapes. *Applied Mathematics and Computation*, 117(2–3):321–350, 2001.

[Reidys and Stadler, 2002] C. M. Reidys and P. F. Stadler. Combinatorial landscapes. *SIAM Review*, 44(2):3–54, 2002.

[Reinelt, 1994] G. Reinelt. *The Traveling Salesman: Computational Solutions for TSP Applications*, volume 840 of *Lecture Notes in Computer Science*. Springer Verlag, Berlin, Germany, 1994.

[Reinelt, 2003] G. Reinelt. TSPLIB. `http://www.iwr.uni-heidelberg.de/groups/comopt/software/TSPLIB95`, 2003. Version visited last on 15 September 2003.

[Reischuk, 1990] K. R. Reischuk. *Einführung in die Komplexitätstheorie*. Teubner Verlag, Stuttgart, Germany, 1990.

[Resende and de Sousa, 2003] M. G. C. Resende and J. P. de Sousa, editors. *Metaheuristics: Computer Decision-Making*. Kluwer Academic Publishers, Boston, MA, USA, 2003.

[Resende and Feo, 1996] M. G. C. Resende and T. A. Feo. A GRASP for satisfiability. In D. S. Johnson and M. A. Trick, editors, *Cliques, Coloring, and Satisfiability: Second DIMACS Implementation Challenge*, volume 26 of DIMACS Series on Discrete Mathematics and Theoretical Computer Science, pages 499–520. American Mathematical Society, Providence, RI, USA, 1996.

[Resende and Ribeiro, 2002] M. G. C. Resende and C. C. Ribeiro. Greedy randomized adaptive search procedures. In F. Glover and G. Kochenberger, editors, *Handbook of Metaheuristics*, pages 219–249. Kluwer Academic Publishers, Norwell, MA, USA, 2002.

[Resende et al., 1997] M. G. C. Resende, L. S. Pitsoulis, and P. M. Pardalos. Approximate solution of weighted MAX-SAT problems using GRASP. In D. Du, J. Gu, and P. M. Pardalos, editors, *Satisfiability problems: Theory and Applications*, volume 35 of *DIMACS Series on Discrete Mathematics and Theoretical Computer Science*, pages 393–405. American Mathematical Society, Providence, RI, USA, 1997.

[Rintanen, 1999a] J. Rintanen. Constructing conditional plans by a theorem-prover. *Journal of Artificial Intelligence Research*, 10:323–352, 1999.

[Rintanen, 1999b] J. Rintanen. Improvements to the evaluation of quantified boolean formulae. In T. Dean, editor, *Proceedings of the Sixteenth International Joint Conference on Artificial Intelligence*, pages 1192–1197. Morgan Kaufmann Publishers, San Francisco, CA, USA, 1999.

[Rish and Frost, 1997] I. Rish and D. Frost. Statistical analysis of backtracking on inconsistent CSPs. In G. Smolka, editor, *Principles and Practice of Constraint Programming – CP 1997*, volume 1330 of *Lecture Notes in Computer Science*, pages 150–163. Springer Verlag, Berlin, Germany, 1997.

[Roberts, 1979] F. S. Roberts. On the mobile radio frequency assignment problem and the traffic light phasing problem. *Annals of the New York Academy of Sciences*, 319:466–483, 1979.

[Robertson and Seymour, 1991] N. Robertson and P. D. Seymour. Graph minors X. Obstructions to tree decompositions. *Journal of Combinatorial Theory, Series B*, 52(2):153–190, 1991.

[Rochet et al., 1997] S. Rochet, G. Venturini, M. Slimane, and E. M. El Kharoubi. A critical and empirical study of epistasis measures for predicting GA performances: A summary. In J.-K. Hao, E. Lutton, E. Ronald, M. Schoenauer, and D. Snyers, editors, *Artificial Evolution*, volume 1363 of *Lecture Notes in Computer Science*, pages 275–286. Springer Verlag, Berlin, Germany, 1997.

[Rohatgi, 1976] V. K. Rohatgi. *An Introduction to Probability Theory and Mathematical Statistics*. John Wiley & Sons, New York, NY, USA, 1976.

[Rohe, 1997] A. Rohe. Parallele Heuristiken für sehr große Traveling Salesman Probleme. Master's thesis, Fachbereich Mathematik, Universität Bonn, Germany, 1997.

[Roli et al., 2001] A. Roli, C. Blum, and M. Dorigo. ACO for maximal constraint satisfaction problems. In *Proceedings of MIC'2001 – Meta–heuristics International Conference*, volume 1, pages 187–191, Porto, Portugal, 2001. Also available as technical report TR/IRIDIA/2001-17, IRIDIA, Université Libre de Bruxelles, Belgium.

[Romeo and Sangiovanni-Vincentelli, 1991] F. Romeo and A. Sangiovanni-Vincentelli. A theoretical framework for simulated annealing. *Algorithmica*, 6(3):302–345, 1991.

[Rosenkrantz et al., 1977] D. J. Rosenkrantz, R. E. Stearns, and P. M. Lewis. An analysis of several heuristics for the traveling salesman problem. *SIAM Journal on Computing*, 6(3):563–581, 1977.

[Rossi et al., 1990] F. Rossi, C. J. Petrie, and V. Dhar. On the equivalence of constraint satisfaction problems. In *Proceedings of the 9th European Conference on Artificial Intelligence*, pages 550–556, Pitman, London, UK, 1990.

[Roth, 1996] D. Roth. On the hardness of approximate reasoning. *Artificial Intelligence*, 82(1–2):273–302, 1996.

[Rozenberg and Salomaa, 1997] G. Rozenberg and A. Salomaa, editors. *Handbook of Formal Languages*. Springer Verlag, Berlin, Germany, 1997.

[Rudolph, 1994] G. Rudolph. Convergence analysis of canonical genetic algorithms. *IEEE Transactions on Neural Networks*, 5(1):96–101, 1994.

[Ruiz and Maroto, to appear] R. Ruiz and C. Maroto. A comprehensive review and evaluation of permutation flowshop heuristics. *European Journal of Operational Research*, to appear.

[Ruml, 2001] W. Ruml. Incomplete tree search using adaptive probing. In B. Nebel, editor, *Proceedings of the 17th International Joint Conference on Artificial Intelligence*, pages 235–241. Morgan Kaufmann Publishers, San Francisco, CA, USA, 2001.

[Russel and Norvig, 2003] S. Russel and P. Norvig. *Artificial Intelligence: A Modern Approach*, second edition. Prentice-Hall, Englewood Cliffs, NJ, USA, 2003.

S   [Sahni and Gonzalez, 1976] S. Sahni and T. Gonzalez. P-complete approximation problems. *Journal of the ACM*, 23(3):555–565, 1976.

[Sait and Youssef, 1999] S. M. Sait and H. Youssef. *Iterative Computer Algorithms with Applications in Engineering*. IEEE Computer Society Press, Los Alamitos, CA, USA, 1999.

[Sampels et al., 2002] M. Sampels, C. Blum, M. Mastrolilli, and O. Rossi-Doria. Meta-heuristics for group shop scheduling. Technical Report IRIDIA/2002-7, IRIDIA, Université Libre de Bruxelles, Belgium, 2002.

[Sandholm et al., 2001] T. Sandholm, S. Suri, A. Gilpin, and D. Levine. CABOB: A fast optimal algorithm for combinatorial auctions. In B. Nebel, editor, *Proceedings of the Seventeenth International Joint Conference on Artificial Intelligence*, pages 1102–1108. Morgan Kaufmann Publishers, San Francisco, CA, USA, 2001.

[Sandholm et al., 2002] T. Sandholm, S. Suri, A. Gilpin, and D. Levine. Winner determination in combinatorial auction generalizations. In C. Castelfranchi and W. L. Johnson, editors, *Proceedings of the First International Joint Conference on Autonomous Agents and Multiagent Systems: Part 1*, pages 69–76. ACM Press, New York, NY, USA, 2002.

[Sandholm, 1999] T. Sandholm. An algorithm for optimal winner determination in combinatorial auctions. In T. Dean, editor, *Proceedings of the Sixteenth International Joint Conference on Artificial Intelligence*, pages 542–547. Morgan Kaufmann Publishers, San Francisco, CA, USA, 1999.

[Sandholm, 2002] T. Sandholm. Algorithms for optimal winner determination in combinatorial auctions. *Artificial Intelligence*, 135(1–2):1–54, 2002.

[Schaerf, 1999] A. Schaerf. A survey of automated timetabling. *Artificial Intelligence Review*, 13(2):87–127, 1999.

[Schiavinotto and Stützle, 2003] T. Schiavinotto and T. Stützle. Search space analysis for the linear ordering problem. In G. R. Raidl et al., editors, *Applications of Evolutionary Computing*, volume 2611 of *Lecture Notes in Computer Science*, pages 322–333. Springer Verlag, Berlin, Germany, 2003.

[Schiex et al., 1995] T. Schiex, H. Fargier, and G. Verfaillie. Valued constraint satisfaction problems: Hard and easy problems. In C. S. Mellish, editor, *Proceedings of the 14th International Joint Conference on Artificial Intelligence*, pages 631–639. Morgan Kaufmann Publishers, San Francisco, CA, USA, 1995.

[Schilham, 2001] R. M. F. Schilham. *Commonalities in Local Search*. PhD thesis, Department of Mathematics and Computer Science, Eindhoven University of Technology, Eindhoven, The Netherlands, 2001.

[Schöning, 1999] U. Schöning. A probabilistic algorithm for k-SAT and constraint satisfaction problems. In *Proceedings of the 40th Annual IEEE Symposium on Foundations of Computer Science*, pages 410–414. IEEE Computer Society Press, Los Alamitos, CA, USA, 1999.

[Schöning, 2002] U. Schöning. A probabilistic algorithm for *k*-SAT based on limited local search and restart. *Algorithmica*, 32(4):615–623, 2002.

[Schreiber and Martin, 1999] G. R. Schreiber and O. C. Martin. Cut size statistics of graph bisection heuristics. *SIAM Journal on Optimization*, 10(1):231–251, 1999.

[Schrijver, 2003] A. Schrijver. On the history of combinatorial optimization. Preprint available at http://www.cwi.nl/~lex, 2003.

[Schuler et al., 2001] R. Schuler, U. Schöning, and O. Watanabe. An Improved Randomized Algorithm for 3-SAT. Technical Report TR-C146, Deptartment of Mathematics and Computer Sciences, Tokyo Institute of Technology, Japan, 2001.

[Schuurmans and Southey, 2000] D. Schuurmans and F. Southey. Local search characteristics of incomplete SAT procedures. In *Proceedings of the Seventeenth National Conference on Artificial Intelligence*, pages 297–302. AAAI Press/The MIT Press, Menlo Park, CA, USA, 2000.

[Schuurmans et al., 2001] D. Schuurmans, F. Southey, and R. C. Holte. The exponentiated subgradient algorithm for heuristic boolean programming. In B. Nebel, editor, *Proceedings of the 17th International Joint Conference on Artificial Intelligence*, pages 334–341. Morgan Kaufmann Publishers, San Francisco, CA, USA, 2001.

[Schwefel, 1981] H.-P. Schwefel. *Numerical Optimization of Computer Models*. John Wiley & Sons, Chichester, UK, 1981.

[Seeman, 1990] N. C. Seeman. De novo design of sequences for nucleic acid structure engineering. *Journal of Biomolecular Structure and Dynamics*, 8:573–581, 1990.

[Selman and Kautz, 1993] B. Selman and H. Kautz. Domain-independent extensions to GSAT: Solving large structured satisfiability problems. In R. Bajcsy, editor, *Proceedings of the 13th International Joint Conference on Artificial Intelligence*, pages 290–295. Morgan Kaufmann Publishers, San Francisco, CA, USA, 1993.

[Selman et al., 1992] B. Selman, H. Levesque, and D. Mitchell. A new method for solving hard satisfiability problems. In *Proceedings of the 10th National Conference on Artificial Intelligence*, pages 440–446. AAAI Press/The MIT Press, Menlo Park, CA, USA, 1992.

[Selman et al., 1994] B. Selman, H. Kautz, and B. Cohen. Noise strategies for improving local search. In *Proceedings of the 12th National Conference on Artificial Intelligence*, pages 337–343. AAAI Press/The MIT Press, Menlo Park, CA, USA, 1994.

[Selman et al., 1997] B. Selman, H. Kautz, and D. McAllester. Ten challenges in propositional reasoning and search. In M. E. Pollack, editor, *Proceedings of the Fifteenth International Joint Conference on Artificial Intelligence*, pages 50–54. Morgan Kaufmann Publishers, San Francisco, CA, USA, 1997.

[Seo and Moon, 2002] D.-I. Seo and B.-R. Moon. Voronoi quantized crossover for the traveling salesman problem. In W. B. Langdon et al., editors, *Proceedings of the Genetic and Evolutionary Computation Conference (GECCO-2002)*, pages 544–552. Morgan Kaufmann Publishers, San Francisco, CA, USA, 2002.

[Serna et al., 1998] M. Serna, L. Trevisan, and F. Xhafa. The parallel approximability of non-boolean constraint satisfaction and restricted integer programming. In *Proceedings of the 15th Symposium on Theoretical Aspects of Computer Science*, volume 1373 of *Lecture Notes in Computer Science*, pages 488–498. Springer Verlag, Berlin, Germany, 1998.

[Shang and Wah, 1997] Y. Shang and B. W. Wah. Discrete Lagrangian-based search for solving MAX-SAT problems. In M. E. Pollack, editor, *Proceedings of the 15th International Joint Conference on Artificial Intelligence*, volume 1, pages 378–383. Morgan Kaufmann Publishers, San Francisco, CA, USA, 1997.

[Shang and Wah, 1998] Y. Shang and B. W. Wah. A discrete Lagrangian-based global-search method for solving satisfiability problems. *Journal of Global Optimization*, 12(1):61–100, 1998.

[Shapiro and Wilk, 1965] S. S. Shapiro and M. B. Wilk. An analysis of variance test for normality (complete samples). *Biometrika*, 52(2–4):591–611, 1965.

[Sheskin, 2000] D. J. Sheskin. *Handbook of Parametric and Nonparametric Statistical Procedures*, second edition. Chapman & Hall / CRC, Boca Raton, Florida, USA, 2000.

[Shonkwiler, 1993] R. Shonkwiler. Parallel genetic algorithms. In S. Forrest, editor, *Proceedings of the Sixth International Conference on Genetic Algorithms (ICGA'95)*, pages 199–205. Morgan Kaufmann Publishers, San Mateo, CA, USA, 1993.

[Siegel et al., 2000] S. Siegel, N. J. Castellan Jr., and N. J. Castellan. *Nonparametric Statistics for the Behavioral Sciences*, second edition. McGraw Hill, New York, NY, USA, 2000.

[Sipser, 1997] M. Sipser. *Introduction to the Theory of Computation*. PWS Publishing Company, Boston, MA, USA, 1997.

[Skorin-Kapov, 1990] J. Skorin-Kapov. Tabu search applied to the quadratic assignment problem. *ORSA Journal on Computing*, 2(1):33–45, 1990.

[Sleator and Tarjan, 1985] D. D. Sleator and R. E. Tarjan. Self-adjusting binary search trees. *Journal of the ACM*, 32(3):652–686, 1985.

[Smith et al., 1996] B. M. Smith, S. C. Brailsford, P. M. Hubbard, and H. P. Williams. The progressive party problem: Integer linear programming and constraint programming compared. *Constraints*, 1(1):119–138, 1996.

[Smith, 1956] W. E. Smith. Various optimizers for single stage production. *Naval Research Logistics Quarterly*, 3:59–66, 1956.

[Smith, 1994] B. M. Smith. Phase transitions and the mushy region in constraint satisfaction problems. In A. G. Cohn, editor, *Proceedings of the 11th European Conference on Artificial Intelligence*, pages 100–104. John Wiley & Sons, Chichester, UK, 1994.

[Smyth et al., 2003] K. Smyth, H. H. Hoos, and T. Stützle. Iterated robust tabu search for MAX-SAT. In Y. Xiang and B. Chaib-draa, editors, *Advances in Artificial Intelligence, 16th Conference of the Canadian Society for Computational Studies of Intelligence*, volume 2671 of *Lecture Notes in Computer Science*, pages 129–144. Springer Verlag, Berlin, Germany, 2003.

[Solnon, 2000] C. Solnon. Solving permutation constraint satisfaction problems with artificial ants. In W. Horn, editor, *Proceedings of the 14th European Conference on Artificial Intelligence*, pages 118–122. IOS Press, Amsterdam, The Netherlands, 2000.

[Solnon, 2002a] C. Solnon. Ants can solve constraint satisfaction problems. *IEEE Transactions on Evolutionary Computation*, 6(4):347–357, 2002.

[Solnon, 2002b] C. Solnon. Boosting ACO with a preprocessing step. In S. Cagnoni et al., editors, *Applications of Evolutionary Computing*, volume 2279 of *Lecture Notes in Computer Science*, pages 163–172. Springer Verlag, Berlin, Germany, 2002.

[Spears, 1993] W. M. Spears. Simulated annealing for hard satisfiability problems. Technical report, Naval Research Laboratory, Washington D.C., USA, 1993.

[Stadler and Flamm, 2003] P. F. Stadler and C. Flamm. Barrier trees on poset-valued landscapes. *Genetic Programming and Evolvable Machines*, 4(1):7–20, 2003.

[Stadler and Happel, 1992] P. F. Stadler and R. Happel. Correlation structure of the landscape of the graph-bipartitioning problem. *Journal of Physics A*, 25(11):3103–3110, 1992.

[Stadler and Krakhofer, 1996] P. F. Stadler and B. Krakhofer. Local minima of p-spin models. *Revista Mexicana de Fisica*, 42:355–363, 1996.

[Stadler and Schnabl, 1992] P. F. Stadler and W. Schnabl. The landscape of the travelling salesman problem. *Physics Letters A*, 161(4):337–344, 1992.

[Stadler, 1995] P. F. Stadler. Towards a theory of landscapes. In R. Lopéz-Peña, R. Capovilla, R. García-Pelayo, H. Waelbroeck, and F. Zertuche, editors, *Complex Systems and Binary Networks*, volume 461 of *Lecture Notes in Physics*, pages 77–163. Springer Verlag, Berlin, Germany, 1995.

[Stadler, 1996] P. F. Stadler. Landscapes and their correlation functions. *Journal of Mathematical Chemistry*, 20(1):1–45, 1996.

[Stadler, 2002] P. F. Stadler. Fitness landscapes. In M. Lässig and A. Valleriani, editors, *Biological Evolution and Statistical Physics*, pages 187–207. Springer Verlag, Berlin, Germany, 2002.

[Stadler, 2003] P. F. Stadler. Spectral landscape theory. In J. P. Crutchfield and P. Schuster, editors, *Evolutionary Dynamics—Exploring the Interplay of Selection, Neutrality, Accident, and Function*, pages 231–272. Oxford University Press, New York, NY, USA, 2003.

[Steinberg, 1961] L. Steinberg. The backboard wiring problem: A placement algorithm. *SIAM Review*, 3:37–50, 1961.

[Steinmann et al., 1997] O. Steinmann, A. Strohmaier, and T. Stützle. Tabu search vs. random walk. In G. Brewka, C. Habel, and B. Nebel, editors, *KI-97: Advances in Artificial Intelligence*, volume 1303 of *Lecture Notes in Artificial Intelligence*, pages 337–348. Springer Verlag, Berlin, Germany, 1997.

[Steuer, 1986] R. E. Steuer. *Multiple Criteria Optimization: Theory, Computation and Application*. John Wiley & Sons, New York, NY, USA, 1986.

[Storer et al., 1992] R. H. Storer, S. D. Wu, and R. Vaccari. New search spaces for sequencing problems with application to job shop scheduling. *Management Science*, 38(10):1495–1509, 1992.

[Stützle and Dorigo, 1999a] T. Stützle and M. Dorigo. ACO algorithms for the quadratic assignment problem. In D. Corne, M. Dorigo, and F. Glover, editors, *New Ideas in Optimization*, pages 33–50. McGraw Hill, London, UK, 1999.

[Stützle and Dorigo, 1999b] T. Stützle and M. Dorigo. ACO algorithms for the traveling salesman problem. In K. Miettinen, M. M. Mäkelä, P. Neittaanmäki, and J. Périaux, editors, *Evolutionary Algorithms in Engineering and Computer Science*, pages 163–183. John Wiley & Sons, Chichester, UK, 1999.

[Stützle and Dorigo, 2002] T. Stützle and M. Dorigo. A short convergence proof for a class of ACO algorithms. *IEEE Transactions on Evolutionary Computation*, 6(4):358–365, 2002.

[Stützle and Hoos, 1996] T. Stützle and H. H. Hoos. Improving the ant system: A detailed report on the $\mathcal{MAX-MIN}$ ant system. Technical Report AIDA–96–12, FG Intellektik, FB Informatik, TU Darmstadt, Germany, August 1996.

[Stützle and Hoos, 1997] T. Stützle and H. H. Hoos. The $\mathcal{MAX-MIN}$ Ant System and local search for the traveling salesman problem. In T. Bäck, Z. Michalewicz, and X. Yao, editors, *Proceedings of the 1997 IEEE International Conference on Evolutionary Computation (ICEC'97)*, pages 309–314. IEEE Press, Piscataway, NJ, USA, 1997.

[Stützle and Hoos, 1999] T. Stützle and H. H. Hoos. $\mathcal{MAX-MIN}$ Ant System and local search for combinatorial optimization problems. In S. Voss, S. Martello, I. H. Osman, and C. Roucairol, editors, *Meta-Heuristics: Advances and Trends in Local Search Paradigms for Optimization*, pages 137–154. Kluwer Academic Publishers, Dordrecht, The Netherlands, 1999.

[Stützle and Hoos, 2000] T. Stützle and H. H. Hoos. $\mathcal{MAX-MIN}$ Ant System. *Future Generation Computer Systems*, 16(8):889–914, 2000.

[Stützle and Hoos, 2001] T. Stützle and H. H. Hoos. Analysing the run-time behaviour of iterated local search for the travelling salesman problem. In P. Hansen and C. C. Ribeiro, editors, *Essays and Surveys on Metaheuristics*, pages 589–611. Kluwer Academic Publishers, Boston, MA, USA, 2001.

[Stützle et al., 2000] T. Stützle, A. Grün, S. Linke, and M. Rüttger. A comparison of nature inspired heuristics on the traveling salesman problem. In M. Schoenauer, K. Deb, G. Rudolph, X. Yao, E. Lutton, J. J. Merelo, and H.-P. Schwefel, editors, *Proceedings of PPSN-VI, Sixth International Conference on Parallel Problem Solving from Nature*, volume 1917 of *Lecture Notes in Computer Science*, pages 661–670. Springer Verlag, Berlin, Germany, 2000.

[Stützle, 1997] T. Stützle. $\mathcal{MAX-MIN}$ Ant System for the quadratic assignment problem. Technical Report AIDA–97–4, FG Intellektik, FB Informatik, TU Darmstadt, Germany, July 1997.

[Stützle, 1998a] T. Stützle. An ant approach to the flow shop problem. In *Proceedings of the 6th European Congress on Intelligent Techniques & Soft Computing (EUFIT'98)*, volume 3, pages 1560–1564. Verlag Mainz, Aachen, Germany, 1998.

[Stützle, 1998b] T. Stützle. Applying iterated local search to the permutation flow shop problem. Technical Report AIDA-98-04, FG Intellektik, TU Darmstadt, FB Informatik, Germany, 1998.

[Stützle, 1998c] T. Stützle. *Local Search Algorithms for Combinatorial Problems — Analysis, Improvements, and New Applications*. PhD thesis, TU Darmstadt, FB Informatik, Darmstadt, Germany, 1998.

[Stützle, 1999] T. Stützle. Iterated local search for the quadratic assignment problem. Technical Report AIDA-99-03, FG Intellektik, FB Informatik, TU Darmstadt, Germany, 1999.

[Suh and Gucht, 1987] J. Y. Suh and D. Van Gucht. Incorporating heuristic information into genetic search. In *Genetic algorithms and their applications: Proceedings of the Second International Conference on Genetic Algorithms*, pages 100–107. Lawrence Erlbaum Associates, Mahwah, NJ, USA, 1987.

[Szwarc et al., 2001] W. Szwarc, A. Grosso, and F. Della Croce. Algorithmic paradoxes of the single machine total tardiness problem. *Journal of Scheduling*, 4(2):93–104, 2001.

T   [Taillard, 1990] É. D. Taillard. Some efficient heuristic methods for the flow shop sequencing problem. *European Journal of Operational Research*, 47(1):65–74, 1990.

[Taillard, 1991] É. D. Taillard. Robust taboo search for the quadratic assignment problem. *Parallel Computing*, 17(4–5):443–455, 1991.

[Taillard, 1993] É. D. Taillard. Benchmarks for basic scheduling problems. *European Journal of Operational Research*, 64(2):278–285, 1993.

[Taillard, 1994] É. D. Taillard. Parallel tabu search techniques for the job shop scheduling problem. *ORSA Journal on Computing*, 6(2):108–117, 1994.

[Taillard, 1995] É. D. Taillard. Comparison of iterative searches for the quadratic assignment problem. *Location Science*, 3(2):87–105, 1995.

[Taillard, 2003] É. D. Taillard. Quadratic assignment instances. `http://ina.eivd.ch/collaborateurs/etd/problemes.dir/qap.dir/qap.html`, 2003. Version visited last on 15 September 2003.

[Tamaki, 2003] H. Tamaki. Alternating cycles contribution: A strategy of tour-merging for the traveling salesman problem. Research Report MPI-I-2003-1-007, Max-Planck-Institut für Informatik, Saarbrücken, Germany, 2003.

[Tanaka and Edwards, 1980] F. Tanaka and S. F. Edwards. Analytic theory of ground state properties of spin glasses: I. Ising spin glass. *Journal Physique France*, 10:2769–2778, 1980.

[ten Eikelder et al., 1996] H. M. M. ten Eikelder, M. G. A. Verhoeven T. V. M. Vossen, and E. H. L. Aarts. A probabilistic analysis of local search. In I. H. Osman and J. P. Kelly, editors, *Metaheuristics: Theory & Applications*, pages 605–618. Kluwer Academic Publishers, Boston, MA, USA, 1996.

[Tesman, 1993a] B. A. Tesman. List T-colorings. *Discrete Applied Mathematics*, 45(3):277–289, 1993.

[Tesman, 1993b] B. A. Tesman. Set T-colorings. *Congressus Numerantium*, 77:229–242, 1990.

[Thompson and Orlin, 1989] P. M. Thompson and J. B. Orlin. The theory of cycle transfers. Working Paper OR 200-89, Operations Research Center, MIT, Cambridge, MA, USA, 1989.

[Thompson and Psaraftis, 1993] P. M. Thompson and H. N. Psaraftis. Cyclic transfer algorithm for multivehicle routing and scheduling problems. *Operations Research*, 41:935–946, 1993.

[Tompkins and Hoos, 2003] D. A. D. Tompkins and H. H. Hoos. Scaling and probabilistic smoothing: Dynamic local search for unweighted MAX-SAT. In Y. Xiang and B. Chaib-draa, editors, *Advances in Artificial Intelligence, 16th Conference of the Canadian Society for Computational Studies of Intelligence*, volume 2671 of *Lecture Notes in Computer Science*, pages 145–159. Springer Verlag, Berlin, Germany, 2003.

[Tsang, 1993] E. Tsang. *Foundations of Constraint Satisfaction*. Academic Press, London, UK, 1993.

[Tulpan and Hoos, 2003] D. C. Tulpan and H. H. Hoos. Hybrid randomised neighbourhoods improve stochastic local search for DNA code design. In Y. Xiang and B. Chaib-draa, editors, *Advances in Artificial Intelligence, 16th Conference of the Canadian Society for Computational Studies of Intelligence*, volume 2671 of *Lecture Notes in Computer Science*, pages 418–433. Springer Verlag, Berlin, Germany, 2003.

[Tulpan et al., 2003] D. C. Tulpan, H. H. Hoos, and A. Condon. Stochastic local search algorithms for DNA word design. In M. Hagiya and A. Ohuchi, editors, *DNA Computing, 8th International Workshop on DNA Based Computers*, volume 2568 of *Lecture Notes in Computer Science*, pages 229–241. Springer Verlag, Berlin, Germany, 2003.

[Turner and Booth, 1987] S. Turner and D. Booth. Comparison of heuristics for flow shop sequencing. *OMEGA*, 15(1):75–85, 1987.

U   [Ulder et al., 1991] N. L. J. Ulder, E. H. L. Aarts, H.-J. Bandelt, P. J. M. van Laarhoven, and E. Pesch. Genetic local search algorithms for the travelling salesman problem. In H.-P. Schwefel and R. Männer, editors, *Parallel Problem Solving from Nature, PPSN I*, volume 496 of *Lecture Notes in Computer Science*, pages 109–116. Springer Verlag, Berlin, Germany, 1991.

V   [Vaessens et al., 1996] R. J. M. Vaessens, E. H. L. Aarts, and J. K. Lenstra. Jop shop scheduling by local search. *INFORMS Journal on Computing*, 8(3):302–317, 1996.

[Vaessens et al., 1998] R. J. M. Vaessens, E. H. L. Aarts, and J. K. Lenstra. A local search template. *Computers & Operations Research*, 25(11):969–979, 1998.

[van Laarhoven and Aarts, 1987] P. J. M. van Laarhoven and E. H. L. Aarts. *Simulated Annealing: Theory and Applications*. D. Reidel Publishing Company, Dordrecht, The Netherlands, 1987.

[Vanbekbergen et al., 1992] P. Vanbekbergen, B. Lin, G. Goossens, and H. De Man. A generalized state assignment theory for transformations on signal transition graphs. In *Proceedings of the IEEE/ACM International Conference on Computer-Aided Design*, pages 112–117. IEEE Computer Society Press, Los Alamitos, CA, USA, 1992.

[Vasko and Wolf, 1987] F. J. Vasko and F. E. Wolf. Optimal selection of ingot sizes via set covering. *Operations Research*, 35:115–121, 1987.

[Vasquez and Hao, 2001] M. Vasquez and J-K. Hao. A hybrid approach for the 0-1 multidimensional knapsack problem. In B. Nebel, editor, *Proceedings of the Seventeenth International Joint Conference on Artificial Intelligence*, pages 328–333. Morgan Kaufmann Publishers, San Francisco, CA, USA, 2001.

[Vazquez and Whitley, 2000] M. Vazquez and D. Whitley. A hybrid genetic algorithm for the quadratic assignment problem. In D. Whitley, D. Goldberg, E. Cantu-Paz, L. Spector, I. Parmee, and H.-G. Beyer, editors, *Proceedings of the Genetic and Evolutionary Computation Conference (GECCO-2000)*, pages 135–142. Morgan Kaufmann Publishers, San Francisco, CA, USA, 2000.

[Voß and Woodruff, 2002] S. Voß and D. L. Woodruff, editors. *Optimization Software Class Libraries*. Kluwer Academic Publishers, Boston, MA, USA, 2002.

[Voß et al., 1999] S. Voß, S. Martello, I. H. Osman, and C. Roucairol, editors. *Meta-Heuristics: Advances and Trends in Local Search Paradigms for Optimization*. Kluwer Academic Publishers, Boston, MA, USA, 1999.

[Voudouris and Tsang, 1995] C. Voudouris and E. Tsang. Guided local search. Technical Report CSM-247, Department of Computer Science, University of Essex, Colchester, UK, 1995.

[Voudouris and Tsang, 1999] C. Voudouris and E. Tsang. Guided local search and its application to the travelling salesman problem. *European Journal of Operational Research*, 113(2):469–499, 1999.

[Voudouris and Tsang, 2002] C. Voudouris and E. Tsang. Guided local search. In F. Glover and G. Kochenberger, editors, *Handbook of Metaheuristics*, pages 185–218. Kluwer Academic Publishers, Norwell, MA, USA, 2002.

[Voudouris, 1997] C. Voudouris. *Guided Local Search for Combinatorial Optimization Problems*. PhD thesis, University of Essex, Department of Computer Science, Colchester, UK, 1997.

W   [Wagner et al., 2003] M. O. Wagner, B. Yannou, S. Kehl, D. Feillet, and J. Eggers. Ergonomic modelling and optimization of the keyboard arrangement with an ant colony algorithm. *Journal of Engineering Design*, 14(2):187–208, 2003.

[Walker, 2003] J. Walker. Hotbits: Genuine random numbers, generated by radioactive decay. `http://www.fourmilab.ch/hotbits`, 2003. Version visited last on 15 September 2003.

[Wallace and Freuder, 1995] R. J. Wallace and E. C. Freuder. Heuristic methods for over-constrained constraint satisfaction problems. In M. Jampel, E. C. Freuder, and M. J. Maher, editors, *Over-Constrained Systems*, volume 1106 of *Lecture Notes in Computer Science*, pages 207–216. Springer Verlag, Berlin, Germany, 1995.

[Wallace, 1994] R. J. Wallace. Directed arc consistency preprocessing as a strategy for maximal constraint satisfaction. In M. Meyer, editor, *Constraint Processing*, volume 923 of *Lecture Notes in Computer Science*, pages 121–138. Springer Verlag, Berlin, Germany, 1994.

[Wallace, 1996a] R. J. Wallace. Analysis of heuristic methods for partial constraint satisfaction problems. In E. C. Freuder, editor, *Proceedings of the Second International Conference on Principles and Practice of Constraint Programming*, volume 1118 of *Lecture Notes in Computer Science*, pages 482–496. Springer Verlag, Berlin, Germany, 1996.

[Wallace, 1996b] R. J. Wallace. Enhancements of branch and bound methods for the maximal constraint satisfaction problem. In *Proceedings of the Thirteenth National Conference on Artificial Intelligence*, volume 1, pages 188–195. AAAI Press/The MIT Press, Menlo Park, CA, USA, 1996.

[Walser et al., 1998] J. P. Walser, R. Iyer, and N. Venkatasubramanyan. An integer local search method with application to capacitated production planning. In *Proceedings of the Fifteenth National Conference on Artificial Intelligence*, pages 373–379. AAAI Press/The MIT Press, Menlo Park, CA, USA, 1998.

[Walser, 1997] J. P. Walser. Solving linear pseudo-boolean constraint problems with local search. In *Proceedings of the Fourteenth National Conference on Artificial Intelligence*, pages 269–274. AAAI Press/The MIT Press, Menlo Park, CA, USA, 1997.

[Walser, 1999] J. P. Walser. *Integer Optimization by Local Search: A Domain-Independent Approach*, volume 1637 of *Lecture Notes in Computer Science*. Springer Verlag, Berlin, Germany, 1999.

[Walser, 2003] J. Walser. Integer local search: WSAT(OIP). `http://www.ps.uni-sb.de/~walser/wsatpb/wsatpb.html`, 2003. Version visited last on 15 September 2003.

[Walsh, 2000] T. Walsh. SAT v CSP. In R. Dechter, editor, *Principles and Practice of Constraint Programming – CP 2000*, volume 1894 of *Lecture Notes in Computer Science*, pages 441–456. Springer Verlag, Berlin, Germany, 2000.

[Walshaw, 2002] C. Walshaw. A multilevel approach to the travelling salesman problem. *Operations Research*, 50(5):862–877, 2002.

[Walters, 1998] T. Walters. Repair and brood selection in the traveling salesman problem. In A. E. Eiben, T. Bäck, M. Schoenauer, and H.-P. Schwefel, editors, *Proceedings of PPSN-V, Fifth International Conference on Parallel Problem Solving from Nature*, volume 1498 of *Lecture Notes in Computer Science*, pages 813–822. Springer Verlag, Berlin, Germany, 1998.

[Watson et al., 2002] J. P. Watson, L. Barbulescu, L. D. Whitley, and A. E. Howe. Contrasting structured and random permutation flow-shop scheduling problems: Search space topology and algorithm performance. *INFORMS Journal on Computing*, 14(2):98–123, 2002.

[Watson et al., 2003] J.-P. Watson, J. C. Beck, A. E. Howe, and L. D. Whitley. Problem difficulty for tabu search in job-shop scheduling. *Artificial Intelligence*, 143(2):189–217, 2003.

[Watson, 2003] J.-P. Watson. Structured versus random benchmarks for the permutation flow-shop scheduling problem. http://www.cs.colostate.edu/sched/generator, 2003. Version visited last on 15 September 2003.

[Wedelin, 1995] D. Wedelin. An algorithm for large scale 0-1 integer programming with application to airline crew scheduling. *Annals of Operations Research*, 57:283–301, 1995.

[Weinberger, 1990] E. D. Weinberger. Correlated and uncorrelated fitness landscapes and how to tell the difference. *Biological Cybernetics*, 63(5):325–336, 1990.

[Weinberger, 1991] E. Weinberger. Local properties of Kauffman's N-k model: A tunably rugged energy landscape. *Physical Review A*, 44(10):6399–6413, 1991.

[Weinberger, 1996] E. D. Weinberger. NP completeness of Kauffman's N-k model, a tuneably rugged fitnes landscapes. Technical Report 96-02-003, Santa Fe Institute, Santa Fe, USA, 1996.

[Werner and Winkler, 1995] F. Werner and A. Winkler. Insertion techniques for the heuristic solution of the job-shop problem. *Discrete Applied Mathematics*, 58(2):191–211, 1995.

[Werner, 1992] F. Werner. On the heuristic solution of the permutation flow shop problem by path algorithms. *Computers & Operations Research*, 20(7):707–722, 1992.

[Whitley et al., 1989] D. Whitley, T. Starkweather, and D. Fuquay. Scheduling problems and travelling salesman: The genetic edge recombination operator. In J. D. Schaffer, editor, *Proceedings of the Third International Conference on Genetic Algorithms (ICGA'89)*, pages 133–140. Morgan Kaufmann Publishers, San Mateo, CA, USA, 1989.

[Whizzkids'97, 1997] Whizzkids'97. http://www.win.tue.nl/whizzkids/1997, 1997. Version visited last on 15 September 2003.

[Widmer and Hertz, 1989] M. Widmer and A. Hertz. A new heuristic method for the flow shop sequencing problem. *European Journal of Operational Research*, 41(2):186–193, 1989.

[Williams Jr. and Dozier, 2001] J. P. Williams Jr. and G. Dozier. A comparison of hill-climbing methods for solving static and recurrent dynamic constraint satisfaction problems. Technical Report CSSE01-01, Auburn University, Department of Computer Science and Software Engineering, Auburn, USA, 2001.

[Winfree et al., 1999] E. Winfree, X. Yang, and N. C. Seeman. Universal computation via self-assembly of DNA: Some theory and experiments. In L. F. Landweber and E. B. Baum, editors, *DNA Based Computers II*, volume 44 of DIMACS Series on Discrete Mathematics and Theoretical Computer Science, Piscataway, NJ, USA, pages 191–213. American Mathematical Society, Providence, RI, USA, 1999.

[Woodruff and Spearman, 1992] D. L. Woodruff and M. L. Spearman. Sequencing and batching for two classes of jobs with deadlines and setup times. *Production and Operations Management*, 1(1):87–102, 1992.

[Wright et al., 2000] A. H. Wright, R. K. Thompson, and J. Zhang. The computational complexity of N-K fitness functions. *IEEE Transactions on Evolutionary Computation*, 4(4):373–379, 2000.

[Wright, 1932] S. Wright. The roles of mutation, inbreeding, crossbreeding and selection in evolution. In D. F. Jones, editor, *International Proceedings of the Sixth International Congress on Genetics*, volume 1, pages 356–366, 1932.

[Wu and Wah, 1999] Z. Wu and B. W. Wah. Trap escaping strategies in discrete Lagrangian methods for solving hard satisfiability and maximum satisfiability problems. In *Proceedings of the Sixteenth National Conference on Artificial Intelligence*, pages 673–678. AAAI Press/The MIT Press, Menlo Park, CA, USA, 1999.

[Wu and Wah, 2000] Z. Wu and B. W. Wah. An efficient global-search strategy in discrete Lagrangian methods for solving hard satisfiability problems. In *Proceedings of the Seventeenth National Conference on Artificial Intelligence*, pages 310–315. AAAI Press/The MIT Press, Menlo Park, CA, USA, 2000.

X   [Xu et al., 1998] J. Xu, S.Y. Chiu, and F. Glover. Fine-tuning a tabu search algorithm with statistical tests. *International Transactions in Operational Research*, 5(4):233–244, 1998.

Y   [Yagiura and Ibaraki, 1998] M. Yagiura and T. Ibaraki. Efficient 2 and 3-flip neighborhoods seach algorithms for the MAX SAT. In W.-L. Hsu and M.-Y. Kao, editors, *Computing and Combinatorics*, volume 1449 of *Lecture Notes in Computer Science*, pages 105–116. Springer Verlag, Berlin, Germany, 1998.

[Yagiura and Ibaraki, 1999] M. Yagiura and T. Ibaraki. Analyses on the 2 and 3-flip neighborhoods for the MAX SAT. *Journal of Combinatorial Optimization*, 3(1):95–114, 1999.

[Yagiura and Ibaraki, 2001] M. Yagiura and T. Ibaraki. Efficient 2 and 3-flip neighborhood search algorithms for the MAX SAT: Experimental evaluation. *Journal of Heuristics*, 7(5):423–442, 2001.

[Yagiura et al., 1999] M. Yagiura, T. Yamaguchi, and T. Ibaraki. A variable depth search algorithm for the generalized assignment problem. In S. Voss, S. Martello, I. H. Osman, and C. Roucairol, editors, *Meta-heuristics: Advances and Trends in Local Search Paradigms for Optimization*, pages 459–471. Kluwer Academic Publishers, Boston, MA, USA, 1999.

[Yagiura et al., 2004] M. Yagiura, M. Kishida, and T. Ibaraki. A 3-flip neighborhood local search for the set covering problem. Technical Report #2004-001, Department of

Applied Mathematics and Physics, Graduate School of Informatics, Kyoto University, Kyoto, Japan, 2004.

[Yamada and Nakano, 1992] T. Yamada and R. Nakano. A genetic algorithm applicable to large-scale job-shop instances. In R. Männer and B. Manderick, editors, *Parallel Problem Solving from Nature 2, PPSN II*, pages 281–290. Elsevier, Amsterdam, The Netherlands, 1992.

[Yamada, 2001] T. Yamada. A pruning pattern list approach to the permutation flowshop scheduling problem. In P. Hansen and C. C. Ribeiro, editors, *Essays and Surveys on Metaheuristics*, pages 641–651. Kluwer Academic Publishers, Boston, MA, USA, 2001.

[Yannakakis, 1990] M. Yannakakis. The analysis of local search problems and their heuristics. In C. Choffrut and T. Lengauer, editors, *STACS 90, 7th Annual Symposium on Theoretical Aspects of Computer Science*, volume 415 of *Lecture Notes in Computer Science*, pages 298–310. Springer Verlag, Berlin, Germany, 1990.

[Yannakakis, 1994] M. Yannakakis. On the approximation of maximum satisfiability. *Journal of Algorithms*, 17(4):475–502, 1994.

[Yannakakis, 1997] M. Yannakakis. Computational complexity. In E. H. L. Aarts and J. K. Lenstra, editors, *Local Search in Combinatorial Optimization*, pages 19–55. John Wiley & Sons, Chichester, UK, 1997.

[Yokoo, 1994] M. Yokoo. Weak-commitment search for solving constraint satisfaction problems. In *Proceedings of the 12th National Conference on Artificial Intelligence*, pages 313–318. AAAI Press/The MIT Press, Menlo Park, CA, USA, 1994.

[Yokoo, 1997] M. Yokoo. Why adding more constraints makes a problem easier for hill-climbing algorithms: Analyzing landscapes of CSPs. In G. Smolka, editor, *Principles and Practice of Constraint Programming – CP 1997*, volume 1330 of *Lecture Notes in Computer Science*, pages 357–370. Springer Verlag, Berlin, Germany, 1997.

[Young et al., 1997] C. Young, D. S. Johnson, D. R. Karger, and M. D. Smith. Near-optimal intraprocedural branch alignment. In *Proceedings of the ACM SIGPLAN'97 Conference on Programming Language Design and Implementation*, pages 183–193, ACM Press, New York, NY, USA, 1997. Published also in SIGPLAN Notices, 32(5).

[Yugami et al., 1994] N. Yugami, Y. Ohta, and H. Hara. Improving repair-based constraint satisfaction methods by value propagation. In *Proceedings of the 12th National Conference on Artificial Intelligence*, pages 344–349. AAAI Press/The MIT Press, Menlo Park, CA, USA, 1994.

Z   [Zachariasen and Dam, 1996] M. Zachariasen and M. Dam. Tabu search on the geometric traveling salesman problem. In I. H. Osman and J. P. Kelly, editors, *Metaheuristics: Theory & Applications*, pages 571–587. Kluwer Academic Publishers, Boston, MA, USA, 1996.

[Zhang and Shin, 1998] B-T. Zhang and S-Y. Shin. Molecular algorithms for efficient and reliable DNA computing. In *Proceedings of the 3rd Annual Genetic Programming Conference*, pages 735–742. Morgan Kaufmann Publishers, San Francisco, CA, USA, 1998.

[Zhang, 1993] W. Zhang. Truncated branch and bound: A case study on the asymmetric TSP. In *Proceedings of the AAAI 1993 Spring Symposium on AI and NP-hard Problems*, pages 160–166, AAAI Press/The MIT Press, Menlo Park, CA, USA, 1993.

[Zhang, 2002] H. Zhang. Generating college conference basketball schedules by a SAT solver. In *Proceedings of the Fifth International Symposium on the Theory and Applications of Satisfiability Testing (SAT 2002)*, pages 281–291, Cincinnati, OH, USA, 2002.

[Zweben and Fox, 1994] M. Zweben and M. Fox, editors. *Intelligent Scheduling*. Morgan Kaufmann Publishers, San Mateo, CA, USA, 1994.

[Zwick, 1999] U. Zwick. Outward rotations: A tool for rounding solutions of semidefinite programming relaxations, with applications to MAX CUT and other problems. In *Proceedings of the Thirty-First Annual ACM Symposium on Theory of Computing*, pages 679–687. ACM Press, New York, NY, USA, 1999.

INDEX

Page numbers in bold indicate where the entry is defined in the text. The letter 'f' indicates a figure, 't' a table and 'g' an entry in the glossary.

Iterative Improvement (II):

determine initial candidate solution s
While s is not a local optimum:
| choose a neighbour s' of s such that
| $\quad g(s') < g(s)$
| $s := s'$

Randomised Iterative Improvement (RII):

determine initial candidate solution s
While termination condition not satisfied:
| With probability wp:
| $\quad$ choose a neighbour s' of s uniformly
| $\qquad$ at random
| Otherwise:
| $\quad$ choose a neighbour s' of s such that
| $\qquad g(s') < g(s)$
| $\qquad$ or, if no such s' exists, choose s'
| $\qquad\quad$ such that $g(s')$ is minimal
| $s := s'$

Variable Neighbourhood Descent (VND):

determine initial candidate solution s
$i := 1$
Repeat:
| choose a most improving neighbour s' of s in N_i
| If $g(s') < g(s)$:
| $\quad s := s'$
| $\quad i := 1$
| Else:
| $\quad i := i + 1$
Until $i > imax$

Note: $N_1, \ldots, N_{imax}$ is a set of neighbourhood relations, typically ordered according to increasing size of the respective local neighbourhoods.

Variable Depth Search (VDS):

determine initial candidate solution s
$\hat{t} := s$
While s is not locally optimal:
| Repeat:
| $\quad$ select best feasible neighbour t
| $\quad$ If $g(t) < g(\hat{t})$: $\hat{t} := t$
| Until construction of complex step
| $\quad$ completed
| $s := \hat{t}$

Simulated Annealing (SA):

determine initial candidate solution s
set initial temperature T according to
$\quad$ annealing schedule
While termination condition not satisfied:
| probabilistically choose a neighbour
| $\quad s'$ of s
| If s' satisfies probabilistic acceptance
| $\quad$ criterion (depending on T):
| $\quad\quad s := s'$
| update T according to annealing schedule

Note: The annealing schedule may keep T constant for a number of search steps.

Tabu Search (TS):

determine initial candidate solution s
While termination criterion is not satisfied:
| determine set N of non-tabu neighbours of s
| choose a best improving solution s' in N
| update tabu attributes based on s'
| $s := s'$

Note: Tabu attributes are associated with solution components.

Dynamic Local Search (DLS):

determine initial candidate solution s
initialise penalties
While termination criterion is not satisfied:
 | compute modified evaluation function g'
 | from g based on penalties
 | perform subsidiary local search on s
 | using evaluation function g'
 | update penalties based on s

Note: Penalties are associated with solution components; the subsidiary local search ends in a local minimum
of g'.

Iterated Local Search (ILS):

determine initial candidate solution s
perform subsidiary local search on s
While termination criterion is not satisfied:
 | $r := s$
 | perform perturbation on s
 | perform subsidiary local search on s
 | based on acceptance criterion, keep s
 | or revert to $s := r$

Note: The search history may additionally influence the
perturbation phase and the acceptance criterion.

Greedy Randomised 'Adaptive' Search Procedure (GRASP):

While termination criterion is not satisfied:
 | generate candidate solution s using
 | subsidiary greedy randomised
 | constructive search
 | perform subsidiary local search on s

Adaptive Iterated Construction Search (AICS):

initialise weights
While termination criterion is not satisfied:
 | generate candidate solution s using subsi-
 | diary randomised constructive search
 | perform subsidiary local search on s
 | adapt weights based on s

Note: The subsidiary constructive search is based
on weights and heuristic information.

Ant Colony Optimisation (ACO):

initialise weights
While termination criterion is not satisfied:
 | generate population sp of candidate
 | solutions using subsidiary
 | randomised constructive search
 | perform subsidiary local search on sp
 | adapt weights based on sp

Note: The subsidiary constructive search uses weights
(pheromone trails) and heuristic information.

Memetic Algorithm (MA):

determine initial population sp
perform subsidiary local search on sp
While termination criterion is not satisfied:
 | generate set spr of new candidate
 | solutions by recombination
 | perform subsidiary local search on spr
 | generate set spm of new candidate
 | solutions from sp and spr by mutation
 | perform subsidiary local search on spm
 | select new population sp from candidate
 | solutions in sp, spr and spm

Notation

$\top$	truth value *true*
$\bot$	truth value *false*
$\neg p$	logical negation of statement p
$p \wedge q$	logical conjunction of p, q (p and q)
$p \vee q$	logical disjunction of p, q (p or q)
$(a_1, a_2, \ldots, a_n)$	ordered list of elements $a_1, a_2, \ldots, a_n$
$\{a_1, a_2, \ldots, a_n\}$	set comprised of elements $a_1, a_2, \ldots, a_n$
$\{x \in S \mid P(x)\}$	set containing all x in set S that satisfy condition $P(x)$
$\#S$	number of elements (i.e., cardinality) of set S
$A \subseteq B$	set A is a subset of B (or equal to B)
$A \subset B$	set A is a strict subset of B
$A \cap B$	intersection of sets A and B
$A \cup B$	union of sets A and B
$A \setminus B$	set containing all elements of A except those in B
$A \times B$	Cartesian product of sets A and B, that is, the set of all pairs (a, b) such that $a \in A$ and $b \in B$
$\bigcup \mathcal{A}$	union over all the sets contained in $\mathcal{A}$, where $\mathcal{A}$ is a set of sets
$\min S$	minimal element of set S
$\max S$	maximal element of set S
$\mathrm{argmin}_{x \in S} f(x)$	set of all $x \in S$ such that $f(x)$ is minimal
$\mathrm{argmin}_{x \in S} f(x)$	set of all $x \in S$ such that $f(x)$ is maximal
$\sum_{s \in S} f(s)$	sum over all elements of set S
$\mathbb{N}$	set of natural numbers (i.e., positive integers)
$\mathbb{R}$	set of real numbers
$\mathbb{R}^+$	set of positive real numbers
$\mathbb{R}_0^+$	set of positive real numbers including zero
$\{a, \ldots, b\}$	set of all integers k such that $a \leq k \leq b$
$[a, b]$	set of all real numbers x such that $a \leq x \leq b$
$[a, b)$	set of all real numbers x such that $a \leq x < b$
$\lfloor x \rfloor$	greatest integer k such that $k \leq x$
$\lceil x \rceil$	least integer k such that $k \geq x$
$f : A \mapsto B$	a function f that maps elements of set A to elements of set B
$\mathcal{D}(S)$	set of all probability distributions over set S
$\mathcal{P}$	class of all problems that can be solved by a deterministic algorithm in polynomial time
$\mathcal{NP}$	class of all problems that can be solved by a nondeterministic algorithm in polynomial time
$ed[m]$	cumulative distribution function of an exponential distribution with median m
$random(S)$	a random element selected from set S according to a uniform probability distribution
$\alpha \mid \beta \mid \gamma$	formal notation for scheduling problems, where α, β and γ represent the machine environment, the job characteristics and the objective function, respectively (for details, see page 422*f*.)

Abbreviations

ACO	Ant Colony Optimisation
AICS	Adaptive Iterated Construction Search
ATSP	Asymmetric Travelling Salesman Problem
BPT	basin partition tree
CAWDP	Combinatorial Auctions Winner Determination Problem
CSP	Constraint Satisfaction Problem
DLS	Dynamic Local Search
DNA-CDP	DNA Code Design Problem
EA	Evolutionary Algorithm
FDA	fitness-distance analysis
FDC	fitness-distance correlation
GA	Genetic Algorithm
GCP	Graph Colouring Problem
GLSM	generalised local search machine
GRASP	Greedy Randomised Adaptive Search Procedures
GSP	Group Shop Scheduling Problem
II	Iterative Improvement
ILS	Iterated Local Search
JSP	Job Shop Scheduling Problem
LVA	(generalised) Las Vegas algorithm
MA	Memetic Algorithm
MAX-CSP	Maximum Constraint Satisfaction Problem
MAX-SAT	Maximum Satisfiability Problem
OLVA	(generalised) Optimisation Las Vegas Algorithm
OSP	Open Shop Scheduling Problem
PCG	plateau connection graph
PFSP	Permutation Flow Shop Scheduling Problem
QAP	Quadratic Assignment Problem
QRTD	qualified solution quality distribution
RTD	run-time distribution
RTQ	solution-quality-dependent run-time statistics
SA	Simulated Annealing
SAT	Propositional Satisfiability Problem
SCP	Set Covering Problem
SLS	Stochastic Local Search
SMTWTP	Single Machine Total Weighted Tardiness Problem
SQD	solution quality distribution
SQT	time-dependent solution quality statistics
TS	Tabu Search
TSP	Travelling Salesman Problem
VDS	Variable Depth Search
VNS	Variable Neighbourhood Search

Printed and bound by CPI Group (UK) Ltd, Croydon, CR0 4YY

03/10/2024

01040344-0001